D1590793

Modeling Hydrologic Change

Statistical Methods

Richard H. McCuen
Department of Civil and Environmental Engineering
University of Maryland

LEWIS PUBLISHERS

A CRC Press Company
Boca Raton London New York Washington, D.C.

Library of Congress Cataloging-in-Publication Data

McCuen, Richard H., 1941
Modeling hydrologic change: statistical methods / Richard H. McCuen.
p. cm.
Includes bibliographical references and index.
ISBN 1-56670-600-9
1. Hydrologic models. 2. Hydrologic—Statistical methods. I. Title.
GB656.2.H9 M33 2002
551.48′01′1—dc21 2002073063
CIP

Catalog record is available from the Library of Congress

Direct all inquiries to CRC Press LLC, 2000 N.W. Corporate Blvd., Boca Raton, Florida 33431.

Visit the CRC Press Web site at www.crcpress.com

Lewis Publishers is an imprint of CRC Press LLC

International Standard Book Number 1-56670-600-9
Library of Congress Card Number 2002073063
Printed in the United States of America 1 2 3 4 5 6 7 8 9 0
Printed on acid-free paper

Preface

Modeling Hydrologic Change: Statistical Methods is about modeling systems where change has affected data that will be used to calibrate and test models of the systems and where models will be used to forecast system responses after change occurs. The focus is not on the hydrology. Instead, hydrology serves as the discipline from which the applications are drawn to illustrate the principles of modeling and the detection of change. All four elements of the modeling process are discussed: conceptualization, formulation, calibration, and verification. Analysis and synthesis are discussed in order to enable both model builders and users to appreciate the importance of both aspects of modeling. The book also focuses on the art and science of modeling.

While modeling techniques may be of great interest to hydrology-oriented professionals, they have value to all disciplines involved in modeling changes. While the book is oriented toward the statistical aspects of modeling, a strong background in statistics is not required. Although the emphasis is on the analysis of temporal and spatial sequences of data, the fundamentals that comprise most of the book are far more applicable. Statistical and modeling methods can be applied to a broad array of problems. This book is not appropriate as a general text for an undergraduate introductory course in probability and statistics. It is intended for advanced undergraduates, graduate students, and practicing professionals.

It includes topics that serve as background material for its central focus and topics related to the graphical and statistical detection of change and the fundamentals of modeling. While Chapters 2, 3, and 5 can be considered foundational, other chapters also introduce basic concepts. Chapters 4 and 6 through 9 are devoted to important graphical and statistical procedures used in modeling. Chapters 10 through 13 provide modeling tools useful in dealing with nonstationary systems.

In Chapter 2, some fundamental time-series concepts are introduced, with a special emphasis on concepts relevant to changing systems. Changes to real systems affect data observations. The different forms that these changes introduce into data are defined and illustrated.

In Chapter 3, basic concepts related to the fundamentals of hypothesis testing are introduced. While most of this material will serve as a review for readers with background in statistical analysis, the chapter includes the basic concepts important to understanding the statistical tests introduced in the middle chapters.

Extreme events contained in measured data are the topics of Chapter 4. They can distort calibrated model parameters and predictions based on models. Thus, their proper assessment and handling are essential in the early stages of modeling.

Frequency analysis is a rank-order statistical method widely used to connect the magnitudes and probabilities of occurrence of a random variable. The basic elements of frequency analysis as applied in hydrology are introduced in Chapter 5.

While statistical methods are important tools for the detection of nonstationarity, they are less effective when not accompanied by graphical analyses. In Chapter 6, the uses of graphical methods in the modeling process are introduced. While graphical analysis alone is inadequate for characterizing hydrologic change, it is a necessary component of the modeling process.

In Chapter 7, the fundamentals of detecting nonhomogeneity in time series are introduced. Special emphasis is placed on selecting the statistical method most sensitive to the types of changes to be evaluated.

Hydrologic change may be evident in the moments of the measured data or more generally in the distribution of the data. The statistical detection of change to moments is discussed in Chapter 8, while the detection of changes in probability distribution is the topic of Chapter 9. Statistical methods sensitive to different types of change are introduced in these chapters.

Chapter 10 covers many fundamentals of model calibration. Basic regression techniques along with advanced topics such as composite modeling and jackknifing are included.

Computer simulation is a valuable tool for modeling expected watershed changes. The manipulation of a model to simulate alternative scenarios of change can be valuable to decision makers. The fundamentals of simulation are presented in Chapter 11.

Sensitivity analysis is an important tool in modeling. It is useful for making error analyses and for assessing the relative importance of causative factors. The mathematical basis of sensitivity analysis and its uses are discussed in Chapter 12.

Chapter 13 presents the role that geographic information systems (GIS) can play in the assessment of hydrologic change. The inclusion of large databases in modeling is discussed. The effects of urbanization on flood frequency analysis are shown.

The Author

Richard H. McCuen, professor of civil engineering at the University of Maryland at College Park, received degrees from Carnegie Mellon University and the Georgia Institute of Technology. He received the Icko Iben Award from the American Water Resource Association and was co-recipient of the 1988 Outstanding Research Award from the American Society of Civil Engineers Water Resources Planning and Management Division. Topics in statistical hydrology and stormwater management are his primary research interest.

He is the author of 17 books and more than 200 professional papers, including *Modeling Hydrologic Change* (CRC Press, 2002); *Hydrologic Analysis and Design, Second Edition* (Prentice-Hall, 1998); *The Elements of Academic Research* (ASCE Press, 1996); *Estimating Debris Volumes for Flood Control* (Lighthouse Publications, 1996; with T.V. Hromadka); and *Dynamic Communication for Engineers* (ASCE Press, 1993; with P. Johnson and C. Davis).

Acknowledgments

Three people contributed greatly to this book. I very much appreciate Glenn Moglen's willingness to contribute the chapter on GIS and its role in modeling change. This book initially was developed as a joint effort with Wilbert O. Thomas, Jr., Baker Engineering, Alexandria, Virginia, but his workload did not permit participation beyond planning and review of earlier material. His insights are appreciated. Finally, the assistance of Dominic Yeh, University of Maryland, for typing the many, many drafts was essential to the completion of the manuscript. His efforts are also very much appreciated.

Richard H. McCuen
College Park, Maryland

Contents

1 Data, Statistics, and Modeling

1.1 INTRODUCTION

Before the 1960s, the availability of hydrologic data was very limited. Short records of streamflow and precipitation were available, but soils data, maps of land cover and use, and data on temporal watershed changes were generally not available. In addition, the availability of computers was limited. These limitations of data and computational firepower restricted the types of hydrologic methods that could be developed for general use. Flood frequency analysis was a staple of hydrologic modelers. It required limited amounts of data and was appropriate for the computational abilities of the era.

Urbanization over the last 40 years, including suburban sprawl, and the resulting increase in flooding necessitated new modeling tools and statistical methods. Watershed change introduced effects into measured hydrologic data that limited the accuracy of the traditional hydrologic computations. For example, frequency analysis assumes a stationary data sequence. The nonstationarity introduced into hydrologic data by watershed change means that T-year (annual) maximum discharge estimates from either predevelopment data or analyses of the nonstationary data would not be accurate estimates of T-year discharges that would occur under the current watershed conditions.

Two statistical advances are required to properly analyze nonstationary data. First, more appropriate statistical tools are necessary to detect the effect of hydrologic change and the characteristics of the change. For example, statistical methods are needed to identify trends, with one class of statistical methods needed for gradual trends and a second class needed for abrupt changes. Second, more sophisticated modeling tools are needed to account for the nonstationarities introduced by watershed change. For example, data measured from an urbanizing watershed would require a model to transform or adjust the data to a common watershed condition before a frequency analysis could be applied to the annual maximum discharge record.

Computers, and advances in software and accessory equipment, have made new, important types of data available that were not generally accessible even a decade ago. For example, maps can be scanned into a computer or taken from a CD-ROM and analyzed to create spatial measures of watershed change. Measured hydrologic data reflect both temporal and spatial changes, and computers make it possible to characterize these changes. The greater availability of data enables and requires more accurate models to account for the spatially and temporally induced nonstationarities in the measured hydrologic data.

1.2 WATERSHED CHANGES

Watersheds, by nature, are dynamic systems; therefore, they are in a constant state of change. While many changes are not detectable over short periods, major storms can cause extreme changes. While the news media may focus on damage to bridges, buildings, or dams, physical features of the watershed such as channel cross-sections may also change drastically during major floods. Thus, natural changes to both the land cover (e.g., afforestation) and the channel (e.g., degradation of the channel bottom) introduce nonstationarities into measured flood records.

In addition to natural watershed changes, the response of a watershed can change drastically due to changes associated with human activities. Changes in agricultural management practices and urban land cover exert significant effects on the responses of watersheds. In-stream changes, such as channelization or the construction of small storage structures, have also had a profound effect on the watershed response to storm rainfall. These changes will also introduce nonstationary effects in measured data.

Changes in hydrologic response due to natural or human-induced causes can change the storage characteristics of the watershed. Agricultural land development often involves both land clearing, which reduces surface roughness and decreases natural storage, and improvements in the hydraulic conveyance of the drainageways. These changes increase the rates of runoff, which decreases the opportunity for infiltration. Thus, peak runoff rates will increase, times to peak will decrease, and volumes of the surface runoff will increase. With less infiltration, baseflow rates will most likely decrease. Increases in flow velocities, which can be major causes of flood damage, accompany the increases in runoff rates, with the higher velocities generally increasing the scour rates of channel bottoms.

Changes in urban land cover reduce both the natural interception and depression storages and the potential for water to infiltrate, especially where the change involves an increase in impervious surfaces. The hydrologic effects of urban land use changes are similar to the effects of agricultural changes. The reduction in both surface storage and roughness increases the runoff rates and decreases computed times of concentration. Because of the reduced potential for infiltration, the volume of surface runoff also increases as a watershed becomes urbanized. The construction of storm sewers and curb-and-gutter streets can also decrease the time to peak and increase peak runoff rates.

In-stream modifications may increase or decrease channel storage and roughness. Channelization (straightening and/or lining) of a stream reach reduces the roughness, which increases flow velocities, and may decrease natural channel storage and/or the flow length (thus increasing channel slope). The net effect is to increase peak runoff rates. Small in-stream detention structures used in urban or rural areas to mitigate the effects of losses of natural storage have the opposite hydrologic effect of channelization. Specifically, the inclusion of in-stream storage decreases peak flow rates of small storms and increases the duration of high-flow conditions. In general, the in-stream detention structure will not significantly change the total volume of runoff but will redistribute the volume over time.

1.3 EFFECT ON FLOOD RECORD

Watershed changes can produce a nonhomogeneous flood record. The nonhomogeneity of the flood record reduces the reliability of the record for making estimates of design discharges. Where ultimate-development discharge estimates are required, an analysis of a nonhomogeneous record that does not account for the nonhomogeneity could result in poor estimates of design discharges.

Watershed changes can introduce an episodic (or abrupt) change or a secular trend in a flood series. An episodic change occurs in a very short period relative to the length of the flood record; channelization of a stream reach would affect the flood record as an episodic change. Once the presence and magnitude of such a change are identified, the flood series can be adjusted to a specified condition.

When a change in watershed processes occurs over a relatively long time period, the continual change in the flood record caused by this watershed change is referred to as a secular change; gradual urbanization would, for example, cause a secular trend. In such cases, the individual events in the flood record reflect different levels of the watershed processes, and the events should be adjusted to reflect a specific watershed state. The effect of the watershed change must be identified and quantified. To achieve this, the type of change must be hypothesized and the cause identified. A model that relates the discharge and a variable that represents the cause of the trend must be developed; such a model may be linear or nonlinear, static or dynamic. The model can then be used to adjust the flood record for a condition that would reflect a specific state; if it is accurate, the method of adjustment would account for the nonhomogeneous nature of the flood record.

Watershed changes that introduce a secular trend are more common in hydrologic analysis than changes due to episodic events. The method of adjustment to account for nonhomogeneity depends on the data available. If a calibrated watershed model and the necessary measured rainfall data are available, then the parameters of the model can be changed to reflect changes in watershed conditions. If the percentage of imperviousness is the only parameter that reflects watershed urbanization, then an appropriate adjustment method must be used in place of the more data-intensive, watershed-modeling approach.

An annual-maximum flood series is characterized by the randomness introduced by the variations of both rainfall and the hydrologic conditions of the watershed when the flood-producing rainfall occurred. The floods of an annual maximum series are produced by storms that vary in intensity, volume, duration, and the spatial characteristics of the storm cells. The extent of these sources of variation is largely responsible for variations in flood magnitudes. Variations in watershed conditions, especially antecedent moisture conditions, also affect variations in flood magnitudes. In a frequency analysis of flood data, variations in both rainfall characteristics and watershed conditions are considered to be random in nature.

Watershed changes, most notably urbanization, also can introduce variation into the floods of an annual maximum series. Urbanization appears to increase the magnitude of floods, especially the smaller floods. Large storms on a nonurbanized watershed saturate the watershed, which then acts as if it is impervious. Therefore, the hydrologic effects of urbanization are less pronounced for the larger storm events. However, unlike the random effects of rainfall and watershed conditions, urbanization is usually considered a deterministic factor. Specifically, urbanization introduces variation into the floods of annual maximum series that is believed to be detectable such that the floods can be adjusted for the effects of urbanization. The difficult task is to isolate the variation introduced by the urbanization in the face of the random variation due to rainfall and watershed conditions. When the random variations are large, it may be difficult to isolate the variation due to urbanization.

Other watershed changes can be important. Channelization and channel-clearing operations can change flood magnitudes because of their effects on hydraulic efficiency. The construction of small, unregulated detention basins can increase the available storage in a way that reduces both the magnitude and variation of flow rates. Intentional deforestation or afforestation are known to have significant hydrologic impacts.

The detection and modeling of trends is not a new problem. Many statistical assessments of business trends have been made. In fact, much of the seminal literature on time series analysis is about business cycles and trends. Haan (1977) briefly discussed fitting of trends in hydrologic data. Davis (1973) used regression techniques for fitting trends in geologic data. More recently, Gilbert (1987) provided an extensive review of tests for detecting trends and approaches for adjusting environmental pollution data. Hirsch, Alexander, and Smith (1991) discussed methods for detecting and estimating trends in water quality data. Systematic changes in data from a broad spectrum of disciplines received considerable professional attention.

1.4 WATERSHED CHANGE AND FREQUENCY ANALYSIS

A variety of watershed changes can affect the magnitude and variability of annual peak discharges and can affect the validity of the flood-frequency estimates. Bulletin 17B, titled "Guidelines for Determining Flood Flow Frequency" by the Interagency Advisory Committee on Water Data (IACWD) (1982; hereafter cited as Bulletin 17B), does not provide appropriate guidance for flood-frequency analyses when watershed changes have affected the magnitude, homogeneity, or randomness of measured annual peak discharges. Analysis techniques are recommended herein for evaluating and adjusting for the effects of watershed changes on the flood-frequency curve at locations where annual peak discharges are available.

Watershed changes do not include the construction of flood-detention structures whose operation can be regulated by human intervention. However, the construction of small in-channel detention ponds with fixed-outlet structures is considered a watershed change. In the latter case, the effect of the flood storage system is consistent from storm to storm and is not subject to nonrandom changes in operational practices as evidenced by human intervention. Many of the examples provided herein relate to urbanization, but the techniques presented are applicable to watershed change due to any factor.

The techniques described in this book should be used only where physical evidence of watershed change exists in a significant portion of the watershed. Guidelines for determining what constitutes a significant portion of the watershed are not universally agreed on but, as an example, Sauer et al. (1983) used watersheds that had more than 15% of the drainage area covered with industrial, commercial, or residential development in a nationwide study of flood characteristics for urban watersheds.

The arbitrary application of the various statistical analyses described in subsequent chapters may provide contradictory results. Each statistical test is based on assumptions that may or may not be applicable to a particular situation. Ultimately, it is the responsibility of the analyst to use hydrologic and engineering judgment in conjunction with the statistical analyses to reach a valid conclusion.

1.5 DETECTION OF NONHOMOGENEITY

Even homogeneous flood records contain a considerable amount of random variation. Outliers exacerbate the difficulty of detecting a trend or episodic variation in a record suspected of representing the effects of watershed change. Separating variation associated with the trend from variation due to outliers and random variation initially requires the detection of the trend so that it can be accurately characterized. This then enables the trend to be modeled and separated from the random variation. Outliers must also be detected; otherwise variation associated with an outlier may be incorrectly attributed to the watershed change or lumped in with the random variation.

The effects of nonhomogeneity due to watershed change should be assessed from both graphical presentations and statistical testing of the data. Graphical analyses have the advantage of allowing the hydrologist to incorporate other information into the decision making, such as physical processes or qualitative data associated with independent variables. The disadvantages of graphical analyses are that they allow for subjective interpretation and that graphing is largely limited to two or three dimensions at a time. In addition, graphical analyses do not provide a calibrated model of the trend, although they may infer one or more reasonable alternatives. In spite of these disadvantages, graphical analysis is still a reasonable first step in analyzing data collected from nonstationary hydrologic processes.

Following the graphical analyses, nonstationary data should be subjected to analyses using statistical tests. Given the large array of available statistical tests, knowing the characteristics and sensitivities of the alternative tests is important. Using a test that is not sensitive to a particular type of nonhomogeneity may result in a failure to detect the nonhomogeneity. For example, some statistical tests are appropriate for detecting abrupt changes while others are useful for detecting gradual trends. The use of an abrupt-change test when the data sequence involves a secular trend over most of the period of record may lead to the incorrect determination that the change did not significantly influence the measured data.

Statistical tests are very useful for analyzing data suspected of being nonhomogeneous; however, tests have their limitations. First, they do not model the effect. Instead, the test is useful only for deciding whether a significant effect exists. Second, statistical tests give only yes-or-no answers, that is, a test will suggest that nonhomogeneity exists in the data or it does not. Thus, the strength of the nonhomogeneity is not qualified. Third, statistical tests require the selection of measures of risk of making incorrect decisions. The procedures for selecting these risk levels are not systematic and often are subjective. Thus, unless the risk measures are selected on the basis of hydrologic considerations, the decisions may defy expectations based on knowledge of the effects of watershed change. In spite of these disadvantages, it is always advisable to subject data suspected of being nonstationary to statistical testing. It provides a measure of credibility to the decision making.

1.6 MODELING OF NONHOMOGENEITY

Assuming that the graphical and statistical analyses for detection suggest that watershed change introduced an effect into measured data, the next task is to model the change. The graphical analyses may have suggested a functional form that would be appropriate. Other sources of data may suggest when the cause of the nonhomogeneity began. For example, construction and building permit data may indicate when urbanization began within a watershed and the growth of development over time. This knowledge may then serve as input to the modeling process.

The following four stages are usually recognized in the development of models: conceptualization, formulation, calibration, and verification. Conceptualization accounts for as much as 80% of the effort in developing models. It involves the composite of all of the thought processes that take place in analyzing a new problem and devising a solution. It is an indefinable mixture of art and science (James, 1970). Science provides a knowledge basis for conceptualization. Included in this base are models developed for previous problems, intellectual abstractions of logical analyses of processes, and mathematical and numerical methodologies, and the hard core of observed hydrologic reality. However, the assembly into an "adequate and efficient" problem solution is an art.

Tremendous freedom in practical conceptualization was gained when electronic computers became everyday working tools. Conceptualization with no hope of activation for prediction is meaningless in applied hydrology. The high-speed capacity of the computer opened up entirely new fields of mathematical and numerical analyses. These new techniques, in turn, allowed wider scope in conceptualization. Formulation, the next stage in model development, no longer placed a severe limitation on conceptualization. It was no longer necessary to gear practical conceptualization to such limited formulations as closed integrals or desk calculator numerics.

The formulation stage of model development means the conversion of concepts to forms for calculation. Formulation seems to imply a mathematical equation or formula. Certainly, equations may be model forms. However, any algorithm or sequence of calculations that converts inputs to outputs is an acceptable formulation. Such a sequence may contain equations or graphical relations or it may include "table look-up." It may also include highly sophisticated finite difference solutions of differential equations. Model formulation is somewhat like the preparation of a detailed flowchart of a problem. In summary, conceptualization involves deciding what effects the model will simulate, while formulation is the phase where functional forms are selected to represent the concepts. Conceptualization involves determining which specific hydrologic processes are important to solving the problem at hand, while formulation provides a structural representation of the selected processes.

A formulated model cannot provide predictions or forecasts. Before a model is ready for use, it must be calibrated for the conditions in which it will be used. For example, a formulated rainfall-runoff model may be equally applicable to humid and semiarid environments because of the physical processes conceptualized and the functional forms selected. However, the model coefficients need numerical values before the model can be used for prediction, and the coefficients of the calibrated rainfall-runoff model for the semiarid environment may be quite different from those used in a model for humid environment. Calibration to the relevant conditions is a necessary task and an important determinant of the accuracy of future predictions.

Failure at any of the first three stages (conceptualization, formulation, and calibration) can be the downfall of a model. This is especially true in the modeling of hydrologic change. Calibration represents the extraction of knowledge from data. An improperly conceptualized or formulated model will not allow even the most sophisticated calibration method to extract the knowledge needed to understand the hydrologic processes modeled. Nonhomogeneous data, such as those from a watershed undergoing change, require methods of analysis that will uncover the continually changing importance of the hydrologic processes involved. Only by detecting and accurately modeling changes in physical processes can a model serve its purposes.

The last stage, testing or validation, is equally important. The primary criterion by which a model is evaluated is rationality. Does it provide accurate predictions under a wide array of conditions, even conditions beyond the range of the data used to calibrate the model? A model must do more than provide predictions. It must be able to provide measures of effects, such as the effects of watershed change. Even a model achieves high goodness-of-fit statistics during calibration may not provide accurate assessments of effects if the fitted coefficients and the functional forms are not rational. Sensitivity analyses and simulation studies are two tools essential to model testing. They enable the model to be tested not only for prediction, but also for its ability to accurately reflect the effects of changes in conditions.

1.7 PROBLEMS

1-1 Discuss how development of a 1-acre parcel of land affects the dominance of the physical watershed processes. Assume that the predevelopment condition is woods, while after development the lot is residential.

1-2 Using Manning's equation, discuss how clearing a stream reach of native vegetation changes the discharge characteristics.

1-3 Would you expect urban development of 10 acres in the upper reaches of a 500-acre watershed to be detectable in a continuous flood record? Explain. Would the effect be noticeable if the 10 acres were near the watershed outlet? Explain.

1-4 Discuss how the effects of an episodic event, such as a hurricane, might influence the flood record of a 200-acre watershed. Assume that the year in which the hurricane occurred was near the middle of the flood record.

1-5 What model components would be necessary for a model to be capable of simulating the effects of watershed change?

1-6 For a 150-acre watershed, how might the installation of numerous small stormwater-detention basins influence the flood record, including both peak discharges and baseflow?

1-7 How might the effects of urbanization on a flood record be detected if specific knowledge of the timing and location of gradual watershed changes were not available?

1-8 Discuss how an outlier due to a hurricane-generated flood might make it difficult to detect the hydrologic effects of gradual urbanization of a watershed.

1-9 What effect would gradual, natural afforestation have on a record of annual-maximum flood hydrographs over a period of 20 years? Discuss effects on the hydrograph characteristics, not only effects on the peak.

2 Introduction to Time Series Modeling

2.1 INTRODUCTION

Time series modeling is the analysis of a temporally distributed sequence of data or the synthesis of a model for prediction in which time is an independent variable. In many cases, time is not actually used to predict the magnitude of a random variable such as peak discharge, but the data are ordered by time. Time series are analyzed for a number of reasons. One might be to detect a trend due to another random variable. For example, an annual maximum flood series may be analyzed to detect an increasing trend due to urban development over all or part of the period of record. Second, time series may be analyzed to formulate and calibrate a model that would describe the time-dependent characteristics of a hydrologic variable. For example, time series of low-flow discharges might be analyzed in order to develop a model of the annual variation of base flow from agricultural watersheds. Third, time series models may be used to predict future values of a time-dependent variable. A continuous simulation model might be used to estimate total maximum daily loads from watersheds undergoing deforestation.

Methods used to analyze time series can also be used to analyze spatial data of hydrologic systems, such as the variation of soil moisture throughout a watershed or the spatial transport of pollutants in a groundwater aquifer. Instead of having measurements spaced in time, data can be location dependent, possibly at some equal interval along a river or down a hill slope. Just as time-dependent data may be temporally correlated, spatial data may be spatially correlated. The extent of the correlation or independence is an important factor in time- and space-series modeling. While the term *time series modeling* suggests that the methods apply to time series, most such modeling techniques can also be applied to space series.

Time and space are not causal variables; they are convenient parameters by which we bring true cause and effect into proper relationships. As an example, evapotranspiration is normally highest in June. This maximum is not caused by the month, but because insolation is highest in June. The seasonal time of June can be used as a model parameter only because it connects evapotranspiration and insolation.

In its most basic form, time series analysis is a bivariate analysis in which time is used as the independent or predictor variable. For example, the annual variation of air temperature can be modeled by a sinusoidal function in which time determines the point on the sinusoid. However, many methods used in time series analysis differ from the bivariate form of regression in that regression assumes independence among the individual measurements. In bivariate regression, the order of the x-y data pairs is not important. Conversely, time series analysis recognizes a time dependence and attempts to use this dependence to improve either the understanding of the underlying physical processes or the accuracy of prediction. More specifically, time series are analyzed to separate the systematic variation from the nonsystematic variation in order to explain the time-dependence characteristics of the data where some of the variation is time dependent. Regression analysis is usually applied to unordered data, while the order in a time series is an important characteristic that must be considered. Actually, it may not be fair to compare regression with time series analysis because regression is a method of calibrating the coefficients of an explicit function, while time series analysis is much broader and refers to an array of data analysis techniques that handle data in which the independent variable is time (or space). The principle of least squares is often used in time series analysis to calibrate the coefficients of explicit time-dependent models.

A time series consists of two general types of variation, systematic and nonsystematic. For example, an upward-sloping trend due to urbanization or the annual variation of air temperature could be modeled as systematic variation. Both types of variation must be analyzed and characterized in order to formulate a model that can be used to predict or synthesize expected values and future events. The objective of the analysis phase of time series modeling is to decompose the data so that the types of variation that make up the time series can be characterized. The objective of the synthesis phase is to formulate a model that reflects the characteristics of the systematic and nonsystematic variations.

Time series modeling that relies on the analysis of data involves four general phases: detection, analysis, synthesis, and verification. For the detection phase, effort is made to identify systematic components, such as secular trends or periodic effects. In this phase, it is also necessary to decide whether the systematic effects are significant, physically and possibly statistically. In the analysis phase, the systematic components are analyzed to identify their characteristics, including magnitudes, form, and duration over which the effect exists. In the synthesis phase, the information from the analysis phase is used to assemble a model of the time series and evaluate its goodness of fit. In the final phase, verification, the model is evaluated using independent data, assessed for rationality, and subjected to a complete sensitivity analysis. Poor judgment in any of the four phases will result in a less-than-optimum model.

2.2 COMPONENTS OF A TIME SERIES

In the decomposition of a time series, five general components may be present, all of which may or may not be present in any single time series. Three components can be characterized as systematic: secular, periodic, and cyclical trends. Episodic events and random variation are components that reflect sources of nonsystematic variation. The process of time series analysis must be viewed as a process of identifying and separating the total variation in measured data into these five components. When a time series has been analyzed and the components accurately characterized, each component present can then be modeled.

2.2.1 Secular Trends

A secular trend is a tendency to increase or decrease continuously for an extended period of time in a systematic manner. The trend can be linear or nonlinear. If urbanization of a watershed occurs over an extended period, the progressive increase in peak discharge characteristics may be viewed as a secular trend. The trend can begin slowly and accelerate upward as urban land development increases with time. The secular trend can occur throughout or only over part of the period of record.

If the secular trend occurs over a short period relative to the length of the time series, it is considered an abrupt change. It may appear almost like an episodic event, with the distinction that a physical cause is associated with the change and the cause is used in the modeling of the change. If the secular trend occurs over a major portion or all of the duration of the time series, it is generally referred to as a gradual change. Secular trends are usually detected by graphical analyses. Filtering techniques can be used to help smooth out random functions. External information, such as news reports or building construction records, can assist in identifying potential periods of secular trends.

Gradual secular trends can be modeled using typical linear and nonlinear functional forms, such as the following:

$$\text{linear: } y = a + bt \tag{2.1a}$$

$$\text{polynomial: } y = a + bt + ct^2 \tag{2.1b}$$

$$\text{power: } y = at^b \tag{2.1c}$$

$$\text{reciprocal: } y = \frac{1}{a + bt} \tag{2.1d}$$

$$\text{exponential: } y = ae^{-bt} \tag{2.1e}$$

$$\text{logistic: } y = \frac{a}{1 + e^{-bt}} \tag{2.1f}$$

in which y is the time series variable; a, b, and c are empirical constants; and t is time scaled to some zero point. In addition to the forms of Equations 2.1, composite or multifunction forms can be used (McCuen, 1993).

Example 2.1

Figure 2.1 shows the annual peak discharges for the northwest branch of the Anacostia River at Hyattsville, Maryland (USGS gaging station 01651000) for water years 1939 to 1988. While some development occurred during the early years of the record, the effect of that development is not evident from the plot. The systematic variation associated with the development is masked by the larger random variation that is inherent to flood peaks that occur under different storm events and when the antecedent soil moisture of the watershed is highly variable. During the early 1960s,

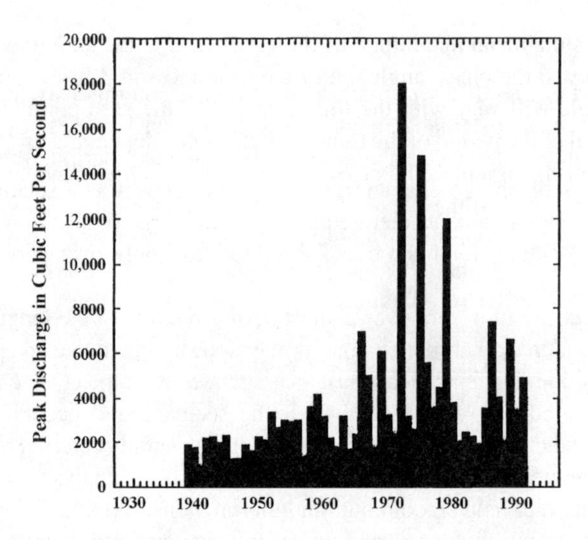

FIGURE 2.1 Annual maximum peak discharge for the Northwest Branch of the Anacostia River near Hyattsville, Maryland.

development increased significantly with the hydrologic effects apparent in Figure 2.1. The peak flows show a marked increase in both the average and variation of the peaks.

To model the secular variation evident in Figure 2.1, a composite model (McCuen, 1993) would need to be fit with a "no-effect" constant used before the mid-1950s and a gradual, nonlinear secular trend for the 1970s and 1980s. After 1980, another "no-effect" flat line may be appropriate. The specific dates of the starts and ends of these three sections of the secular trend should be based on records indicating when the levels of significant development started and ended. The logistic model of Equation 2.1f may be a reasonable model to represent the middle portion of the secular trend.

2.2.2 Periodic and Cyclical Variations

Periodic trends are common in hydrologic time series. Rainfall, runoff, and evaporation rates often show periodic trends over an annual period. Air temperature shows distinct periodic behavior. Seasonal trends may also be apparent in hydrologic data and may be detected using graphical analyses. Filtering methods may be helpful to reduce the visual effects of random variations. Appropriate statistical tests can be used to test the significance of the periodicity. The association of an apparent periodic or cyclical trend with a physical cause is generally more important than the results of a statistical test. Once a periodic trend has been shown, a functional form can be used to represent the trend. Quite frequently, one or more sine functions are used to represent the trend:

$$f(t) = A\sin(2\pi f_0 t + \theta) + \overline{Y} \tag{2.2}$$

in which $\bar{Y}$ is s the mean magnitude of the variable, A is the amplitude of the trend, f_0 the frequency, θ the phase angle, and t the time measured from some zero point. The phase angle will vary with the time selected as the zero point. The frequency is the reciprocal of the period of the trend, with the units depending on the dimensions of the time-varying variable. The phase angle is necessary to adjust the trend so that the sine function crosses the mean of the trend at the appropriate time. The values of A, f_0, and θ can be optimized using a numerical optimization method. In some cases, f_0 may be set by the nature of the variable, such as the reciprocal of 1 year, 12 months, or 365 days for an annual cycle.

Unlike periodic trends, cyclical trends occur irregularly. Business cycles are classic examples. Cyclical trends are less common in hydrology, but cyclical behavior of some climatic factors has been proposed. Sunspot activity is cyclical.

Example 2.2

Figure 2.2 shows elevation of the Great Salt Lake surface for the water years 1989 to 1994. The plot reveals a secular trend, probably due to decreased precipitation in the region, periodic or cyclical variation in each year, and a small degree of random variation. While the secular decline is fairly constant for the first 4 years, the slope of the trend appears to decline during the last 2 years. Therefore, a decreasing nonlinear function, such as an exponential, may be appropriate as a model representation of the secular trend.

The cyclical component of the time series is not an exact periodic function. The peaks occur in different months, likely linked to the timing of the spring snowmelt. The peak occurs as early as April (1989 and 1992) and as late as July (1991). Peaks also occur in May (1994) and June (1990, 1993). While the timing of the maximum amplitude of the cyclical waves is likely related to the temperature cycle, it may be appropriate to model the cyclical variation evident in Figure 2.2 using a periodic function (Equation 2.2). This would introduce some error, but since the actual month

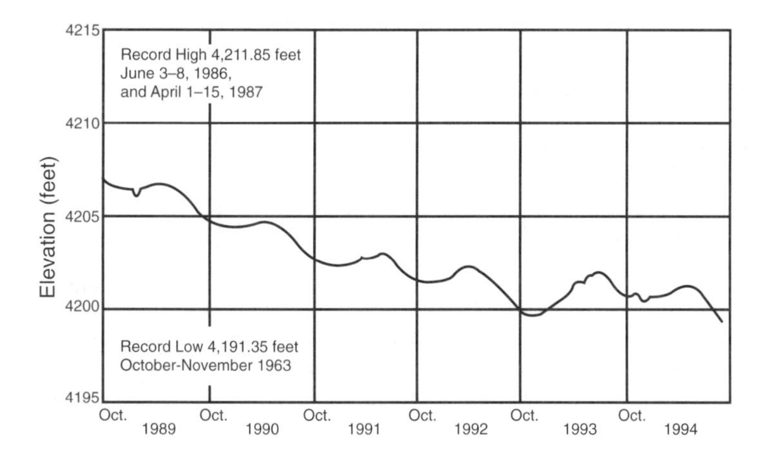

FIGURE 2.2 Variation of water surface elevation of Great Salt Lake (October 1988 to August 1994).

of the peak cannot be precisely predicted in advance, the simplification of a constant 12-month period may be necessary for a simple model. If a more complex model is needed, the time of the peak may depend on another variable. If the dependency can be modeled, better accuracy may be achieved.

2.2.3 Episodic Variation

Episodic variation results from "one-shot" events. Over a long period, only one or two such events may occur. Extreme meteorological events, such as monsoons or hurricanes, may cause episodic variation in hydrological data. The change in the location of a recording gage may also act as an episodic event. A cause of the variation may or may not be known. If the cause can be quantified and used to estimate the magnitude and timing of the variation, then it is treated as an abrupt secular effect. Urbanization of a small watershed may appear as an episodic event if the time to urbanize is very small relative to the period of record. The failure of an upstream dam may produce an unusually large peak discharge that may need to be modeled as an episodic event. If knowledge of the cause cannot help predict the magnitude, then it is necessary to treat it as random variation.

The identification of an episodic event often is made with graphical analyses and usually requires supplementary information. Although extreme changes may appear in a time series, one should be cautious about labeling a variation as an episodic event without supporting data. It must be remembered that extreme events can be observed in any set of measurements on a random variable. If the supporting data do not provide the basis for evaluating the characteristics of the episodic event, one must characterize the remaining components of the time series and use the residual to define the characteristics of the episodic event. It is also necessary to distinguish between an episodic event and a large random variation.

Example 2.3

Figure 2.3 shows the time series of the annual maximum discharges for the Saddle River at Lodi, New Jersey (USGS gaging station 01391500), from 1924 to 1988. The watershed was channelized in 1968, which is evident from the episodic change. The characteristics of the entire series and the parts of the series before and after the channelization are summarized below.

		Discharge (cfs)		Logarithms		
Series	n	Mean	Standard Deviation	Mean	Standard Deviation	Skew
Total	65	1660	923	3.155	0.2432	0.0
Pre	44	1202	587	3.037	0.1928	0.2
Post	21	2620	746	3.402	0.1212	0.1

The discharge series after completion of the project has a higher average discharge than prior to the channelization. The project reduced the roughness of the channel and increased the slope, both of which contributed to the higher average flow rate.

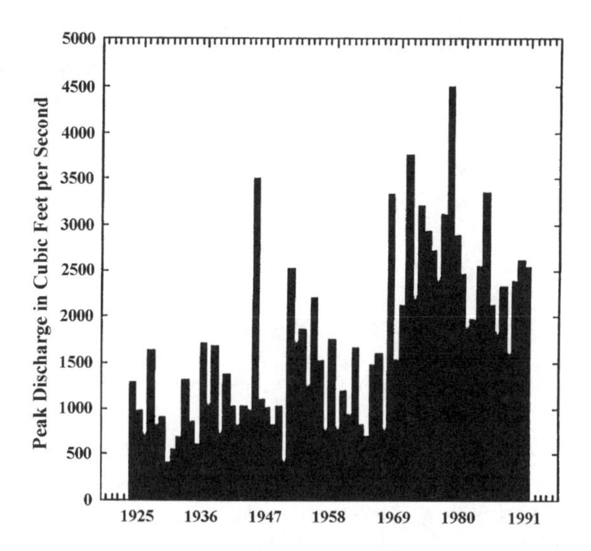

FIGURE 2.3 Annual maximum peak discharges for Saddle River at Lodi, New Jersey.

The reduction in variance of the logarithms is due to the removal of pockets of natural storage that would affect small flow rates more than the larger flow rates. The skew is essentially unchanged after channelization.

The flood frequency characteristics (i.e., moments) before channelization are much different than those after channelization, and different modeling would be necessary. For a flood frequency analysis, separate analyses would need to be made for the two periods of record. The log-Pearson type-III models for pre- and post-channelization are:

$$x = 3.037 + 0.1928K$$

and

$$x = 3.402 + 0.1212K,$$

respectively, in which K is the log-Pearson deviate for the skew and exceedance probability. For developing a simulation model of the annual maximum discharges, defining the stochastic properties for each section of the record would be necessary.

2.2.4 Random Variation

Random fluctuations within a time series are often a significant source of variation. This source of variation results from physical occurrences that are not measurable; these are sometimes called environmental factors since they are considered to be uncontrolled or unmeasured characteristics of the physical processes that drive the system. Examples of such physical processes are antecedent moisture levels, small amounts of snowmelt runoff that contribute to the overall flow, and the amount of vegetal cover in the watershed at the times of the events.

The objective of the analysis phase is to characterize the random variation. Generally, the characteristics of random variation require the modeling of the secular, periodic, cyclical, and episodic variations, subtracting these effects from the measured time series, and then fitting a known probability function and the values of its parameters to the residuals. The normal distribution is often used to represent the random fluctuations, with a zero mean and a scale parameter equal to the standard error of the residuals.

The distribution selected for modeling random variation can be identified using a frequency analysis. Statistical hypothesis tests can be used to verify the assumed population. For example, the chi-square goodness-of-fit test is useful for large samples, while the Kolmogorov–Smirnov one-sample test can be used for small samples. These methods are discussed in Chapter 9.

2.3 MOVING-AVERAGE FILTERING

Moving-average filtering is a computational technique for reducing the effects of nonsystematic variations. It is based on the premise that the systematic components of a time series exhibit autocorrelation (i.e., correlation between adjacent and nearby measurements) while the random fluctuations are not autocorrelated. Therefore, the averaging of adjacent measurements will eliminate the random fluctuations, with the remaining variation converging to a description of the systematic trend.

The moving-average computation uses a weighted average of adjacent observations. The averaging of adjacent measurements eliminates some of the total variation in the measured data. Hopefully, the variation smoothed out or lost is random rather than a portion of the systematic variation. Moving-average filtering produces a new time series that should reflect the systematic trend. Given a time series Y, the filtered series $\hat{Y}$ is derived by:

$$\hat{Y}_i = \sum_{j=1}^{m} w_j Y_{i+j-0.5(m+1)} \qquad \text{for } i = 0.5(m+1), \ldots, n - \frac{1}{2}(m-1) \tag{2.3}$$

in which m is the number of observations used to compute the filtered value (i.e., the smoothing interval), and w_j is the weight applied to value j of the series Y. The smoothing interval is generally an odd integer, with 0.5 $(m - 1)$ values of Y before observation i and 0.5 $(m - 1)$ values of Y after observation i used to estimate the smoothed value $\hat{Y}$. A total of $(m - 1)$ observations is lost; that is, while the length of the measured time series equals n, the smoothed series, $\hat{Y}$, only has $n - m + 1$ values. The simplest weighting scheme would be the arithmetic mean (i.e., $w_j = 1/m$):

$$\hat{Y}_t = \frac{1}{m} \sum_{j=1}^{m} Y_{i+j-0.5(m-1)} \tag{2.4}$$

Other weighting schemes often give the greatest weight to the central point in the interval, with successively smaller weights given to points farther removed from the central point. For example, if weights of 0.25 were applied to the two adjacent time periods and a weight of 0.5 to the value at the time of interest, then the moving-average filter would have the form:

$$\hat{Y}_t = 0.25Y_{t-1} + 0.5Y_t + 0.25Y_{t+1} \tag{2.5}$$

Moving-average filtering has several disadvantages. First, $m - 1$ observations are lost, which may be a serious limitation for short record lengths. Second, a moving-average filter is not itself a mathematical representation, and thus forecasting with the filter is not possible; a functional form must still be calibrated to forecast any systematic trend identified by the filtering. Third, the choice of the smoothing interval is not always obvious, and it is often necessary to try several intervals to identify the best separation of systematic and nonsystematic variation. Fourth, if the smoothing interval is not properly selected, it is possible to eliminate both systematic and nonsystematic variation.

Filter characteristics are important in properly identifying systematic variation. As the length of the filter is increased, an increasingly larger portion of the systematic variation will be eliminated along with the nonsystematic variation. For example, if a moving-average filter is applied to a sine curve that does not include any random variation, the smoothed series will also be a sine curve with an amplitude that is smaller than that of the time series. When the smoothing interval equals the period of the sine curve, the entire systematic variation will be eliminated, with the smoothed series equal to the mean of the series (i.e., $\bar{Y}$ of Equation 2.2). Generally, the moving-average filter is applied to a time series using progressively longer intervals. Each smoothed series is interpreted, and decisions are made based on the knowledge gained from all analyses.

A moving-average filter can be used to identify a trend or a cycle. A smoothed series may make it easier to identify the form of the trend or the period of the cycle to be fitted. A model can then be developed to represent the systematic component and the model coefficients evaluated with an analytical or numerical optimization method.

The mean square variation of a time series is a measure of the information content of the data. The mean square variation is usually standardized to a variance by dividing by the number of degrees of freedom, which equals $n - 1$, where n is the number of observations in the time series. As a series is smoothed, the variance will decrease. Generally speaking, if the nonsystematic variation is small relative to the signal, smoothing will only reduce the variation by a small amount. When the length of the smoothing interval is increased and smoothing begins to remove variation associated with the signal, then the variance of the smoothed series begins to decrease at a faster rate relative to the variance of the raw data. Thus, a precipitous drop in the variance with increasing smoothing intervals may be an indication that the smoothing process is eliminating some of the systematic variation. When computing the raw-data variance to compare with the variance of the smoothed series, it is common to only use the observations of the measured data that correspond to points on the smoothed series, rather than the variance of the entire time series. This series is called the truncated series. The ratio of the variances of the smoothed series to the truncated series is a useful indicator of the amount of variance reduction associated with smoothing.

Example 2.4

Consider the following time series with a record length of 8:

$$Y = \{13, 13, 22, 22, 22, 31, 31, 34\} \tag{2.6}$$

While a general upward trend is apparent, the data appear to resemble a series of step functions rather than a predominantly linear trend. Applying a moving-average filter with equal weights of one-third for a smoothing interval of three yields the following smoothed series:

$$\hat{Y} = \{16, 19, 22, 25, 28, 32\} \tag{2.7}$$

While two observations are lost, one at each end, the smoothed series still shows a distinctly linear trend. Of course, if the physical processes would suggest a step function, then the smoothed series would not be rational. However, if a linear trend were plausible, then the smoothed series suggests a rational model structure for the data of Equation 2.6. The model should be calibrated from the data of Equation 2.6, not the smoothed series of Equation 2.7. The nonsystematic variation can be assessed by computing the differences between the smoothed and measured series, that is, $e_i = \hat{Y}_i - Y_i$:

$$e_3 = \{3, -3, 0, 3, -3, 1\}$$

The differences suggest a pattern; however, it is not strong enough, given the small record length, to conclude that the data of Equation 2.6 includes a second systematic component.

The variance of the series of Equation 2.6 is 64.29, and the variance of the smoothed series of Equation 2.7 is 34.67. The truncated portion of Equation 2.6 has a variance of 45.90. Therefore, the ratio of the smoothed series to the truncated series is 0.76. The residuals have a variance of 7.37, which is 16% of the variance of the truncated series. Therefore, the variation in the residuals relative to the variation of the smoothed series is small, and thus the filtering probably eliminated random variation.

A moving-average filtering with a smoothing interval of five produces the following series and residual series:

$$\hat{Y}_5 = \{18.4, 22.0, 25.6, 28.0\} \tag{2.8a}$$

and

$$e_5 = \{-3.6, 0.0, 3.6, -3.0\}. \tag{2.8b}$$

The variance of the truncated series is 20.25, while the variances of $\hat{Y}_5$ and e_5 are 17.64 and 10.89, respectively. Thus, the variance ratios are 0.87 and 0.53. While the variance ratio for the smoothed series (Equation 2.8a) is actually larger than that of the smoothed series of Equation 2.7, the variance ratio of the residuals has increased greatly from that of the e_3 series. Therefore, these results suggest that the smoothing based on an interval of five eliminates too much variation from the series, even though the smoothed series of Equations 2.7 and 2.8a are nearly identical. The five-point smoothing reduces the sample size too much to allow confidence in the accuracy of the smoothed series.

Example 2.5

Consider the following record:

$$X = \{35, 36, 51, 41, 21, 19, 23, 27, 45, 47, 50, 58, 42, 47, 37, 36, 51, 59, 77, 70\} \tag{2.9}$$

The sample of 20 shows a slightly upper trend for the latter part of the record, an up-and-down variation that is suggestive of a periodic or cyclical component, and considerable random scatter. A three-point moving-average analysis with equal weights yields the following smoothed series:

$$\hat{X} = \{41, 43, 38, 27, 21, 23, 32, 40, 47, 52, 50, 49. 42, 40, 41, 49, 62, 69\} \tag{2.10}$$

The smoothed values for the first half of the series are relatively low, while the data seem to increase thereafter. The second point in the smoothed series is a local high and is eight time steps before the local high of 52, which is eight time steps before the apparent local peak of 69. The low point of 21 is nine time steps before the local low point of 40. These highs and lows at reasonably regular intervals suggest a periodic or cyclical component. A five-point moving-average yields the following smoothed series:

$$\begin{aligned}\tilde{\nabla}_5 = \{&36.8, 33.6, 31.0, 26.2, 27.0, 32.2, 38.4, 45.4, 48.4,\\ &48.8, 46.8, 44.0, 42.6, 46.0, 52.0, 58.6\}\end{aligned} \tag{2.11}$$

This smoothed series exhibits an up-and-down shape with an initial decline followed by an increase over six time intervals, a short dip over three time intervals, and then a final steep increase. While the length of the smoothed series (16 time intervals) is short, the irregularity of the local peaks and troughs does not suggest a periodic component, but possibly a cyclical component.

If a smoothing function with weights of 0.25, 0.50, and 0.25 is applied to the sequence of Equation 2.9, the smoothed series is

$$\tilde{\nabla}_3 = \{40, 45, 39, 26, 20, 23, 30, 41, 47, 51, 52, 47, 43, 39, 40, 49, 62, 71\} \tag{2.12}$$

TABLE 2.1
Variances of Series, Example 2.5

Smoothing Interval	Series Length	Variance of Truncated Series, S_t^2	Variance of Smoothed Series, S_s^2	Variance of Residual Series, S_e^2	$\frac{S_s^2}{S_t^2}$	$\frac{S_e^2}{S_t^2}$
3	18	217.8	150.8	25.9	0.692	0.119
5	16	161.8	87.2	49.9	0.539	0.308

The same general trends are present in Equation 2.12 and Equation 2.10.

To assess the magnitude of the nonsystematic variation, the smoothed series of Equation 2.10 and the actual series of Equation 2.9 can be used to compute the residuals, e:

$$e_3 = \{5, -8\ -3, 6, 2, 0, 5, -5, 0, 2, -8, 7, -5, 3, 5, -2, 3, -8\}. \quad (2.13)$$

These appear to be randomly distributed with a mean of −0.06 and a standard deviation of 5.093. The residuals for the smoothed series based on a five-point smoothing interval follow:

$$e_5 = \{-14.2, -7.4, 10.0, 7.2, 4.0, 5.2, -6.6, -1.6, -1.6, -9.2, 4.8, -3.0, 5.6, 10.0, 1.0, -0.4\} \quad (2.14)$$

The variances for the truncated series, the smoothed series, and the residual series are shown in Table 2.1, along with the ratios of the variances of both the smoothed and residual series to the truncated series. The large decrease in the percentage of variance of the five-interval smoothing suggests that some of the systematic variation is being removed along with the error variation.

In summary, the smoothed time series of Equation 2.12 appears to consist of a secular trend for the last half of the series, a periodic component, and random variation. While the record length is short, these observations could fit a model to the original series.

Example 2.6

Consider a time series based on the sine function:

$$Y(t) = 20 + 10 \sin t \quad (2.15)$$

in which t is an angle measured in degrees. Assume that the time series is available at an increment of 30 degrees (see column 2 of Table 2.2). For this case, the time series is entirely deterministic, with the values not corrupted with random variation. Columns 3 to 6 of Table 2.2 give the smoothed series for smoothing intervals of 3, 5, 7, and 9. Whereas the actual series $Y(t)$ varies from 10 to 30, the smoothed series vary from 10.89 to 29.11, from 12.54 to 27.46, from 14.67 to 25.33, and from 16.96 to 23.03, respectively. As the length of the smoothing interval increases, the amplitude of the smoothed series decreases. At a smoothing interval of 15, the smoothed series would be a horizontal line, with all values equal to 20.

TABLE 2.2
Characteristics of Smoothed Time Series

t (deg)	$Y(t)$	Smoothed Series with a Smoothing Interval of			
		3	5	7	9
0	20	—	—	—	—
30	25	24.55	—	—	—
60	28.66	27.89	26.46	—	—
90	30	29.11	27.46	25.33	—
120	28.66	27.89	26.46	24.62	22.63
150	25	24.55	23.73	22.67	21.52
180	20	20.00	20.00	20.00	20.00
210	15	15.45	16.27	17.33	18.48
240	11.34	12.11	13.54	15.38	17.37
270	10	10.89	12.54	14.67	16.96
300	11.34	12.11	13.54	15.38	17.37
330	15	15.45	16.27	17.33	18.48
360	20	20.00	20.00	20.00	20.00
30	25	24.55	23.73	22.67	21.52
60	28.66	27.89	26.46	24.62	22.63
90	30	29.11	27.46	25.33	23.03
120	28.66	27.89	26.46	24.62	—
150	25	24.55	23.73	—	—
180	20	20.00	—	—	—
210	15	—	—	—	—

This example illustrates the dilemma of moving-average filtering. The process reduces the total variation by smoothing both random and systematic variation. Smoothing of random variation is desirable, while eliminating part of the systematic variation is not. As the smoothing interval is increased, more of the systematic variation is smoothed out. However, if the smoothing interval is too short, an insufficient amount of the nonsystematic variation will be smoothed to allow identification of the signal.

To model time series in order to circumvent this dilemma, it is a good practice to develop several smoothed series for different smoothing intervals and evaluate each in an attempt to select the smoothed series that appears to provide the best definition of the signal. Developing general rules for determining the smoothing interval is difficult to impossible.

Example 2.7

A common problem in hydrologic modeling is evaluation of the effects of urban development on runoff characteristics, especially peak discharge. It is difficult to determine the hydrologic effects of urbanization over time because development occurs gradually in a large watershed and other factors can cause variation in runoff characteristics, such as variation in storm event rainfall or antecedent moisture conditions.

TABLE 2.3
Annual Flood Series and Smoothed Series for Pond Creek Watershed, 1945–1968

Year	Annual Maximum (cfs)	Smoothed Series (cfs)
1945	2000	
1946	1740	
1947	1460	
1948	2060	1720
1949	1530	1640
1950	1590	1580
1951	1690	1460
1952	1420	1360
1953	1330	1380
1954	607	1480
1955	1380	1610
1956	1660	1870
1957	2290	2040
1958	2590	2390
1959	3260	2560
1960	2490	2800
1961	3080	3620
1962	2520	3860
1963	3360	4020
1964	8020	4130
1965	4310	4300
1966	4380	
1967	6220	
1968	4320	

This example demonstrates the use of moving-average smoothing for detecting a secular trend in data. The data consist of the annual flood series from 1945 through 1968 for the Pond Creek watershed, a 64-square-mile watershed in north-central Kentucky. Between 1946 and 1966, the percentage of urbanization increased from 2.3 to 13.3, while the degree of channelization increased from 18.6% to 56.7% with most of the changes occurring after 1954. The annual flood series for the 24-year period is shown in Table 2.3.

The data were subjected to a moving-average smoothing with a smoothing interval of 7 years. Shorter intervals were attempted but did not show the secular trend as well as the 7-year interval. The smoothed series is shown in Figure 2.4. The smoothed series has a length of 18 years because three values are lost at each end of the series for a smoothing interval of 7 years. In Figure 2.4, it is evident that the smoothed series contains a trend. Relatively little variation exists in the smoothed series before the mid-1950s; this variation can be considered random. After urbanization became significant in the mid-1950s, the flood peaks appear to have increased,

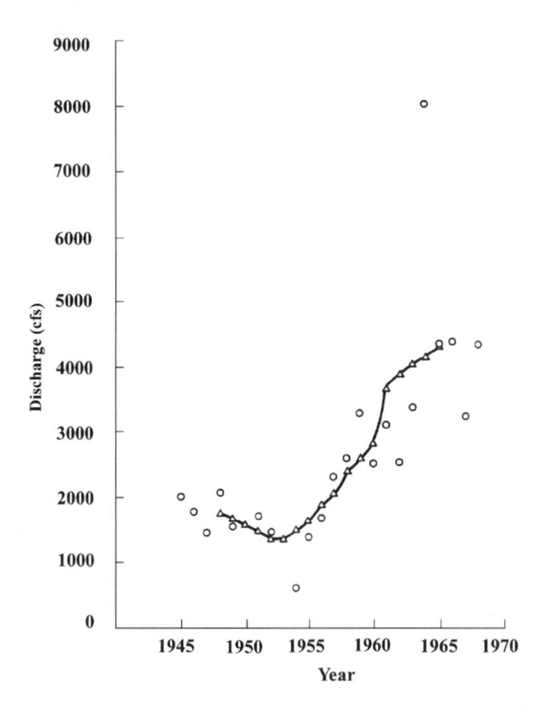

FIGURE 2.4 Annual flood series and smoothed series for Pond Creek Watershed, 1945–1968. Key: ○, annual flood series; Δ, smoothed series.

as is evident from the nearly linear upward trend in the smoothed series. It appears that the most appropriate model would be a composite (McCuen, 1993) with zero-sloped lines from 1945 to 1954 and from 1964 to 1968. A variable-intercept power model might be used for the period from 1954 to 1963. Fitting this formulation yielded the following calibrated model:

$$Q_i = \begin{cases} 1592 & \text{for} \quad t \le 10 \\ 400t^{0.6}e^{0.072(t-10)} & \text{for } 10 < t < 20 \\ 4959 & \text{for} \quad t \ge 10 \end{cases}$$

where $t = 1$, 10, 20, and 24 for 1945, 1954, 1964, and 1968, respectively. This model could be used to show the effect of urbanization in the annual maximum discharges from 1945–1968. The model has a correlation coefficient of 0.85 with Pond Creek data.

It is important to emphasize two points. First, the moving-average smoothing does not provide a forecast equation. After the systematic trend has been identified, it would be necessary to fit a representative equation to the data. Second, the trend evident in the smoothed series may not actually be the result of urban development. Some chance that it is due either to randomness or to another causal factor, such as increased rainfall, exists. Therefore, once a trend has been found to exist and a model structure hypothesized, a reason for the trend should be identified. The reasons may suggest a form of model to be used to represent the data.

Example 2.8

To formulate an accurate water-yield time series model, it is necessary to determine the variation of water yield over time. The moving-average filtering method can be used to identify the components of a water-yield time series. For a watershed that has not undergone significant changes in land use, one would not expect a secular trend. Therefore, one would expect the dominant element of the systematic variation to be a periodic trend that correlates with the seasonal variation in meteorological conditions. For such a component, identifying the period (1 year) and the amplitude and phase of the periodic component would be necessary. A moving-average filtering of the time series will enable the modeler to get reasonably accurate initial estimates of these elements so that a formal model structure can be fit using numerical optimization. For locations where two flooding seasons occur per year, two periodic trends may be indicated by moving-average filtering. Once the periodic trends are identified and subtracted from the time series, frequency analysis methods can be used to identify the population underlying the nonsystematic variations of the residuals.

A record of monthly runoff data (March 1944 to October 1950; $n = 80$) for the Chestuee Creek near Dial, Georgia, was subjected to a moving-average filtering. Both the time series and the smoothed series for a smoothing interval of 3 months are shown in Figure 2.5. As the smoothing interval was increased, the variation decreased significantly. Figure 2.5 shows that the monthly water-yield data are characterized by a dominant annual cycle, with a fairly constant base flow at the trough between each pair of peaks. The smoothed series closely approximates the actual water yield during the dry period of the year. However, there is considerable nonsystematic variation around the peaks. For example, the smoothed peaks for 1944, 1946, 1947, and 1949 are nearly equal, even though the actual monthly peaks for those years show considerable variation.

The smoothed series suggests that the systematic variation can be represented by a sine function with a mean of about 2.0 in., a period of 12 months, and a phase angle of about 3 months. The following model was calibrated with the 80 months of runoff depth:

$$Q_t = 1.955 + 1.632 \sin(2\pi t/12 + 2.076 \text{ radians})$$

in which t is the month number ($t = 1$ for October, $t = 4$ for January, $t = 12$ for September), and Q_t is the depth (inches). The model had a standard error of 1.27 in. and a correlation coefficient of 0.68. The predicted values show a slight underprediction in the spring months and a slight overprediction in the fall months. The residuals suggest that the nonsystematic variation has a nonconstant variance and thus the population of the nonsystematic variation is time dependent. This further suggests that separate analyses may need to be conducted for each month.

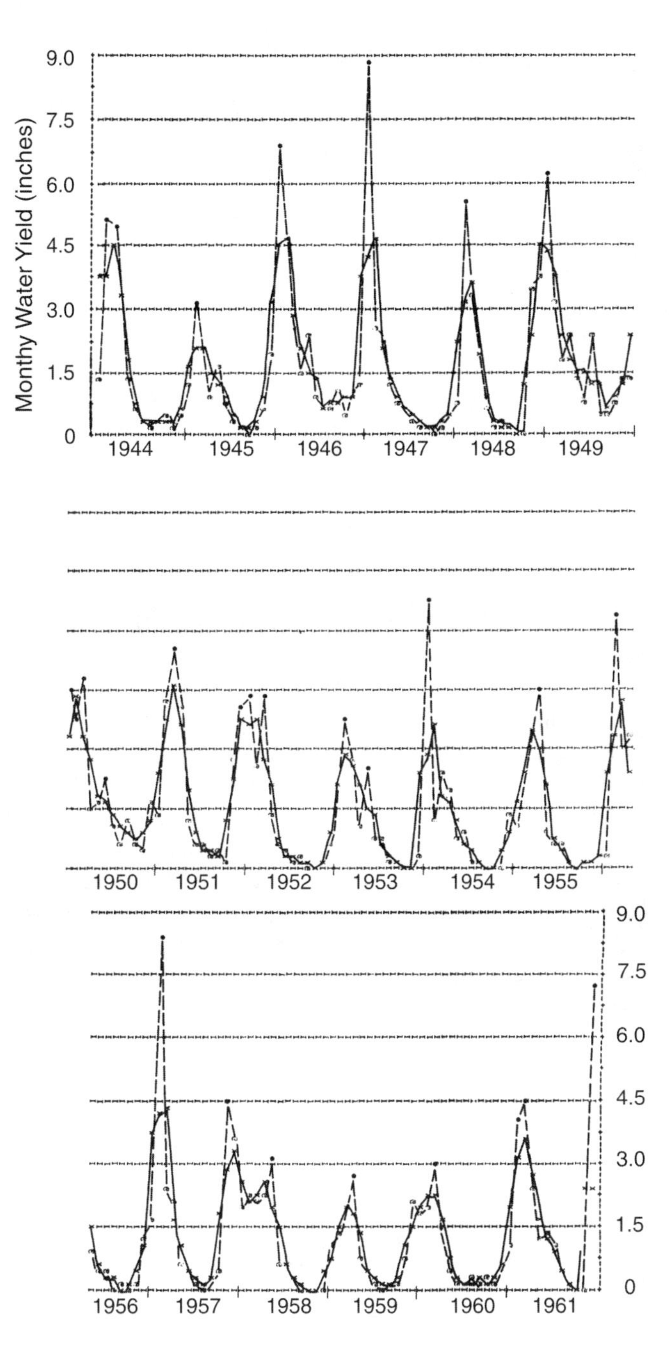

FIGURE 2.5 Monthly water yield and moving-average filtered series for Chestuee Creek Watershed, 1944–1961. Key: Solid line, moving-average filtered series; dashed line, monthly water yield.

2.4 AUTOCORRELATION ANALYSIS

Many time and space series are characterized by high autocorrelation. That is, adjacent values in a series may be correlated. For example, when the daily streamflow from a large river system is high above the mean flow, high flows should be expected for 1 or more days afterward; very little chance of having a flow of 100 cfs exists if the flow on the preceding day was 10,000 cfs. Similarly, measurements of soil moisture down the slope of a hill will probably be spatially correlated, possibly with the soil moisture increasing steadily down the slope. Serial correlation is sometimes called *persistence*. As will be indicated later, monthly water yields on large rivers usually show high correlations between flows measured in adjacent months. If this serial correlation is sufficiently high, one can take advantage of it in forecasting water yield. The depletion of snow cover during the spring melt is another hydrologic variable that exhibits serial correlation for intervals up to a week or more. The strength of the autocorrelation depends on the size of the watershed, the amount of snow cover, and the amounts and time distributions of precipitation during the melt season. Many snowmelt forecast models use this serial correlation improve forecast accuracy.

The computational objective of autocorrelation analysis is to analyze a time series to determine the degree of correlation between adjacent values. High values followed by high values and low values followed by low values suggest high autocorrelation. The strength of the correlation will depend on the time interval between the individual measured values. Actually, the analysis usually examines the changes in correlation as the separation distance increases. The separation distance is called the *lag* and is denoted by the letter tau, τ. Thus, the correlation coefficient computed with adjacent values is referred to as a lag-1 autocorrelation. The correlation between values separated by two time intervals is called the lag-2 autocorrelation.

A plot of the autocorrelation coefficient versus lag is called the *correlogram*. Computationally, the correlogram is computed by finding the value of the Pearson product-moment correlation coefficient for lags from one time unit to a maximum lag of approximately 10% of the record length. The autocorrelation function is computed as follows:

$$R(\tau)=\frac{\sum_{i=1}^{N-\tau} X_i X_{i+\tau}-\frac{1}{N-\tau}\left(\sum_{i=1}^{N-\tau} X_i\right)\left(\sum_{i=\tau+1}^{N} X_i\right)}{\left[\sum_{i=1}^{N-\tau} X_i^2-\frac{1}{N-\tau}\left(\sum_{i=1}^{N-\tau} X_i\right)^2\right]^{0.5}\left[\sum_{i=\tau+1}^{N} X_i^2-\frac{1}{N-\tau}\left(\sum_{i=\tau+1}^{N} X_i\right)^2\right]^{0.5}} \tag{2.16}$$

Obviously, for τ equal to zero, $R(\tau)$ equals 1. The graph of $R(\tau)$ versus τ is called the *correlogram*. As τ increases, the number of values used to compute $R(\tau)$ decreases, and the correlogram may begin to oscillate. The oscillations usually reflect random variation and are generally not meaningful. This is the reason for the empirical rule of thumb that suggests limiting the maximum value of τ to approximately 10% of N; for large sample sizes, this limitation may not be important.

TABLE 2.4
Analysis of Time Series

(1)	(2)	(3) Offset	(4)	(5)	(6)	(7)	(8)	(9)	(10)
t	X_t	X_{t+1}	$X_t X_{t+1}$	X_t^2	X_{t+1}^2	X_{t+2}	$X_t X_{t+2}$	X_t^2	X_{t+2}^2
1	4	—	—	—	—	—	—	—	—
2	3	4	12	9	16	—	—	—	—
3	5	3	15	25	9	4	20	25	16
4	4	5	20	16	25	3	12	16	9
5	6	4	24	36	16	5	30	36	25
6	5	6	30	25	36	4	20	25	16
7	8	5	40	64	25	6	48	64	36
Total	35	27	141	175	127	22	130	166	102

Autocorrelation in a time or space series can be caused by a secular trend or a periodic variation. The correlogram is useful for understanding the data and looking for the type of systematic variation that caused the autocorrelation. A strong secular trend will produce a correlogram characterized by high autocorrelation for small lags, with the autocorrelation decreasing slightly with increasing lag. A periodic component will be evident from a correlogram with a peak occurring at the lag that corresponds to the period of the component. For example, a time series on a monthly time interval that includes an annual cycle will have a correlogram with a spike at the 12-month lag. If both annual and semi-annual periodic components are present, the correlogram will have local peaks for both periods. In the presence of periodic components, the correlogram usually decreases fairly rapidly toward a correlation of zero, with spikes at the lags that reflect the periodic components.

Example 2.9

Consider the time series X_t given in column 2 of Table 2.4, which has a sample size of 7. Column 3 shows the values of X offset by one time period. The summations at the bottom of columns 3 to 6 and a partial summation of column 2 are entered into Equation 2.16 to compute the lag-1 autocorrelation coefficient:

$$R(1) = \frac{141 - (31)(27)/6}{[175 - (31)^2/6]^{0.5}[127 - (27)^2/6]^{0.5}} = 0.166 \tag{2.17}$$

The lag-2 autocorrelation coefficient is computed by offsetting the values in column 2 by two time increments (see column 7). The summations at the bottom of columns 7 to 10 and a partial summation of column 2 yield the lag-2 correlation coefficient:

$$R(2) = \frac{130 - 28(22)/5}{[166 - (28)^2/5]^{0.5}[102 - (22)^2/5]^{0.5}} = 0.983 \tag{2.18}$$

TABLE 2.5
Serial Correlation Coefficients, (*R*) as a Function of Time Lag (τ) in Years for Pond Creek Data

τ	*R*	τ	*R*
0	1.00	4	0.43
1	0.61	5	0.33
2	0.55	6	0.24
3	0.43		

These two autocorrelation coefficients can be verified graphically by plotting X_{t+1} and X_{t+2} versus X_t.

Example 2.10

The correlogram was computed using the annual maximum flood series for the Pond Creek watershed (column 2 of Table 2.3). The correlation coefficients are given in Table 2.5. The lag-1 correlation, 0.61, which is statistically significant at the 0.5% level, indicates that about 36% of the variation in an annual maximum discharge can be explained by the value from the previous year. Of course, this is an irrational interpretation because the annual maximum events are actually independent of each other. Instead, the serial correlation of 0.61 is indicative of the secular trend in the annual maximum series because of the increase in urbanization. Although correlograms are commonly used to detect periodic components in time series, this example indicates that secular trends can also cause significant serial correlations. The significant lag-1 correlation should not be viewed as an indication that a forecast model with time as the predictor variable can be calibrated to assist in predicting annual maximum floods. The significant correlation only indicates that a third variable (i.e., urbanization) that changes systematically with time causes the variation in the annual maximum flood data. If a forecast model is developed, the amount of impervious land cover could be used as a predictor variable.

The lag-2 correlation, while less than the lag-1 correlation, is also significant. The closeness of the lag-2 correlation to the lag-1 correlation and the gradual decrease of the lag-3, lag-4, and lag-5 autocorrelations suggest a secular trend rather than a periodic component.

Example 2.11

The Chestuee Creek water-yield data were also subject to a serial correlation analysis. The correlogram is shown in Figure 2.6 and the correlation coefficients in Table 2.6. The correlogram shows a periodic trend, with high values for lags in multiples of 12 months and low values starting at lag-6 and a period of 12 months. The lag-1 serial coefficient of 0.47, which corresponds to an explained variance of about 22%, indicates that monthly variation is not very predictable. This suggests that a lag-1 forecast model would not provide highly accurate predictions of monthly water yield for the

TABLE 2.6
Serial Correlation Coefficients (*R*) as a Function of Time Lag (τ) in Months for Chestuee Creek Monthly Water-Yield Data

τ	$R(\tau)$	τ	$R(\tau)$	τ	$R(\tau)$	τ	$R(\tau)$
0	1.00	14	0.18	28	–0.24	42	–0.41
1	0.44	15	–0.03	29	–0.38	43	–0.36
2	0.19	16	–0.26	30	–0.39	44	–0.27
3	–0.08	17	–0.35	31	–0.38	45	–0.12
4	–0.24	18	–0.41	32	–0.23	46	0.13
5	–0.34	19	–0.37	33	–0.03	47	0.33
6	–0.37	20	–0.24	34	0.21	48	0.49
7	–0.34	21	–0.03	35	0.42	49	0.49
8	–0.17	22	0.27	36	0.47	50	0.22
9	0.01	23	0.39	37	0.47	51	–0.04
10	0.23	24	0.43	38	0.20	52	–0.23
11	0.39	25	0.41	39	–0.06	53	–0.37
12	0.47	26	0.20	40	–0.22	54	–0.39
13	0.32	27	–0.01	41	–0.36		

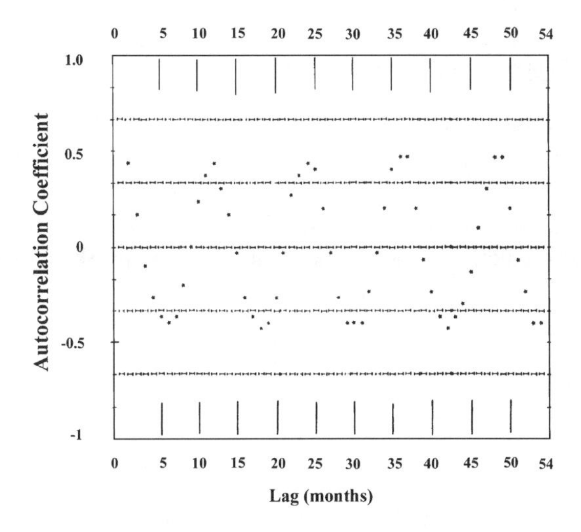

FIGURE 2.6 Correlogram for Chestuee Creek monthly streamflow.

watershed. Figure 2.6 suggests that the low lag-1 autocorrelation coefficient is due to large variation during the few months of high flows. From Figure 2.6, one would expect a lag-1 autoregressive model to provide reasonably accurate estimates during months with low flows, but probably not during months of high flows.

The lag-1 autoregression analysis was performed, with the following result:

$$Y_{t+1} = 0.9846 + 0.4591Y_t \tag{2.19}$$

The model of Equation 2.19 yields a correlation coefficient of 0.47 and a standard error ratio (S_e/S_y) of 0.883, both of which indicate that the model will not provide highly accurate estimates of Y_{t+1}; the stochastic component would be significant. The residuals were stored in a data file and subjected to a normal frequency analysis. The standardized skew of −2.66 indicates that the residuals cannot be accurately represented with a normal distribution. A frequency plot would show that the negative residuals depart significantly from the normal population curve. In addition, it is interesting that there were 56 positive residuals and only 23 negative residuals, rather than the expected equal split for a symmetric distribution such as the normal distribution.

In Example 2.6, a sine function was calibrated with the same data. It provided better goodness of fit than the autoregressive model of Equation 2.19. The sine function resulted in a correlation coefficient of 0.68, which represents an increase of 24% in explained variance. The sine function has the advantage of using three fitting coefficients, versus two for Equation 2.19. The sine function is, therefore, more flexible. The advantage of the autoregressive model is that, if a value is especially large or small, the model can at least attempt to predict a similar value for the next time period. The sine curve predicts the same value every year for any month. These results should not suggest that the sine function will always provide better goodness of fit than an autoregressive model.

2.5 CROSS-CORRELATION ANALYSIS

In some situations, one may wish to compute or forecast values for one time series using values from a second time series. For example, we may wish to use rainfall for time t (i.e., P_t) to predict runoff at time t or time $t + 1$ (i.e., RO_t or RO_{t+1}). Or we may wish to predict the runoff at a downstream site (i.e., Y_t) from the runoff at an upstream site for the preceding time period (i.e., X_{t-1}). The first objective of a cross-correlation analysis is to identify the significance of the correlation, and thus the predictability, between the two time series. If the cross-correlation coefficient suggests poor prediction accuracy, then the relationship is unlikely to be any more worthwhile than the autocorrelation model.

Cross-correlation analysis is computationally similar to autocorrelation analysis except that two time or space series are involved rather than one series offset from itself. Cross-correlation coefficients can be plotted against lag to produce a cross-correlogram. Distinct differences exist between auto- and cross-correlation. First, while the autocorrelation coefficient for lag-0 must be 1, the cross-correlation coefficient for lag-0 can take on any value between −1 and 1. Second, the peak of the cross-correlogram for two time series may peak at a value other than lag-0, especially when a physical cause for the lag between the two series exists. For example, the cross-correlogram between stream gages on the same river would probably show a peak on the cross-correlogram corresponding to the time lag most closely equal to the travel time of flow between the two gages. A third distinguishing characteristic of the cross-correlogram is that one may need to compute the correlations for both positive and negative lags.

While the autocorrelation function is a mirror image about lag $\tau = 0$, the cross-correlation function is not a mirror image. However, in some cases, only one side of the cross-correlogram is relevant because of the nature of the physical processes involved. As an example, consider the time series of rainfall X and runoff Y, which are measured at a time increment of 1 day. The rainfall on May 10, denoted as X_t, can influence the runoff on May 10, May 11, or May 12, which are denoted as Y_t, Y_{t+1}, and Y_{t+2}, respectively. Thus, positive lags can be computed appropriately. However, the rainfall on May 10 cannot influence the runoff on May 9 (Y_{t-1}) or May 8 (Y_{t-2}). Therefore, it would be incorrect to compute the cross-correlation coefficients for negative lags since they are not physically rational.

The cross-correlation coefficient $R_c(\tau)$ can be computed by modifying Equation 2.16 by substituting $Y_{i+\tau}$ for $X_{i+\tau}$:

$$R_c(\tau) = \frac{\sum_{i=1}^{N-|\tau|} X_i Y_{i+\tau} - \frac{1}{N-|\tau|}\left(\sum_{i=1}^{N-|\tau|} X_i\right)\left(\sum_{i=\tau+1}^{N} Y_i\right)}{\left[\sum_{i=1}^{N-|\tau|} X_i^2 - \frac{1}{N-|\tau|}\left(\sum_{i=1}^{N-|\tau|} X_i\right)^2\right]^{0.5}\left[\sum_{i=1+|\tau|}^{N} Y_i^2 - \frac{1}{N-|\tau|}\left(\sum_{i-1+|\tau|}^{N} Y_i\right)^2\right]^{0.5}} \tag{2.20}$$

It is important to note that the absolute value of τ is used in Equation 2.20. A plot of $R_t(\tau)$ versus τ is the *cross-correlogram.*

Example 2.12

Table 2.7 shows depths (in.) of rainfall X_t and runoff Y_t for a 7-month period. The lag-0 cross correlation coefficient is computed with Equation 2.20:

$$R_c(0) = \frac{51.56 - 26.1(13.4)/7}{[101.83 - (26.1)^2/7]^{0.5}[26.48 - (13.4)^2/7]^{0.5}} = 0.8258 \tag{2.21}$$

TABLE 2.7
Analysis of Cross-Correlation

Month	X_t	Y_t	Y_{t+1}	Y_{t+2}	X_tY_t	X_t^2	Y_t^2
April	5.0	2.5	2.1	2.0	12.50	25.00	6.25
May	4.8	2.1	2.0	1.3	10.08	23.04	4.41
June	3.7	2.0	1.3	1.7	7.40	13.69	4.00
July	2.8	1.3	1.7	2.0	3.64	7.84	1.69
August	3.6	1.7	2.0	1.8	6.12	12.96	2.89
September	3.3	2.0	1.8	—	6.60	10.89	4.00
October	2.9	1.8	—	—	5.22	8.41	3.24
Total	26.1	13.4	10.9	9.8	51.56	101.83	26.48

Since May rainfall cannot influence April runoff, the appropriate analysis is to lag runoff such that Y_t is correlated with X_{t-1}. Thus, the lag-1 cross-correlation coefficient is:

$$R_c(1) = \frac{42.81 - (23.2)(10.9)/6}{[93.42 - (23.2)^2/6]^{0.5}[20.23 - (10.9)^2/6]^{0.5}} = 0.5260 \tag{2.22}$$

The lag-2 cross-correlation coefficient relates April rainfall and June runoff and other values of X_{t-2} and Y_t:

$$R_c(2) = \frac{34.61 - (19.9)(8.8)/5}{[82.53 - (19.9)^2/5]^{0.5}[15.82 - (8.8)^2/5]^{0.5}} = -0.3939 \tag{2.23}$$

The cross-correlogram shows a trend typical of rainfall-runoff data. The lag correlation coefficient decreases with increasing lag. A descending trend in a cross-correlogram often reflects a periodic trend in monthly rainfall and is especially evident in large watersheds.

Example 2.13

In northern regions, floods may be the result of snowmelt runoff. In the springtime, a watershed may be partially or totally covered by snow of varying depth. The SCA (snow-covered area) is the fraction of the area covered by snow. When the temperature increases above freezing, the snow begins to melt and the SCA decreases. The variable used to reflect the rise in temperature is the degree-day factor, F_T, which is the number of degrees of temperature above some threshold, such as 32°F. As the F_T increases, the snow should melt, and SCA should decrease. Therefore, the relationship between F_T and SCA is indirect.

Figure 2.7 shows one season (1979) of record for the Conejos River near Magote, Colorado, which is a 282-mi^2 watershed. The SCA was 100% until May 9 (day 9 in Figure 2.7). It decreased steadily until reaching a value of 0 on day 51. Table 2.8 shows the autocorrelation and regression coefficients for lag-0 to lag-9. The smooth decline of SCA in Figure 2.7 is evident in the high autocorrelation coefficients even for lag-9.

The degree-day factor is also shown in Figure 2.7 for the same period of record. The nine-point moving-average series is also shown for F_T since the measured values include a significant amount of random variation. The smoothed series shows that F_T is relatively flat for the first week followed by a steep rise for more than a week, and then a leveling off followed by a more gradual upward trend about the time that the snow covered area is approaching zero.

A cross-correlation analysis was made between the SCA and F_T. The F_T on day t could influence the SCA on day $t + \tau$ but not on day $t - \tau$. Therefore, the cross-correlation analysis is presented only for the positive lags (see Table 2.9). The results yield correlations of approximately –0.7, where the negative sign only indicates that SCA decreases as F_T increases, which is rational. The correlation for lag-9 is the

TABLE 2.8
Autocorrelation and Autoregression Analysis for SCA

Lag	Serial R	Se (Standard Error)	Se/Sy	Autoregression Coefficients b_0	b_1
0	1.000	0.0000	0.000	0.0	1.0
1	0.999	0.0126	0.035	–0.0194	0.9989
2	0.998	0.0243	0.068	–0.0385	0.9959
3	0.995	0.0356	0.101	–0.0575	0.9919
4	0.991	0.0465	0.134	–0.0765	0.9870
5	0.986	0.0568	0.166	–0.0953	0.9810
6	0.980	0.0666	0.198	–0.1136	0.9736
7	0.973	0.0758	0.229	–0.1317	0.9651
8	0.966	0.0843	0.259	–0.1496	0.9555
9	0.957	0.0920	0.289	–0.1669	0.9445

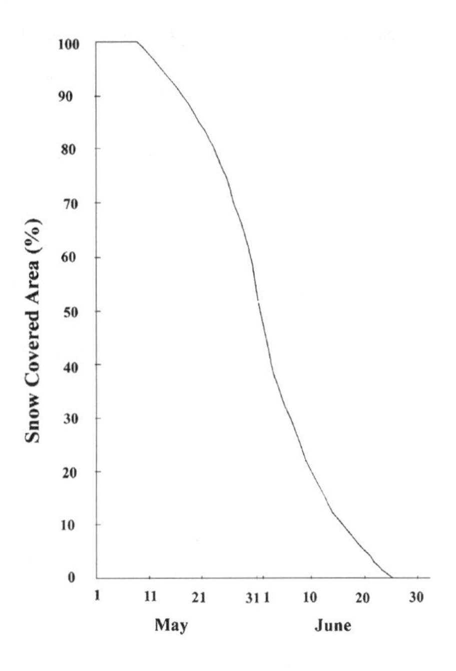

FIGURE 2.7 Snow-covered area versus date from Conejos River near Magote, Colorado (zone 2).

largest, but this only reflects the increase in the correlation as the sample size decreases. The most rational correlation is for lag-0. While this indicates an explained variation of 50%, the standard error ratio suggests the potential prediction accuracy of the cross-regression model is not good.

TABLE 2.9
Cross-Correlation and Cross-Regression Analyses between SCA and Degree-Day Factor F_T

Lag	R_c	Se (Standard Error)	Se/Sy	Cross-Regression Coefficients b_0	b_1
0	−0.721	0.252	0.693	0.956	−0.0395
1	−0.709	0.253	0.706	0.935	−0.0394
2	−0.698	0.254	0.716	0.914	−0.0392
3	−0.691	0.254	0.723	0.891	−0.0387
4	−0.685	0.252	0.728	0.867	−0.0381
5	−0.678	0.251	0.735	0.845	−0.0379
6	−0.671	0.249	0.741	0.822	−0.0376
7	−0.669	0.246	0.743	0.799	−0.0371
8	−0.686	0.236	0.728	0.784	−0.0369
9	−0.731	0.217	0.683	0.788	−0.0382

2.6 IDENTIFICATION OF RANDOM COMPONENTS

For some types of problems, it is only necessary to identify systematic components of the time series. For example, in developing a water-yield model that relates the volume of runoff for any month to the rainfall depths for the current and previous months, it is only necessary to identify the systematic variation between RO_t and both P_t and P_{t-1}. The resulting water-yield model would then be based solely on the deterministic elements of the time series. For other types of problems, most notably simulation, it is necessary to analyze the time series and identify and characterize both the systematic and nonsystematic components.

Assuming that the nonsystematic variation is due entirely to random variation (i.e., noise), it is usually sufficient to identify a probability function that can be used to represent the random variation. Both the function and values of the parameters of the function need to be identified. As an initial step, a histogram of the errors can be plotted to suggest possible distribution that could reasonably represent the random component. For the most likely candidate, a probability function can be tested using a frequency analysis of the errors defined by subtracting all systematic variation from the measured time series. In some cases, it is possible to calibrate simultaneously both the coefficients of the deterministic and random components using a numerical optimization strategy (McCuen, 1993). A hypothesis test on the distribution can be made using one of the methods discussed in Chapter 9. Such a test serves to support the findings of the graphical and frequency analyses.

2.7 AUTOREGRESSION AND CROSS-REGRESSION MODELS

When values of a single time series exhibit significant serial correlation, autoregression models are commonly used to predict future values. When values of two time series exhibit significant cross-correlation, cross-regression models can be used to predict future values of one time series from values of the second series. Autoregression and cross-regression models include both deterministic and stochastic components.

2.7.1 Deterministic Component

If the autocorrelogram indicates that Y_t and Y_{t+1} are correlated, then a lag-1 autoregression model can be used; the deterministic component for a stationary process could have the linear form

$$Y_{t+1} = \bar{Y} + b(Y_t - \bar{Y}) = a + bY_t \tag{2.24}$$

in which a is the autoregression intercept, b is the autoregression slope coefficient, and $\bar{Y}$ is the mean of Y. If both the lag-1 and lag-2 correlations are significant, then a two-lag model with the following deterministic component can be used:

$$\hat{Y}_{t+1} = \bar{Y} + b(Y_t - \bar{Y}) + c(Y_{t-1} - \bar{Y}) \tag{2.25}$$

The problem with the model depicted in Equation 2.25 is that Y_t and Y_{t-1} are also correlated and their intercorrelation may produce an irrational sign on b or c if a multiple regression analysis is used to evaluate b and c. If the moving-average filter identifies a linear trend, the trend can be represented by the following element:

$$\bar{Y} + d(t - t_0) \tag{2.26}$$

in which d is a fitting coefficient and t_0 is a time-offset coefficient; both d and t_0 must be fit by data analysis techniques. If the filtering identifies a periodic component, a model component such as Equation 2.2 can be used. When a trend or periodic component is used with an autoregressive element, the coefficients must be fit using numerical optimization (McCuen, 1993).

2.7.2 Stochastic Element

Once the deterministic component has been calibrated using measured data, the residuals (e_i) can be computed:

$$e_i = \hat{Y}_i - Y_i \tag{2.27}$$

The distribution of e_i is evaluated, possibly using frequency analysis or a visual assessment of a histogram. From the probability distribution of e_i, including estimates of the parameters, the stochastic component of the model can be formulated. A commonly used function is:

$$z_i S_e = z_i S_y (1 - R^2)^{0.5} \tag{2.28}$$

in which z_i is the random variate for the distribution of e_i, S_e is the standard error of estimate, S_y is the standard deviation of Y, and R is the autocorrelation coefficient.

2.7.3 Cross-Regression Models

Cross-regression models can be developed from autoregression models by substituting X_t for Y_t. Therefore, a stochastic lag-1 cross-regression model between rainfall for time t (P_t) and runoff for time $t + 1$ (RO_{t+1}) is

$$\hat{RO}_{t+1} = \overline{RO} + b(P_i - \bar{P}) + z_i \quad S_{RO}(1 - R_1^2)^{0.5} \tag{2.29}$$

in which $\overline{RO}$ is the mean runoff, b is the regression coefficient obtained by regressing the RO data on the P data, $\bar{P}$ is the mean rainfall, S_{RO} is the standard deviation of the runoff, and R_1 is the lag-1 cross-correlation coefficient. In some cases, Equation 2.29 uses logarithmic statistics so that the runoff values follow a lognormal distribution.

Example 2.14

The data of Example 2.13 was also used to develop a autoregression model for SCA. If a lag-1 model is used because the lag-1 autocorrelation coefficient is the largest, then Equation 2.24 becomes:

$$Y_{t+1} = -0.0194 + 0.9989Y_t \tag{2.30}$$

This yields a correlation coefficient of 0.999 and a standard error ratio of 0.035, both of which indicate very good accuracy. The standard error of estimate (Table 2.7) is 0.0126. This means that the stochastic component with assumption of a normal distribution would have the form of Equation 2.28, zS_e. Thus, the stochastic model of the process would be:

$$y_{t+1} = -0.0194 + 0.9989y_t + 0.0126z_t \tag{2.31}$$

Assuming that SCA would begin at a value of 1.0 in any year, a sequence could be generated using random normal deviates and Equation 2.31. The computed values may need to be constrained so they do not exceed 1 and are not smaller than 0.

2.8 PROBLEMS

2-1 Develop the structure of a random model for representing a flood series in which the mid-section has a linear secular trend (Equation 2.1a) and the two end portions are values of random processes with different means and standard deviations.

2-2 If Equation 2.1c is used to represent the secular trend of the annual percent imperviousness of a watershed for a 30-year record, what influences the magnitudes of the model parameters a and b?

2-3 If Equation 2.1e is used to represent the secular trend of daily SCA for a watershed, what factors influence the magnitudes of model parameters a and b?

2-4 Transform Equation 2.1d into a linear form by which the coefficients a and b can be calibrated using analytical least squares.

2-5 Discuss the pitfall of using the quadratic model of Equation 2.1b to represent a secular trend when the coefficients a, b, and c must be fitted with least squares.

2-6 A time series consists of the daily values over the length of the day (i.e., potential hours of sunshine). If the model depicted by Equation 2.2 is fitted to the 365 daily values at a particular latitude, what factors influence the magnitudes of the coefficients?

2-7 Propose a model structure to represent a random temporal sequence of annual maximum discharges for which the river was channelized during a year near the middle of the flood series.

2-8 Discuss the use of frequency analysis in assessing the characteristics of errors in a temporal model.

2-9 Discuss the implications of using a moving-average filter in which the weights do not sum to 1.

2-10 Develop a moving-average filter that has a set of weights that are applied to Y_t, Y_{t-1}, Y_{t-2}, and Y_{t-3} to estimate Y_t such that weight $w_j = 2w_{j-1}$ for each weight in the smoothing equation and the sum of the weights is equal to 1.

2-11 Discuss the effect on smoothing of increasing (a) the number of smoothing weights (m of Equation 2.3) and (b) the peakedness of the smoothing weights for a given m.

2-12 The smoothing weights w_j of Equation 2.3 can be positive or negative. What is the effect of negative weights on the smoothed series?

2-13 What effect does moving-average filtering have on periodic or cyclic systematic variation?

2-14 Use a moving-average filter with a smoothing interval of three and equal weights to smooth the following series: $X = \{5, 8, 6, 9, 5, 13, 16, 10, 12, 17, 11\}$. Discuss the nature of any systematic variation.

2-15 Using a moving-average filter with a smoothing interval of three and equals weights, smooth the following series: $Y = \{5, 8, 4, 9, 6, 15, 20, 17, 22, 31, 30\}$. Discuss the nature of any systematic variation.

2-16 Consider a 10-year series of monthly water-quality measurements that involve an annual periodic component, a linear trend downward due to a reduction in the amounts of fertilizer applied to fields, and random variation. Discuss how variation in each component would make it difficult to assess the downward linear trend.

2-17 Use three-point and five-point moving-average filters to smooth the daily degree-day factor (July and August, 1979) for the Conejos River near Magote, Colorado. Characterize the systematic and nonsystematic variation. $D = \{25.45, 24.57, 24.95, 23.57, 24.49, 24.99, 24.87, 27.34, 27.57, 29.57, 29.40, 30.69, 29.84, 29.25, 28.13, 25.83, 27.95, 25.14, 24.93, 25.19, 27.07, 26.16, 26.78, 29.57, 27.52, 27.61, 28.23, 27.81, 27.93, 27.99, 24.28, 24.49, 27.16, 25.95, 26.72, 28.13, 29.30, 27.92, 26.98, 24.60, 24.22, 23.34, 24.98, 21.95, 22.60, 19.37, 20.10, 20.28, 17.90, 17.28, 17.04, 17.45, 22.28, 23.49, 23.40, 21.28, 22.25, 19.80, 23.63, 22.13, 21.72, 23.28\}$.

2-18 Use three-point and five-point moving-average filters to smooth the daily discharge (July 1979) for the Conejos River near Magote, Colorado. Characterize the systematic and nonsystematic variation. Q = {2235, 2200, 2090, 1735, 1640, 1530, 1560, 1530, 1490, 1310, 1050, 1120, 1070, 990, 960, 1060, 985, 795, 720, 616, 566, 561, 690, 530, 489, 480, 476, 468, 468, 460, 384}.

2-19 Using Equation 2.16, discuss how large nonsystematic variation can mask the variation of a linear trend in a lag-1 autocorrelation coefficient.

2-20 Discuss the implications of each of the following correlograms in terms of modeling the underlying time series: (a) $R(\tau)$ = {1.0, 0.95, 0.90, 0.80}; (b) $R(\tau)$ = {1.0, 0.2, 0.1, -0.2}; and (c) $R(\tau)$ = {1.0, 0.5, 0.0, 0.1}.

2-21 What would be the general shape of a correlogram for each of the following types of systematic variation: (a) linear secular trend; (b) sinusoidal trend; (c) cyclical trend with an irregular period; (d) an exponential, decreasing secular trend; (e) an episodic event; and (f) random variation.

2-22 Compute the correlogram for the degree-day data of Problem 2-17 for 0 to 6 lags. Discuss the implications to modeling the underlying process.

2-23 Compute the correlogram for the discharge data of Problem 2-18 for 0 to 5 lags. Discuss the implications to modeling the underlying process.

2-24 Compute the correlogram for the following snow covered area for 0 to 6 lags. Discuss the implications to modeling the underlying process. X = {0.970, 0.960, 0.950, 0.938, 0.925, 0.915, 0.900, 0.890, 0.879, 0.863, 0.850, 0.839, 0.820, 0.809, 0.794, 0.780, 0.760, 0.749, 0.730, 0.710, 0.695, 0.675, 0.658, 0.640, 0.620, 0.600, 0.580, 0.560, 0.540, 0.520}.

2-25 Compute the cross-correlogram for the following daily degree-day factors X and daily streamflow Y data for −5 to +5 lags. Discuss the relevance of positive and negative lags. Discuss the implications to modeling the underlying process. X = {20.55, 22.30, 20.86, 20.24, 22.11, 22.36, 21.67, 19.11, 17.42, 17.74, 17.55, 17.11, 16.87, 15.42, 14.75, 12.92, 16.12, 14.43, 13.12, 11.49, 12.36, 18.12, 19.55, 19.93, 17.12, 16.93, 12.86, 17.61, 16.11, 15.24, 19.12}. Y = {324, 288, 278, 288, 271, 257, 229, 218, 257, 368, 336, 328, 288, 268, 299, 320, 344, 292, 212, 177, 177, 158, 151, 142, 131, 122, 151, 125, 109, 109, 109}.

2-26 How might the distance between a rain gage and a streamflow gage affect the characteristics of a cross-correlogram?

2-27 Compute the cross-correlogram for the following daily SCA X and daily streamflow Y data for 0 to 5 lags. Discuss the implications of the cross-correlogram to modeling the relationship. X = {0.500, 0.478, 0.455, 0.435, 0.415, 0.395, 0.375, 0.355, 0.335, 0.312, 0.299, 0.280, 0.260, 0.245, 0.220, 0.211, 0.200, 0.183, 0.169, 0.159, 0.140, 0.129, 0.115, 0.100, 0.090, 0.079, 0.069, 0.055, 0.045, 0.033, 0.025}. Y = {2057, 1934, 1827, 1723, 1626, 1555, 1499, 1420, 1350, 1306, 1273, 1250, 1225, 1167, 1116, 1041, 934, 863, 788, 719, 687, 614, 562, 517, 498, 477, 434, 432, 567, 524, 465}.

3 Statistical Hypothesis Testing

3.1 INTRODUCTION

In the absence of a reliable theoretical model, empirical evidence is often an alternative for decision making. An intermediate step in decision making is reducing a set of observations on one or more random variables to descriptive statistics. Examples of frequently used descriptive statistics include the moments (i.e., mean and variance) of a random variable and the correlation coefficient of two random variables.

Statistical hypothesis testing is a tool for making decisions about descriptive statistics in a systematic manner. Based on concepts of probability and statistical theory, it provides a means of incorporating the concept of risk into the assessment of alternative decisions. More importantly, it enables statistical theory to assist in decision making. A systematic analysis based on theoretical knowledge inserts a measure of objectivity into the decision making.

It may be enlightening to introduce hypothesis testing in terms of populations and samples. Data are measured in the field or in a laboratory. These represent samples of data, and descriptive statistics computed from the measured data are sample estimators. However, decisions should be made using the true population, which unfortunately is rarely known. When using the empirical approach in decision making, the data analyst is interested in extrapolating from a data sample the statements about the population from which the individual observations that make up the sample were obtained. Since the population is not known, it is necessary to use the sample data to identify a likely population. The assumed population is then used to make predictions or forecasts. Thus, hypothesis tests combine statistical theory and sample information to make inferences about populations or parameters of a population. The first step is to formulate hypotheses that reflect the alternative decisions.

Because of the inherent variability in a random sample of data, a sample statistic will usually differ from the corresponding parameter of the underlying population. The difference cannot be known for a specific sample because the population is not known. However, theory can suggest the distribution of the statistic from which probability statements can be made about the difference. The difference between the sample and population values is assumed to be the result of chance, and the degree of difference between a sample value and the population value is a reflection of the sampling variation. Rarely does the result of a pre-election day poll match exactly the election result, even though the method of polling may adhere to the proper methods of sampling. The margin of error is the best assessment of the sampling variation. As another example, one would not expect the mean of five random-grab samples of the dissolved oxygen concentration in a stream to exactly equal the true mean dissolved-oxygen concentration. Some difference between a sample estimate of the mean and the population mean should be expected. Although some differences may be acceptable, at some point the difference becomes so large that it is unlikely to be the result of chance. The theoretical basis of a hypothesis test allows one to determine the difference that is likely to result from chance, at least within the expectations of statistical theory.

If a sufficiently large number of samples could be obtained from a population and the value of the statistic of interest computed for each sample, the characteristics (i.e., mean, variance, probability density function) of the statistic could be estimated empirically. The mean of the values is the expected value. The variance of the values indicates the sampling error of the statistic. The probability function defines the sampling distribution of the statistic. Knowledge of the sampling distribution of the parameter provides the basis for making decisions. Fortunately, the theoretical sampling distributions of many population parameters, such as the mean and variance, are known from theoretical models, and inferences about these population parameters can be made when sampled data are available to approximate unknown values of parameters.

Given the appropriate hypotheses and a theoretical model that defines sampling distribution, an investigator can select a decision rule that specifies sample statistics likely to arise from the sampling distribution for each hypothesis included in the analysis. The theoretical sampling distribution is thus used to develop the probability statements needed for decision making.

Example 3.1

Consider Table 3.1. The individual values were sampled randomly from a standard normal population that has a mean of 0 and a standard deviation of 1. The values vary from −3.246 to 3.591. While many of the 200 values range from −1 to +1, a good portion fall outside these bounds.

The data are divided into 40 samples of 5, and the 40 means, standard deviations, and variances are computed for each sample of 5 (see Tables 3.2, 3.3, and 3.4, respectively). Even though the population mean is equal to 0.0, none of the 40 sample means is the same. The 40 values show a range from −0.793 to +1.412. The sample values vary with the spread reflective of the sampling variation of the mean. Similarly, the sample standard deviations (Table 3.3) and variances (Table 3.4) show considerable variation; none of the values equals the corresponding population value of 1. Again, the variation of the sample values reflects the sampling variation of the statistics. The basic statistics question is whether or not any of the sample statistics (e.g., mean, standard deviation, variance) are significantly different from the true population values that are known. The answer requires knowledge of basic concepts of statistical theory.

Theory indicates that the mean of a sample of values drawn from a normal population with mean μ and standard deviation on σ has an underlying normal population with mean μ and standard deviation $\sigma / \sqrt{n}$. Similarly, statistical theory

TABLE 3.1
Forty Random Samples of Five Observations on a Standard Normal Distribution, N(0, 1)

0.048	1.040	–0.111	–0.120	1.396	–0.393	–0.220	0.422	0.233	0.197
–0.521	–0.563	–0.116	–0.512	–0.518	–2.194	2.261	0.461	–1.533	–1.836
–1.407	–0.213	0.948	–0.073	–1.474	–0.236	–0.649	1.555	1.285	–0.747
1.822	0.898	–0.691	0.972	–0.011	0.517	0.808	2.651	–0.650	0.592
1.346	–0.137	0.952	1.467	–0.352	0.309	0.578	–1.881	–0.488	–0.329
0.420	–1.085	–1.578	–0.125	1.337	0.169	0.551	–0.745	–0.588	1.810
–1.760	–1.868	0.677	0.545	1.465	0.572	–0.770	0.655	–0.574	1.262
–0.959	0.061	–1.260	–0.573	–0.646	–0.697	–0.026	–1.115	3.591	–0.519
0.561	–0.534	–0.730	–1.172	–0.261	–0.049	0.173	0.027	1.138	0.524
–0.717	0.254	0.421	–1.891	2.592	–1.443	–0.061	–2.520	–0.497	0.909
–2.097	–0.180	–1.298	–0.647	0.159	0.769	–0.735	–0.343	0.966	0.595
0.443	–0.191	0.705	0.420	–0.486	–1.038	–0.396	1.406	0.327	1.198
0.481	0.161	–0.044	–0.864	–0.587	–0.037	–1.304	–1.544	0.946	–0.344
–2.219	–0.123	–0.260	0.680	0.224	–1.217	0.052	0.174	0.692	–1.068
1.723	–0.215	–0.158	0.369	1.073	–2.442	–0.472	2.060	–3.246	–1.020
–0.937	1.253	0.321	–0.541	–0.648	0.265	1.487	–0.554	1.890	0.499
–0.568	–0.146	0.285	1.337	–0.840	0.361	–0.468	0.746	0.470	0.171
–1.717	–1.293	–0.556	–0.545	1.344	0.320	–0.087	0.418	1.076	1.669
–0.151	–0.266	0.920	–2.370	0.484	–1.915	–0.268	0.718	2.075	–0.975
2.278	–1.819	0.245	–0.163	0.980	–1.629	–0.094	–0.573	1.548	–0.896

TABLE 3.2
Sample Means

0.258	0.205	0.196	0.347	–0.246	–0.399	0.556	0.642	–0.231	–0.425
–0.491	–0.634	–0.694	–0.643	0.897	–0.290	–0.027	–0.740	0.614	0.797
–0.334	–0.110	–0.211	–0.008	0.077	–0.793	–0.571	0.351	–0.063	–0.128
–0.219	–0.454	0.243	–0.456	0.264	–0.520	0.114	0.151	1.412	0.094

TABLE 3.3
Sample Standard Deviations

1.328	0.717	0.727	0.833	1.128	1.071	1.121	1.682	1.055	0.939
0.977	0.867	1.151	0.938	1.333	0.792	0.481	1.209	1.818	0.875
1.744	0.155	0.717	0.696	0.667	1.222	0.498	1.426	1.798	1.001
1.510	1.184	0.525	1.321	0.972	1.148	0.783	0.665	0.649	1.092

TABLE 3.4
Sample Variances

1.764	0.514	0.529	0.694	1.272	1.147	1.257	2.829	1.113	0.882
0.955	0.752	1.325	0.880	1.777	0.627	0.231	1.462	3.305	0.766
3.042	0.024	0.514	0.484	0.445	1.493	0.248	2.033	3.233	1.002
2.280	1.402	0.276	1.745	0.945	1.318	0.613	0.442	0.421	1.192

indicates that if S^2 is the variance of a random sample of size n taken from a normal population that has the variance σ^2, then:

$$\chi^2 = \frac{(n-1)S^2}{\sigma^2} \tag{3.1}$$

is the value of a random variable that has a chi-square distribution with degrees of freedom $\upsilon = n - 1$.

Figure 3.1 compares the sample and population distributions. Figure 3.1(a) shows the distributions of the 200 sample values of the random variable z and the standard normal distribution, which is the underlying population. For samples of five from the stated population, the underlying distribution of the mean is also a normal distribution with a mean of 0 but it has a standard deviation of $1/\sqrt{5}$ rather than 1. The frequency distribution for the 40 sample means and the distribution of the population are shown in Figure 3.1(b). Differences in the sample and population distributions for both Figures 3.1(a) and 3.1(b) are due to sampling variation and the relatively small samples, both the size of each sample (i.e., five) and the number of samples (i.e., 40). As the sample size would increase towards infinity, the distribution of sample means would approach the population distribution. Figure 3.1(c) shows the sample frequency histogram and the distribution of the underlying population for the chi-square statistic of Equation 3.1. Again, the difference in the two distributions reflects sampling variation. Samples much larger than 40 would show less difference.

This example illustrates a fundamental concept of statistical analysis, namely sampling variation. The example indicates that individual values of a sample statistic can be quite unlike the underlying population value; however, most sample values of a statistic are close to the population value.

3.2 PROCEDURE FOR TESTING HYPOTHESES

How can one decide whether a sample statistic is likely to have come from a specified population? Knowledge of the theoretical sampling distribution of a test statistic based on the statistic of interest can be used to test a stated hypothesis. The test of a hypothesis leads to a determination whether a stated hypothesis is valid. Tests are

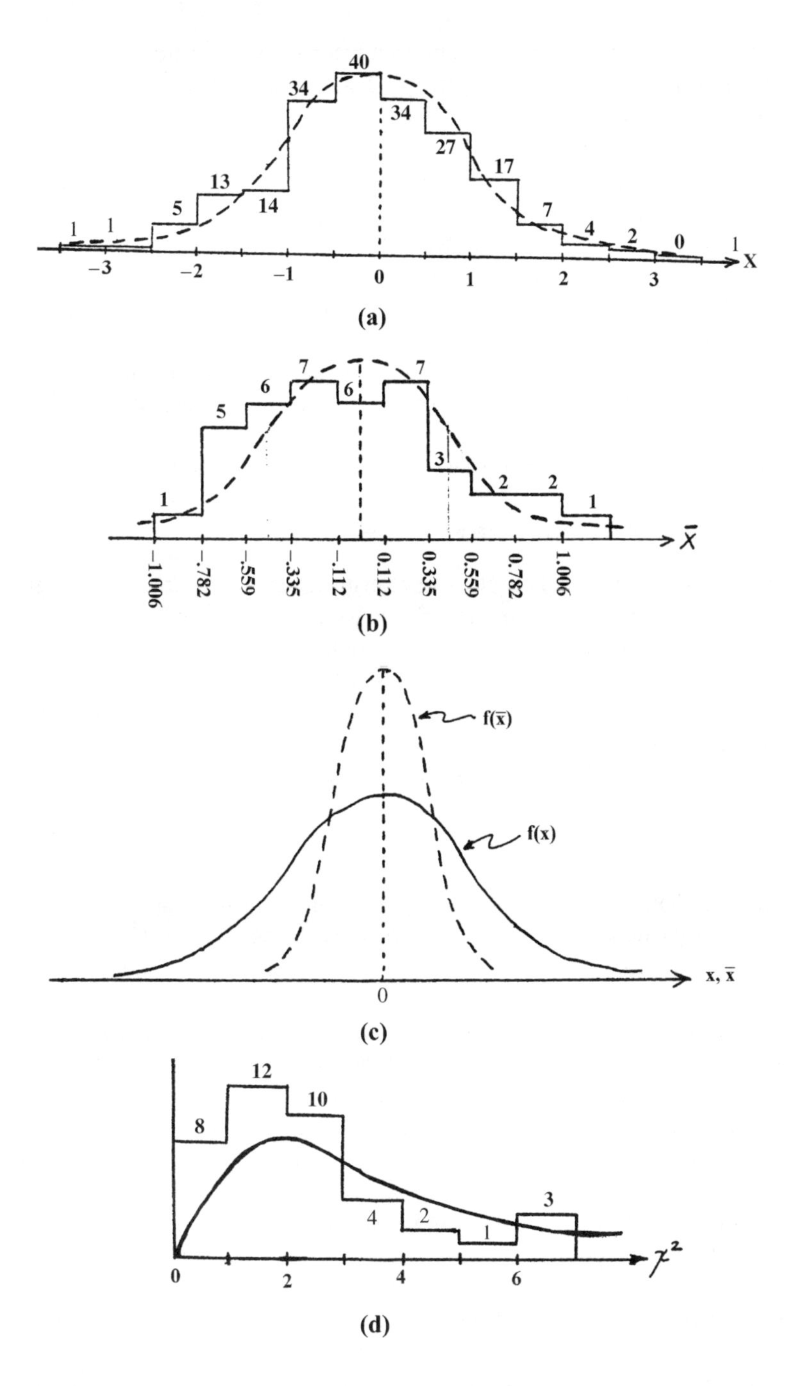

FIGURE 3.1 Based on the data of Table 3.1: (a) the distribution of the random sample values; (b) the distribution of the sample means; (c) distributions of the populations of X and the mean $\bar{X}$; and (d) chi-square distribution of the variance: sample and population.

available for almost every statistic, and each test follows the same basic steps. The following six steps can be used to perform a statistical analysis of a hypothesis:

1. Formulate hypotheses.
2. Select the appropriate statistical model (theorem) that identifies the test statistic and its distribution.
3. Specify the level of significance, which is a measure of risk.
4. Collect a sample of data and compute an estimate of the test statistic.
5. Obtain the critical value of the test statistic, which defines the region of rejection.
6. Compare the computed value of the test statistic (step 4) with the critical value (step 5) and make a decision by selecting the appropriate hypothesis.

Each of these six steps will be discussed in more detail in the following sections.

3.2.1 Step 1: Formulation of Hypotheses

Hypothesis testing represents a class of statistical techniques that are designed to extrapolate information from samples of data to make inferences about populations. The first step is to formulate two hypotheses for testing. The hypotheses will depend on the problem under investigation. Specifically, if the objective is to make inferences about a single population, the hypotheses will be statements indicating that a random variable has or does not have a specific distribution with specific values of the population parameters. If the objective is to compare two or more specific parameters, such as the means of two samples, the hypotheses will be statements formulated to indicate the absence or presence of differences between two means. Note that the hypotheses are composed of statements that involve population distributions or parameters; hypotheses should not be expressed in terms of sample statistics.

The first hypothesis is called the *null hypothesis*, denoted by H_0, and is always formulated to indicate that a difference does not exist. The second or *alternative hypothesis* is formulated to indicate that a difference does exist. Both are expressed in terms of populations or population parameters. The alternative hypothesis is denoted by either H_1 or H_A. The null and alternative hypotheses should be expressed in words and in mathematical terms and should represent mutually exclusive conditions. Thus, when a statistical analysis of sampled data suggests that the null hypothesis should be rejected, the alternative hypothesis is assumed to be correct. Some are more cautious in their interpretations and decide that failure to reject the null hypothesis implies only that it can be accepted.

While the null hypothesis is always expressed as an equality, the alternative hypothesis can be a statement of inequality ($\neq$), less than ($<$), or greater than ($>$). The selection depends on the problem. If standards for a water quality index indicated that a stream was polluted when the index was greater than some value, the H_A would be expressed as a greater-than statement. If the mean dissolved oxygen was not supposed to be lower than some standard, the H_A would be a less-than statement. If a direction is not physically meaningful, such as when the mean should not be significantly less than or significantly greater than some value, then a two-tailed

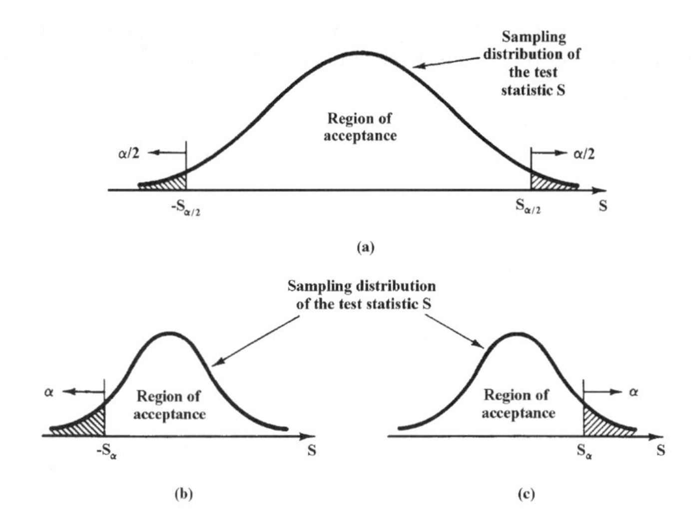

FIGURE 3.2 Representation of the region of rejection (cross-hatched area), region of acceptance, and the critical value (S_α): (a) H_A: $\mu \neq \mu_0$; (b) H_A: $\mu < \mu_0$; (c) H_A: $\mu > \mu_0$.

inequality statement is used for H_A. The statement of the alternative hypothesis is important in steps 5 and 6. The three possible alternative hypotheses are illustrated in Figure 3.2.

3.2.2 Step 2: Test Statistic and Its Sampling Distribution

The two hypotheses of step 1 allow an equality or a difference between specified populations or parameters. To test the hypotheses, it is necessary to identify the test statistic that reflects the difference suggested by the alternative hypothesis. The specific test statistic is generally the result of known statistical theory. The sample value of a test statistic will vary from one sample to the next because of sampling variation. Therefore, the test statistic is a random variable and has a sampling distribution. A hypothesis test should be based on a theoretical model that defines the sampling distribution of the test statistic and its parameters. Based on the distribution of the test statistic, probability statements about computed sample values may be made.

Theoretical models are available for all of the more frequently used hypothesis tests. In cases where theoretical models are not available, approximations have usually been developed. In any case, a model or theorem that specifies the test statistic, its distribution, and its parameters must be identified in order to make a hypothesis test.

3.2.3 Step 3: Level of Significance

Two hypotheses were formulated in step 1; in step 2, a test statistic and its distribution were selected to reflect the problem for which the hypotheses were formulated. In step 4, data will be collected to test the hypotheses. Before data collection, it is necessary to provide a probabilistic framework for accepting or rejecting the null

TABLE 3.5
Decision Table for Hypothesis Testing

Decision	Situation: H_0 is true	Situation: H_0 is false
Accept H_0	Correct decision	Incorrect decision: type II error
Reject H_0	Incorrect decision: type I error	Correct decision

hypothesis and subsequently making a decision; the framework will reflect the allowance for the variation that can be expected in a sample of data. Table 3.5 shows the situations that could exist in the population, but are unknown (i.e., H_0 is true or false) and the decisions that the data could suggest (i.e., accept or reject H_0). The decision table suggests two types of error:

Type I error: reject H_0 when, in fact, H_0 is true.
Type II error: accept H_0 when, in fact, H_0 is false.

These two incorrect decisions are not independent; for a given sample size, the magnitude of one type of error increases as the magnitude of the other type of error is decreased. While both types of errors are important, the decision process most often considers only one of the errors, specifically the type I error.

The level of significance, which is usually the primary element of the decision process in hypothesis testing, represents the probability of making a type I error and is denoted by the Greek lower-case letter alpha, α. The probability of a type II error is denoted by the Greek lower-case letter beta, β. The two possible incorrect decisions are not independent. The level of significance should not be made exceptionally small, because the probability of making a type II error will then be increased. Selection of the level of significance should, therefore, be based on a rational analysis of the physical system being studied. Specifically, one would expect the level of significance to be different when considering a case involving the loss of human life and a case involving minor property damage. However, the value chosen for α is often based on convention and the availability of statistical tables; values for α of 0.05 and 0.01 are selected frequently and the arbitrary nature of this traditional means of specifying α should be recognized.

Because α and β are not independent, it is necessary to consider the implications of both types of errors in selecting a level of significance. The concept of the power of a statistical test is important when discussing a type II error. The power is defined as the probability of rejecting H_0 when, in fact, it is false:

$$\text{Power} = 1 - \beta \tag{3.2}$$

For some hypotheses, more than one theorem and test statistic are available, with alternatives usually based on different assumptions. The theorems will produce different powers, and when the assumptions are valid, the test that has the highest power for a given level of significance is generally preferred.

3.2.4 Step 4: Data Analysis

After obtaining the necessary data, the sample is used to provide an estimate of the test statistic. In most cases, the data are also used to provide estimates of the parameters required to define the sampling distribution of the test statistic. Many tests require computing statistics called *degrees of freedom* in order to define the sampling distribution of the test statistic.

3.2.5 Step 5: Region of Rejection

The region of rejection consists of values of the test statistic that are unlikely to occur when the null hypothesis is true, as shown in the cross-hatched areas in Figure 3.2. Extreme values of the test statistic are least likely to occur when the null hypothesis is true. Thus, the region of rejection usually lies in one or both tails of the distribution of the test statistic. The location of the region of rejection depends on the statement of the alternative hypothesis. The region of acceptance consists of all values of the test statistic that are likely if the null hypothesis is true.

The critical value of the test statistic is defined as the value that separates the region of rejection from the region of acceptance. The critical value of the test statistic depends on (1) the statement of the alternative hypothesis, (2) the distribution of the test statistic, (3) the level of significance, and (4) characteristics of the sample or data. These components represent the first four steps of a hypothesis test. Values of the critical test statistics are usually given in tables.

The region of rejection may consist of values in both tails or in only one tail of the distribution as suggested by Figure 3.2. Whether the problem is two-tailed, one-tailed lower, or one-tailed upper will depend on the statement of the underlying problem. The decision is not based on statistics, but rather is determined by the nature of the problem tested. Although the region of rejection should be defined in terms of values of the test statistic, it is often pictorially associated with an area of the sampling distribution that is equal to the level of significance. The region of rejection, region of acceptance, and the critical value are shown in Figure 3.2 for both two-tailed and one-tailed tests. For a two-tailed test, it is standard practice to define the critical values such that one-half of α is in each tail. For a symmetric distribution, such as the normal or t, the two critical values will have the same magnitude and different signs. For a nonsymmetric distribution such as the chi-square, values will be obtained from the table such that one-half of α is in each tail; magnitudes will be different.

Some computer programs avoid dealing with the level of significance as part of the output and instead compute and print the rejection probability. The rejection probability is the area in the tail of the distribution beyond the computed value of the test statistic. This concept is best illustrated by way of examples. Assume a software package is used to analyze a set of data and prints out a computed value of the test statistic z of 1.92 and a rejection probability of 0.0274. This means that approximately 2.74% of the area under the probability distribution of z lies beyond a value of 1.92. To use this information for making a one-tailed upper test, the null hypothesis would be rejected for any level of significance larger than 2.74% and accepted for any level of significance below 2.74%. For a 5% level, H_0 is rejected, while for a 1% level of significance, the H_0 is accepted. Printing the rejection probability places the decision in the hands of the reader of the output.

3.2.6 Step 6: Select Appropriate Hypothesis

A decision whether to accept the null hypothesis depends on a comparison of the computed value (step 4) of the test statistic and the critical value (step 5). The null hypothesis is rejected when the computed value lies in the region of rejection. Rejection of the null hypothesis implies acceptance of the alternative hypothesis. When a computed value of the test statistic lies in the region of rejection, two explanations are possible. The sampling procedure many have produced an extreme value purely by chance; although this is very unlikely, it corresponds to the type I error of Table 3.5. Because the probability of this event is relatively small, this explanation is usually rejected. The extreme value of the test statistic may have occurred because the null hypothesis was false; this explanation is most often accepted and forms the basis for statistical inference.

The decision for most hypothesis tests can be summarized in a table such as the following:

If H_A is	Then reject H_0 if
$P \neq P_0$	$S > S_{\alpha/2}$ or $S < S_{1-\alpha/2}$
$P < P_0$	$S < S_{1-\alpha}$
$P > P_0$	$S > S_\alpha$

where P is the parameter tested against a standard value, P_0; S is the computed value of the test statistic; and $S_{\alpha/2}$ and $S_{1-\alpha/2}$ are the tabled values for the population and have an area of $\alpha/2$ in the respective tails.

Example 3.2

Consider the comparison of runoff volumes from two watersheds that are similar in drainage area and other important characteristics such as slope, but differ in the extent of development. On one watershed, small pockets of land have been developed. The hydrologist wants to know whether the small amount of development is sufficient to increase storm runoff. The watersheds are located near each other and are likely to experience the same rainfall distributions. While rainfall characteristics are not measured, the total storm runoff volumes are measured.

The statement of the problem suggests that two means will be compared, one for a developed watershed population μ_d and one for an undeveloped watershed population μ_μ. The hydrologist believes that the case where μ_d is less than μ_μ is not rational and prepares to test the following hypotheses:

$$H_0\colon \mu_d = \mu_\mu \tag{3.3a}$$

$$H_A\colon \mu_d > \mu_\mu \tag{3.3b}$$

A one-sided test of two means will be made, with the statement of the alternative hypothesis determined by the problem statement.

Several theorems are available for comparing two means, and the hydrologist will select the most appropriate one for the data expected to be collected in step 4. For example, one theorem assumes equal variances that are unknown, while another theorem assumes variances that are known and do not have to be equal. A third theorem assumes unequal and unknown variances. The theorem should be specified before the data are collected. In step 3, the level of significance needs to be specified. The implications of the two types of error are:

Type I: Conclude that H_0 is false when it is true and wrongly assume that even spotty development can increase runoff volumes. This might lead to the requirement for unnecessary BMPs.

Type II: Conclude that H_0 is true when it is not and wrongly assume that spotty development does not increase runoff volumes. This might allow increases in runoff volumes to enter small streams and ultimately cause erosion problems.

Assume that the local government concludes that the implications of a type II error are more significant than those of the type I error. They would, therefore, want to make β small, which may mean selecting a level of significance that is larger than the traditional 5%.

While the data have not been collected, the problem statement has been transformed into a research hypothesis (step 1), the relevant statistical theory has been identified (step 2), and the risk of sampling errors has been considered (step 3). It is generally considered incorrect experimental practice to collect and peruse the data prior to establishing the first three steps of the test.

Step 4 is data collection. Generally, the largest sample size that is practical to collect should be obtained. Accuracy is assumed to improve with increasing sample size. Once the data are collected and organized, the test statistic and parameters identified in the theorem are computed. It may also be necessary to check any assumptions specified in the theorem. For the case of the runoff volumes, the sample size may be limited by the number of storms that occur during the period allotted to the experiment.

In step 5, the critical value would be obtained from the appropriate table. The value may depend on parameters, such as the sample size, from step 4. The critical value would also depend on the statement of the alternative hypothesis (step 1) and the level of significance (step 3). The critical value and the statement of the alternative hypothesis would define the region of rejection.

In step 6, the computed value of the test statistic is compared with the critical value. If it lies in the region of rejection, the hydrologist might assume that the null hypothesis is not correct and that spotty development in a watershed causes increases in runoff volumes. The value of the level of significance would indicate the probability that the null hypothesis was falsely rejected.

3.3 RELATIONSHIPS AMONG HYPOTHESIS TEST PARAMETERS

The purpose for using a statistical hypothesis test is to make a systematic decision. The following four decision parameters are inherent to every hypothesis test, although only two are generally given explicit consideration: sample size n, level of significance α, power of test P, and decision criterion C. Generally, n and α are the two parameters selected for making the test, with a value of 5% often used for α. However, any two of the four can be selected, and whenever two parameters are specified, the other two are uniquely set. Each parameter plays a role in the decision:

- n: The sample size is an indication of the accuracy of the statistic, that is, the magnitude of its standard error, with the accuracy increasing with sample size.
- α: The probability of making a type I error decision, that is, the consumer's risk.
- P: A measure of the probability of making a type II error decision, that is, the producer's risk. Note that Equation 3.2 shows the relationship between P and β.
- C: The criterion value that separates the regions of acceptance and rejection.

To understand the relationship of these four parameters, it is necessary to introduce two new concepts: the region of uncertainty and the rejection hypothesis denoted as H_r. The decision process includes three hypotheses: null, alternative, and rejection. The rejection hypothesis is established to reflect the condition where the null hypothesis is truly incorrect and should be rejected. The null and rejection hypotheses can be represented by probability density functions. The region between the distribution of the test statistic when H_0 is true and the distribution when H_r is true is the region of uncertainty (see Figure 3.3).

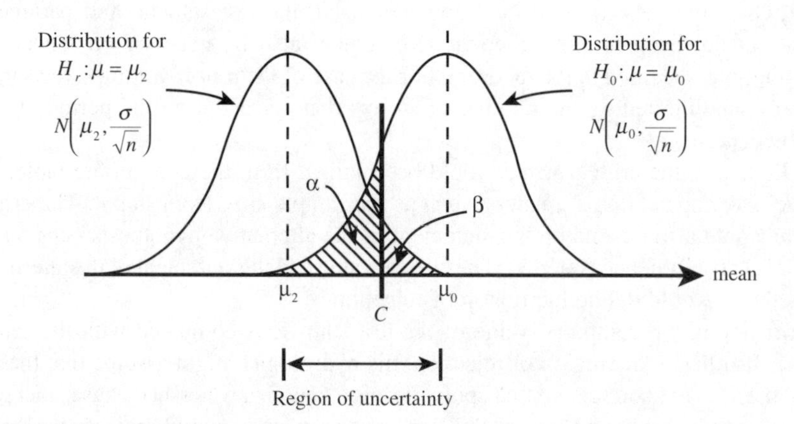

FIGURE 3.3 Relationship between type I and II errors and critical test statistic C.

Consider the case of a hypothesis test on a mean against some standard μ_0, with the null hypothesis H_0: $\mu = \mu_0$ and the one-tailed alternative hypothesis, H_A: $\mu < \mu_0$. If this hypothesis is true, then the test statistic is normally distributed with mean μ_0 and standard deviation $\sigma / \sqrt{n}$, where σ is the population standard deviation, which is assumed to be known. This means that, if H_0 is true, a sample value of the mean is likely to fall within the distribution shown on the right side in Figure 3.3. If the null hypothesis is false and the rejection hypothesis H_r: $\mu = \mu_2$ is true, then the test statistic has a normal distribution with mean μ_2 and standard deviation $\sigma / \sqrt{n}$. This means that, if H_r is true, a sample value of the mean is likely to fall within the distribution shown on the left side in Figure 3.3. The region between μ_0 and μ_2 is the region of uncertainty.

If H_0 is correct and the mean has the distribution shown on the right, then the level of significance indicates a portion of the lower tail where type I errors are most likely to occur. If the H_r rejection hypothesis is correct, then the value of β indicates the portion of the distribution for H_r where type II errors are most likely to occur, which is in the upper tail of the H_r distribution. The variation within each of the two distributions depends on the sample size, with the spread decreasing as the sample size is increased. This reflects the greater confidence in the computed value of the mean as the sample size increases. The cross-hatched area indicated with α represents the probability that the null hypothesis should be rejected if H_0 is true. The cross-hatched area indicated with β reflects the probability that the null hypothesis will be accepted when the mean is distributed by the H_r distribution. The decision criterion C, which serves as the boundary for both the α and β regions, is the value of the test statistic below which a computed value will indicate rejection of the null hypothesis.

As indicated, when two of the four parameters are set, values for the other two parameters are uniquely established. Each of the four parameters has statistical implications and a physical-system association. The statistical implications of the parameters follow:

n: A measure of the error variance of the statistical parameter, with the variance decreasing as the sample size increases.
C: The separation line between the regions of acceptance and rejection.
α: The probability of a type I error.
β: The probability of a type II error and a measure of the statistical power of the test (see Equation 3.2).

The physical implications of the four parameters are:

n: The quantity of empirical evidence that characterizes the underlying population of the test statistic.
C: The decision criterion in units of the decision variable.
α: The consumer's risk of a wrong decision.
β: The producer's risk of a wrong decision.

Consider the case where n and C are set. If μ_0, μ_2, and σ are known based on the characteristics of the physical system, then α and β, both shown in Figure 3.3, are determined as follows:

$$\alpha = P(\mu < C \mid H_0 \text{ is true}) = P\left(z < \frac{C - \mu_0}{\sigma / \sqrt{n}}\right) \tag{3.4a}$$

$$\beta = P(\mu > C \mid H_2 \text{ is true}) = P\left(z > \frac{C - \mu_2}{\sigma / \sqrt{n}}\right) \tag{3.4b}$$

As an alternative, if C and β are the unknowns, then they are determined by:

$$C = \mu_0 + z(\sigma / \sqrt{n}) \tag{3.5a}$$

$$\beta = P(\mu > C \mid H_2 \text{ is true}) = P\left(z > \frac{C - \mu_2}{\sigma / \sqrt{n}}\right) \tag{3.5b}$$

Note that Figure 3.3 could be restructured, as when the region of rejection is for a one-tailed upper test; in this case, the normal distribution of μ for the rejection hypothesis H_2 would be to the right of the normal distribution for H_0.

Example 3.3

Consider the hypothetical case of a state that wants to establish a criterion on a water pollutant. A small sample is required, with the exact size set by practicality and cost concerns. Assume that five independent grab samples are considered cost effective and sufficiently accurate for making decisions. Extensive laboratory tests suggest that the variation in five measurements in conditions similar to those where the test will be applied is ±0.2 mg/L. State water quality specialists believe that conditions are acceptably safe at 2.6 mg/L but problems occur when the concentration begins to exceed 3.0 mg/L. They require the test on the mean to use a 5% level of significance. Based on these conditions, they seek to determine the decision criterion C to be used in the field and the type II error probability.

Figure 3.4 shows the normal distributions for the null hypothesis H_0: $\mu = 2.6$ mg/L and the rejection hypothesis H_r: $\mu = 3.0$ mg/L. A one-sided test is appropriate, with H_A: $\mu > 2.6$ mg/L because it is not a relevant pollution problem if the level of the pollutant is below the safe concentration of 2.6 mg/L. The distribution of the mean for the null hypothesis is $N(2.6, 0.2/\sqrt{5})$. Therefore, the decision criterion can be calculated by:

$$C = \mu_0 + z_\alpha(\sigma / \sqrt{n}) = 2.6 + 1.645(0.2 / \sqrt{5}) = 2.747 \text{mg/L} \tag{3.6}$$

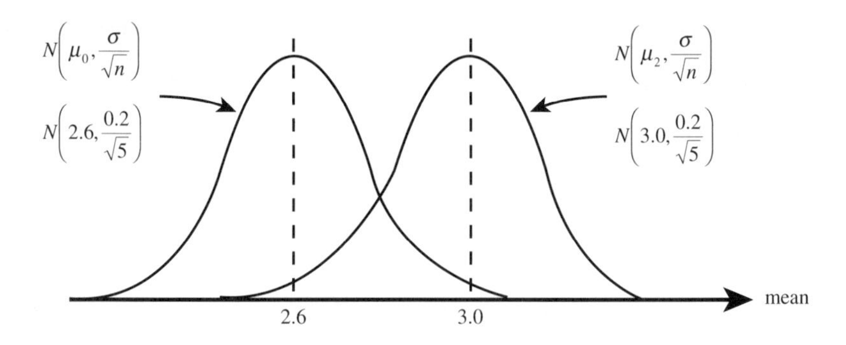

FIGURE 3.4 Distributions of the mean.

Therefore, the probability of a type II error is found from:

$$\beta = P(\mu < 2.747 \mid \mu_2 = 3\,\text{mg/L}) = P\left(z < \frac{2.747 - 3.0}{0.2/\sqrt{5}}\right) = P(z < -2.829) = 0.0023 \tag{3.7}$$

Consequently, the state establishes the guideline that five samples are to be taken and the null hypothesis is rejected if the mean of the sample of five exceeds 2.75 mg/L. Note that since the criterion C of 2.75 is slightly larger than 2.747, α is theoretically slightly below 5% and β is slightly above 0.23%.

3.4 PARAMETRIC AND NONPARAMETRIC TESTS

A parametric test is based on theory or concepts that require specific conditions about the underlying population and/or its parameters from which sample information will be obtained. For example, it might assume that the sample data are drawn from a normal population or that the variance of the population is known. The accuracy of decisions based on parametric tests depends on the extent to which these assumptions are met. Parametric tests also require that the random variable on which the values are measured be at least on an interval scale. Parametric tests cannot be applied with data measured on the nominal or ordinal scale.

A nonparametric test is based on theory or concepts that have not required the sample data to be drawn from a certain population or have conditions placed on the parameters of the population. Nonparametric tests, in contrast to parametric tests, do not require that the random variables be measured on the interval or ratio scales. Many nonparametric tests are applicable to random variables measured on the nominal or ordinal scale, and very often the nonparametric tests require that interval-scale measurements be transformed to ranks. This does not mean that the application of nonparametric tests makes no assumptions about the data. Many nonparametric tests make assumptions such as data independence or that the random variable is continuously distributed. The primary difference between the assumptions made for the two classes of tests is that those made for nonparametric tests are not as restrictive as those made for parametric tests, such as complete specification of the underlying population.

3.4.1 Disadvantages of Nonparametric Tests

Nonparametric tests have applicability to a wide variety of situations and, therefore, represent an important array of statistical decision tools. However, they have a few disadvantages.

1. Many nonparametric test statistics are based on ranks or counts, which are often integers. Unlike test statistics that are continuously distributed, rank-based statistics are discrete, and therefore, it is not possible to obtain critical values for exact levels of significance, such as 5% or 1%. Alternatively, parametric test statistics are continuously distributed and critical values for specific levels of significance can be obtained.

   Consider the following hypothetical example. Assume that the test statistic T can only take on integer values, and small values (i.e., near zero) are unlikely to occur if the null hypothesis is true. Assume that the cumulative probability distribution $F(T)$ for small values of T is as follows:

T	0	1	2	3	4
$F(T)$	0.003	0.008	0.021	0.064	0.122

   If a 5% level of significance was of interest and a rejection probability greater than 5% was considered undesirable, then a critical value of 2 would need to be used. Unfortunately, this will provide a decision that is conservative with respect to T because use of a value of 2 indicates that the probability of a type I error is 2.1%. If a value of 3 were used for T, then the rejection probability would be 6.4%, which would mean that the test would not meet the desired 5% criterion. Similarly, if a 1% rejection probability were of interest, a critical value of 2 could not be used, but a critical value of 1 would yield a smaller than desired rejection probability.
2. Using ranks or integer scores often results in tied values. These are more troublesome to deal with in nonparametric tests than in parametric tests. Ties are much rarer with continuously distributed random variables. For some nonparametric tests, dealing with ties is not straightforward, with several alternatives having been proposed. In some cases, the method of handling ties distorts the rejection probability.
3. The most common criticism of nonparametric tests is that if the assumptions underlying a parametric alternative are met, the parametric test will always be more powerful statistically than the nonparametric test. This is a valid criticism, but the counterargument states that it is difficult to know whether the assumptions underlying the parametric alternative have been met, and if they have not been met, the nonparametric test may, in reality, be the better test.

3.4.2 Advantages of Nonparametric Tests

1. Small samples are common and in such cases, the assumptions of a parametric test must be met exactly in order for the decision to be accurate. Since it is extremely difficult to evaluate the extent to which parametric test assumptions have been met, nonparametric tests are generally preferred for cases of small sample sizes.
2. While parametric tests are limited to random variables on interval or ratio scales, nonparametric tests can be used for random variables measured on nominal and ordinal scales. When measurement on an interval scale is highly imprecise, nonparametric tests may yield more accurate decisions than parametric alternatives.
3. It is easier to detect violations of the assumptions of a nonparametric test than it is to detect violations for parametric tests. The assumptions for nonparametric tests are usually less stringent and play a smaller role in calculation of the test statistic.
4. When assumptions of the parametric test are violated, the level of significance used to make the decision will not be a precise measure of the rejection probability. However, since nonparametric tests are less dependent on the adequacy of the assumptions, the probabilities are usually exact.

3.5 PROBLEMS

3-1 What are the characteristics of a null hypothesis and an alternative hypothesis?

3-2 Why is it necessary to state the null hypothesis as a finding of no significant difference (i.e., an equality) when the objective of the research may be to show a difference?

3-3 Why are hypotheses stated in terms of the population parameters rather than sample values?

3-4 What four factors influence the critical value of a test statistic? Show pictorially how each factor affects the critical value.

3-5 Define the region of rejection in the following terms:
- (a) Values of the test statistic
- (b) Proportions of the area of the probability density function of the test statistic
- (c) The region of acceptance
- (d) The critical value(s) of the test statistic

3-6 What factors contribute to sample variation? Discuss the effect of each factor on the magnitude of the sampling variation.

3-7 Graphical analyses show the sampling distribution of the mean for samples of 25 drawn from a normal population with $\mu = 8$ and $\sigma^2 = 1.2$. Is it likely that a sample of 25 from this population would have a mean of 9? Explain.

3-8 If a sample of 9 has a standard deviation of 3, it is likely that the sample is from a normal population with a variance of 3? Explain.

3-9 From a research standpoint, why should the first three steps of a hypothesis test be made before data are collected and reviewed?

3-10 Distinguish between the sampling distribution of the random variable and the sampling distribution of the test statistic in the various steps of a hypothesis test.

3-11 Develop one-tailed upper, one-tailed lower, and two-tailed hypotheses related to the hydrologic effects of afforestation.

3-12 Develop one-tailed upper, one-tailed lower, and two-tailed hypotheses related to hydrologic effects of clearing vegetation from a stream channel.

3-13 Assume that the following null hypothesis needs to be tested: H_0: the mean stream scour rate of a restored stream is the same as the mean rate prior to stream restoration. What is an appropriate alternative hypothesis? What are the implications of type I and type II errors?

3-14 Assume that the following null hypothesis needs to be tested: H_0: the average baseflow from a small forested watershed is the same as the average baseflow from an agricultural watershed. What is an appropriate alternative hypothesis? What are the implications of type I and type II errors?

3-15 What is wrong with always using a 5% level of significance for hypothesis testing related to watershed change?

3-16 Discuss why it might be best to conclude that a decision cannot be made when the rejection probability is in the range from 1% to 5%.

3-17 What nonstatistical factors should be considered in setting a level of significance?

3-18 Explain the distribution for H_r: $\mu = \mu 2$ of Figure 3.3. What is its implication with respect to sample values of the random variable and the test statistic?

3-19 Discuss the advantages of nonparametric and parametric tests.

4 Outlier Detection

4.1 INTRODUCTION

After plotting a frequency histogram, an event that is much larger or much smaller than the remainder of the sample may be evident. This will initially be labeled as an *extreme event*. Some data samples may contain more than one extreme event. Extreme events can create problems in data analysis and modeling. For example, an extremely large value can cause the sample mean and standard deviation to be much larger than the population values. In bivariate analysis (i.e., X vs. Y), an extreme point can adversely influence the sample value of a correlation coefficient; it can also distort the coefficients of the regression line, thus suggesting an effect that may not reflect the true relationship between the two variables.

Having subjectively decided that one or more values in a sample are extreme events, the values should be objectively evaluated. The intent is to to assess whether the extreme event is likely to have occurred if the sample were correctly obtained from the assumed population. Statistical theory in the form of a hypothesis test can be used to make a decision. If the statistical test indicates that the observed extreme event is unlikely to have occurred during sampling from the assumed population, the extreme event is called an *outlier*. An outlier is a measured value that, according to a statistical test, is unlikely to have been drawn from the same population as the remainder of the sample data.

Having determined that an extreme event is an outlier, the question arises: What can be done with the value? If the value is kept in the sample, then it may distort values or relationships computed from the sample. Eliminating a data point proven to be an outlier should yield more accurate statistics and relationships with other variables. However, some professionals oppose censoring (eliminating the statistically proven outlier from the sample). Their argument is that if the value was measured, it could have occurred and it is incorrect to censor it from the sample. Both are legitimate arguments. While the general consensus is toward allowing proven outliers to be censored, every effort should be made to find the reason that such an extreme event occurred. If the data point is the result of sampling from another population, it seems reasonable to analyze separately the samples from the two populations. Examples of multiple populations in hydrology are hurricane versus nonhurricane floods, snowmelt floods versus nonsnowmelt floods, and channel erosion from cohesive and noncohesive beds.

While a number of methods have been proposed, the Dixon–Thompson and Rosner methods are commonly used alternatives. Chauvenet's method, a third test for outliers, is simple to apply. These methods define a test statistic as the ratio of two deviations and assume that sampling is from a normal population. If one suspects that a lognormal population underlies the sampling, the tests can be applied to the logarithms of the measured data. Rosner's method is only valid for samples larger than 25. The Dixon–Thompson test can be applied for smaller samples. Chauvenet's method can be used with any sample size. Two methods for use with the log-Pearson type III distribution are also presented here.

The literature contains a number of hypothesis tests for outlier detection. Some are only valid for testing a single value while others can test for multiple outliers. Some are valid for all sample sizes while others are only valid for large sample sizes. The most important discriminating factor is probably the assumed distribution of the population. Some outlier tests assume that the sampled data came from a normal probability distribution, while others are valid for the Pearson type III distribution. The underlying distribution is an important factor in selecting the most appropriate test. A decision based on a test may be invalid if the data are sampled from a distribution that is not the same one as assumed in the development of the critical values of the test. Such application may identify more or fewer outliers than actually exist in the sampled data when the assumption of the underlying population is incorrect. While robustness is not discussed here, the underlying distribution assumptions should be seriously considered when selecting a method.

Hydrologic data are often assumed to have come from a lognormal or log-Pearson type III distribution. In such cases, the logarithms of the sampled data are obtained and the test performed on the logarithms. Sample points identified as outliers in the analysis of the logarithms may not be the same points identified under the assumption of the nontransformed distribution. For this reason, the specification of the distribution is an important decision.

4.2 CHAUVENET'S METHOD

Chauvenet's method can be used to test a single value in a sample to determine whether it is from a different population as the remainder of the sample points. The critical values are based on the assumption of a normal distribution. The data point to be tested is the sample value (X_0) that deviates the most from the mean. The point may lie above or below the mean. The hypotheses for this test follow:

H_0: All sample points are from the same normal population (4.1a)

H_A: The most extreme sample point in the sample is unlikely to have come from the normal population from which the remainder of the sample points were drawn. (4.1b)

Rejection of the null hypothesis suggests that the extreme sample value tested is an outlier. It is also possible that the value was detected as an outlier because the test assumed normality when, in fact, the data were from a highly skewed distribution.

The test statistic for Chauvenet's method requires the sample mean $\bar{X}$ and standard deviation S to be computed, which are estimates of the corresponding population parameters μ and σ, respectively. Then the most extreme value in the sample, X_0, is identified and used to compute the standard normal deviate Z:

$$Z = \frac{X_0 - \bar{X}}{S} \quad (4.2)$$

Chauvenet defined the region of rejection by assuming that an observation is unlikely to be from the normal distribution $N(\mu, \sigma)$ if it is in the tails of the distribution with the probability p of $1/(2n)$ divided equally between the upper and lower tails of the standard normal distribution. If the absolute value of Z computed from Equation 4.2 exceeds the value $Z_{p/2}$, then the null hypothesis can be rejected. Values of $Z_{p/2}$ can be obtained directly from the standard normal table by entering with the probability of $p/2$. The following examples illustrate the computations of critical values as a function of the sample size n:

n	$p = 0.5/n$	$p/2$	$Z_{p/2}$
10	0.050	0.025	1.96
20	0.025	0.0125	2.24
30	0.010	0.005	2.575

This is a two-tailed test and cannot be applied as a one-tailed test if Chauvenet's criterion of $p = 1/(2n)$ is applied. It should only be used to test for one outlier and not reapplied to a second observation if the most extreme event is deemed an outlier. It seems that this test has an inherent weakness in that it does not allow the level of significance to be specified. The probability varies with the sample size, so the level of significance is set by the sample size. The method is easy to apply.

Example 4.1

Consider the following annual maximum flood series for the Floyd River at James, Iowa (USGS gaging station 066005) for the period from 1935 to 1973.

71,500	7440	4520	2260	1360
20,600	7170	4050	2060	1330
17,300	6280	3810	2000	1300
15,100	6250	3570	1920	970
13,900	5660	3240	1720	829
13,400	5320	2940	1460	726
8320	4840	2870	1400	318
7500	4740	2710	1390	

The mean ($\bar{X}$) and standard deviation (S_x) are 6771 cfs 11,696 cfs, respectively. The log mean ($\bar{Y}$) and log standard deviation (S_y) are 3.5553 and 0.46418, respectively. Given the mean of 6771 cfs, the highest flow of 71,500 cfs is more deviant from the mean than the lowest flow of 318 cfs. Therefore, it is the flow selected to test as an outlier. Applying Chauvenet's statistic (Equation 4.2) to the largest flood gives:

$$z = \frac{71,500 - 6771}{11,696} = 5.53 \tag{4.3}$$

For a sample size of 39, the probability used to make the test is 1/78 = 0.01282. For a two-tailed test with one-half of the probability in each tail, the critical normal deviate is ±2.49. Since the computed value exceeds the critical value, the largest flood is considered an outlier. The result assumes that the remaining 38 values are from the same normal population. Since 71,500 is further from the log mean than 318, the largest flow can be tested for being an outlier. Applying Chauvenet's statistic to the logarithms of the data gives:

$$z = \frac{\log(71,500) - 3.5553}{0.46418} = 2.798 \tag{4.4}$$

Again, the computed test statistic exceeds the critical value for a 1.28% probability, which implies that the largest value is an outlier when the remaining values are from a lognormal population.

Example 4.2

The following are daily mean values of the suspended sediment (mg/L) from Brandywine Creek at Chadds Ford, Pennsylvania.

160	45	20	16	14	12	10	10	8	8	7	6
115	40	20	16	14	11	10	9	8	8	7	6
60	38	19	15	13	11	10	9	8	8	7	6
60	35	17	15	12	10	10	9	8	8	6	5
46	20	16	15	12	10	10	8	8	7	6	

The mean and standard deviation, both untransformed and with a log transform, are as follows:

					Logarithms		
n	X	Mean	Standard Deviation	Coefficient of Variation	Mean	Standard Deviation	Coefficient of Variation
57	60	15.12	12.81	0.85	1.0832	0.2642	0.24
58	115	16.84	18.25	1.08	1.1000	0.2916	0.27
59	160	19.27	25.98	1.35	1.1187	0.3229	0.29

The values indicate that even for a moderately sized sample, a single value can significantly change the moments. For a test of the largest value, the z value is

$$z = \frac{160 - 19.27}{25.98} = 5.417 \tag{4.5}$$

For the logarithms, the z value is

$$z = \frac{\log(160) - 1.1187}{0.3229} = 3.361 \tag{4.6}$$

For a sample size of 59, Chauvenet's probability is 1/(2(59)) = 0.00847. For a two-tailed test, the $z_{\alpha/2}$ value is ±2.63. Therefore, the null hypothesis must be rejected for the untransformed and the log-transformed data. The sample value is considered an outlier.

4.3 DIXON–THOMPSON TEST

The Dixon–Thompson outlier test has the advantage of suitability for samples as small as three. It uses the same null hypothesis as Chauvenet's method (see Equations 4.1). It is only valid for testing one outlier. It can be used as a one-tailed or two-tailed test and is preferable to Chauvenet's test when the direction of the outlier is specified prior to collecting the data. For example, if only low outliers are important to the problem, then it is preferable to use the Dixon–Thompson test for low outliers rather than Chauvenet's test, with the chance that the largest value deviates from the mean more than the lowest value does.

To conduct the Dixon–Thompson test, the data are ranked from smallest to largest, with the smallest denoted as X_1 and the largest denoted as X_n. The subscript indicates the rank of the value from smallest to largest. The test statistic R and critical value R_c depend on the sample size. Table 4.1 gives the equations (Equations 4.3 to 4.6) used to compute R and corresponding critical values R_c for 5%, 2.5% and 1% levels of significance. The null hypothesis is rejected if R is greater than R_c.

Example 4.3

The Dixon–Thompson test can be applied to the flood series for San Emigdio Creek, California, to test the largest value as an outlier: (6690, 538, 500, 340, 262, 234, 212, 208, 171, 159, 150, 145, 118, 94, 92, 60, 58, 54, 40, 38, 28, 27, 26, 25). In this case, the value to be examined is prespecified so it is appropriate to use a one-tailed upper test. The equation from Table 4.1 gives:

$$R = \frac{X_n - X_{n-2}}{X_n - X_3} = \frac{6690 - 500}{6690 - 27} = 0.929 \tag{4.7}$$

The critical values obtained from Table 4.1 are 0.405, 0.443, and 0.485 for levels of significance of 5%, 2.5%, and 1%, respectively. Thus, the largest flow is rejected and considered an outlier by the Dixon–Thompson test.

It is important when applying this test to specify the hypotheses (one-tailed lower, one-tailed upper, or two-tailed) and the distribution (normal or lognormal) prior to collecting the data. The data should not be collected and examined prior to specifying the hypotheses and distribution.

TABLE 4.1
Test Statistics and Critical Values for Dixon–Thompson Outlier Test

Sample Size	Low Outlier Test Statistic	High Outlier Test Statistic	Equation	m	Critical Value 5%	2.5%	1%
3 to 7	$R=\frac{X_2-X_1}{X_n-X_1}$	$R=\frac{X_n-X_{n-1}}{X_n-X_1}$	(4.3)	3	0.943	0.970	0.988
				4	0.765	0.829	0.889
				5	0.641	0.707	0.777
				6	0.560	0.626	0.693
				7	0.503	0.562	0.630
8 to 10	$R=\frac{X_2-X_1}{X_{n-1}-X_1}$	$R=\frac{X_n-X_{n-1}}{X_n-X_2}$	(4.4)	8	0.549	0.610	0.675
				9	0.506	0.565	0.630
				10	0.472	0.528	0.590
11 to 13	$R=\frac{X_3-X_1}{X_{n-1}-X_1}$	$R=\frac{X_n-X_{n-2}}{X_n-X_2}$	(4.5)	11	0.570	0.617	0.670
				12	0.540	0.586	0.637
				13	0.515	0.560	0.610
14 to 25	$R=\frac{X_3-X_1}{X_{n-2}-X_1}$	$R=\frac{X_n-X_{n-2}}{X_n-X_3}$	(4.6)	14	0.538	0.583	0.632
				15	0.518	0.562	0.611
				16	0.499	0.542	0.590
				17	0.482	0.525	0.574
				18	0.467	0.509	0.556
				19	0.455	0.495	0.541
				20	0.444	0.482	0.528
				21	0.431	0.470	0.516
				22	0.422	0.461	0.506
				23	0.414	0.452	0.494
				24	0.405	0.443	0.485
				25	0.397	0.435	0.480

Example 4.4

The following data measured from 18 debris basins in southern California were subjected to a Dixon–Thompson test.

70000	103980	124070	69000	119200	15000
67050	63000	73040	80000	112070	20300
33990	66410	57140	16000	3010	19040

The debris volumes are expressed in cubic yards per square mile. The goal was to evaluate the data for either a low or high outlier while assuming a normal distribution. The mean and standard deviation are 61,794 and 37,726 yd^3/mi^2, respectively. The smallest measured volume (3010 yd^3/mi^2) was noticeably smaller than the remainder of the data, so it was tested as a low outlier:

$$R=\frac{16000-3010}{112070-3010}=0.119$$

The critical value computed with Table 4.1 is 0.467 for a 5% level of significance. The results indicate that the smallest value is not significantly different from rates that can be expected under the assumption of a normal population. Therefore, the lowest value is not an outlier if the normal distribution is assumed.

4.4 ROSNER'S OUTLIER TEST

Rosner (1975; 1983) devised a two-tailed outlier test that can be used to detect multiple outliers. The test is applicable to normal distributions. Rosner's test is an iterative approach to testing for as many as k outliers; thus, k tests are performed. At step m, where $m = 1, 2, \ldots k$, the null and alternative hypotheses are:

H_0: All values in the sample of size $n - m + 1$ are from the same normal population. (4.8a)

H_A: The m most extreme events are unlikely to have come from the same normal population as the remainder of the sample of size $n - m$. (4.8b)

Prior to collecting the data, the value of k should be specified. Once the data are collected, the following procedure is used:

1. Compute the sample mean $\bar{X}_{(1)}$ and standard deviation $S_{(1)}$ based on the n sample values.
2. The sample value $X_{(1)}$ that deviates the most from the mean is identified and used to computed the sample value of the test statistic $R_{(1)}$:

$$R_{(1)} = \frac{\left|X_{(1)} - \bar{X}_{(1)}\right|}{S_{(1)}} \tag{4.9}$$

   Note that $X_{(1)}$ can be either the largest or smallest value in the sample of size $R_{(1)}$.
3. The sample value $X_{(1)}$ is removed from the sample, and the mean $\bar{X}_{(2)}$ and standard deviation $S_{(2)}$ are computed from the remaining $n - 1$ values.
4. The sample value $X_{(2)}$ that deviates the most from the mean $\bar{X}_{(2)}$ is identified and used to compute the sample value of the test statistic $R_{(2)}$:

$$R_{(2)} = \frac{\left|X_{(2)} - \bar{X}_{(2)}\right|}{S_{(2)}} \tag{4.10}$$

   The value $X_{(2)}$ can be either the largest or smallest value in the sample of size $n - 1$.

5. The process of removing the $(m - 1)$th value and computing the mean $\overline{X}_{(m)}$ and standard deviation $S_{(m)}$ is continued, with the test statistic $R_{(m)}$ computed with the value $X_{(m)}$ that deviates the most from the mean:

$$R_{(m)} = \frac{\left| X_{(m)} - \overline{X}_{(m)} \right|}{S_{(m)}} \tag{4.11}$$

Step 5 is repeated until k values have been examined and the test statistic $R_{(k)}$ computed. This yields k values of $R_{(m)}$. To this point, none of the values have been tested for statistical significance. Only the sample values $R_{(m)}$ of the test statistic have been computed. The test procedure then continues, but in an order reverse to the order in which the computed values of the test statistic $R_{(m)}$ were computed. Approximate values of the critical test statistic, denoted as $R_{c(m)}$, are obtained from the following polynomials as a function of both the sample size n and the value of m:

$$\begin{aligned} R_c = {} & 2.295 + 0.02734n - 0.0002175n^2 - 0.03786m + 0.0009356nm \\ & - 0.0000793m^2 - 0.00006973nm^2 + 0.000008374m^3 \\ & + 0.000003943nm^3 \quad \text{for } 25 \le n \le 50 \text{ and } 1 \le m \le 10 \end{aligned} \tag{4.12a}$$

$$\begin{aligned} R_c = {} & 2.8613 + 0.006641n - 0.00001521n^2 - 0.003914m + 0.001207nm \\ & - 0.0006349m^2 - 0.00002405nm^2 + 0.00003133m^3 \\ & + 0.00000163nm^3 \quad \text{for } 51 \le n \le 200 \text{ and } 1 \le m \le 10 \end{aligned} \tag{4.12b}$$

These equations provide estimates of the critical value for a 5% level of significance with an accuracy of about 0.01. Tables are available for other levels of significance and sample sizes. Using either Equation 4.12a or 4.12b, the critical value can be estimated for the sample size n and each value of m from 1 to k. Note that when using Equations 4.12, the sample size remains equal to that for the entire sample n and is not reduced as each value $X_{(m)}$ is removed from the sample; however, for computing $R_{c(m)}$ the value of m changes with each step. Having computed estimates of R_c for each step, with each denoted as $R_{c(m)}$, the sample values of $R_{(m)}$ are compared with the critical values $R_{c(m)}$. The testing continues with the following steps.

6. The values $R_{(k)}$ and $R_{c(k)}$ are compared. If $R_{(k)}$ is greater than $R_{c(k)}$, then $X_{(k)}$ is considered to be an outlier and all of the other extreme values $X_{(k-1)}$ to $X_{(1)}$ are also considered outliers regardless of the values of R and R_c for these values. If $R_{(k)}$ is less than $R_{c(k)}$, than $X_{(k)}$ is not an outlier and the testing process continues.
7. The values of $R_{(k-1)}$ and $R_{c(k-1)}$ are compared. If $R_{(k-1)}$ is greater than $R_{c(k-1)}$, then $X_{(k-1)}$ is considered to be an outlier. All remaining untested values of $X_{(k-2)}$ to $X_{(1)}$ are also considered to be outliers. If $R_{(k-1)}$ is less than $R_{c(k-1)}$, then $X_{(k-1)}$ is not an outlier and the testing process continues.
8. Until an outlier is identified, the process continues with $R_{(m)}$ compared with $R_{c(m)}$. When an outlier is identified, all of the more deviant extreme

values are considered outliers. The process stops when an outlier is identified.

Example 4.5

A small hypothetical sample will be used to illustrate the computational steps of Rosner's method. Assume that the sample consists of five values: 44, 21, 16, 14, and 5. Note that Rosner's test requires a sample size of 25 or more; this use of a sample of five is intended only to illustrate the steps of the procedure. Also assume that before the data were collected, a decision was made to evaluate the sample for three outliers. The mean and standard deviation for the sample of five are 20 and 14.61, respectively. Since the value of 44 deviates from the mean of 20 more than the value of 5, it is used to compute the test statistic $R_{(1)}$:

$$R_{(1)} = \frac{|44 - 20|}{14.61} = 1.643 \tag{4.13a}$$

The value of 44 is then eliminated from the sample and the moments recomputed: $\bar{X}_{(2)} = 14$ and $S_{(2)} = 6.683$. Since the 5 deviates more from 14 than 21 does, it is used to compute the test statistic $R_{(2)}$:

$$R_{(2)} = \frac{|5 - 14|}{6.683} = 1.347 \tag{4.13b}$$

Since k is 3, then the sample point of 5 is removed from the sample and the moments recomputed: $\bar{X}_{(3)} = 17$ and $S_{(3)} = 3.606$. Since the largest sample value of 21 deviates from more than does the lowest value of 14, it is tested next:

$$R_{(3)} = \frac{|21 - 17|}{3.606} = 1.109 \tag{4.13c}$$

Since k was set at 3, the first phase of the test is complete. The second phase tests the significance of the three computed values of $R_{(m)}$.

Generally, critical values are obtained from a table of exact values or estimated using Equations 4.12. Since this is a computational example only and the sample size is only five, for which Rosner's method is not applicable, values will be assumed. The value of $R_{(3)}$ is tested first. If the critical value is less than 1.109, then the sample value of 21 is considered as an outlier; therefore, the values $X_{(1)} = 44$ and $X_{(2)} = 5$ would also be outliers. If $R_{c(3)}$ is greater than 1.109, then $X_{(3)} = 21$ is not an outlier and the procedure moves to test $R_{(2)}$. Assume $X_{(3)}$ was not an outlier. If the critical value for testing $R_{(2)}$ is less than 1.347, then both $X_{(2)}$ and $X_{(1)}$ are outliers; otherwise, $X_{(2)}$ is not an outlier and $R_{(1)}$ should be checked. Assume $X_{(2)}$ is not an outlier. If $R_{c(1)}$ is less than 1.643, then $X_{(1)}$ is an outlier; otherwise, it is not an outlier. Note that $R_{c(1)}$, $R_{c(2)}$, and $R_{c(3)}$ will have different values because m is changing.

Example 4.6

Extreme events are not uncommon in flood series and can exert significant effect on computed frequency curves. Rosner's test was applied to the 44-year annual maximum flood record for the Pond Creek watershed (see Appendix Table B.1) to test for a maximum of three outliers. Since logarithmic transforms are common in hydrologic analysis, Rosner's test was applied to the logarithms of the Pond Creek data. The largest value is 8020 (log 3.904) and the smallest value is 607 (log 2.783). The latter log value is furthest from the log mean of 3.414, so it is the first value tested. The test statistic value $R_{(1)}$ is:

$$R_{(1)} = \frac{|\log(607) - 3.4141|}{0.2048} = 3.081$$

The discharge of 607 is removed from the sample and the logarithmic moments recomputed (see Table 4.2). Because the log of the largest discharge is now the most extreme point, it is used to compute $R_{(1)}$:

$$R_{(2)} = \frac{\log(8020) - 3.429}{0.1823} = 2.606$$

The discharge of 8020 is then removed from the sample and the logarithmic moments recomputed. The third value, $R_{(3)}$, follows:

$$R_{(3)} = \frac{\log(5970) - 3.418}{0.1685} = 2.125$$

Since k was set at 3, the process moves to the testing phase.

The critical value $R_{c(3)}$ for $m = 3$ and $n = 44$ is obtained from a table as 3.058. Since $R_{o(3)}$ is less than the critical value, 5970 is not an outlier from a lognormal population.

TABLE 4.2
Rosner's Test of Logarithms of the Pond Creek Annual Maximum Series (n = 44)

m	$n - m + 1$	Mean	Standard Deviation	Sample Value	Computed Test Statistic	Critical Test Statistic	Decision
1	44	3.41445	0.20479	2.78319	3.081	3.077	[a]
2	43	3.42913	0.18229	3.90417	2.606	3.068	Accept H_0
3	42	3.41782	0.16854	3.77597	2.125	3.058	Accept H_0

[a] Reject H_0; this sample value is an outlier.

TABLE 4.3
Rosner's Test of Pond Creek Annual Maximum Series ($n = 44$)

m	$n-m+1$	Mean	Standard Deviation	Sample Value	Computed Test Statistic	CriticalTest Statistic	Decision
1	44	2881.5	1364.7	8020.	3.765	3.077	[a]
2	43	2762.0	1124.0	5970.	2.854	3.068	Accept H_0
3	42	2685.6	1018.4	5180.	2.449	3.058	Accept H_0

[a] Reject H_0; this sample value is an outlier.

Therefore, $R_{(2)}$ is tested. The tabled value of 3.068 is greater than $R_{(2)}$; therefore, the H_0 is accepted, and the process moves the check $R_{(1)}$. The critical value for $m = 1$ and $n = 44$ is 3.077. Since $R_{(1)}$ is greater than $R_{c(1)}$, H_0 is rejected, so $X_{(1)}$ of 607 is considered to be an outlier. Therefore, Rosner's test of the logarithms concludes that only the lowest discharge of 607 cfs is an outlier if the data were sampled from a lognormal population.

The data were also analyzed for $k = 3$ without a logarithmic transform (see Table 4.3). The lowest discharge of 607 cfs is never furthest from the mean, so it was not considered an extreme event. The three largest discharges were furthest from the central tendency. The largest discharge of 8020 cfs failed the test and is considered an outlier if the data were from a normal population.

The same data were analyzed with and without a logarithmic transform. The resulting decisions were quite different. One analysis concluded that the sample included a single low outlier while the other analysis identified a single high outlier; however, different underlying populations were assumed. This illustrates the importance to the decision of whether to make a nonlinear data transformation.

Example 4.7

The first predictor variable of the sediment yield data (see Appendix Table B.3) was subjected to Rosner's test, with seven events tested, that is, $k = 7$ (see Table 4.4). The three largest were identified as outliers. Since the data were mostly values near 0, all seven extreme events were the largest flows. For the first four of the seven tested (i.e., $m = 4, 5, 6, 7$), the computed value of $R_{(m)}$ was less than the critical value $R_{c(m)}$. Therefore, they were assumed to be part of normal distribution. For the third largest value, the computed value of the test statistic ($R = 3.38$) was larger than the critical value ($R_{c(3)} = 2.98$), so the null hypothesis was rejected. Thus, the two values larger than this value must also be considered outliers. Even though the computed value of the test statistic (2.93) for the next to the largest event is less than the corresponding critical value (2.99), the 1.14 event is still considered an outlier because the next smaller value of 1.126 caused the null hypothesis to be rejected. It would not be rational for 1.126 to be an outlier if 1.14 was not. Once Rosner's null hypothesis is rejected, all more extreme values in either tail of the distribution are assumed to be outliers.

TABLE 4.4
Rosner's Test of Precipitation–Temperature Ratio of Sediment Data

m	$n - m + 1$	Mean	Standard Deviation	Sample Value	Computed Test Statistic	Critical Test Statistic	Decision
7	31	0.2457	0.1403	0.536	2.069	2.922	Accept H_0
6	32	0.2591	0.1573	0.673	2.631	2.938	Accept H_0
5	33	0.2732	0.1748	0.725	2.585	2.953	Accept H_0
4	34	0.2896	0.1971	0.833	2.757	2.966	Accept H_0
3	35	0.3135	0.2402	1.126	3.383	2.979	[a]
2	36	0.3365	0.2739	1.140	2.934	2.992	[b]
1	37	0.3660	0.3242	1.428	3.275	3.004	[b]

[a] Reject H_0; this sample value is an outlier.
[b] Sample values that are more extreme than a proven outlier are also outliers.

4.5 LOG-PEARSON TYPE III OUTLIER DETECTION: BULLETIN 17B

Outliers that may be found at either or both ends of a flood frequency curve are measured values that appear to be from a longer period of record or a different population. Short flood records often contain one or more events that appear to be extreme even though they would not appear to be outliers if the record were longer. Similarly, one or two snowmelt events in a series of rainfall events represent a different population and may cause such events to appear as outliers. Outliers are evident when one or more data points do not follow the trend of the remaining data plotted on a frequency curve.

Bulletin 17B (Interagency Advisory Committee on Water Data, 1982) presents criteria based on a one-sided test to detect outliers at a 10% significance level. If the station skew based on logarithms of the data exceeds 0.4, tests are applied for high outliers first; if it is below –0.4, low outliers are considered first. If the station skew is ±0.4, both high and low outliers are tested before any data are censored. The high and low outliers are determined with the following equations, respectively:

$$Y_L = \bar{Y} + K_N S_Y \tag{4.14}$$

$$Y_L = \bar{Y} - K_N S_Y \tag{4.15}$$

where Y_L is the log of high or low outlier limit, $\bar{Y}$ is the mean of the log of the sample flows, S_Y is the standard deviation of the logs of the sample flows, and K_N is the critical deviate taken from Table 4.5.

If a sample is found to contain high outliers, the peak flows should be checked against historical data and data from nearby stations. Bulletin 17B recommends that

TABLE 4.5
Outlier Test Deviates (K_N) at 10% Significance Level (from Bulletin 17B)

Sample Size	K_N Value	Sample Size	K_N Value	Sample Size	K_N Value	Sample Size	K_N Value
10	2.036	45	2.727	80	2.940	115	3.064
11	2.088	46	2.736	81	2.945	116	3.067
12	2.134	47	2.744	82	2.949	117	3.070
13	2.175	48	2.753	83	2.953	118	3.073
14	2.213	49	2.760	84	2.957	119	3.075
15	2.247	50	2.768	85	2.961	120	3.078
16	2.279	51	2.775	86	2.966	121	3.081
17	2.309	52	2.783	87	2.970	122	3.083
18	2.335	53	2.790	88	2.973	123	3.086
19	2.361	54	2.798	89	2.977	124	3.089
20	2.385	55	2.804	90	2.981	125	3.092
21	2.408	56	2.811	91	2.984	126	3.095
22	2.429	57	2.818	92	2.989	127	3.097
23	2.448	58	2.824	93	2.993	128	3.100
24	2.467	59	2.831	94	2.996	129	3.102
25	2.486	60	2.837	95	3.000	130	3.104
26	2.502	61	2.842	96	3.003	131	3.107
27	2.519	62	2.849	97	3.006	132	3.109
28	2.534	63	2.854	98	3.011	133	3.112
29	2.549	64	2.860	99	3.014	134	3.114
30	2.563	65	2.866	100	3.017	135	3.116
31	2.577	66	2.871	101	3.021	136	3.119
32	2.591	67	2.877	102	3.024	137	3.122
33	2.604	68	2.883	103	3.027	138	3.124
34	2.616	69	2.888	104	3.030	139	3.126
35	2.628	70	2.893	105	3.033	140	3.129
36	2.639	71	2.897	106	3.037	141	3.131
37	2.650	72	2.903	107	3.040	142	3.133
38	2.661	73	2.908	108	3.043	143	3.135
39	2.671	74	2.912	109	3.046	144	3.138
40	2.682	75	2.917	110	3.049	145	3.140
41	2.692	76	2.922	111	3.052	146	3.142
42	2.700	77	2.927	112	3.055	147	3.144
43	2.710	78	2.931	113	3.058	148	3.146
44	2.719	79	2.935	114	3.061	149	3.148

high outliers be adjusted for historical information or retained in the sample as a systematic peak. High outliers should not be discarded unless the peak flow is shown to be seriously in error. If a high outlier is adjusted based on historical data, the mean and standard deviation of the log distribution should be recomputed for the adjusted data before testing for low outliers.

To test for low outliers, the low outlier threshold Y_L of Equation 4.15 is computed. The corresponding discharge $Y_L = 10^{Y_L}$ is then computed. If any discharges in the flood series are less than X_L, they are considered low outliers and should be deleted from the sample. The moments should be recomputed and a conditional probability adjustment applied. Thus, low outliers can be censored from the data regardless of the availability of historic data.

Example 4.8

To illustrate this criteria for outlier detection, Equations 4.14 and 4.15 are applied to the 43-year record for the Medina River, which has a log mean of 2.0912 and a log standard deviation of 0.3941. From Table 4.5, $K_N = 2.710$. Testing first for high outliers, Equation 4.14 yields:

$$Y_L = 2.0912 + 2.710(0.3941) = 3.1592 \tag{4.16}$$

and

$$X_L = 10^{3.1592} = 1443\ m^3/s \tag{4.17}$$

The flood record does not include any flows that exceed this amount, so high outliers are not present in the data.

Now testing for low outliers, Equation 4.15 gives

$$Y_L = 2.0912 - 2.710(0.3941) = 1.0232 \tag{4.18}$$

and

$$X_L = 10^{1.0232} = 11\ m^3/s \tag{4.19}$$

The Medina River flood record does not include any flows that are less than this critical value. Therefore, the entire sample should be used in the log-Pearson III frequency analysis.

4.6 PEARSON TYPE III OUTLIER DETECTION

The guidelines for treating extreme events in records of annual peak discharges presented in Bulletin 17B are limited. The critical values for low and high outlier thresholds assume a normal distribution (zero skew), consider only one level of significance (10%), and do not provide a separate test to evaluate the possibility of more than one outlier. In addition, the critical deviates in Bulletin 17B are presented in a tabular format and do not appear to increase consistently with sample size. Specifically, the slope of the tabular values presented in Bulletin 17B is slightly irregular, which suggests some inaccuracy in developing critical values.

A broader outlier test was developed by Spencer and McCuen (1996), with the procedure summarized as follows:

1. Obtain the sample mean ($\bar{X}$), standard deviation (S), and skew (g).
2. Before examining the data, decide the number of outliers that can be expected. The method can handle one, two, or three outliers. Generally, information beyond the flood record will be used to decide on the number of values to be tested as outliers. For example, if two hurricanes are known to have occurred during the period of record, it may be of interest to deal with only nonhurricane generated floods in the analysis; therefore, a test for two outliers would be selected a priori. As another example, low outliers may occur during drought years, so information about drought occurrences could be used to suggest the number of outliers for which the test should be conducted.
3. Obtain the critical deviate K_N of Equation 4.14 using the following composite model:

$$K_N = c_1 n^2 + c_2 n + c_3 \quad \text{for } 10 \le n \le 15 \tag{4.20a}$$

$$K_N = c_4 + c_5 e^{-c_6 n} n^{c_7} \quad \text{for } 16 \le n \le 89 \tag{4.20b}$$

$$K_N = c_4 + c_5 e^{-90 c_6} 90^{c_7} \left[1 + (n - 90) \left(\frac{c_7}{90} - c_6 \right) \right] \quad \text{for } 90 \le n \le 150 \tag{4.20c}$$

Values for the coefficients c_i ($i = 1, 2, \ldots, 7$) are obtained from Tables 4.6 to 4.11. Critical deviates are given only for skews of –1, –0.5, 0, 0.5,

TABLE 4.6
Coefficients for Single Outlier Test (Sample Size from 10 to 15)

Skew for High Outliers	10% Level of Significance	5% Level of Significance	1% Level of Significance	Skew for Low Outliers
−1.0	C(1) = −0.0007001	C(1) = −0.0007001	C(1) = −0.0002334	1.0
	C(2) = 0.02790	C(2) = 0.02790	C(2) = 0.01963	
	C(3) = 1.475	C(3) = 1.599	C(3) = 1.883	
−0.5	C(1) = −0.0006666	C(1) = −0.001767	C(1) = −0.001733	0.5
	C(2) = 0.008335	C(2) = 0.07337	C(2) = 0.07913	
	C(3) = 1.713	C(3) = 1.435	C(3) = 1.638	
0	C(1) = −0.001733	C(1) = −0.001700	C(1) = −0.001667	0
	C(2) = 0.08313	C(2) = 0.0879	C(2) = 0.1047	
	C(3) = 1.381	C(3) = 1.466	C(3) = 1.517	
0.5	C(1) = −0.001700	C(1) = −0.002900	C(1) = −0.001533	−0.5
	C(2) = 0.09790	C(2) = 0.1363	C(2) = 0.1207	
	C(3) = 1.386	C(3) = 1.258	C(3) = 1.484	
1.0	C(1) = −0.002367	C(1) = −0.003267	C(1) = −0.001900	−1.0
	C(2) = 0.1296	C(2) = 0.1619	C(2) = 0.1493	
	C(3) = 1.274	C(3) = 1.172	C(3) = 1.334	

TABLE 4.7
Coefficients for Single Outlier Test (Sample Size from 16 to 150)

Skew for High Outliers	10% Level of Significance	5% Level of Significance	1% Level of Significance	Skew for Low Outliers
−1.0	C(4) = 0.4976	C(4) = 0.7022	C(4) = 1.325	1.0
	C(5) = 1.040	C(5) = 1.050	C(5) = 0.9722	
	C(6) = 0.0006810	C(6) = 0.0005446	C(6) = −0.0000016	
	C(7) = 0.07019	C(7) = 0.04113	C(7) = −0.07090	
−0.5	C(4) = 0.2917	C(4) = 0.2490	C(4) = 0.4411	0.5
	C(5) = 1.125	C(5) = 1.349	C(5) = 1.455	
	C(6) = 0.001137	C(6) = 0.0009226	C(6) = 0.001201	
	C(7) = 0.1598	C(7) = 0.1310	C(7) = 0.1236	
0	C(4) = 0.4236	C(4) = 0.5683	C(4) = 0.7647	0.0
	C(5) = 0.9423	C(5) = 0.9117	C(5) = 0.9002	
	C(6) = 0.001669	C(6) = 0.001909	C(6) = 0.002537	
	C(7) = 0.2547	C(7) = 0.2710	C(7) = 0.3000	
0.5	C(4) = 0.2551	C(4) = 0.2735	C(4) = 0.2270	−0.5
	C(5) = 0.9987	C(5) = 1.051	C(5) = 1.127	
	C(6) = 0.002022	C(6) = 0.002098	C(6) = 0.002689	
	C(7) = 0.3081	C(7) = 0.3146	C(7) = 0.3424	
1.0	C(4) = 0.3575	C(4) = 0.2578	C(4) = 0.02386	−1.0
	C(5) = 0.8896	C(5) = 1.021	C(5) = 1.168	
	C(6) = 0.002379	C(6) = 0.002220	C(6) = 0.002833	
	C(7) = 0.3709	C(7) = 0.3613	C(7) = 0.3862	

and 1. Critical deviates for samples with other skews can be determined by interpolation. For example, if the sample skew is between 0 and 0.5, the critical deviates for skews of 0 and 0.5 should be computed, with the deviate for the desired skew obtained by linear interpolation.

Example 4.9

The following is the annual maximum flood series for Oak Creek near Mojave, California, for the 1958–1984 period.

1740	97	22	8.2	3.6	0.43
750	82	12	6.9	3.1	0.40
235	44	11	6.3	3.0	
230	36	9.8	6.2	2.0	
165	29	9.2	4.0	0.8	

The annual series will be checked for one high outlier under the assumption of a log-Pearson III distribution.

TABLE 4.8
Coefficients for Consecutive Outlier Test (Two Outliers, Sample Size from 10 to 15)

Skew for High Outliers	10% Level of Significance	5% Level of Significance	1% Level of Significance	Skew for Low Outliers
−1.0	C(1) = 0.0000333	C(1) = −0.001300	C(1) = −0.0003668	1.0
	C(2) = 0.009768	C(2) = 0.04310	C(2) = 0.01757	
	C(3) = 1.468	C(3) = 1.392	C(3) = 1.815	
−0.5	C(1) = −0.0005667	C(1) = −0.0007001	C(1) = −0.002033	0.5
	C(2) = 0.03597	C(2) = 0.03990	C(2) = 0.07420	
	C(3) = 1.376	C(3) = 1.488	C(3) = 1.537	
0.0	C(1) = −0.001167	C(1) = −0.001300	C(1) = −0.003733	0.0
	C(2) = 0.06317	C(2) = 0.06710	C(2) = 0.1351	
	C(3) = 1.278	C(3) = 1.390	C(3) = 1.204	
0.5	C(1) = −0.001367	C(1) = −0.0009334	C(1) = −0.003467	−0.5
	C(2) = 0.07857	C(2) = 0.07254	C(2) = 0.1493	
	C(3) = 1.263	C(3) = 1.417	C(3) = 1.138	
1.0	C(1) = −0.001333	C(1) = −0.002200	C(1) = −0.003700	−1.0
	C(2) = 0.08933	C(2) = 0.1164	C(2) = 0.1749	
	C(3) = 1.275	C(3) = 1.224	C(3) = 0.9960	

The log statistics are $\bar{Y} = 1.1836$, $S_y = 0.9247$, and $g = 0.4$. For testing one outlier, values of the critical deviate are computed with Equation 4.20b and the coefficient of Table 4.7, with the following results:

$$\text{For 10\%: } K_N = 0.2888 + 0.9874e^{-0.001951(27)}\,(27)^{0.2974} = 2.7853$$

$$\text{For 5\%: } K_N = 0.3325 + 1.0231e^{-0.00206(27)}\,(27)^{0.30588} = 2.9845$$

$$\text{For 1\%: } K_N = 0.3345 + 1.0816e^{-0.002659(27)}\,(27)^{0.3339} = 3.3605$$

The coefficients were interpolated for a skew of 0.4. The critical deviates are used with Equation 4.14:

$$\overline{Y_U} = 1.1836 + 0.9247K_N$$

which yields X_U values of 5756, 8813, and 19,630 cfs for 10%, 5%, and 1% probabilities, respectively. The largest flow of 1740 cfs is considerably below the computed bounds. Therefore, it is not large enough to be considered an outlier.

TABLE 4.9
Coefficients for Consecutive Outlier Test (Two Outliers, Sample Size from 16 to 150)

Skew for High Outliers	10% Level of Significance	5% Level of Significance	1% Level of Significance	Skew for Low Outliers
−1.0	C(4) = 0.5054	C(4) = 0.6063	C(4) = 0.9588	1.0
	C(5) = 0.9508	C(5) = 1.050	C(5) = 1.146	
	C(6) = 0.0003294	C(6) = 0.0001929	C(6) = −0.0000166	
	C(7) = 0.06331	C(7) = 0.03125	C(7) = −0.03885	
−0.5	C(4) = 0.2690	C(4) = 0.4534	C(4) = 0.6562	0.5
	C(5) = 1.050	C(5) = 1.034	C(5) = 1.178	
	C(6) = 0.0007525	C(6) = 0.0007993	C(6) = 0.0006414	
	C(7) = 0.1422	C(7) = 0.1354	C(7) = 0.1011	
0.0	C(4) = 0.3447	C(4) = 0.4336	C(4) = 0.5882	0.0
	C(5) = 0.8814	C(5) = 0.9126	C(5) = 1.033	
	C(6) = 0.001343	C(6) = 0.001487	C(6) = 0.001483	
	C(7) = 0.2319	C(7) = 0.2316	C(7) = 0.2143	
0.5	C(4) = 0.3614	C(4) = 0.4545	C(4) = 0.6567	−0.5
	C(5) = 0.8247	C(5) = 0.8403	C(5) = 0.8728	
	C(6) = 0.001680	C(6) = 0.001904	C(6) = 0.002059	
	C(7) = 0.2922	C(7) = 0.2996	C(7) = 0.3067	
1.0	C(4) = 0.2889	C(4) = 0.3092	C(4) = 0.3940	−1.0
	C(5) = 0.8676	C(5) = 0.9314	C(5) = 0.9905	
	C(6) = 0.001838	C(6) = 0.001941	C(6) = 0.002308	
	C(7) = 0.3232	C(7) = 0.3235	C(7) = 0.3397	

TABLE 4.10
Coefficients for Consecutive Outlier Test (Three Outliers, Sample Size from 10 to 15)

Skew for High Outliers	10% Level of Significance	5% Level of Significance	1% Level of Significance	Skew for Low Outliers
−1.0	C(1) = −0.0005001	C(1) = −0.0007001	C(1) = 0.0008665	1.0
	C(2) = 0.02150	C(2) = 0.02290	C(2) = −0.02106	
	C(3) = 1.352	C(3) = 1.490	C(3) = 2.036	
−0.5	C(1) = −0.0004667	C(1) = −0.001400	C(1) = −0.001233	0.5
	C(2) = 0.03027	C(2) = 0.05080	C(2) = 0.04763	
	C(3) = 1.346	C(3) = 1.365	C(3) = 1.632	
0.0	C(1) = −0.001733	C(1) = −0.001033	C(1) = −0.002067	0.0
	C(2) = 0.07213	C(2) = 0.05323	C(2) = 0.08647	
	C(3) = 1.138	C(3) = 1.394	C(3) = 1.396	
0.5	C(1) = 0.003233	C(1) = −0.001800	C(1) = −0.003067	−0.5
	C(2) = 0.1206	C(2) = 0.08560	C(2) = 0.1285	
	C(3) = 0.8910	C(3) = 1.237	C(3) = 1.149	
1.0	C(1) = 0.003267	C(1) = −0.003667	C(1) = −0.003700	−1.0
	C(2) = 0.1329	C(2) = 0.1467	C(2) = 0.1639	
	C(3) = 0.8720	C(3) = 0.9030	C(3) = 0.9410	

TABLE 4.11
Coefficients for Consecutive Outlier Test (Three Outliers, Sample Size from 16 to 150)

Skew for High Outliers	10% Level of Significance	5% Level of Significance	1% Level of Significance	Skew for Low Outliers
−1.0	C(4) = 0.4599	C(4) = 0.6026	C(4) = 1.041	1.0
	C(5) = 0.9492	C(5) = 0.9996	C(5) = 0.9610	
	C(6) = 0.0001685	C(6) = 0.0000281	C(6) = 0.0001729	
	C(7) = 0.05781	C(7) = 0.02747	C(7) = −0.03357	
−0.5	C(4) = 0.3350	C(4) = 0.4161	C(4) = 0.5637	0.5
	C(5) = 0.9498	C(5) = 1.015	C(5) = 1.251	
	C(6) = 0.0005051	C(6) = 0.0005468	C(6) = 0.0002952	
	C(7) = 0.1351	C(7) = 0.1217	C(7) = 0.07192	
0.0	C(4) = 0.2828	C(4) = 0.3055	C(4) = 0.5341	0.0
	C(5) = 0.9033	C(5) = 1.013	C(5) = 1.086	
	C(6) = 0.0009574	C(6) = 0.0008973	C(6) = 0.0009050	
	C(7) = 0.2047	C(7) = 0.1867	C(7) = 0.1709	
0.5	C(4) = 0.4382	C(4) = 0.5662	C(4) = 0.7590	−0.5
	C(5) = 0.7295	C(5) = 0.7392	C(5) = 0.7746	
	C(6) = 0.001466	C(6) = 0.001492	C(6) = 0.001770	
	C(7) = 0.2839	C(7) = 0.2826	C(7) = 0.2875	
1.0	C(4) = 0.2557	C(4) = 0.3122	C(4) = 0.4608	−1.0
	C(5) = 0.8650	C(5) = 0.9135	C(5) = 0.9581	
	C(6) = 0.001510	C(6) = 0.001557	C(6) = 0.001856	
	C(7) = 0.2948	C(7) = 0.2926	C(7) = 0.3038	

4.7 PROBLEMS

4-1 Discuss the pros and cons of censoring extreme events.

4-2 An extreme event in a flood series may be the result of an event from a different population (e.g., one hurricane event in a record of nonhurricane events) or the result of sampling variation. How would the cause be distinguished? Should these two types of extreme events be treated differently?

4-3 Test the following data for the Little Patuxent River, Maryland, using Chauvenet's method: 12400, 1880, 1110, 5370, 1910, 3670, 4680, 2780, 1490, 1430, 680, 1230, 1230, 5070, and 1040.

4-4 Use Chauvenet's method to test the following data for Seneca Creek, Maryland, for an outlier: 26100, 3020, 3160, 16000, 4900, 7850, 16000, 3260, 3010, 3620, 1070, 4950, 7410, 8250, and 2270.

4-5 The following data are the annual maximum discharges for the Grey River near Alpine, Wyoming. Test the data for outliers using Chauvenet's method: 4210, 2010, 5010, 4290, 3720, 2920, 2500, 2110, 3110, 2420, 4280, 3860, 3150, 4050, 3260, 7230, 5170, 2550, 5220, 3650, 3590, and 650.

4-6 The following data are the annual maximum discharges for San Emigdio Creek, California: 6690, 538, 500, 340, 262, 234, 212, 208, 171, 159, 150, 145, 118, 94, 92, 60, 58, 54, 40, 38, 28, 27, 26, 25, 24, 17, 16, 8, and 1.23. Test the data for outliers using Chauvenet's method.

4-7 The following data are the annual maximum discharges for Oak Creek near Mojave, California: 1740, 750, 235, 230, 165, 97, 82, 44, 36, 29, 22, 12, 11, 9.8, 9.2, 8.2, 6.9, 6.3, 6.2, 4.0, 3.6, 3.1, 3.0, 2.0, 0.8, 0.43, and 0.4. Test the data for an outlier using Chauvenet's method.

4-8 Using the data from Problem 4-7, test the logarithms of the data for an outlier using Chauvenet's method.

4-9 When using the Dixon–Thompson method to test for a low outlier with a sample size of 5, how much larger must X_2 be in comparison with X_1 for X_1 to be considered a low outlier? Assume $X_n - X_1 = 10$.

4-10 For a sample size of ten, with the Dixon–Thompson test, what minimum value of X_{10} is necessary to reject the largest sample value as a high outlier if $X_2 = 4.3$ and $X_9 = 11.4$?

4-11 For the following data, use the Dixon–Thompson test to assess whether or not the largest value is an outlier: 3, 6, 7, 9, 26.

4-12 For the data of Problem 4-4, use the Dixon–Thompson test to assess whether the largest value is an outlier.

4-13 For the data of Problem 4-3, use the Dixon–Thompson test to assess whether the largest value is an outlier.

4-14 For the data of Problem 4-5, use the Dixon–Thompson test to assess whether the smallest value is an outlier.

4-15 Use Rosner's test to test for up to three outliers for the following annual minimum flow in a small river: 28.2, 12.6, 9.4, 7.9, 7.8, 7.8, 7.6, 7.5, 7.5, 7.5, 7.3, 7.2, 7.2, 7.1, 6.8, 6.6, 6.6, 6.5, 6.5, 6.4, 6.3, 6.1, 6.1, 6.0, 5.4, 3.1, and 0.7.

4-16 Use Rosner's test to test the data of Problem 4-6 for up to four outliers.

4-17 Use Rosner's test to test the data of Problem 4-7 for up to five outliers.

4-18 Critique the rationale of the Rosner test assumption that once an outlier is detected, all other extreme events are considered outliers.

4-19 Discuss the application of outlier tests to logarithms, rather than the actual data, to recorded events declared as outliers based on decisions made with logarithms.

4-20 For a test on a data set with $n = 30$, the mean ($X_{(3)}$) and standard deviation ($S_{(3)}$) are 48 and 12, respectively. When checking the third most extreme event in the sample, what minimum value would be considered a high outlier?

4-21 The following data are the annual maximum series for Joshua Creek, California (1959–1989). Using Rosner's test, test for as many as ten outliers: 4500, 2660, 2540, 470, 246, 180, 113, 86, 43, 28, 20, 17, 13, 13, 11, 9, 9, 4, 4, 3, 1, 1, 0, 0, 0, 0, 0, 0, 0, 0, and 0.

4-22 For the data of Problem 4-15, use the Spencer–McCuen test to assess whether or not the largest event is a high outlier for a log-Pearson III distribution.

4-23 For the data of Problem 4-6, use the Spencer–McCuen test to assess whether or not the largest event is a high outlier for a log-Pearson III distribution.

5 Statistical Frequency Analysis

5.1 INTRODUCTION

Univariate frequency analysis is widely used for analyzing hydrologic data, including rainfall characteristics, peak discharge series, and low flow records. It is primarily used to estimate exceedance probabilities and variable magnitudes. A basic assumption of frequency analysis is that the vector of data was measured from a temporally or spatially homogeneous system. If measured data are significantly nonhomogeneous, the estimated probabilities or magnitudes will be inaccurate. Thus, changes such as climate or watershed alterations render the data unfit for frequency analysis and other modeling methods.

If changes to the physical processes that influence the data are suspected, the data vector should be subjected to statistical tests to decide whether the nonstationarity is significant. If the change had a significant effect on the measured data, it may be necessary to adjust the data before subjecting it to frequency analysis. Thus, the detection of the effects of change, the identification of the nature of any change detected, and the appropriate adjustment of the data are prerequisite steps required before a frequency model can be used to make probability or magnitude estimates.

5.2 FREQUENCY ANALYSIS AND SYNTHESIS

Design problems such as the delineation of flood profiles require estimates of discharge rates. A number of methods of estimating peak discharge rates are available. They fall into two basic groups, one used at sites where gaged stream-flow records are available (gaged) and the other at sites where such records are not available (ungaged).

Statistical frequency analysis is the most common procedure for the analysis of flood data at a gaged location. It is a general procedure that can be applied to any type of data. Because it is so widely used with flood data, the method is sometimes designated *flood frequency analysis*. However, statistical frequency analysis can also be applied to other hydrologic variables such as rainfall data for the development of intensity-duration-frequency curves and low-flow discharges for use in water quality control. The variable could also be the mean annual rainfall, the peak discharge, the 7-day low flow, or a water quality parameter. Therefore, the topic will be treated in both general and specific terms.

5.2.1 Population versus Sample

In frequency modeling, it is important to distinguish between the population and the sample. Frequency modeling is a statistical method that deals with a single random variable and thus is classified as a univariate method. The goal of univariate prediction is to make estimates of probabilities or magnitudes of random variables. A first step is to identify the population. The objective of univariate data analysis is to use sample information to determine the appropriate population density function, with the probability density function (PDF) being the univariate model from which probability statements can be made. The input requirements for frequency modeling include a data series and a probability distribution assumed to describe the occurrence of the random variable. The data series could include the largest instantaneous peak discharge to occur each year of the record. The probability distribution could be the normal distribution. Analysis is the process of using the sample information to estimate the population. The population consists of a mathematical model that is a function of one or more parameters. For example, the normal distribution is a function of two parameters: the mean μ and standard deviation σ. In addition to identifying the correct PDF, it is necessary to quantify the parameters of the PDF. The population consists of both the probability distribution function and the parameters.

A frequently used procedure called the *method of moments* equates characteristics of the sample (e.g., sample moments) to characteristics of the population (e.g., population parameters). It is important to note that estimates of probability and magnitudes are made using the assumed population and not the data sample; the sample is used only in identifying and verifying the population.

5.2.2 Analysis versus Synthesis

As with many hydrologic methods that have statistical bases, the terms *analysis* and *synthesis* apply to the statistical frequency method. Frequency analysis is "breaking down" data in a way that leads to a mathematical or graphical model of the relationship between flood magnitude and its probability of occurrence. Conversely, synthesis refers to the estimation of (1) a value of the random variable X for some selected exceedance probability or (2) the exceedance probability for a selected value of the random variable X. In other words, analysis is the derivation of a model that can represent the relation between a random variable and its likelihood of occurrence, while synthesis is using the resulting relation for purposes of estimation.

It is important to point out that frequency analysis may actually be part of a more elaborate problem of synthesis. Specifically, separate frequency analyses can be performed at a large number of sites within a region and the value of the random variable X for a selected exceedance probability determined for each site; these values can then be used to develop a regression model using the random variable X as the criterion or dependent variable. As an example, regression equations that relate peak discharges of a selected exceedance probability for a number of sites to watershed characteristics are widely used in hydrologic design. This process is called *regionalization*. These equations are derived by (1) making a frequency analysis of annual maximum discharges at a number (n) of stream gage stations in a region;

(2) selecting the value of the peak discharge from each of the n frequency curves for a selected exceedance probability, say the 100-year flood; and (3) developing the regression equation relating the n values of peak discharge to watershed characteristics for the same n watersheds.

5.2.3 Probability Paper

Frequency analysis is a common task in hydrologic studies. A frequency analysis usually produces a graph of the value of a single hydrologic variable versus the probability of its occurrence. The computed graph represents the best estimate of the statistical population from which the sample of data was drawn.

Since frequency analyses are often presented graphically, a special type of graph paper, which is called probability paper, is required. The paper has two axes. The ordinate is used to plot the value of the random variable, that is, the magnitude, and the probability of its occurrence is given on the abscissa. The probability scale will vary depending on the probability distribution used. In hydrology, the normal and Gumbel extreme-value distributions are the two PDFs used most frequently to define the probability scale. Figure 5.1 is on normal probability paper. The probability scale represents the cumulative normal distribution. The scale at the top of the graph is the exceedance probability, that is, the probability that the random variable will be equaled or exceeded in one time period. It varies from 99.99% to 0.01%. The lower scale is the nonexceedance probability, which is the probability that the corresponding value of the random variable will not be exceeded in any one time period. This scale extends from 0.01% to 99.99%. The ordinate of probability paper is used for

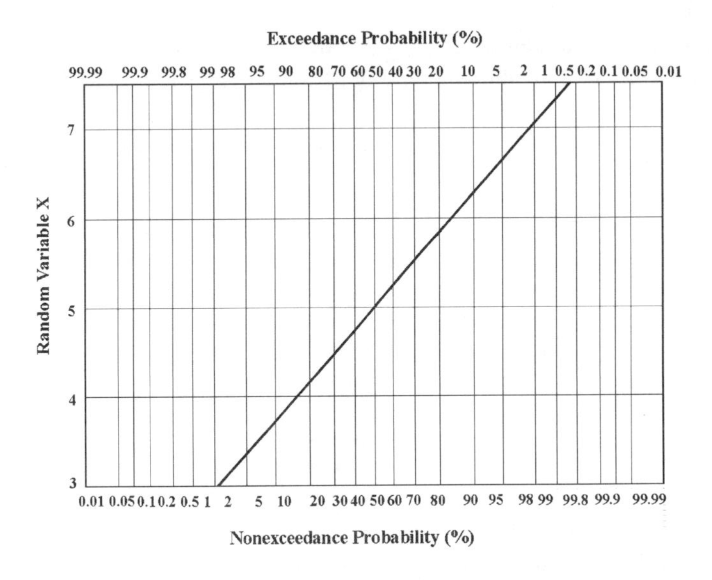

FIGURE 5.1 Frequency curve for a normal population with $\mu = 5$ and $\sigma = 1$.

the random variable, such as peak discharge. The example shown in Figure 5.1 has an arithmetic scale. Lognormal probability paper is also available, with the scale for the random variable in logarithmic form. Gumbel and log-Gumbel papers can also be obtained and used to describe the probabilistic behavior of random variables that follow these probability distributions.

A frequency curve provides a probabilistic description of the likelihood of occurrence or nonoccurrence of a variable. Figure 5.1 shows a frequency curve, with the value of the random variable X versus its probability of occurrence. The upper probability scale gives the probability that X will be exceeded in one time period, while the lower probability scale gives the probability that X will not be exceeded. For the frequency curve of Figure 5.1, the probability that X will be greater than 7 in one time period is 0.023 and the probability that X will not be greater than 7 in one time period is 0.977.

Although a unique probability plotting paper could be developed for each probability distribution, papers for the normal and extreme value distributions are the most frequently used. The probability paper is presented as a cumulative distribution function. If the sample of data is from the distribution function used to scale the probability paper, the data will follow the pattern of the population line when properly plotted on the paper. If the data do not follow the population line, then (1) the sample is from a different population or (2) sampling variation produced a nonrepresentative sample. In most cases, the former reason is assumed, especially when the sample size is reasonably large.

5.2.4 Mathematical Model

As an alternative to a graphical solution using probability paper, a frequency analysis may be conducted using a mathematical model. A model that is commonly used in hydrology for normal, lognormal, and log-Pearson Type III analyses has the form

$$X = \bar{X} + KS \tag{5.1}$$

in which X is the value of the random variable having mean $\bar{X}$ and standard deviation S, and K is a frequency factor. Depending on the underlying population, the specific value of K reflects the probability of occurrence of the value X. Equation 5.1 can be rearranged to solve for K when X, $\bar{X}$, and S are known and an estimate of the probability of X occurring is necessary:

$$K = \frac{X - \bar{X}}{S} \tag{5.2}$$

In summary, Equation 5.1 is used when the probability is known and an estimation of the magnitude is needed, while Equation 5.2 is used when the magnitude is known and the probability is needed.

5.2.5 Procedure

In a broad sense, frequency analysis can be divided into two phases: deriving the population curve and plotting the data to evaluate the goodness of fit. The following procedure is often used to derive the frequency curve to represent the population:

1. Hypothesize the underlying density function.
2. Obtain a sample and compute the sample moments.
3. Equate the sample moments and the parameters of the proposed density function.
4. Construct a frequency curve that represents the underlying population.

This procedure is referred to as method-of-moments estimation because the sample moments are used to provide numerical values for the parameters of the assumed population. The computed frequency curve representing the population can then be used to estimate magnitudes for a given return period or probabilities for specified values of the random variable. Both the graphical frequency curve and the mathematical model of Equation 5.1 are the population.

It is important to recognize that it is not necessary to plot the data points in order to make probability statements about the random variable. While the four steps listed above lead to an estimate of the population frequency curve, the data should be plotted to ensure that the population curve is a good representation of the data. The plotting of the data is a somewhat separate part of a frequency analysis; its purpose is to assess the quality of the fit rather than act as a part of the estimation process.

5.2.6 Sample Moments

For the random variable X, the sample mean ($\bar{X}$), standard deviation (S), and standardized skew (g) are, respectively, computed by:

$$\bar{X} = \frac{1}{n}\sum_{i=1}^{n} X_i \tag{5.3a}$$

$$S = \left[\frac{1}{n-1}\sum_{i=1}^{n}(X_i - \bar{X})^2\right]^{0.5} \tag{5.3b}$$

$$g = \frac{n\sum_{i=1}^{n}(X_i - \bar{X})^3}{(n-1)(n-2)S^3} \tag{5.3c}$$

For use in frequency analyses where the skew is used, Equation 5.3c represents a standardized value of the skew. Equations 5.3 can also be used when the data are transformed by taking the logarithms. In this case, the log transformation should be done before computing the moments.

5.2.7 Plotting Position Formulas

It is important to note that it is not necessary to plot the data before probability statements can be made using the frequency curve; however, the data should be plotted to determine how well they agree with the fitted curve of the assumed population. A rank-order method is used to plot the data. This involves ordering the data from the largest event to the smallest event, assigning a rank of 1 to the largest event and a rank of n to the smallest event, and using the rank (i) of the event to obtain a probability plotting position; numerous plotting position formulas are available. Bulletin 17B (Interagency Advisory Committee on Water Data, 1982) provides the following generalized equation for computing plotting position probabilities:

$$P_i = \frac{i - a}{n - a - n + 1} \tag{5.4}$$

where a and b are constants that depend on the probability distribution. An example is $a = b = 0$ for the uniform distribution. Numerous formulas have been proposed, including the following:

Weibull: $$P_i = \frac{i}{n+1} \tag{5.5a}$$

Hazen: $$P_i = \frac{2i-1}{2n} = \frac{i-0.5}{n} \tag{5.5b}$$

Cunnane: $$P_i = \frac{i-0.4}{n+0.2} \tag{5.5c}$$

in which i is the rank of the event, n is the sample size, and p_i values give the exceedance probabilities for an event with rank i. The data are plotted by placing a point for each value of the random variable at the intersection of the value of the random variable and the value of the exceedance probability at the top of the graph. The plotted data should approximate the population line if the assumed population model is a reasonable assumption. The various plotting position formulas provide different probability estimates, especially in the tails of the distributions. The following summary shows computed probabilities for each rank for a sample of nine using the plotting position formulas of Equations 5.5.

	$n = 9$				$n = 99$		
Rank	p_w	p_h	p_c	**Rank**	p_w	p_h	p_c
1	0.1	0.05	0.065	1	0.01	0.005	0.006
2	0.2	0.15	0.174	2	0.02	0.015	0.016
3	0.3	0.25	0.283	.			
4	0.4	0.35	0.391	.			
5	0.5	0.45	0.500	.			
6	0.6	0.55	0.609	98	0.98	0.985	0.984
7	0.7	0.65	0.717	99	0.99	0.995	0.994
8	0.8	0.75	0.826				
9	0.9	0.85	0.935				

The Hazen formula gives smaller probabilities for all ranks than the Weibull and Cunnane formulas. The probabilities for the Cunnane formula are more dispersed than either of the others. For a sample size of 99, the same trends exist as for $n = 9$.

5.2.8 Return Period

The concept of return period is used to describe the likelihood of flood magnitudes. The return period is the reciprocal of the exceedance probability, that is, $p = 1/T$. Just as a 25-year rainfall has a probability of 0.04 of occurring in any one year, a 25-year flood has a probability of 0.04 of occurring in any one year. It is incorrect to believe that a 25-year event will not occur again for another 25 years. Two 25-year events can occur in consecutive years. Then again, a period of 100 years may pass before a second 25-year event occurs.

Does a 25-year rainfall cause a 25-year flood magnitude? Some hydrologic models make this assumption; however, it is unlikely to be the case in actuality. It is a reasonable assumption for modeling because models are based on the average of expectation or on-the-average behavior. In actuality, a 25-year flood magnitude will not occur if a 25-year rainfall occurs on a dry watershed. Similarly, a 50-year flood could occur from a 25-year rainfall if the watershed was saturated. Modeling often assumes that a T-year rainfall on a watershed that exists in a T-year hydrologic condition will produce a T-year flood.

5.3 POPULATION MODELS

Step 1 of the frequency analysis procedure indicates that it is necessary to select a model to represent the population. Any probability distribution can serve as the model, but the lognormal and log-Pearson Type III distributions are the most widely used in hydrologic analysis. They are introduced subsequent sections, along with the normal distribution or basic model.

5.3.1 Normal Distribution

Commercially available normal probability paper is commonly used in hydrology. Following the general procedure outlined above, the specific steps used to develop a curve for a normal population are as follows:

1. Assume that the random variable has a normal distribution with population parameters μ and σ.
2. Compute the sample moments $\bar{X}$ and S (the skew is not needed).
3. For normal distribution, the parameters and sample moments are related by $\mu = \bar{X}$ and $\sigma = S$.
4. A curve is fitted as a straight line with $(\bar{X} - S)$ plotted at an exceedance probability of 0.8413 and $(\bar{X} + S)$ at an exceedance probability of 0.1587.

The frequency curve of Figure 5.1 is an example for a normal distribution with a mean of 5 and a standard deviation of 1. It is important to note that the curve passes through the two points: $(\bar{X} - S,\ 0.8413)$ and $(\bar{X} + S,\ 0.1587)$. It also passes through the point defined by the mean and a probability of 0.5. Two other points that could be used are $(\bar{X} + 2S, 0.0228)$ and $(\bar{X} - 2S,\ 0.9772)$. Using the points farther removed from the mean has the advantage that inaccuracies in the line drawn to represent the population will be smaller than when using more interior points.

The sample values should then be plotted (see Section 5.2.7) to decide whether the measured values closely approximate the population. If the data provide a reasonable fit to the line, one can assume that the underlying population is the normal distribution and the sample mean and standard deviation are reasonable estimates of the location and scale parameters, respectively. A poor fit indicates that the normal distribution is not appropriate, that the sample statistics are not good estimators of the population parameters, or both.

When using a frequency curve, it is common to discuss the likelihood of events in terms of exceedance frequency, exceedance probability, or the return period (T) related to the exceedance probability (p) by $p = 1/T$, or $T = 1/p$. Thus, an event with an exceedance probability of 0.01 should be expected to occur 1 time in 100. In many cases, a time unit is attached to the return period. For example, if the data represent annual floods at a location, the basic time unit is 1 year. The return period for an event with an exceedance probability of 0.01 would be the 100-year event (i.e., $T = 1/0.01 = 100$); similarly, the 25-year event has an exceedance probability of 0.04 (i.e., $p = 1/25 = 0.04$). It is important to emphasize that two T-year events will not necessarily occur exactly T years apart. They can occur in successive years or may be spaced three times T years apart. On average, the events will be spaced T years apart. Thus, in a long period, say 10,000 years, we would expect $10{,}000/T$ events to occur. In any single 10,000-year period, we may observe more or fewer occurrences than the mean ($10{,}000/T$).

Estimation with normal frequency curve—For normal distribution, estimation may involve finding a probability corresponding to a specified value of the random variable or finding the value of the random variable for a given probability. Both problems can be solved using graphical analysis or the mathematical models of

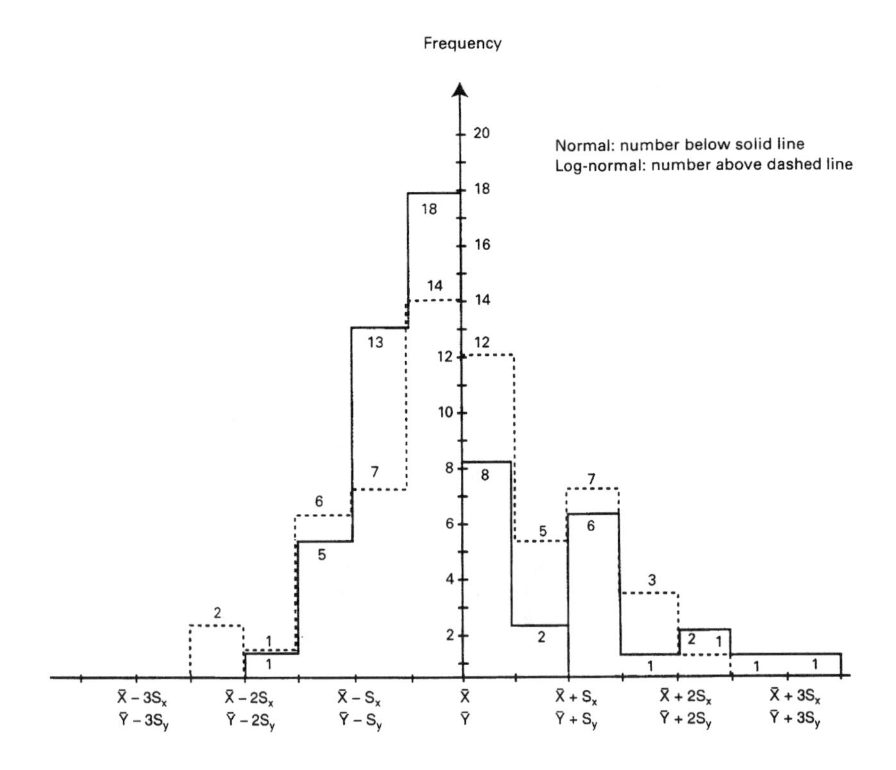

FIGURE 5.2 Frequency histograms of the annual maximum flood series (solid line) and logarithms (dashed line) based on mean (and for logarithms) and standard deviations (S_x and for logarithms S_y): Piscataquis River near Dover-Foxcroft, Maine.

Equations 5.1 and 5.2. A graphical analysis estimation involves simply entering the probability and finding the corresponding value of the random variable or entering the value of the random variable and finding the corresponding exceedance probability. In both cases, the fitted line (population) is used. The accuracy of the estimated value will be influenced by the accuracy used in drawing the line or graph.

Example 5.1

Figure 5.2 shows a frequency histogram for the data in Table 5.1. The sample consists of 58 annual maximum instantaneous discharges, with a mean of 8620 ft^3/sec, a standard deviation of 4128 ft^3/sec, and a standardized skew of 1.14. In spite of the large skew, the normal frequency curve was fitted using the procedure of the preceding section. Figure 5.3 shows the cumulative normal distribution using the sample mean and the standard deviation as estimates of the location and scale parameters. The population line was drawn by plotting $X + S = 12{,}748$ at $p = 15.87\%$ and $\bar{X} - S = 4492$ at $p = 84.13\%$, using the upper scale for the probabilities. The data were plotted using the Weibull plotting position formula (Equation 5.5a). The data do not provide a reasonable fit to the population; they show a significant skew with an

TABLE 5.1
Frequency Analysis of Peak Discharge Data: Piscataquis River

Rank	Weibull Probability	Random Variable	Logarithm of Variable
1	0.0169	21500	4.332438
2	0.0339	19300	4.285557
3	0.0508	17400	4.240549
4	0.0678	17400	4.240549
5	0.0847	15200	4.181844
6	0.1017	14600	4.164353
7	0.1186	13700	4.136721
8	0.1356	13500	4.130334
9	0.1525	13300	4.123852
10	0.1695	13200	4.120574
11	0.1864	12900	4.110590
12	0.2034	11600	4.064458
13	0.2203	11100	4.045323
14	0.2373	10400	4.017034
15	0.2542	10400	4.017034
16	0.2712	10100	4.004322
17	0.2881	9640	3.984077
18	0.3051	9560	3.980458
19	0.3220	9310	3.968950
20	0.3390	8850	3.946943
21	0.3559	8690	3.939020
22	0.3729	8600	3.934499
23	0.3898	8350	3.921686
24	0.4068	8110	3.909021
25	0.4237	8040	3.905256
26	0.4407	8040	3.905256
27	0.4576	8040	3.905256
28	0.4746	8040	3.905256
29	0.4915	7780	3.890980
30	0.5085	7600	3.880814
31	0.5254	7420	3.870404
32	0.5424	7380	3.868056
33	0.5593	7190	3.856729
34	0.5763	7190	3.856729
35	0.5932	7130	3.853090
36	0.6102	6970	3.843233
37	0.6271	6930	3.840733
38	0.6441	6870	3.836957
39	0.6610	6750	3.829304
40	0.6780	6350	3.802774
41	0.6949	6240	3.795185

TABLE 5.1
Frequency Analysis of Peak Discharge Data: Piscataquis River (*Continued*)

Rank	Weibull Probability	Random Variable	Logarithm of Variable
42	0.7119	6200	3.792392
43	0.7288	6100	3.785330
44	0.7458	5960	3.775246
45	0.7627	5590	3.747412
46	0.7797	5300	3.724276
47	0.7966	5250	3.720159
48	0.8136	5150	3.711807
49	0.8305	5140	3.710963
50	0.8475	4710	3.673021
51	0.8644	4680	3.670246
52	0.8814	4570	3.659916
53	0.8983	4110	3.613842
54	0.9153	4010	3.603144
55	0.9322	4010	3.603144
56	0.9492	3100	3.491362
57	0.9661	2990	3.475671
58	0.9831	2410	3.382017
Mean		8620	3.889416
Standard deviation		4128	0.203080
Standardized skew		1.14	−0.066

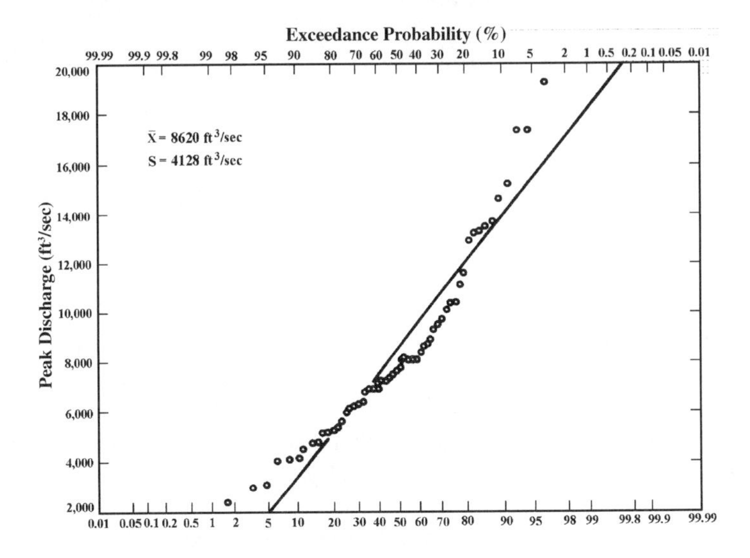

FIGURE 5.3 Piscataquis River near Dover-Foxcroft, Maine.

especially poor fit to the tails of the distribution (i.e., high and low exceedance probabilities). Because of the poor fit, the line shown in Figure 5.3 should not be used to make probability statements about the future occurrences of floods; for example, the normal distribution (i.e., the line) suggests a 1% chance flood magnitude of slightly more than 18,000 ft^3/sec. However, if a line was drawn subjectively through the trend of the points, the flood would be considerably larger, say about 23,000 ft^3/sec.

The 100-year flood is estimated by entering with a probability of 1% and finding the corresponding flood magnitude. Probabilities can be also estimated. For example, if a levee system at this site would be overtopped at a magnitude of 16,000 ft^3/sec, the curve indicates a corresponding probability of about 4%, which is the 25-year flood.

To estimate probabilities or flood magnitudes using the mathematical model, Equation 5.1 becomes $X = \overline{X} + zS$ because the frequency factor K of Equation 5.1 becomes the standard normal deviate z for a normal distribution, where values of z are from Appendix Table A.1. To find the value of the random variable X, estimates of $\overline{X}$ and S must be known and the value of z obtained from Appendix Table A.1 for any probability. To find the probability for a given value of the random variable X, Equation 5.2 is used to solve for the frequency factor z (which is K in Equation 5.2); the probability is then obtained from Table A.1 using the computed value of z. For example, the value of z from Table A.1 for a probability of 0.01 (i.e., the 100-year event) is 2.327; thus, the flood magnitude is:

$$X = \overline{X} + zS = 8620 + 2.327(4128) = 18.226 \text{ ft}^3\text{/sec}$$

which agrees with the value obtained from the graphical analysis. For a discharge of 16,000 ft^3/sec, the corresponding z value is:

$$z = \frac{X - \overline{X}}{S} = \frac{16{,}000 - 8{,}620}{4{,}128} = 1.788$$

Appendix Table A.1 indicates the probability is 0.0377, which agrees with the graphical estimate of about 4%.

5.3.2 Lognormal Distribution

When a poor fit to observed data is obtained, a different distribution function should be considered. For example, when the data demonstrate a concave, upward curve, as in Figure 5.3, it is reasonable to try a lognormal distribution or an extreme value distribution. It may be preferable to fit with a distribution that requires an estimate of the skew coefficient, such as a log-Pearson Type III distribution. However, sample estimates of the skew coefficient may be inaccurate for small samples.

The same procedure used for fitting the normal distribution can be used to fit the lognormal distribution. The underlying population is assumed to be lognormal. The data must first be transformed to logarithms, $Y = \log X$. This transformation creates a new random variable Y. The mean and standard deviation of the logarithms are computed and used as the parameters of the population; it is important to recognize that the logarithm of the mean does not equal the mean of the logarithms, which is also true for the standard deviation. Thus, the logarithms of the mean and standard deviation should not be used as parameters; the mean and standard deviation of the logarithms should be computed and used as the parameters. Either natural or base-10 logarithms may be used, although the latter is more common in hydrology. The population line is defined by plotting the straight line on arithmetic probability paper between the points ($\bar{Y} + S_y$, 0.1587) and ($\bar{Y} - S_y$, 0.8413), where $\bar{Y}$ and S_y are the mean and standard deviation of the logarithms, respectively. In plotting the data, either the logarithms can be plotted on an arithmetic scale or the untransformed data can be plotted on a logarithmic scale.

When using a frequency curve for a lognormal distribution, the value of the random variable Y and the moments of the logarithms ($\bar{Y}$ and S_y) are related by the equation:

$$Y = \bar{Y} + zS_y \tag{5.6}$$

in which z is the value of the standardized normal variate; values of z and corresponding probabilities can be found in Table A.1. Equation 5.6 can be used to estimate either flood magnitudes for a given exceedance probability or an exceedance probability for a specific discharge. To find a discharge for a specific exceedance probability the standard normal deviate z is obtained from Table A.1 and used in Equation 5.6 to compute the discharge. To find the exceedance probability for a given discharge Y, Equation 5.6 is rearranged by solving for z. With values of Y, $\bar{Y}$, and S_y, a value of z is computed and used with Table A.1 to compute the exceedance probability. Of course, the same values of both Y and the probability can be obtained directly from the frequency curve.

Example 5.2

The peak discharge data for the Piscataquis River were transformed by taking the logarithm of each of the 58 values. The moments of the logarithms are as follows: $\bar{Y} = 3.8894$, $S_y = 0.20308$, and $g = -0.07$. Figure 5.4 is a histogram of the logarithms. In comparison to the histogram of Figure 5.2, the logarithms of the sample data are less skewed. While a skew of –0.07 would usually be rounded to –0.1, it is sufficiently close to 0 such that the discharges can be represented with a lognormal distribution. The frequency curve is shown in Figure 5.5. To plot the lognormal population curve, the following two points were used: $\bar{Y} - S_y = 3.686$ at $p = 84.13\%$ and $\bar{Y} + S_y = 4.092$ at $p = 15.87\%$. The data points were plotted on Figure 5.5 using the Weibull formula and show a much closer agreement with the population line in comparison to the points for the normal distribution in Figure 5.3. It is reasonable to assume that the measured peak discharge rates can be represented by a lognormal distribution and that the future flood behavior of the watershed can be described statistically using a lognormal distribution.

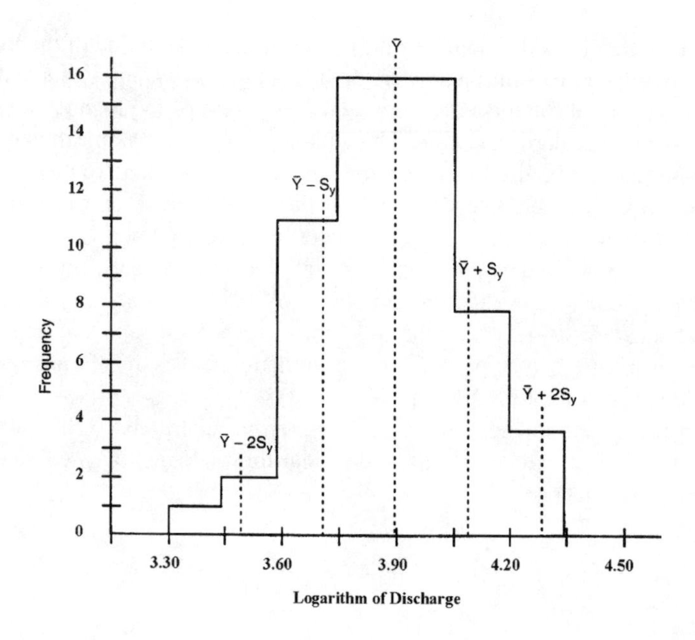

FIGURE 5.4 Histogram of logarithms of annual maximum series: Piscataquis River.

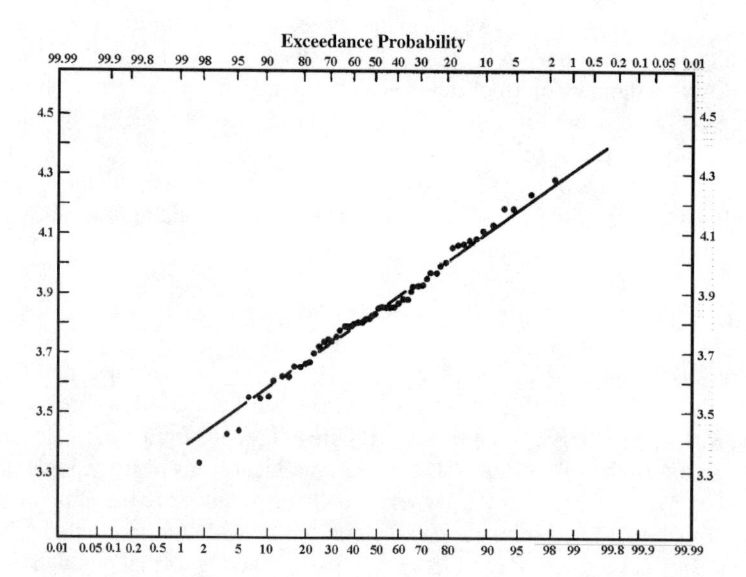

FIGURE 5.5 Frequency curve for the logarithms of the annual maximum discharge.

If one were interested in the probability that a flood of 20,000 ft^3/sec would be exceeded in a given year, the logarithm of 20,000 (4.301) would be entered on the discharge axis and followed to the assumed population line. Reading the exceedance probability corresponding to that point on the frequency curve, a flood of 20,000 ft^3/sec has a 1.7% chance of being equaled or exceeded in any one year. It can also be interpreted that over the span of 1,000 years, a flood discharge of 20,000 ft^3/sec would be exceeded in 17 of those years; it is important to understand that this is an average. In any period of 1,000 years, a value of 20,000 ft^3/sec may be exceeded or reached less frequently than in 17 of the 1,000 years, but on average 17 exceedances would occur in 1,000 years.

The probability of a discharge of 20,000 ft^3/sec can also be estimated mathematically. The standard normal deviate is

$$z = \frac{\log(20,000) - \bar{Y}}{S_y} = \frac{4.301 - 3.8894}{0.20308} = 2.027 \tag{5.7a}$$

The value of z is entered into Table A.1, which yields a probability of 0.9786. Since the exceedance probability is of interest, this is subtracted from 1, which yields a value of 0.0214. This corresponds to a 47-year flood. The difference between the mathematical estimate of 2.1% and the graphical estimate of 1.7% is due to the error in the graph. The computed value of 2.1% should be used.

The frequency curve can also be used to estimate flood magnitudes for selected probabilities. Flood magnitude is found by entering the figure with the exceedance probability, moving vertically to the frequency curve, and finally moving horizontally to flood magnitude. For example, the 100-year flood for the Piscataquis River can be found by starting with an exceedance probability of 1%, moving to the curve of Figure 5.5 and then to the ordinate, which indicates a logarithm of about 4.3586 or a discharge of 22,800 ft^3/sec. Discharges for other exceedance probabilities can be found that way or by using the mathematical model of Equation 5.1. In addition to the graphical estimate, Equation 5.6 can be used to obtain a more exact estimate. For an exceedance probability of 0.01, a z value of 2.327 is obtained from Table A.1. Thus, the logarithm is

$$Y = \bar{Y} + zS_y = 3.8894 + 2.327(0.20308) = 4.3620 \tag{5.7b}$$

Taking the antilogarithm yields a discharge of 23,013 ft^3/sec.

5.3.3 Log-Pearson Type III Distribution

Normal and lognormal frequency analyses were introduced because they are easy to understand and have a variety of uses. The statistical distribution most commonly used in hydrology in the United States is the log-Pearson Type III (LP3) because it was recommended by the U.S. Water Resources Council in Bulletin 17B (Interagency Advisory Committee on Water Data, 1982). The Pearson Type III is a PDF. It is widely accepted because it is easy to apply when the parameters are estimated using the method of moments and it usually provides a good fit to measured data. LP3 analysis requires a logarithmic transformation of data; specifically, the common logarithms are used as the variates, and the Pearson Type III distribution is used as the PDF.

While LP3 analyses have been made for most stream gage sites in the United States and can be obtained from the U.S. Geological Survey, a brief description of the analysis procedure is provided here. Although the method of analysis presented here will follow the procedure recommended by the Water Resources Council, you should consult Bulletin 17B when performing an analysis because a number of options and adjustments cited in the bulletin are not discussed here. This section only provides sufficient detail so that a basic frequency analysis can be made and properly interpreted.

Bulletin 17B provides details for the analysis of three types of data: a systematic gage record, regional data, and historic information for the site. The systematic record consists of the annual maximum flood record. It is not necessary for the record to be continuous as long as the missing part of the record is not the result of flood experience, such as the destruction of the stream gage during a large flood. Regional information includes a generalized skew coefficient, a weighting procedure for handling independent estimates, and a means of correlating a short systematic record with a longer systematic record from a nearby stream gaging station. Historic records, such as high-water marks and newspaper accounts of flooding that occurred before installation of the gage, can be used to augment information at the site.

The procedure for fitting an LP3 curve with a measured systematic record is similar to the procedure used for the normal and lognormal analyses described earlier. The steps of the Bulletin 17B procedure for analyzing a systematic record based on a method-of-moments analysis are as follows:

1. Create a series that consists of the logarithms Y_i of the annual maximum flood series x_i.
2. Using Equations 5.3 compute the sample mean, $\overline{Y}$, standard deviation, S_y, and standardized skew, g_s, of the logarithms created in step 1.
3. For selected values of the exceedance probability (p), obtain values of the standardized variate K from Appendix Table A.5 (round the skew to the nearest tenth).
4. Determine the values of the LP3 curve for the exceedance probabilities selected in Step 3 using the equation

$$Y = \overline{Y} + KS_y \tag{5.8}$$

in which y is the logarithmic value of the LP3 curve.
5. Use the antilogarithms of the Y_j values to plot the LP3 frequency curve.

After determining the LP3 population curve, the data can be plotted to determine adequacy of the curve. The Weibull plotting position is commonly used. Confidence limits can also be placed on the curve; a procedure for computing confidence intervals is discussed in Bulletin 17B. In step 3, it is necessary to select two or more points to compute and plot the LP3 curve. If the absolute value of the skew is small, the line will be nearly straight and only a few points are necessary to draw it accurately. When the absolute value of the skew is large, more points must be used

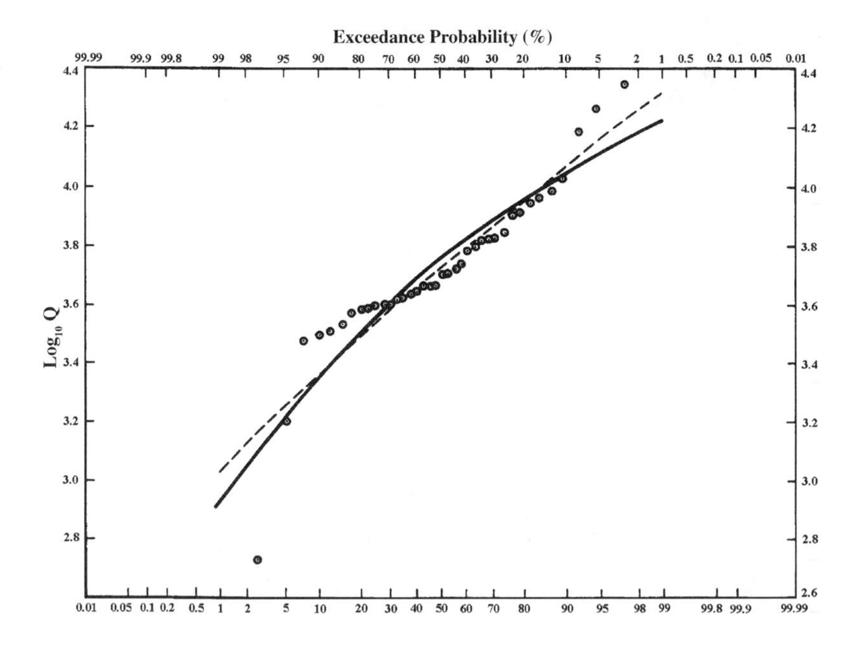

FIGURE 5.6 Log-Pearson Type III frequency curve for Back Creek near Jones Spring, West Virginia, with station skew (—) and weighted skew (- - -).

because of the greater curvature. When selecting exceedance probabilities to compute the LP3 curve, it is common to include 0.5, 0.2, 0.1, 0.04, 0.02, 0.01, and 0.002 because these correspond to return periods that are usually of interest.

Example 5.3

Data for the Back Creek near Jones Springs, West Virginia (USGS gaging station 016140), are given in Table 5.2. Based on the 38 years of record (1929–1931 and 1939–1973), the mean, standard deviation, and skew of the common logarithms are 3.722, 0.2804, and –0.731, respectively; the skew will be rounded to –0.7. Table 5.3 shows the K values of Equation 5.8 for selected values of the exceedance probability (p); these values were obtained from Table A.5 using p and the sample skew of –0.7. Equation 5.8 was used to compute the logarithms of the LP3 discharges for the selected exceedance probabilities (Table 5.3).

The logarithms of the discharges were then plotted versus the exceedance probabilities, as shown in Figure 5.6. The rank of each event is also shown in Table 5.2 and was used to compute the exceedance probability using the Weibull plotting position formula (Equation 5.4a). The logarithms of the measured data were plotted versus the exceedance probability (Figure 5.6). The data show a reasonable fit to the frequency curve, although the fit is not especially good for the few highest and the lowest measured discharges; the points with exceedance probabilities between 80% and 92% suggest a poor fit. Given that we can only make subjective assessments of

TABLE 5.2
Annual Maximum Floods for Back Creek

Year	Q	Log Q	Rank	p
1929	8750	3.9420	7	0.179
1930	15500	4.1903	3	0.077
1931	4060	3.6085	27	0.692
1939	6300	3.7993	14	0.359
1940	3130	3.4955	35	0.897
1941	4160	3.6191	26	0.667
1942	6700	3.8261	12	0.308
1943	22400	4.3502	1	0.026
1944	3880	3.5888	30	0.769
1945	8050	3.9058	9	0.231
1946	4020	3.6042	28	0.718
1947	1600	3.2041	37	0.949
1948	4460	3.6493	23	0.590
1949	4230	3.6263	25	0.641
1950	3010	3.4786	36	0.923
1951	9150	3.9614	6	0.154
1952	5100	3.7076	19	0.487
1953	9820	3.9921	5	0.128
1954	6200	3.7924	15	0.385
1955	10700	4.0294	4	0.103
1956	3880	3.5888	31	0.795
1957	3420	3.5340	33	0.846
1958	3240	3.5105	34	0.872
1959	6800	3.8325	11	0.282
1960	3740	3.5729	32	0.821
1961	4700	3.6721	20	0.513
1962	4380	3.6415	24	0.615
1963	5190	3.7152	18	0.462
1964	3960	3.5977	29	0.744
1965	5600	3.7482	16	0.410
1966	4670	3.6693	21	0.538
1967	7080	3.8500	10	0.256
1968	4640	3.6665	22	0.564
1969	536	2.7292	38	0.974
1970	6680	3.8248	13	0.333
1971	8360	3.9222	8	0.205
1972	18700	4.2718	2	0.051
1973	5210	3.7168	17	0.436

Source: Interagency Advisory Committee on Water Data, Guidelines for Determining Flood-Flow Frequency, Bulletin 17B, U.S. Geological Survey, Office of Water Data Coordination, Reston, VA, 1982.

TABLE 5.3
Computation of Log-Pearson Type III Frequency Curve for Back Creek Near Jones Springs, West Virginia

p	K	$+KS_y$	Q(ft^3/sec)
0.99	−2.82359	2.9303	852
0.90	−1.33294	3.3482	2,230
0.70	−0.42851	3.6018	3,998
0.50	0.11578	3.7545	5,682
0.20	0.85703	3.9623	9,169
0.10	1.18347	4.0538	11,320
0.04	1.48852	4.1394	13,784
0.02	1.66325	4.1884	15,430
0.01	1.80621	4.2285	16,922

the goodness of fit, the computed frequency curve appears to be a reasonable estimate of the population.

The curve of Figure 5.6 can be used to estimate flood discharges for selected exceedance probabilities or exceedance probabilities for selected discharges. For example, the 100-year flood ($p = 0.01$) equals 16,992 ft^3/sec (i.e., log $Q = 4.2285$ from Table 5.3). The exceedance probability for a discharge of 10,000 ft^3/sec (i.e., log $Q = 4$) is approximately 0.155 from Figure 5.6; this corresponds to the 6-year event.

5.4 ADJUSTING FLOOD RECORD FOR URBANIZATION

A statistical flood frequency analysis is based on the assumption of a homogeneous annual flood record. Significant changes in land use lead to nonhomogeneity of flood characteristics, thus violating the assumptions that underline frequency analysis. A flood frequency analysis based on a nonhomogeneous record will produce inaccurate estimates. The effects of nonhomogeneity must be estimated before computation of frequency analysis so that the flood record can be adjusted.

Urbanization is a primary cause of nonhomogeneity of flood records. Although the problem has been recognized for decades, few attempts to develop a systematic procedure for adjuisting flood records have been made. Multiparameter watershed models have been used for this purpose; however, a single model or procedure for adjustment has not been widely accepted by the professional community. Comparisons of methods for adjusting records have not been made.

5.4.1 Effects of Urbanization

A number of models that allow assessment of the effects of urbanization on peak discharges are available. Some models provide bases for accounting for urbanization, but it is difficult to develop a general statement of the effects of urbanization from these models. For example, with the rational method, urban development affects the runoff coefficient and the time of concentration. Thus, it is not possible to make a general statement that a 5% increase in imperviousness will cause an $x\%$ increase in the peak discharge for a specific return period. Other models are not so constrained. A number of regression equations are available that include the percentage of imperviousness as a predictor variable. They make it possible to develop a general statement on the effects of urbanization. Sarma, Delleur, and Rao (1969) provided one such example:

$$q_p = 484.1 A^{0.723}(1+U)^{1.516} P_E^{1.113} T_R^{-0.403} \tag{5.9}$$

in which A is the drainage area (mi^2), U is the fraction of imperviousness, P_E is the volume of excess rainfall (in.), T_R is the duration of rainfall excess (hours), and q_p is the peak discharge (ft^3/sec). Since the model has the power-model form, the specific effect of urbanization depends on the values of the other predictor variables (A, P_E, and T_R). However, the relative sensitivity of Equation 5.9 can be used as a measure of the effect of urbanization. The relative sensitivity (S_R) is given by:

$$S_R = \frac{\partial q_p}{\partial U} \cdot \frac{U}{q_p} \tag{5.10}$$

Evaluation of Equation 5.10 yields a relative sensitivity of 1.516. Thus, a 1% change in U will cause a change of 1.516% in the peak discharge. This estimate is an average effect since it is independent of both the value of U and the return period.

Based on the work of Carter (1961) and Anderson (1970), Dunne and Leopold (1978) provided the following equation for estimating the effect of urbanization:

$$f = 1 + 0.015U \tag{5.11}$$

in which f is a factor that gives the relative increase in peak discharge for a percent imperviousness of U. The following is a summary of the effect of urbanization based on the model of Equation 5.11:

U	0	10	20	30	40	50	100
f	1	1.15	1.3	1.45	1.6	1.75	2.5

Thus, a 1% increase in U will increase the peak discharge by 1.5%, which is the same effect shown by Equation 5.10.

The Soil Conservation Service (SCS) provided an adjustment for urbanization for the first edition of the TR-55 (1975) chart method. The adjustment depended on the percentages of imperviousness, the hydraulic length modified, and the runoff curve number (*CN*). Although the adjustment did not specifically include return period as a factor, the method incorporates the return period through rainfall input. Table 5.4 shows adjustment factors for imperviousness and the hydraulic length modified.

TABLE 5.4
Adjustment Factors for Urbanization

SCS Chart Method					USGS Urban Equations			
CN	*U*	*f*	f^2	R_s	*T*	*U*	f_1	R_s
70	20	1.13	1.28	0.018	2-yr	20	1.70	0.016
	25	1.17	1.37	0.019		25	1.78	0.016
	30	1.21	1.46	0.025		30	1.86	0.018
	35	1.26	1.59	0.026		35	1.95	0.020
	40	1.31	1.72	—		40	2.05	—
80	20	1.10	1.21	0.013	100-yr	20	1.23	0.010
	25	1.13	1.28	0.014		25	1.28	0.008
	30	1.16	1.35	0.019		30	1.32	0.008
	35	1.20	1.44	0.015		35	1.36	0.010
	40	1.23	1.51	—		40	1.41	—

Assuming that these changes occur in the same direct proportion, the effect of urbanization on peak discharges is the square of the factor. Approximate measures of the effect of changes in the adjustment factor from change in *U* are also given in Table 5.4 (R_s); these values of R_s represent the change in peak discharge due to the peak factors provided in the first edition of TR-55. Additional effects of urban development on the peak discharge would be reflected in change in the *CN*. However, the relative sensitivities of the SCS chart method suggest a change in peak discharge of 1.3% to 2.6% for a 1% change in urbanization, which represents the combined effect of changes in imperviousness and modifications of the hydraulic length.

USGS urban peak discharge equations (Sauer et al., 1981) provide an alternative for assessing effects of urbanization. Figures 5.7 and 5.8 show the ratios of urban-to-rural discharges as functions of the percentages of imperviousness and basin development factors. For the 2-year event, the ratios range from 1 to 4.5; 4.5 indicates complete development. For the 100-year event, the ratio has a maximum value of 2.7. For purposes of illustration and assuming basin development occurs in direct proportion to changes in imperviousness, the values of Table 5.4 (R_s) show the effects of urbanization on peak discharge. The average changes in peak discharge due to a 1% change in urbanization are 1.75 and 0.9% for the 2- and 100-year events, respectively. While the methods discussed reveal an effect of about 1.5%, the USGS equations suggest that the effect is slightly higher for more frequent storm events and slightly lower for less frequent events. This is generally considered rational.

Rantz (1971) devised a method for assessing the effects of urbanization on peak discharges using simulated data of James (1965) for the San Francisco Bay area. Urbanization is characterized by two variables: the percentages of channels sewered and basins developed. The percentage of basins developed is approximately twice the percentage of imperviousness. Table 5.5 shows the relative sensitivity of the peak discharge to the percent imperviousness and the combined effect of the percentages of channels sewered and basins developed. For urbanization as measured by the percentage change in imperviousness, the mean relative sensitivities are 2.6%, 1.7%,

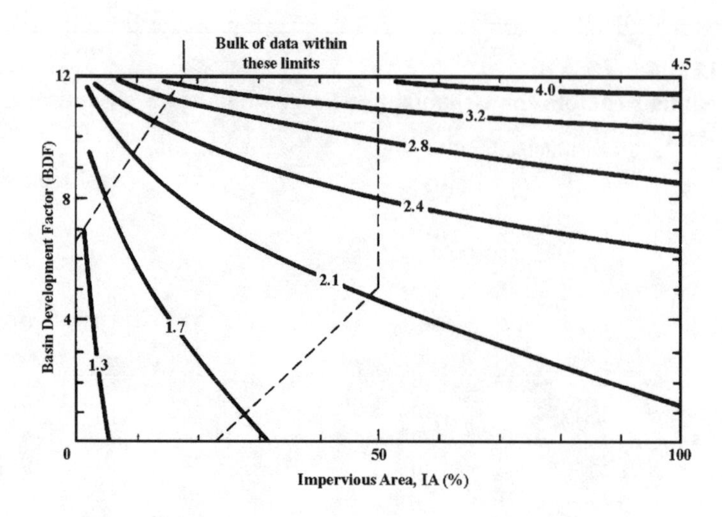

FIGURE 5.7 Ratio of the urban-to-rural 2-year peak discharge as a function of basin development factor and impervious area.

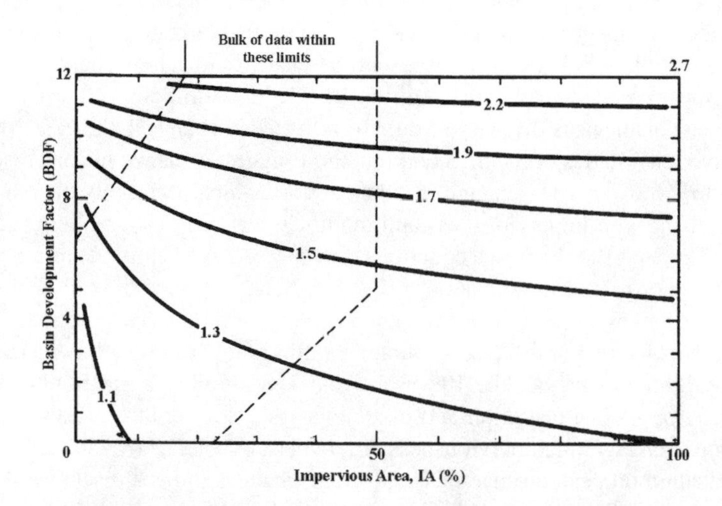

FIGURE 5.8 Ratio of the urban-to-rural 100-year peak discharge as a function of basin development factor and impervious area.

and 1.2% for the 2-, 10-, and 100-year events, respectively. These values are larger by about 30% to 50% than the values computed from the USGS urban equations. When both the percentages of channels sewered and basins developed are used as indices of development, the relative sensitivities are considerably higher. The mean relative sensitivities are 7.1, 5.1, and 3.5% for the 2-, 10-, and 100-year events, respectively. These values are much larger than the values suggested by the other methods discussed above. Thus, the effects of urbanization vary considerably, which complicates the adjustment of measured flow data to produce a homogeneous flood series.

TABLE 5.5
Effect on Peak Discharge of (a) the Percentage of Imperviousness (U) and (b) the Combined Effect of Urban Development (D)

		T = 2-yr		T = 10-yr		T = 100-yr	
		f	R_s	f	R_s	f	R_s
(a)	U (%)						
	10	1.22	0.025	1.13	0.015	1.08	0.011
	20	1.47	0.025	1.28	0.017	1.19	0.012
	30	1.72	0.026	1.45	0.018	1.31	0.013
	40	1.98	0.029	1.63	0.018	1.44	0.012
	50	2.27		1.81		1.56	
(b)	D (%)						
	10	1.35	0.040	1.18	0.022	1.15	0.010
	20	1.75	0.060	1.40	0.040	1.25	0.025
	30	2.35	0.085	1.80	0.050	1.50	0.050
	40	3.20	0.100	2.30	0.092	2.00	0.055
	50	4.20		3.22		2.55	

5.4.2 Method for Adjusting Flood Record

The literature does not identify a single method considered best for adjusting a flood record. Each method depends on the data used to calibrate the prediction process. The databases used to calibrate the methods are very sparse. However, the sensitivities suggest that a 1% increase in imperviousness causes an increase in peak discharge of about 1% to 2.5% with the former value for the 100-year event and the latter for the 2-year event. However, considerable variation is evident at any return period.

Based on the general trends of the data, a method of adjusting a flood record was developed. Figure 5.9 shows the peak adjustment factor as a function of the exceedance probability for percentages of imperviousness up to 60%. The greatest effect is for the more frequent events and the highest percentage of imperviousness. Given the return period of a flood peak for a nonurbanized watershed, the effect of an increase in imperviousness can be assessed by multiplying the discharge by the peak adjustment factor for the return period and percentage of imperviousness.

Where it is necessary to adjust a discharge from a partially urbanized watershed to a discharge for another condition, the discharge can be divided by the peak adjustment factor for the existing condition and the resulting "rural" discharge multiplied by the peak adjustment factor for the second watershed condition. The first operation (division) adjusts the discharge to a magnitude representative of a nonurbanized condition. The adjustment method of Figure 5.9 requires an exceedance probability. For a flood record, the best estimate of the probability is obtained from a plotting position formula. The following procedure can be used to adjust a

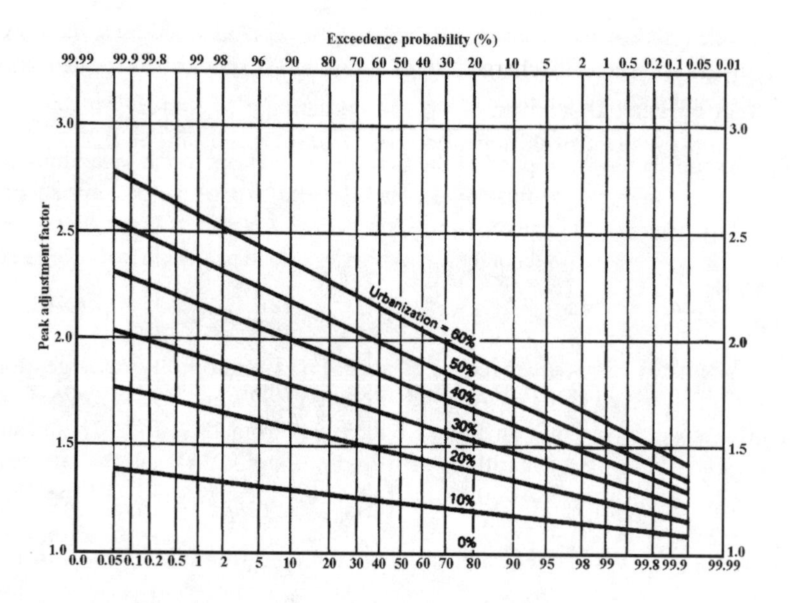

FIGURE 5.9 Peak adjustment factors for urbanizing watersheds.

flood record for which the individual flood events have occurred on a watershed undergoing continuous changes in levels of urbanization.

1. Identify the percentage of imperviousness for each event in the flood record and the percentage of imperviousness for which an adjusted flood record is needed.
2. Compute the rank (i) and exceedance probability (p) for each event in the flood record (a plotting position formula can be used to compute the probability).
3. Using exceedance probability and the actual percentage of imperviousness, find from Figure 5.9 the peak adjustment factor (f_1) to transform the measured peak from the level of imperviousness to a nonurbanized condition.
4. Using the exceedance probability and the percentage of imperviousness for which a flood series is needed, find from Figure 5.9 the peak adjustment factor (f_2) necessary to transform the nonurbanized peak to a discharge for the desired level of imperviousness.
5. Compute the adjusted discharge (Q_a) by:

$$Q_a = \frac{f_2}{f_1} Q \tag{5.12}$$

   in which Q is the measured discharge.
6. Repeat steps 3, 4, and 5 for each event in the flood record and rank the adjusted series.

7. If the ranks of the adjusted discharges differ considerably from the ranks of the measured discharges, steps 2 through 6 should be repeated until the ranks do not change.

This procedure should be applied to the flood peaks and not to the logarithms of the flood peaks, even when the adjusted series will be used to compute a lognormal or log-Pearson Type III frequency curve. The peak adjustment factors of Figure 5.9 are based on methods that show the effect of imperviousness on flood peaks, not on the logarithms.

Example 5.4

Table 5.6 contains the 48-year record of annual maximum peak discharges for the Rubio Wash watershed in Los Angeles. Between 1929 and 1964, the percent of impervious cover also shown in Table 5.6 increased from 18 to 40%. The mean and standard deviation of the logarithms of the record are 3.2517 and 0.1910, respectively. The station skew was –0.53, and the map skew was –0.45. Therefore, a weighted skew of –0.5 was used.

The procedure was used to adjust the flood record from actual levels of imperviousness for the period from 1929 to 1963 to current impervious cover conditions. For example, while the peak discharges for 1931 and 1945 occurred when the percent cover was 19% and 34%, respectively, the values were adjusted to a common percentage of 40%, which is the watershed state after 1964. Three iterations of adjustments were required. The iterative process is required because the return period for some of the earlier events changed considerably from the measured record; for example, the rank of the 1930 peak changed from 30 to 22 on the first trial, and the rank of the 1933 event went from 20 to 14. Because of such changes in the rank, the exceedance probabilities change and thus the adjustment factors, which depend on the exceedance probabilities, change. After the second adjustment is made, the rank of the events did not change, so the process is complete. The adjusted series is given in the last part of Table 5.6.

The adjusted series has a mean and standard deviation of 3.2800 and 0.1785, respectively. As expected, the mean increased because earlier events occurred when less impervious cover existed and the standard deviation decreased because the measured data include both natural variation and variation due to different levels of imperviousness. The adjustment corrected for the latter variation. The adjusted flood frequency curve will generally be higher than the curve for the measured series but will have a shallower slope. The higher curve reflects the effect of greater imperviousness (40%). The lower slope reflects the single level of imperviousness of the adjusted series. The computations for the adjusted and unadjusted flood frequency curves are given in Table 5.7. The percent increases in the 2-, 5-, 10-, 15-, 50-, and 100-year flood magnitudes appear in Table 5.7. The change is minor because the imperviousness did not change after 1964 and changes from 1942 to 1964 were minor (i.e., 10%). Most larger storm events occurred after the watershed reached developed condition. The adjusted series represents annual flooding for a constant urbanization condition (40% imperviousness). Of course, the adjusted series is not measured and its accuracy depends on the representativeness of Figure 5.9 for measuring the effects of urbanization.

TABLE 5.6
Adjustment of Rubio Wash Annual Flood Record for Urbanization

Measured Series					Ordered Data			
Year	Urbanization (%)	Annual Peak	Rank	Exceedance Probability	Rank	Annual Peak	Year	Exceedance Probability
1929	18	661	47	0.9592	1	3700	1970	0.0204
1930	18	1690	30	0.6122	2	3180	1974	0.0408
1931	19	798	46	0.9388	3	3166	1972	0.0612
1932	20	1510	34	0.6939	4	3020	1951	0.0816
1933	20	2071	20	0.4082	5	2980	1956	0.1020
1934	21	1680	31	0.6327	6	2890	1968	0.1224
1935	21	1370	35	0.7143	7	2781	1958	0.1429
1936	22	1181	40	0.8163	8	2780	1942	0.1633
1937	23	2400	14	0.2857	9	2740	1957	0.1837
1938	25	1720	29	0.5918	10	2650	1946	0.2041
1939	26	1000	43	0.8776	11	2610	1976	0.2245
1940	28	1940	26	0.5306	12	2540	1969	0.2449
1941	29	1201	38	0.7755	13	2460	1967	0.2653
1942	30	2780	8	0.1633	14	2400	1937	0.2857
1943	31	1930	27	0.5510	15	2310	1953	0.3061
1944	33	1780	28	0.5714	16	2300	1965	0.3265
1945	34	1630	32	0.6531	17	2290	1950	0.3469
1946	34	2650	10	0.2041	18	2200	1952	0.3673
1947	35	2090	19	0.3878	19	2090	1947	0.3878
1948	36	530	48	0.9796	20	2071	1933	0.4082
1949	37	1060	42	0.8571	21	2070	1975	0.4286
1950	38	2290	17	0.3469	22	2041	1966	0.4490
1951	38	3020	4	0.0816	23	2040	1964	0.4694
1952	39	2200	18	0.3673	24	1985	1973	0.4898
1953	39	2310	15	0.3061	25	1970	1955	0.5102
1954	39	1290	36	0.7347	26	1940	1940	0.5306
1955	39	1970	25	0.5102	27	1930	1943	0.5510
1956	39	2980	5	0.1020	28	1780	1944	0.5714
1957	39	2740	9	0.1837	29	1720	1938	0.5918
1958	39	2781	7	0.1429	30	1690	1930	0.6122
1959	39	985	44	0.8980	31	1680	1934	0.6327
1960	39	902	45	0.9184	32	1630	1945	0.6531
1961	39	1200	39	0.7959	33	1570	1963	0.6735
1962	39	1180	41	0.8367	34	1510	1932	0.6939
1963	39	1570	33	0.6735	35	1370	1935	0.7143
1964	40	2040	23	0.4694	36	1290	1954	0.7347
1965	40	2300	16	0.3265	37	1240	1971	0.7551
1966	40	2041	22	0.4490	38	1201	1941	0.7755
1967	40	2460	13	0.2653	39	1200	1961	0.7959
1968	40	2890	6	0.1224	40	1181	1936	0.8163
1969	40	2540	12	0.2449	41	1180	1962	0.8367

TABLE 5.6
Adjustment of Rubio Wash Annual Flood Record for Urbanization (*Continued*)

Measured Series					Ordered Data			
Year	Urbanization (%)	Annual Peak	Rank	Exceedance Probability	Rank	Annual Peak	Year	Exceedance Probability
1970	40	3700	1	0.0204	42	1060	1949	0.8571
1971	40	1240	37	0.7551	43	1000	1939	0.8776
1972	40	3166	3	0.0612	44	985	1959	0.8980
1973	40	1985	24	0.4898	45	902	1960	0.9184
1974	40	3180	2	0.0408	46	798	1931	0.9388
1975	40	2070	21	0.4286	47	661	1929	0.9592
1976	40	2610	11	0.2245	48	530	1948	0.9796

Iteration 1

			Correction Factor			Adjusted Series	
Year	Urbanization (%)	Measured Peak	Existing	Ultimate	Peak	Rank	Exceedance Probability
1929	18	661	1.560	2.075	879.3	47	0.9592
1930	18	1690	1.434	1.846	2175.6	22	0.4490
1931	19	798	1.573	2.044	1037.1	44	0.8980
1932	20	1510	1.503	1.881	1889.3	32	0.6531
1933	20	2071	1.433	1.765	2551.4	13	0.2653
1934	21	1680	1.506	1.855	2069.4	25	0.5102
1935	21	1370	1.528	1.890	1694.8	34	0.6939
1936	22	1181	1.581	1.943	1451.4	36	0.7347
1937	23	2400	1.448	1.713	2838.1	8	0.1633
1938	25	1720	1.568	1.838	2016.7	28	0.5714
1939	26	1000	1.690	1.984	1173.8	42	0.8571
1940	28	1940	1.603	1.814	2194.3	20	0.4082
1941	29	1201	1.703	1.920	1354.0	37	0.7551
1942	30	2780	1.508	1.648	3037.7	5	0.1020
1943	31	1930	1.663	1.822	2114.5	23	0.4694
1944	33	1780	1.705	1.830	1910.3	31	0.6327
1945	34	1630	1.752	1.863	1733.5	33	0.6735
1946	34	2650	1.585	1.672	2795.3	10	0.2041
1947	35	2090	1.675	1.757	2191.6	21	0.4286
1948	36	530	2.027	2.123	555.1	48	0.9796
1949	37	1060	1.907	1.969	1094.7	43	0.8776
1950	38	2290	1.708	1.740	2332.7	16	0.3265
1951	38	3020	1.557	1.583	3068.9	4	0.0816
1952	39	2200	1.732	1.748	2220.5	19	0.3878
1953	39	2310	1.706	1.722	2331.0	17	0.3469
1954	39	1290	1.881	1.900	1303.2	38	0.7755
1955	39	1970	1.788	1.806	1989.1	29	0.5918

(Continued)

TABLE 5.6
Iteration 1 (*Continued*)

Year	Urbanization (%)	Measured Peak	Correction Factor		Adjusted Series		
			Existing	Ultimate	Peak	Rank	Exceedance Probability
1956	39	2980	1.589	1.602	3004.4	6	0.1224
1957	39	2740	1.646	1.660	2763.7	11	0.2245
1958	39	2781	1.620	1.634	2804.5	9	0.1837
1959	39	985	1.979	2.001	995.7	45	0.9184
1960	39	902	1.999	2.020	911.9	46	0.9388
1961	39	1200	1.911	1.931	1212.5	40	0.8163
1962	39	1180	1.935	1.956	1192.5	41	0.8367
1963	39	1570	1.853	1.872	1585.9	35	0.7143
1964	40	2040	1.790	1.790	2040.0	27	0.5510
1965	40	2300	1.731	1.731	2300.0	18	0.3673
1966	40	2041	1.781	1.781	2041.0	26	0.5306
1967	40	2460	1.703	1.703	2460.0	15	0.3061
1968	40	2890	1.619	1.619	2890.0	7	0.1429
1969	40	2540	1.693	1.693	2540.0	14	0.2857
1970	40	3700	1.480	1.480	3700.0	1	0.0204
1971	40	1240	1.910	1.910	1240.0	39	0.7959
1972	40	3166	1.559	1.559	3166.0	3	0.0612
1973	40	1985	1.798	1.798	1985.0	30	0.6122
1974	40	3180	1.528	1.528	3180.0	2	0.0408
1975	40	2070	1.773	1.773	2070.0	24	0.4898
1976	40	2610	1.683	1.683	2610.0	12	0.2449

Iteration 2

Year	Urbanization (%)	Measured Peak	Correction Factor		Adjusted Series		
			Existing	Ultimate	Peak	Rank	Exceedance Probability
1929	18	661	1.560	2.075	879.3	47	0.9592
1930	18	1690	1.399	1.781	2152.6	22	0.4490
1931	19	798	1.548	2.001	1031.6	44	0.8980
1932	20	1510	1.493	1.863	1885.1	32	0.6531
1933	20	2071	1.395	1.703	2528.5	14	0.2857
1934	21	1680	1.475	1.806	2056.7	25	0.5102
1935	21	1370	1.522	1.881	1692.9	34	0.6939
1936	22	1181	1.553	1.900	1444.6	36	0.7347
1937	23	2400	1.405	1.648	2814.1	8	0.1633
1938	25	1720	1.562	1.830	2015.1	28	0.5714
1939	26	1000	1.680	1.969	1172.4	42	0.8571
1940	28	1940	1.567	1.765	2185.4	21	0.4286

TABLE 5.6
Iteration 2 (*Continued*)

Year	Urbanization (%)	Measured Peak	Correction Factor		Adjusted Series		
			Existing	Ultimate	Peak	Rank	Exceedance Probability
1941	29	1201	1.695	1.910	1353.1	37	0.7551
1942	30	2780	1.472	1.602	3025.8	5	0.1020
1943	31	1930	1.637	1.790	2110.2	23	0.4694
1944	33	1780	1.726	1.855	1912.5	31	0.6327
1945	34	1630	1.760	1.872	1734.1	33	0.6735
1946	34	2650	1.585	1.672	2795.3	10	0.2041
1947	35	2090	1.690	1.773	2192.9	20	0.4082
1948	36	530	2.027	2.123	555.1	48	0.9796
1949	37	1060	1.921	1.984	1094.9	43	0.8776
1950	38	2290	1.699	1.731	2332.4	16	0.3265
1951	38	3020	1.557	1.583	3068.9	4	0.0816
1952	39	2200	1.741	1.757	2220.6	19	0.3878
1953	39	2310	1.724	1.740	2331.3	17	0.3469
1954	39	1290	1.901	1.920	1303.4	38	0.7755
1955	39	1970	1.820	1.838	1989.5	29	0.5918
1956	39	2980	1.606	1.619	3004.8	6	0.1224
1957	39	2740	1.668	1.683	2764.2	11	0.2245
1958	39	2781	1.646	1.660	2805.1	9	0.1837
1959	39	985	1.999	2.020	995.8	45	0.9184
1960	39	902	2.022	2.044	912.0	46	0.9388
1961	39	1200	1.923	1.943	1212.6	40	0.8163
1962	39	1180	1.935	1.956	1192.5	41	0.8367
1963	39	1570	1.871	1.890	1586.0	35	0.7143
1964	40	2040	1.822	1.822	2040.0	27	0.5510
1965	40	2300	1.748	1.748	2300.0	18	0.3673
1966	40	2041	1.814	1.814	2041.0	26	0.5306
1967	40	2460	1.722	1.722	2460.0	15	0.3061
1968	40	2890	1.634	1.634	2890.0	7	0.1429
1969	40	2540	1.713	1.713	2540.0	13	0.2653
1970	40	3700	1.480	1.480	3700.0	1	0.0204
1971	40	1240	1.931	1.931	1240.0	39	0.7959
1972	40	3166	1.559	1.559	3166.0	3	0.0612
1973	40	1985	1.846	1.846	1985.0	30	0.6122
1974	40	3180	1.528	1.528	3180.0	2	0.0408
1975	40	2070	1.798	1.798	2070.0	24	0.4898
1976	40	2610	1.693	1.693	2610.0	12	0.2449

TABLE 5.6
Iteration 3

Year	Urbanization (%)	Measured Peak	Correction Factor		Adjusted Series		
			Existing	Ultimate	Peak	Rank	Exceedance Probability
1929	18	661	1.560	2.075	879.3	47	0.9592
1930	18	1690	1.399	1.781	2152.6	22	0.4490
1931	19	798	1.548	2.001	1031.6	44	0.8980
1932	20	1510	1.493	1.863	1885.1	32	0.6531
1933	20	2071	1.401	1.713	2532.2	14	0.2857
1934	21	1680	1.475	1.806	2056.7	25	0.5102
1935	21	1370	1.522	1.881	1692.9	34	0.6939
1936	22	1181	1.553	1.900	1444.6	36	0.7347
1937	23	2400	1.405	1.648	2814.1	8	0.1633
1938	25	1720	1.562	1.830	2015.1	28	0.5714
1939	26	1000	1.680	1.969	1172.4	42	0.8571
1940	28	1940	1.573	1.773	2186.9	21	0.4286
1941	29	1201	1.695	1.910	1353.1	37	0.7551
1942	30	2780	1.472	1.602	3025.8	5	0.1020
1943	31	1930	1.637	1.790	2110.2	23	0.4694
1944	33	1780	1.726	1.855	1912.5	31	0.6327
1945	34	1630	1.760	1.872	1734.1	33	0.6735
1946	34	2650	1.585	1.672	2795.3	10	0.2041
1947	35	2090	1.683	1.765	2192.2	20	0.4082
1948	36	530	2.027	2.123	555.1	48	0.9796
1949	37	1060	1.921	1.984	1094.9	43	0.8776
1950	38	2290	1.699	1.731	2332.4	16	0.3265
1951	38	3020	1.557	1.583	3068.9	4	0.0816
1952	39	2200	1.741	1.757	2220.6	19	0.3878
1953	39	2310	1.724	1.740	2331.3	17	0.3469
1954	39	1290	1.901	1.920	1303.4	38	0.7755
1955	39	1970	1.820	1.838	1989.5	29	0.5918
1956	39	2980	1.606	1.619	3004.8	6	0.1224
1957	39	2740	1.668	1.683	2764.2	11	0.2245
1958	39	2781	1.646	1.660	2805.1	9	0.1837
1959	39	985	1.999	2.020	995.8	45	0.9184
1960	39	902	2.022	2.044	912.0	46	0.9388
1961	39	1200	1.923	1.943	1212.6	40	0.8163
1962	39	1180	1.935	1.956	1192.5	41	0.8367
1963	39	1570	1.871	1.890	1586.0	35	0.7143
1964	40	2040	1.822	1.822	2040.0	27	0.5510
1965	40	2300	1.748	1.748	2300.0	18	0.3673
1966	40	2041	1.814	1.814	2041.0	26	0.5306
1967	40	2460	1.722	1.722	2460.0	15	0.3061
1968	40	2890	1.634	1.634	2890.0	7	0.1429
1969	40	2540	1.703	1.703	2540.0	13	0.2653

TABLE 5.6
Iteration 3 (*Continued*)

Year	Urbanization (%)	Measured Peak	Correction Factor Existing	Correction Factor Ultimate	Adjusted Series Peak	Adjusted Series Rank	Adjusted Series Exceedance Probability
1970	40	3700	1.480	1.480	3700.0	1	0.0204
1971	40	1240	1.931	1.931	1240.0	39	0.7959
1972	40	3166	1.559	1.559	3166.0	3	0.0612
1973	40	1985	1.846	1.846	1985.0	30	0.6122
1974	40	3180	1.528	1.528	3180.0	2	0.0408
1975	40	2070	1.798	1.798	2070.0	24	0.4898
1976	40	2610	1.693	1.693	2610.0	12	0.2449

TABLE 5.7
Computation of Flood Frequency Curves for Rubio Wash Watershed in Actual (Q) and Ultimate Development (Q_a) Conditions

(1) p	(2) K	(3) $\log_{10}Q$	(4) Q (ft³/sec)	(5) $\log_{10}Q_a$	(6) Q_a (ft³/sec)	(7)
0.99	−2.68572	2.7386	548	2.8006	632	
0.90	−1.32309	2.9989	998	3.0439	1106	
0.70	−0.45812	3.1642	1459	3.1983	1579	
0.50	0.08302	3.2676	1852	3.2949	1972	0.065
0.20	0.85653	3.4153	2602	3.4329	2710	0.041
0.10	1.21618	3.4840	3048	3.4971	3142	0.031
0.04	1.56740	3.5511	3557	3.5598	3629	0.020
0.02	1.77716	3.5912	3901	3.5973	3956	0.014
0.01	1.95472	3.6251	4218	3.6290	4256	0.009

Note: p, exceedance probability; K, LP3 variate for $g_w = -0.5$; (3), $\log_{10}Q = \bar{Y} + KS_y$; (4), $Q = 10^{**}\log_{10}Q$; (5), $\log_{10}Q_a = \bar{Y}_a + KS_{ya}$; (6), $Q_a = 10^{**}\log_{10}Q_a$.

5.4.3 Testing Significance of Urbanization

A basic assumption of frequency analysis is that all values in the record were sampled from the same population. In hydrologic terms, this assumption implies that the watershed has not undergone a systematic change. Obviously, the state of a watershed is continually changing. Seasonal variations of land cover and soil moisture conditions are important hydrologically. These sources of variation along with the variation introduced into floods by the storm-to-storm variation in rainfall cause the variation in the annual maximum flood series. These factors are usually considered random variations and do not violate the assumption of watershed homogeneity.

Major land use changes, including urban development, afforestation, and deforestation, introduce variations into a flood record that violate the assumption of homogeneity. If the systematic variations are hydrologically significant, they can introduce considerable variation into a record such that flood magnitudes or probabilities estimated from a flood frequency analysis will not be accurate indicators of flooding for the changed watershed condition. For example, if urban development increased from 0 to 50% imperviousness over the duration of a record and flood magnitudes will be computed from a flood frequency analysis for the current 50% imperviousness, estimates from the frequency curve will not be representative because many floods used to derive the frequency curve occurred under conditions of much less urban development.

Where a flood record is suspected of lacking homogeneity, the data should be tested before flood frequency analysis. Statistical tests are intended for this purpose, although their results should not be accepted unless a hydrologic basis for the result is strongly suspected. Before a flood record is adjusted, the results of both hydrologic and statistical assessments should suggest the need for adjustment.

Several statistical methods can test for nonhomogeneity. The selection of a test is not arbitrary. The type of nonhomogeneity should be considered in selecting a test. Urban development can occur abruptly or gradually. An abrupt hydrologic change occurs over a small portion of the record, for example, a 40-year flood record collected from 1950 to 1989 on a small watershed urbanized from 1962 to 1966 would be suitable for analysis with a method appropriate for abrupt change. Conversely, if the urbanization occurred from 1955 to 1982, a statistical method sensitive to gradual change would be more appropriate. A test designed for one type of nonhomogeneity and used with another type may not detect hydrologic effect.

If a flood record has been analyzed and found to lack homogeneity, the flood magnitudes should be adjusted prior to making frequency analysis. Flood magnitudes computed from an adjusted series can be significantly different from magnitudes computed from the unadjusted series.

5.5 PROBLEMS

5-1 Given a mean and standard deviation of 3500 and 2000 ft^3/sec, respectively, find the 2-, 10-, and 100-year peak floods for a normal distribution.

5-2 Given a mean and standard deviation of 1000 and 600 ft^3/sec, respectively, find the 5- and 25-year peak floods for a normal distribution.

5-3 Assuming a normal distribution with a mean and standard deviation of 2400 and 1200 ft^3/sec, find the exceedance probability and return period for flood magnitudes of 3600 and 5200 ft^3/sec.

5-4 Assuming a normal distribution with a mean and standard deviation of 5700 and 1300 ft^3/sec, respectively, find the exceedance probability and return period for flood magnitudes of 7400 and 8700 ft^3/sec.

5-5 For a sample of 60, compute the Weibull, Hazen, and Cunnane plotting probabilities and the corresponding return periods for the two largest and two smallest ranks. Comment on the results.

5-6 For sample sizes of 5, 20, and 50 compute the Weibull, Hazen, and Cunnane plotting probabilities and the corresponding return periods for the largest and smallest ranks. Discuss the magnitude of the differences as a function of sample size.

5-7 For the random variable X of Figure 5.1, find the following probabilities: (a) $P(X > 4.5)$; (b) $P(X < 6.2)$; (c) $P(X > 6.7)$; and (d) $P(X = 7.1)$.

5-8 For the random variable X of Figure 5.1, find the following probabilities: (a) $P(X > 5.3)$; (b) $P(X < 6.5)$; (c) $P(X > 7.1)$; and (d) $P(4.3 < X < 6.6)$.

5-9 For the random variable X of Figure 5.1, find X_o for each of the following probabilities: (a) $P(X > X_O) = 0.1$; (b) $P(X > X_O) = 0.8$; (c) $P(X < X_O) = 0.05$; and (d) $P(X < X_O) = 0.98$.

5-10 If the random variable X of Figure 5.1 is the peak rate of runoff in a stream (in ft^3/sec × 10^3), find the following: (a) $P(X > 7040$ ft^3/sec); (b) $P(X > 3270$ ft^3/sec); (c) $P(X < 3360$ ft^3/sec); and (d) $P(X < 6540$ ft^3/sec).

5-11 If the random variable X of Figure 5.1 is the annual sediment yield (in tons/acre × 10^1), find the following: (a) $P(X > 40$ tons/acre/year); (b) $P(X > 67.3$ tons/acre/year); and (c) $P(X < 32.7$ tons/acre/year).

5-12 If the random variable X of Figure 5.1 is the peak flow rate (in ft^3/sec × 10) from a 20-acre urban watershed, what is (a) the 10-year flood; (b) the 25-year flood; and (c) the 100-year flood?

5-13 Assuming a normal distribution, make a frequency analysis of the January rainfall (P) on Wildcat Creek for 1954–1971. Plot the data using the Weibull formula. Based on the frequency curve, estimate: (a) the 100-year January rainfall and (b) the probability that the January rainfall in any one year will exceed 7 in.

Year	*P*	Year	*P*	Year	*P*	Year	*P*
1954	6.43	1958	3.59	1962	4.56	1967	5.19
1955	5.01	1959	3.99	1963	5.26	1968	6.55
1956	2.13	1960	8.62	1964	8.51	1969	5.19
1957	4.49	1961	2.55	1965	3.93	1970	4.29
				1966	6.20	1971	5.43

5-14 Assuming a normal distribution, make a frequency analysis of the July evaporation (E) on Wildcat Creek for 1954–1971. Plot the data using the Weibull formula. Based on the frequency curve, estimate the following: (a) the 50-year July evaporation and (b) the probability that the July evaporation in any one year will exceed 10 in.

Year	*E*	Year	*E*	Year	*E*	Year	*E*
1954	8.30	1958	8.39	1962	8.05	1967	6.16
1955	7.74	1959	7.94	1963	8.03	1968	7.44
1956	6.84	1960	9.24	1964	6.91	1969	9.19
1957	8.45	1961	8.79	1965	7.50	1970	9.24
				1966	6.20	1971	5.43

5-15 Assuming a normal distribution, make a frequency analysis of the annual maximum flood series on the Grey River at Reservoir near Alpine, Wyoming. Plot the data using the Weibull formula. Estimate (a) the 20-year annual maximum; (b) the exceedance probability and return period for an event of 5500 cfs; and (c) the index ratio of the 100-year flood to the 2-year flood.

4210	3720	3110	3150	2550
2010	2920	2420	4050	5220
5010	2500	4280	3260	3650
4290	2110	3860	7230	3590
			5170	650

5-16 Given a mean and standard deviation of the base-10 logarithms of 0.36 and 0.15, respectively, find the 10-, 50-, and 100-year magnitudes assuming a lognormal distribution.

5-17 Given a mean and standard deviation of the base-10 logarithms of –0.43 and 0.12, respectively, find the 5-, 25-, and 100-year magnitudes assuming a lognormal distribution.

5-18 Assuming a lognormal distribution with a mean and standard deviation of –0.89 and 0.10, respectively, find the exceedance probability and return period for rainfall intensities of 0.22 in./hr and 0.28 in./hr.

5-19 The following data are annual 15-minute peak rainfall intensities I (in./hr) for 9 years of record. Compute and plot the $\log_{10}$-normal frequency curve and the data. Use the Weibull formula. Using both the curve and the mathematical equation, estimate (a) the 25-year, 15-minute peak rainfall intensity; (b) the return period for an intensity of 7 in./hr; and (c) the probability that the annual maximum 15-minute rainfall intensity will be between 4 and 6 in./hr.

Year	I	Year	I
1972	3.16	1977	2.24
1973	2.29	1978	4.37
1974	4.07	1979	6.03
1975	4.57	1980	2.75
1976	2.82		

5-20 Assuming a lognormal distribution, make a frequency analysis of the total annual runoff Q (in.) from Wildcat Creek for the period 1954–1970. Plot the data using the Weibull plotting position formula. Based on the frequency curve, estimate the following: (a) The 50-year annual runoff; (b) the probability that the total annual runoff in any one year will be less than 5 in.; and (c) the probability that the total annual runoff will exceed 15 in.

5-21 Assuming a lognormal distribution, make a frequency analysis of the mean August temperature (7°F) at Wildcat Creek for the 1954–1971 period. Plot the data using the Weibull plotting position formula. Based on the frequency curve, estimate: (a) the August temperature that can be expected to be exceeded once in 10 years; (b) the probability that the mean August temperature will exceed 83°F in any one year; and (c) the probability that the mean August temperature will not exceed 72°F.

82.5	79.5	78.9	78.0	76.6	76.2
80.1	78.9	74.1	76.1	74.6	79.1
80.4	80.6	75.7	77.7	79.6	76.7

5-22 Using the data of Problem 5-15, perform a lognormal frequency analysis. Make the three estimations indicated in Problem 5-15 and compare the results for the normal and lognormal analyses.

5-23 Assuming a log-Pearson Type III distribution, make a frequency analysis of the total annual rainfall at Wildcat Creek for 1954–1970. Plot the data using the Weibull formula. Based on the frequency curve, estimate (a) the 100-year annual rainfall and (b) the probability that the total annual rainfall in any one year will exceed 50 in.

Year	*P*	Year	*P*	Year	*P*	Year	*P*
1954	34.67	1958	42.39	1962	56.22	1966	57.60
1955	36.46	1959	42.57	1963	56.00	1967	64.17
1956	50.27	1960	43.81	1964	69.18	1968	50.28
1957	44.10	1961	58.06	1965	49.25	1969	51.91
						1970	38.23

5-24 Compute a log-Pearson Type III frequency analysis for the data of Problem 5-20. Compare the fit with the lognormal and the estimates of parts (a), (b), and (c).

5-25 Using the data of Problem 5-15, perform a log-Pearson Type III analysis. Make the three estimations indicated in Problem 5-15, and compare the results for the normal, lognormal (Problem 5-22), and LP3 analyses.

5-26 Obtain a stream flow record of at least 20 years from the U.S. Geological Survey Water Supply Papers. Find the maximum instantaneous peak discharge for each year of record. Perform a frequency analysis for a log-Pearson Type III distribution and estimate the 2-year (Q_2), 10-year (Q_{10}), and 100-year (Q_{100}) peak discharges. Next, compute the index ratios of the Q_{10}/Q_2 and Q_{100}/Q_2.

5-27 Make a log-Pearson Type III analysis of the data of Problem 5-14. Based on the frequency curve, estimate (a) the 50-year July evaporation and (b) the probability that the July evaporation in any one year will exceed 10 in.

5-28 The following data are the annual maximum series (Q_p) and the percentage of impervious area (I) for the Alhambra Wash watershed for 1930 through 1977. Using the method of Figure 5.9, adjust the flood series to ultimate development of 50%. Perform LP3 analyses on both the adjusted and unadjusted series and evaluate the effect on the estimated 2-, 10-, 25-, and 100-year floods.

Year	I%	Q_p (cfs)	Year	I%	Q_p (cfs)	Year	I%	Q_p (cfs)
1930	21	1870	1946	35	1600	1962	45	2560
1931	21	1530	1947	37	3810	1963	45	2215
1932	22	1120	1948	39	2670	1964	45	2210
1933	22	1850	1949	41	758	1965	45	3730
1934	23	4890	1950	43	1630	1966	45	3520
1935	23	2280	1951	45	1620	1967	45	3550
1936	24	1700	1952	45	3811	1968	45	3480
1937	24	2470	1953	45	3140	1969	45	3980
1938	25	5010	1954	45	2140	1970	45	3430
1939	25	2480	1955	45	1980	1971	45	4040
1940	26	1280	1956	45	4550	1972	46	2000
1941	27	2080	1957	45	3090	1973	46	4450
1942	29	2320	1958	45	4830	1974	46	4330
1943	30	4480	1959	45	3170	1975	46	6000
1944	32	1860	1960	45	1710	1976	46	1820
1945	33	2220	1961	45	1480	1977	46	1770

5-29 Residents in a community at the discharge point of a 240-mi^2 watershed believe that recent increases in peak discharge rates are due to deforestation by a logging company in recent years. Use the Spearman test to analyze the annual maximum discharges (q_p) and an average fraction of forest cover (f) for the watershed.

Year	q_p	f	Year	q_p	f	Year	q_p	f
1982	8000	53	1987	12200	54	1992	5800	46
1983	8800	56	1988	5700	51	1993	14300	44
1984	7400	57	1989	9400	50	1994	11600	43
1985	6700	58	1990	14200	49	1995	10400	42
1986	11100	55	1991	7600	47			

6 Graphical Detection of Nonhomogeneity

6.1 INTRODUCTION

The preparation phase of data analysis involves compilation, preliminary organization, and hypothesis formulation. All available physiographic, climatic, and hydrologic data should be compiled. While a criterion variable, such as annual maximum discharge, is often of primary interest, other hydrologic data can be studied to decide whether a change in the criterion variable occurred. The analysis of daily flows, flow volumes, and low flow magnitudes may be useful for detecting watershed change. Physiographic data, such as land use or channel changes, are useful for assigning responsibility to changes and developing a method that can be used to adjust the flood record. Climatic data, such as rainfall volumes, reflect the extent to which the change in the annual flood series is climate related. If physiographic or climatic data do not suggest a significant watershed change, it may not be necessary to apply trend tests to the flood data. The variation in the annual flood series may simply be a function of random climatic variability, and this hypothesis can be evaluated by applying univariate trend tests to the sequence of data.

Graphical analyses are often the first step in data analyses. They are preludes to quantitative analyses on which decisions can be based. Graphical analyses should always be considered initial steps, not conclusive steps. They can be misleading when not accompanied by quantitative analyses. However, failure to graph data may prevent the detection of a trend or the nature of the trend. Graphical analyses should be used in conjunction with other quantitative methods. They will be discussed in this chapter and quantitative analyses will be discussed in other chapters.

6.2 GRAPHICAL ANALYSES

After compilation of data, several preliminary analyses can be made in preparing to test for and, if necessary, adjust for the effects of watershed changes. Three general types of analyses can be made. First, one or more graphical analyses of the series can be made, including the standard frequency analysis (e.g., plotting several Pearson Type III frequency curves for different time periods). The purpose of graphical analysis is to study the data to identify the ways watershed changes affected the flood series. For example, does the central tendency change with time? Did the variance of the data change? Did the watershed changes affect only part of the temporal series, thus producing a mixed population series? Graphical analyses can provide some insight into characteristics of the changes and suggest the best path for detecting the effects and adjusting the series.

Graphical methods that can be initially used to understand the data include plots of data versus time, ranking the annual event versus the water year, the number of occurrences above a threshold versus water year, and histograms or empirical cumulative probability plots of the data for two or more periods of the record. Where untransformed data are characterized by considerable random variation, the logarithms of the data can be plotted to assist in detecting the effects of watershed change. In most cases, several plots should be made, as different types of plots will identify different characteristics of the data.

6.2.1 Univariate Histograms

Graphical analyses are often the first steps in analyzing data. Univariate graphical analyses in the form of histograms help identify the distribution of the random variable being analyzed. A frequency histogram is a tabulation or plot of the frequency of occurrence versus selected intervals of the continuous random variable. It is the equivalent of a bar graph used for graphing discrete random variables. The effectiveness of a graphical analysis in identifying characteristics of a random variable or its probability density function depends on the sample size and interval selected to plot the abscissa. For small samples, it is difficult to separate the data into a sufficient number of groups to provide a meaningful indication of data characteristics. With small samples, the impressions of the data will be very sensitive to the cell boundaries and widths selected for the histogram. It is generally wise to try several sets of cell boundaries and widths to ensure accurate assessments of the data. The following are general guidelines for constructing frequency histograms:

1. Set the minimum value (X_m) as (a) the smallest sample value or (b) a physically limiting value, such as zero.
2. Set the maximum value (X_x) as (a) the largest sample value or (b) an upper limit considered the largest value expected.
3. Select the number of intervals (k), which is usually about 5 for small samples and a maximum of about 20 for large samples. For moderate size samples, the following empirical equation can be used to estimate the number of cells:

$$k = 1 + 3.3 \log_{10}(n) \tag{6.1}$$

4. Compute the approximate cell width (w) where $w = (X_x - X_m)/k$.
5. Round the computed value of w to a convenient value w_0.
6. Set the upper bound (B_i) for cell i using the minimum value X_m, and the cell width w_0:

$$B_i = X_m + i\, w_0 \quad \text{for } i = 1, 2, \ldots, k \tag{6.2}$$

7. Using the sample data, compute the sample frequencies for each cell.

In addition to the cell width w_0, assessments of the data characteristics can be influenced by the scale used as the ordinate. For example, a histogram where all frequencies are 10 to 15 per cell will appear quite different when the ordinate is scaled from 10 to 15 and from 0 to 15. The former scale suggests the cell frequencies are varied. The latter suggests a relatively uniform set of frequencies. This can skew the viewer's impression of the data characteristics.

Histograms provide a pictorial representation of the data. They provide for assessing the central tendency of the data; the range and spread of the data; the symmetry (skewness) of the data; the existence of extreme events, which can then be checked for being outliers; and approximate sample probabilities. Frequency histograms can be transformed to relative frequency or probability histograms by dividing the frequencies of every cell by the sample size.

Example 6.1

Consider the 38-year discharge record of Table 6.1. To achieve an average frequency of five per cell will require a histogram with no more than seven cells. The use of more cells would produce cells with low frequencies and invite problems in characterizing the data.

Figure 6.1(a) shows a nine-cell histogram based on a cell width of 50 cfs. With an average of 4.2 floods per cell, only four of the nine cells have frequencies of five or more. The histogram is multimodal and does not suggest an underlying distribution. The cell with a one-count in the middle would discount the use of a normal or lognormal distribution.

Figure 6.1(b) shows a histogram of the same data but with a cell width of 100 cfs. With only five cells, the average frequency is 7.6. Except for the 500–600 cfs cell, the data appear to follow a uniform distribution. However, with only five cells, it is difficult to have confidence in the shape of the distribution.

Figure 6.1(c) also shows the frequency histogram with a cell width of 100 cfs, but the lowest cell bound is 550 cfs rather than the 500 cfs used in Figure 6.1(b). The histogram of Figure 6.1(c) is characterized by one high-count cell, with the other cells having nearly the same count. The histogram might suggest a lognormal distribution.

The important observation about the histograms of Figure 6.1 is that, even with a sample size of 38, it is difficult to characterize the data. When graphing such data, several cell widths and cell bound delineations should be tried. The three histograms of Figure 6.1 could lead to different interpretations. While the data could be transformed using logarithms, the same problems would exist. The frequencies in each cell would be limited because of the sample size.

TABLE 6.1
Annual Maximum Discharge Record

654	967	583	690	957	814	871	859	843	837
714	725	917	708	618	685	941	822	883	766
827	693	660	902	672	612	742	703	731	
637	810	981	646	992	734	565	678	962	

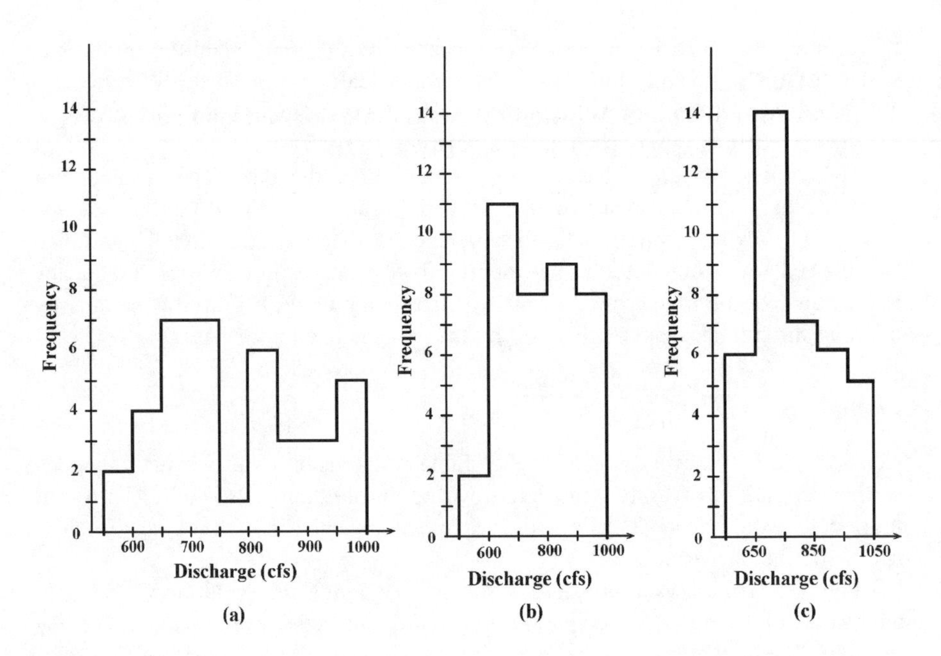

FIGURE 6.1 Frequency histograms for annual maximum discharge record of Table 6.1: effects of cell width.

Example 6.2

Table 6.2 includes the measured annual maximum flood series for the Rubio Wash Watershed for the 1929–1976 period (n = 48). During the period of record, the percentage of imperviousness increased from 18% to 40%; thus, the record is nonhomogeneous. Using the method of Section 5.4.2, the measured series was adjusted to a homogeneous record based on 40% imperviousness (Table 6.2). The histograms for the two series are shown in Figure 6.2. Since the homogeneous series has large discharge rates, it appears shifted to the right. The adjusted series shows a more bell-shaped profile than the measured series. However, the small sample size allows the one high-frequency cell for the adjusted series (Figure 6.2b) to dominate the profile of the histogram.

In summary, even though the flood record increases to 48 annual maximum discharges, it is difficult to use the graphical analysis alone to identify the underlying population. The analyses suggest that the adjusted series is different from the measured but nonhomogeneous series.

6.2.2 Bivariate Graphical Analysis

In addition to univariate graphing with histograms and frequency plots, graphs of related variables can be helpful in understanding data, such as flood peaks versus the level of urbanization or percent forest cover. The first step in examining a

TABLE 6.2
Measured (Y) and Adjusted (X) Annual Maximum Flood Series for the Rubio Wash Watershed, 1929–1976

Year	Y	X	Year	Y	X	Year	Y	X
1929	661	879	1945	1630	1734	1961	1200	1213
1930	1690	2153	1946	2650	2795	1962	1180	1193
1931	798	1032	1947	2090	2192	1963	1570	1586
1932	1510	1885	1948	530	555	1964	2040	2040
1933	2071	2532	1949	1060	1095	1965	2300	2300
1934	1680	2057	1950	2290	2332	1966	2041	2041
1935	1370	1693	1951	3020	3069	1967	2460	2460
1936	1181	1445	1952	2200	2221	1968	2890	2890
1937	2400	2814	1953	2310	2331	1969	2540	2540
1938	1720	2015	1954	1290	1303	1970	3700	3700
1939	1000	1172	1955	1970	1990	1971	1240	1240
1940	1940	2186	1956	2980	3005	1972	3166	3166
1941	1201	1353	1957	2740	2764	1973	1985	1985
1942	2780	3026	1958	2781	2805	1974	3180	3180
1943	1930	2110	1959	985	996	1975	2070	2070
1944	1780	1912	1960	902	912	1976	2610	2610

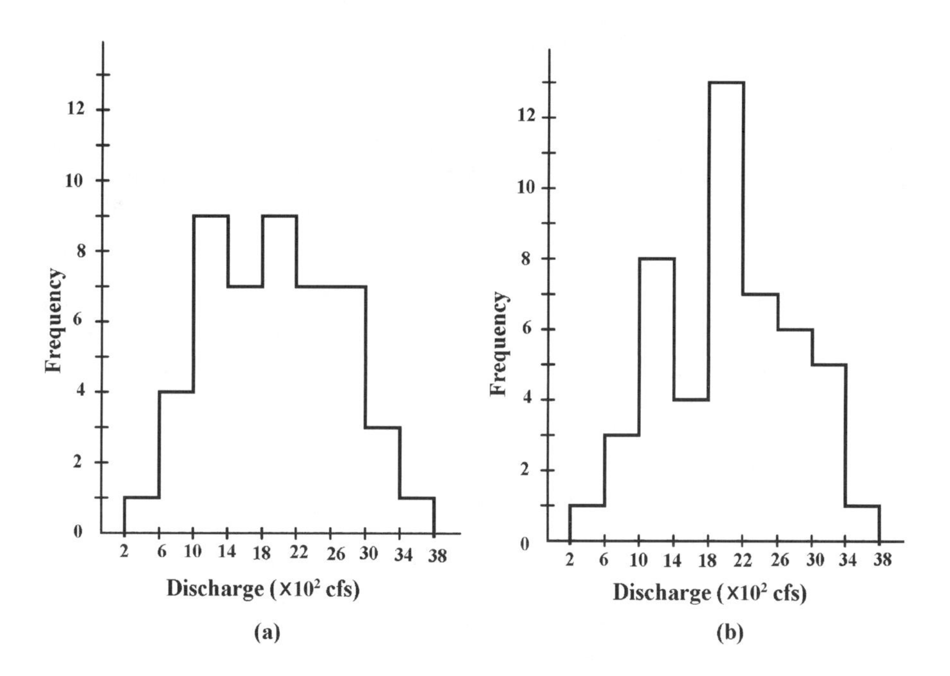

FIGURE 6.2 Frequency histograms for annual maximum discharges for Rubio Wash, California: (a) nonhomogeneous series; and (b) homogeneous series.

relationship of two variables is to perform a graphical analysis. Visual inspection of the graphed data can identify:

1. The degree of common variation, which is an indication of the degree to which the two variables are related
2. The range and distribution of the sample data points
3. The presence of extreme events
4. The form of the relationship between the two variables (linear, power, exponential)
5. The type of relationship (direct or indirect)

All these factors are of importance in the statistical analysis of sample data and decision making.

When variables show a high degree of association, one assumes that a causal relationship exists. If a physical reason suggests that a causal relationship exists, the association demonstrated by the sample data provides empirical support for the assumed relationship. Systematic variation implies that when the value of one of the random variables changes, the value of the other variable will change predictably, that is, an increase in the value of one variable occurs when the value of another variable increases. For example, a graph of the mean annual discharge against the percentage of imperviousness may show an increasing trend.

If the change in the one variable is highly predictable from a given change in the other variable, a high degree of common variation exists. Figure 6.3 shows graphs of different samples of data for two variables having different degrees of common variation. In Figures 6.3(a) and (e), the degrees of common variation are very high; thus the variables are said to be correlated. In Figure 6.3(c), the two variables are not correlated because, as the value of X is increased, it is not certain whether Y will increase or decrease. In Figures 6.3(b) and (d), the degree of correlation is moderate; in Figure 6.3(b), it is evident that Y will increase as X is increased, but the exact change in Y for a change in X is difficult to estimate. A more quantitative discussion of the concept of common variation appears later in this chapter.

It is important to use a graphical analysis to identify the range and distribution of the sample data points so that the stability of the relationship can be assessed and so that one can assess the ability of the data sample to represent the distribution of the population. If the range of the data is limited, a fitted relationship may not be stable; that is, it may not apply to the distribution of the population. Figure 6.4 shows a case where the range of the sample is much smaller than the expected range of the population. If an attempt is made to use the sample to project the relationship between the two random variables, a small change in the slope of the relationship will cause a large change in the predicted estimate of Y for values of X at the extremes of the range of the population. A graph of two random variables might alert an investigator to a sample in which the range of the sample data may cause stability problems in a derived relationship between two random variables, especially when the relationship will be extrapolated beyond the range of the sample data.

It is important to identify extreme events in a sample of data for several reasons. First, extreme events can dominate a computed relationship between two variables.

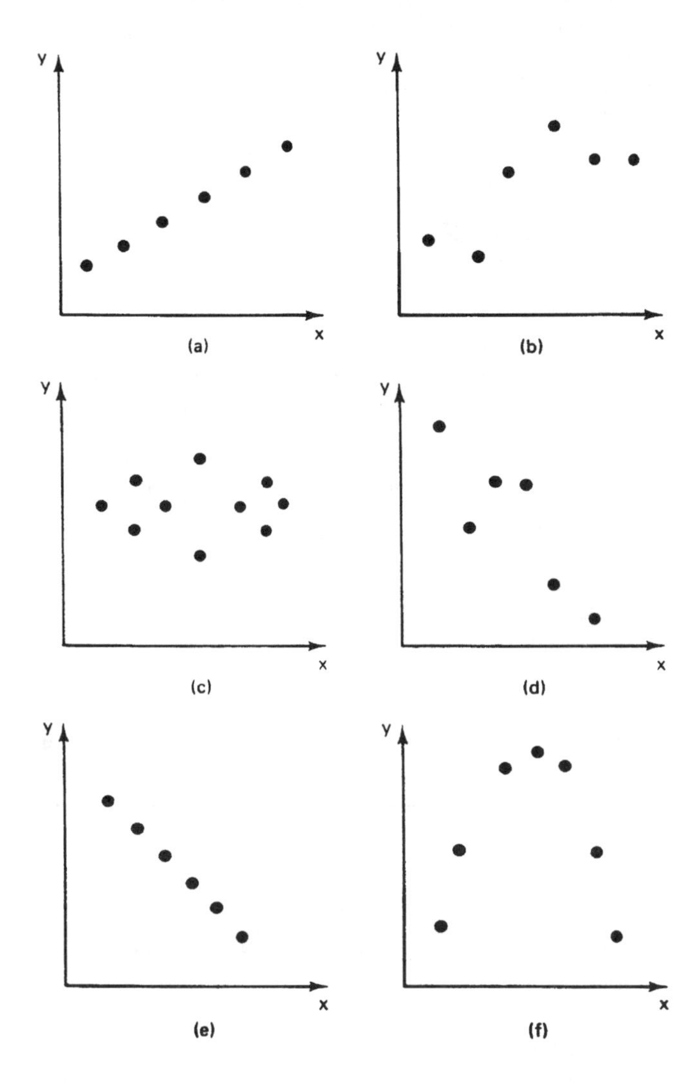

FIGURE 6.3 Different degrees of correlation between two random variables (X and Y): (a) $R = 1.0$; (b) $R = 0.5$; (c) $R = 0.0$; (d) $R = -0.5$; (e) $R = -1.0$; (f) $R = 0.3$.

For example, in Figure 6.5(a), the extreme point suggests a high correlation between X and Y and the cluster of points acts like a single observation. In Figure 6.5(b), the extreme point causes a poor correlation between the two random variables. Since the cluster of points has the same mean value of Y as the value of Y of the extreme point, the data of Figure 6.5(b) suggest that a change in X is not associated with a change in Y. A correlation coefficient is more sensitive to an extreme point when sample size is small. An extreme event may be due to errors in recording or plotting the data or a legitimate observation in the tail of the distribution. Therefore, an extreme event must be identified and its cause determined. Otherwise, it will not be possible to properly interpret the results of correlation analysis.

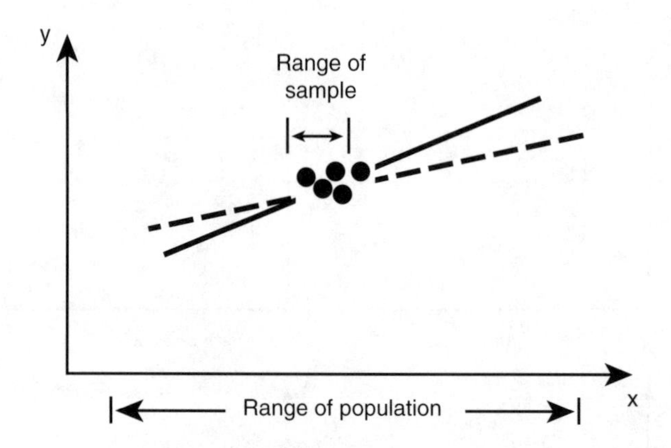

FIGURE 6.4 Instability in the relationship between two random variables.

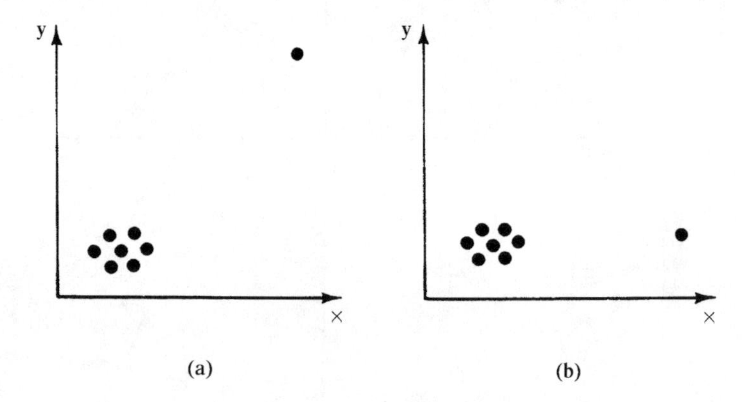

FIGURE 6.5 Effect of an extreme event in a data sample on correlation: (a) high correlation; and (b) low correlation.

Relationships can be linear or nonlinear. Since the statistical methods to be used for the two forms of a relationship differ, it is important to identify the form. In addition, the most frequently used correlation coefficient depends on a linear relationship between the two random variables; thus low correlation may result for a nonlinear relationship even when a strong relationship is obvious. For example, the bivariate relationship of Figure 6.3(f) suggests a predictable trend in the relationship between Y and X; however, the correlation coefficient will be low, and is certainly not as high as that in Figure 6.3(a).

Graphs relating pairs of variables can be used to identify the type of the relationship. Linear trends can be either direct or indirect, with an indirect relationship indicating a decrease in Y as X increases. This information is useful for checking the rationality of the relationship, especially when dealing with data sets that include more than two variables. A variable that is not dominant in the physical relationship may demonstrate a physically irrational relationship with another variable because of the values of the other variables affecting the physical relationship.

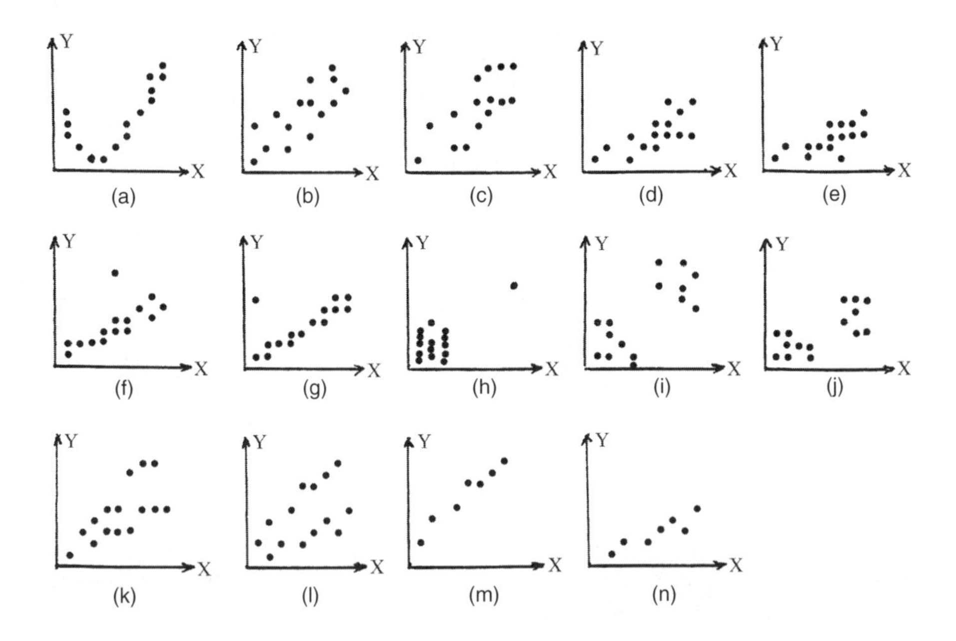

FIGURE 6.6 Graphical assessment of bivariate plots.

Consider the *X-Y* graphs of Figure 6.6. If the Pearson correlation coefficient did not assume linearity, a very high correlation (near 1) could be expected for Figure 6.6(a). The data suggest a high degree of systematic variation, but because the relationship is nonlinear, the correlation coefficient is only 0.7 or 49% explained variance. Figures 6.6(b) and 6.6(c) show far more nonsystematic variation than Figure 6.6(a), but have the same correlation (0.7) because the trend is more linear.

The graphs of Figures 6.6(d) and 6.6(e) show less nonsystemic variation than seen in Figures 6.6(b) and 6.6(c), but they have the same correlation of 0.7 because the total variation is less. Thus, the ratio of the variation explained by the linear trend to the total variation is the same in all four graphs.

Figures 6.6(f), (g), and (h) show single events that deviate from the remainder of the sample points. In Figure 6.6(f), the deviant point lies at about the mean of the *X* values but is outside the range of the *Y* values. The deviant point in Figure 6.6(g) lies at the upper end of the *Y* values and at the lower end of the *X* values. Both points are located away from the general linear trend shown by the other sample points. In Figure 6.6(h), the deviant point falls beyond the ranges of the sample values of both *X* and *Y* but the one deviant point creates a linear trend. The correlations for the three graphs are the same, 0.7, in spite of the positioning of the deviant points.

Figures 6.6(i) and 6.6(j) show two clusters of points. The two clusters in Figure 6.6(i) show greater internal variation than those in Figure 6.6(j) but they are more dispersed along the y-axis. Thus, the two graphs have the same correlation of 0.7 and show that the correlation depends on both the slope of the relationship and the amount of nonsystematic variation or scatter.

All the graphs in Figures 6.6(a) through 6.6(j) have the same correlation coefficient of 0.7 despite the dissimilar patterns of points. This leads to several important observations about bivariate graphs. Both graphs and computed correlation coefficients can be very misleading. Either one alone can lead to poor modeling. It is necessary to graph the data and, if the trend is somewhat linear, compute the correlation coefficient. Second, the correlation coefficient is a single-valued index that cannot reflect all circumstances such as clustering of points, extreme deviant points, nonlinearity, and random versus systematic scatter. Third, the correlation coefficient may not be adequate to suggest a model form, as the data of Figure 6.6(a) obviously need a different model form than needed by the data of Figure 6.6(i).

Bivariate graphs and correlation coefficients also suffer from the effects of other variables. Specifically, the apparent random scatter in an *X*-*Y* graph may be due to a third variable, suggesting that *X* and *Y* are not related. Consider Figures 6.6(k) and 6.6(l). Both show considerable random scatter, but if they are viewed as data for different levels of a third variable, the degree of linear association between *Y* and *X* is considerably better. Figure 6.6(k) with a correlation of 0.7 shows a smaller effect of a second variable than does Figure 6.6(l), which has a correlation of 0.5. However, if the data of Figure 6.6(l) is separated for the two levels of the second predictor variable, the correlation between *Y* and *X* is much better. Figures 6.6(m) and 6.6(n) show the data of Figure 6.6(l) separated into values for the two levels of the second predictor variable. The correlations for Figures 6.6(m) and 6.6(n) are 0.98 and 0.9, respectively. Figures 6.6(k) through 6.6(n) show the importance of considering the effects of other predictor variables when evaluating bivariate plots.

Graphing is an important modeling tool, but it cannot be used alone. Numerical indicators such as correlation coefficients must supplement the information extracted from graphical analyses.

Example 6.3

Table 6.3 contains data for 22 watersheds in the Western Coastal Plain of Maryland and Virginia. The data includes the drainage area (*A*, mi^2), the percentage of forest cover (F), and the 10-year log-Pearson type III discharge (*Q*, cfs). The correlation matrix for the three variables follows.

A	F	Q	
1.000	−0.362	0.933	*A*
	1.000	−0.407	*F*
		1.000	*Q*

The area and discharge are highly correlated, while the correlation between forest cover and discharge is only moderate. Both correlations are rational, as peak discharge should increase with area and decrease with forest cover. The relatively high correlation between peak discharge and area suggests a strong linear relationship. The moderate correlation between peak discharge and forest cover may be the result of a nonlinear relationship or lack of common variation. This cannot be known without plotting the data.

TABLE 6.3
Data Matrix

A (mi²)	F (%)	Q (10-year discharge, cfs)	n (record length)	SN (USGS station number)
0.30	25	54	11	661430
1.19	19	334	10	594445
1.70	96	679	10	496080
2.18	68	193	25	668300
2.30	82	325	14	660900
2.82	69	418	9	668200
3.85	46	440	19	594600
6.73	83	192	11	594800
6.82	85	400	24	661800
6.92	70	548	35	590500
6.98	66	1100	18	661600
8.50	70	350	42	590000
10.4	66	974	25	661000
18.5	60	3010	21	661050
24.0	82	2820	44	661500
24.3	22	3050	36	496000
26.8	23	3240	10	495500
28.0	69	813	39	669000
30.2	47	1590	25	594500
39.5	42	2470	25	653600
54.8	69	4200	37	658000
98.4	22	7730	42	594000

Figure 6.7(a) shows the plot of drainage area versus peak discharge. The graph shows a cluster of points near the origin and a single point at the upper end of the graph. A second point at an area of about 55 mi² is also influential in defining the trend of the data. While dispersed, a second cluster of seven points contributes to the nonsystematic variation. In spite of the points that deviate from a line connecting the cluster of points near the origin and the points for the two largest watersheds, the strong positive association suggested by the correlation coefficient of 0.933 is evident.

Figure 6.7(b) shows the data when peak discharge is plotted against forest cover. The data show considerably more scatter than in Figure 6.7(a). The one point for the largest watershed might suggest a negative relationship, but points for the remainder of the data do not suggest a relationship between forest cover and discharge. Thus, the low correlation of –0.407 reflects a lack of systematic variation rather than a strong nonlinear relationship.

In Figure 6.7(b), the drainage area rounded to an integer is plotted next to the corresponding point. It is evident that higher discharges are associated with larger watersheds. In Figure 6.7(a), the percentage of forest cover is plotted next to the corresponding point. A relationship between discharge and forest cover is not evident. This reflects the lower correlation of discharge and forest cover.

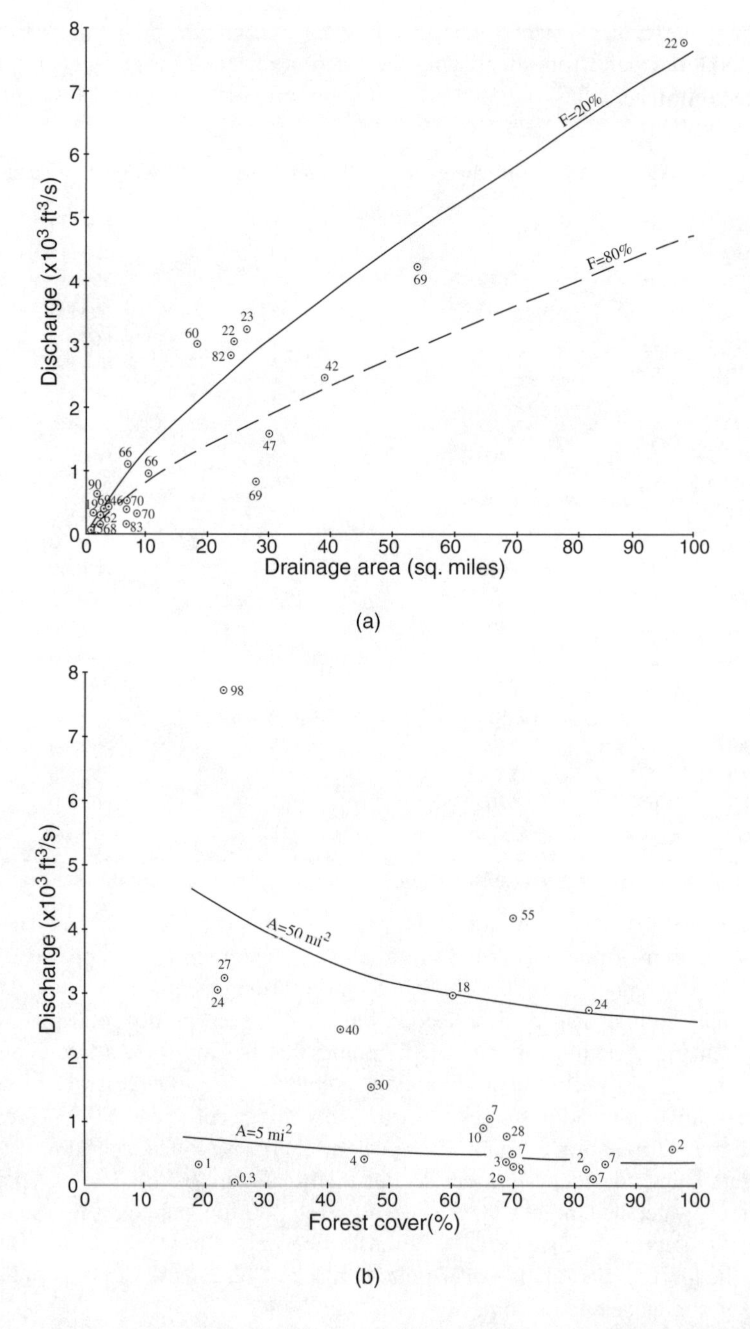

FIGURE 6.7 Graphical representation of a multiple regression analysis for peak discharge (Q) as a function of drainage area (A) and percent forest cover (F): (a) Q versus A with values of F indicated; and (b) Q versus F with values of A indicated.

Since the percentage of forest cover is not a dominant predictor variable, its effect is not evident from identifying the fraction of forest cover by each point on the graph. Instead, a log-transformed multiple regression power model was fit to the data, with the following result:

$$\hat{Q} = 653.7A^{0.7603}F^{-0.3508} \tag{6.3}$$

Figure 6.7(a) shows equations for forest covers of 20% and 80%. The effect of forest cover is not as strong as for drainage area. This illustrates the general rule that the effects of nondominant variables are more difficult to discern than the effects of more important variables. When the equation is plotted in Figure 6.7(b), the effect of drainage area is more evident.

6.3 COMPILATION OF CAUSAL INFORMATION

In addition to graphical analyses of the annual maximum series, compilation and summary analyses of causal information can help formulate hypotheses to be tested and plan adjustment methods. For example, if one suspects that urban development or deforestation contributed to changes in flood characteristics, values that are indicators of these causal factors, such as the percent of imperviousness or percent forest cover, should be obtained for each year of the record. This information may not be readily available and may require a diligent search of files or inquiries to organizations, groups, or individuals responsible for the watershed changes. If such data are available, it may be useful to plot them to assess the extent of systematic variation.

For episodic changes, such as the construction of levees, channelization, or the installation of small in-stream detention reservoirs, the date or dates of the changes should be noted. It might also be of value to obtain important climatic or meteorological data, such as rainfall volumes. The causal information may be of value in hypothesizing the type of nonhomogeneity that exists, selecting methods for detecting changes, or adjusting the flood series.

In the Bulletin 17B flood-frequency environment (Interagency Advisory Committee on Water Data, 1983), the logarithms of the annual peak discharges are assumed to fit a Pearson Type III frequency distribution. All calculations and frequency plots in the bulletin are based on the logarithms. Most of the statistical tests to detect nonhomogeneity are nonparametric and the results of the tests would be the same for the logarithms as for the untransformed flood data. However, some graphs and quantitative analyses will yield different results when the data are transformed.

Example 6.4

Figure 6.8 is a plot of the annual peak discharges versus water year for the Northwest Branch of the Anacostia River at Hyattsville, Maryland (USGS gaging station 01651000) for 1939 to 1988. During the period of record, the watershed was subject

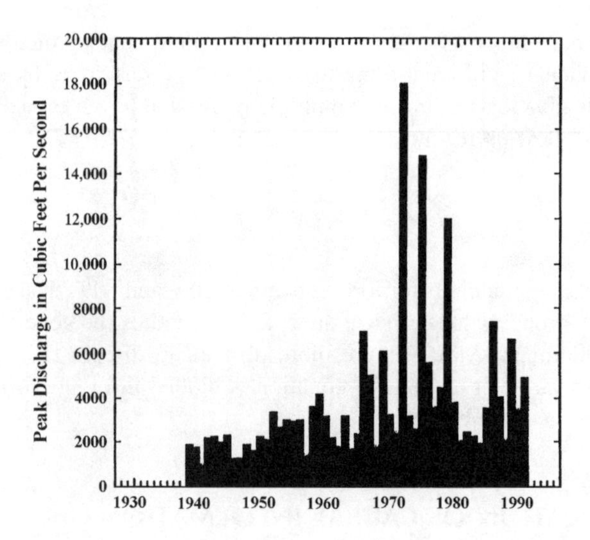

FIGURE 6.8 Annual maximum peak discharge for the Northwest Branch of the Anacostia River near Hyattsville, Maryland.

to considerable urbanization. While considerable scatter is evident, the secular or gradual trend in peak discharge is evident in Figure 6.8. Before the mid-1960s, most of the annual maximums were less than 4000 cfs. The increase in urbanization produced a nonstationary annual maximum series characterized by substantial increases in both the central tendency and spread of the data. While the three large peaks during the 1970s may be partly influenced by large rainfalls, the increasing trend over the duration of record is evident. The large scatter in the latter part of the record makes it difficult to recommend a functional form for the effect of the urbanization.

Example 6.5

Figures 6.9 and 6.10 show a plot of the annual maximum peak discharges and a plot of the rank of the annual maximum peak discharges versus the water year for the Saddle River at Lodi, New Jersey (USGS gaging station 01391500) from 1924 to 1988. The watershed was channelized in 1968, which is evident from the episodic change in flood peak characteristics evident in Figure 6.9. In the prechannelization period, the flood peaks show considerable scatter, from many of the lowest peaks of record to one of the largest peaks of record. After channelization, the peaks are generally larger, but less scatter is evident. The channelization evidently caused the peaks to bunch around the mean because of the loss of natural storage that dominated in the earlier period.

As an alternative to plotting discharges or logarithms of discharges, the discharges can be transformed to ranks and the ranks can be plotted versus time, that is, water year. One advantage of graphing the rank rather than the discharge is that

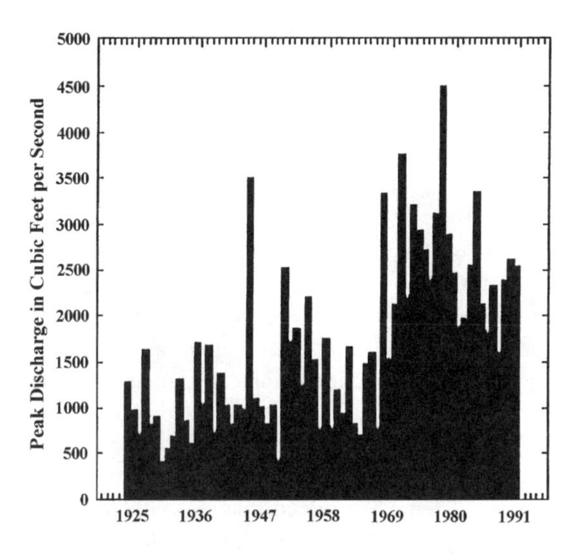

FIGURE 6.9 Annual maximum peak discharges for Saddle River at Lodi, New Jersey.

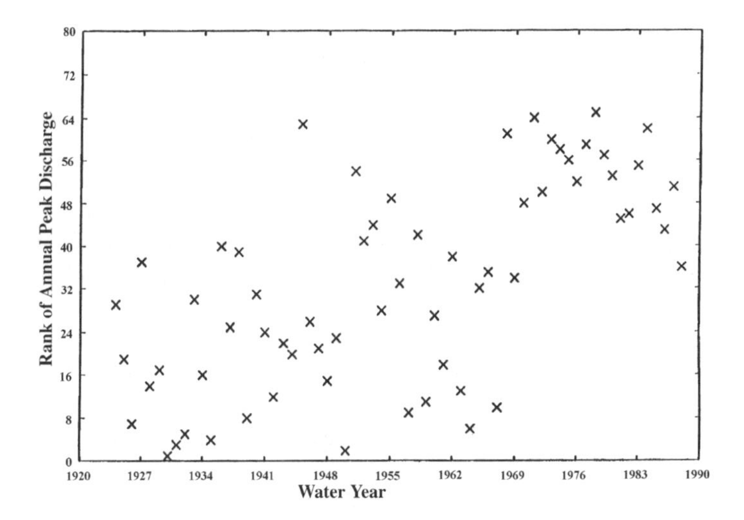

FIGURE 6.10 Rank of annual maximum peak discharges for Saddle River at Lodi, New Jersey.

extreme events in the series do not cause misperception of the significance of nonhomogeneity. As evident from Figure 6.10, the peaks occurring after channelization show little variation in rank. The plot of the ranks shows much less scatter than the plot of the discharges. However, if the goal was to provide functional representation of the discharges, the plot of the ranks would not be helpful. The rank plot is only useful for detecting general trends.

6.4 SUPPORTING COMPUTATIONAL ANALYSES

Graphical analyses are very useful but systematic changes or trends are often difficult to detect because of the dominance of random variation. For example, in Figure 6.8 (Northwest Branch of the Anacostia River near Hyattsville), the data show considerable scatter that may make it difficult to characterize the nature of the secular trend. A number of methods are available for analyzing data to separate systematic and random variations for the purpose of evaluating the type of modeling effort to use. Moving-average filtering, which was introduced in Chapter 2, is one method that can be used to reduce the effect of random variation in order to detect secular (systematic) trends. It is also useful in detecting abrupt or episodic change.

Moving-average filtering is a computational data-analysis technique for reducing the effects of nonsystematic variations. As indicated in Chapter 2, the method is based on the premise that the systematic component of a time series exhibits some autocorrelation between adjacent and nearby measurements while the random fluctuations are not autocorrelated. Such autocorrelation could result from the hydrologic effects of watershed change. Therefore, the averaging of adjacent measurements will eliminate random fluctuations, with the resulting data converging to a description of the systematic trend.

Moving-average filtering has several disadvantages. First, $2k$ observations are lost, which may be a very limiting disadvantage for short record lengths. Second, a moving-average filter is not a mathematical representation, and thus forecasting with the filter is not possible; a structural form must still be calibrated to forecast any systematic trend identified by the filtering. Third, the choice of the smoothing interval is not always obvious, and it is often necessary to try several values in order to provide the best separation of systematic and nonsystematic variation. Fourth, if the smoothing interval is not properly selected, it is possible to eliminate both the systematic and the nonsystematic variation. Fifth, extreme points in a data set can cause large fluctuations in the smoothed series just as it does in the measured series.

One disadvantage in applying a moving-average filter is that it is necessary to specify the filter length (i.e., smoothing interval). An objective criterion for selecting m is not generally accepted, so in practice, it is necessary to perform successive analyses using different smoothing intervals and select the one that enables the systematic effects to be most accurately assessed. Based on the data sets presented herein, a smoothing interval of seven or nine appears to be best for detecting systematic changes in annual flood series that were subject to watershed change.

Example 6.6

The use of moving-average smoothing for detecting a secular trend is illustrated using data for 1945 through 1968 for two adjacent watersheds in north-central Kentucky, about 50 miles south of Louisville. The data include the annual flood series for Pond Creek, a 64-square-mile watershed, and the north fork of the Nolin River at Hodgenville, which has an area of 36.4 square miles. From 1945 to 1968, urbanization in the Pond Creek watershed increased from 2.3 to 13.3% while

TABLE 6.4
Annual Flood Series for Pond Creek (Q_{p1}) and North Fork of Nolin River (Q_{p2}) Watersheds

Water Year	Q_{p1} (ft^3/s)	Q_{p2} (ft^3/s)
1945	2000	4390
1946	1740	3550
1947	1460	2470
1948	2060	6560
1949	1530	5170
1950	1590	4720
1951	1690	2720
1952	1420	5290
1953	1330	6580
1954	607	548
1955	1380	6840
1956	1660	3810
1957	2290	6510
1958	2590	8300
1959	3260	7310
1960	2490	1640
1961	3080	4970
1962	2520	2220
1963	3360	2100
1964	8020	8860
1965	4310	2300
1966	4380	4280
1967	3220	7900
1968	4320	5000
Median	2175	4845

channelization increased from 18.6 to 56.7%. Most changes occurred between 1954 and 1965. The Nolin River watershed served as a control since change was minimal for 1945 through 1968. The annual flood series for Pond Creek and the North Fork for the 24-year period is given in Table 6.4.

The data for Pond Creek were subjected to a moving-average smoothing with a smoothing interval of 7 years. Shorter smoothing intervals were tried but did not show the secular trend as well as the 7-year interval. The smoothed series shown in Figure 6.11 has a length of 18 years because six values ($2k = 6$) are lost during smoothing. A visual inspection of Figure 6.11 indicates a trend in the smoothed series. Plots similar to Figure 6.11 simply indicate that a trend might be prevalent and show a need for further statistical tests to determine whether the apparent trend is significant. Relatively little variation in the smoothed series is evident before the mid-1950s;

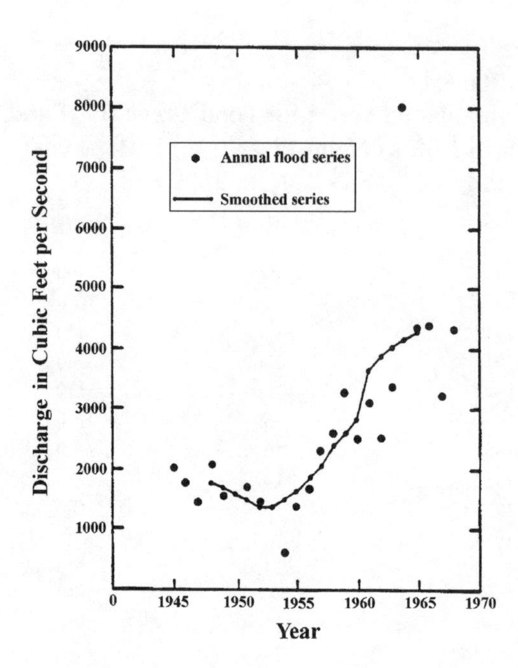

FIGURE 6.11 Annual-maximum flood series and smoothed series for Pond Creek near Louisville, Kentucky.

the variation in this portion of the smoothed series can probably be attributed to climatic variation. As watershed development increased in the mid-1950s, the flood peaks increased, as indicated by the upward trend after 1955.

Figure 6.12 is a plot of the annual maximum series versus water year for the North Fork of the Nolin River along with the moving-average smoothed series for a smoothing interval of 7 years. In contrast to the smoothed series for Pond Creek, an upward secular trend is not evident in the annual maximum series of the Nolin River. The discharges in the latter part of the smoothed series in Figure 6.12 are similar in magnitude to the discharges in the early years of the record. The slight increase in the discharges in the middle of the smoothed series possibly reflects either higher rainfall events of that period or random variation. Rainfall data associated with the annual peak discharges should be examined to determine whether a hypothesis of higher rainfall is true. Thus, Figure 6.12 probably reflects a flood series in which a secular trend, such as that caused by watershed change, is not present.

Example 6.7

The moving-average filter was used in the previous example to identify a systematic trend that may have resulted from watershed change. Filtering can also be used to assess an association between two time series. This is illustrated using an annual maximum flood series and the series of storm event rainfalls that produced the flood peaks in the annual series. Data were available for a 41-year period (1949–1989) during which time the watershed underwent some urbanization. Figure 6.13 shows

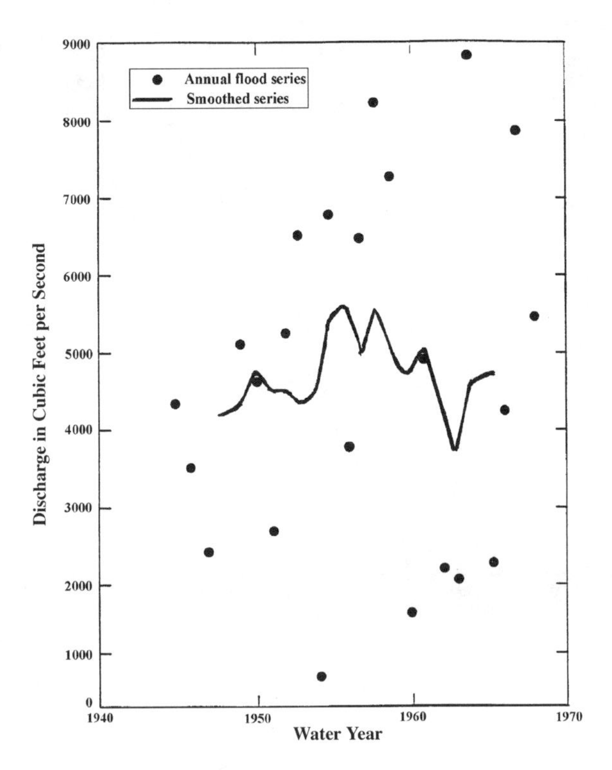

FIGURE 6.12 Annual-maximum flood series and smoothed series for north fork of Nolin River at Hodgenville, Kentucky.

the two smoothed series computed using a smoothing interval of 9 years, with 4 years lost from each end. The similarities in the fluctuations in the two smoothed series suggest that variation in the rainfall is a dominant factor in the variation in the annual maximums. All of the increase in the peak discharges that started in 1968 should not be attributed to urbanization. In the example in Figure 6.13, the annual peak discharges must be adjusted for the effect of increasing storm event rainfalls before the effect of urbanization can be evaluated.

6.5 PROBLEMS

6-1 Discuss the advantages and disadvantages of histograms.

6-2 Create frequency and probability histograms for the following data. Assess the likely distribution from which the data were sampled.

05	57	23	06	26	23	08	66	16	11	75	28	81
37	78	16	06	57	12	46	22	90	97	78	67	39
23	71	15	08	82	64	87	29	01	20	46	72	05
42	67	98	41	67	44	28	71	45	08	19	47	76
06	83	03	84	32	62	83	27	48	83	09	19	84

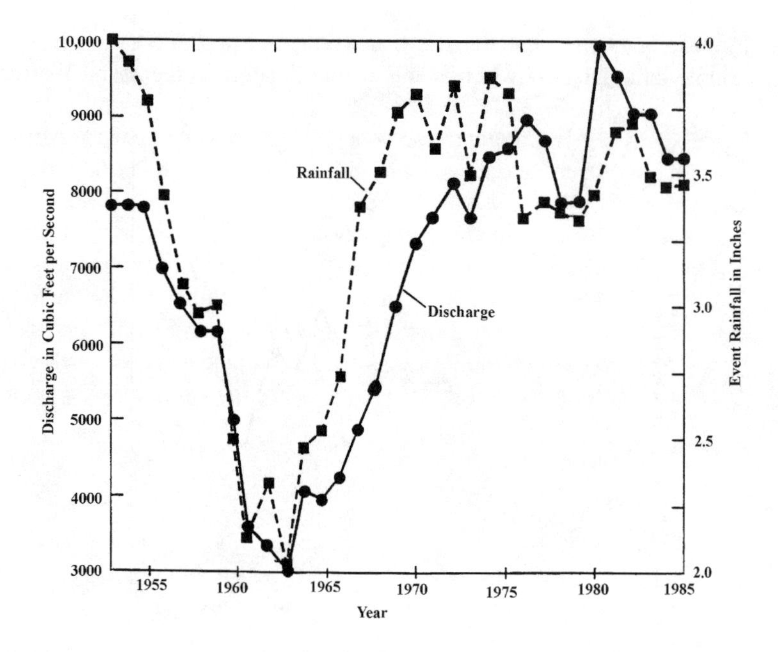

FIGURE 6.13 An example of a moving-average filtered series for annual-maximum peak discharges and associated event rainfall.

6-3 Create frequency and probability histograms for the following data. Assess the likely distribution from which the data were sampled.

4.05	5.04	3.89	3.88	5.40	3.61	3.78	4.42	4.23
3.48	3.44	3.88	3.48	3.48	1.81	6.26	4.46	2.47
2.59	3.79	4.95	3.93	2.53	3.76	3.35	5.56	5.28
5.82	4.90	3.31	4.97	3.99	4.52	4.81	6.65	3.35
5.35	3.86	4.95	5.50	3.65	4.31	4.58	2.12	3.51
4.42	2.91	2.42	3.88	5.34	4.17	4.55	3.25	3.41
2.24	2.13	4.68	4.54	5.47	4.57	3.23	4.66	3.42

6-4 Discuss the uses of bivariate graphical analyses and indicate the characteristics of graphs that will illustrate each use.

6-5 Provide an example of a bivariate graph that contains a sample value that would be considered a univariate outlier in the Y direction, but not in the X direction. Assume that a univariate outlier would be detected with a test from Chapter 4.

6-6 Provide a bivariate graph that contains a sample value that would not be considered an outlier in either the X or Y direction, but would be an outlier in the bivariate space.

6-7 Propose an outlier test statistic that would be appropriate for bivariate sample data. Discuss whether this would depend on the X and Y directions.

6-8 Discuss why it is important to use both correlation and graphical analyses in evaluating bivariate data and why the use of each without the other is inadequate.

6-9 The following annual maximum discharges (Q) occurred during a period of rapid urbanization, as measured by the percent imperviousness (I). (a) Graph the data and assess the importance of the imperviousness to the magnitude of the discharges. (b) Compute the correlation coefficient between Q and I. (c) Would it make sense to compute the autocorrelogram for Q and the cross-correlogram for Q and I? (d) Discuss the potential interpretation.

I	8	12	17	20	25	31	34	41
Q	140	710	420	380	530	960	580	670

6-10 The following table shows monthly rainfall (P, in.) and corresponding runoff (Q, in.) for a 145-acre watershed. Compute the correlation coefficient and graph the data. Interpret the results from statistical and hydrologic standpoints.

P	3.78	2.94	5.17	4.33	4.60	5.25	6.41	3.37	7.29
Q	1.57	0.86	2.20	2.01	1.72	2.51	2.62	1.52	3.65

6-11 The following time series represents the annual summer baseflow over 21 years for a moderate-sized watershed that has undergone a land use change. Characterize the effect of that land use change. Q = {8, 5, 9, 7, 12, 8, 7, 18, 14, 20, 22, 20, 23, 28, 31, 26, 34, 25, 30, 32, 27}.

6-12 The following is a time series of sediment trap efficiency for 18 storm events for a wetland (T, %): {74, 44, 44, 51, 65, 49, 52, 55, 35, 59, 41, 61, 37, 92, 73, 36, 72, 40}. (a) Graph the data and assess the degree of trend in the data. (b) Compute and interpret the autocorrelation coefficient. (c) The following are storm event rainfall depths (P, in.) corresponding to the trap efficiencies: {0.43, 1.55, 0.49, 1.25, 0.25, 1.78, 0.90, 0.44, 1.38, 0.63, 2.02, 0.78, 1.64, 0.27, 0.66, 1.88, 0.51, 0.98}. Graph T versus P and discuss the importance of P as a causative variable in the time series of T. (d) Compute the lag-0 cross-correlation coefficient between T and P, and interpret the result.

6-13 Discuss the use of moving-average filtering on the criterion variable of a bivariate relationship.

6-14 The following is a 13-year record of the annual load of nitrogen-based fertilizer (X_t) applied to a field during the growing season and the annual average concentration of nitrogen measured in a nearby stream (Y_t). (a) Perform a moving-average filtering of Y_t using the following filter: $\hat{Y}_t = 0.6Y_t + 0.3Y_{t-1} + 0.1Y_{t-2}$. Graph both X_t versus Y_t and X_t versus $\hat{Y}_t$, and discuss the benefit or information loss of smoothing. (b) Would the filter $\hat{Y}_t = 0.25Y_{t+1} + 0.5Y_t + 0.25Y_{t-1}$ be preferred?

X_t	64	56	54	45	48	42	38	39	34	36	35	31	29
Y_t	280	300	210	200	240	200	170	210	150	160	180	150	150

7 Statistical Detection of Nonhomogeneity

7.1 INTRODUCTION

Data independent of the flood record may suggest that a flood record may not be stationary. Knowing that changes in land cover occurred during the period of record will necessitate assessing the effect of the land cover change on the peaks in the record. Statistical hypothesis testing is the fundamental approach for analyzing a flood record for nonhomogeneity. Statistical testing only suggests whether a flood record has been affected; it does not quantify the effect. Statistical tests have been used in flood frequency and hydrologic analyses for the detection of nonhomogeneity (Natural Environment Research Council, 1975; Hirsch, Slack, and Smith, 1982; Pilon and Harvey, 1992; Helsel and Hirsch, 1992).

The runs test can be used to test for nonhomogeneity due to a trend or an episodic event. The Kendall test tests for nonhomogeneity associated with a trend. Correlation analyses can also be applied to a flood series to test for serial independence, with significance tests applied to assess whether an observed dependency is significant; the Pearson test and the Spearman test are commonly used to test for serial correlation. If a nonhomogeneity is thought to be episodic, separate flood frequency analyses can be done to detect differences in characteristics, with standard techniques used to assess the significance of the differences. The Mann–Whitney test is useful for detecting nonhomogeneity associated with an episodic event.

Four of these tests (all but the Pearson test) are classified as nonparametric. They tests can be applied directly to the discharges in the annual maximum series without making a logarithmic transform. The exact same solution results when the test is applied to the logarithms and to the untransformed data with all four tests. This is not true for the Pearson test, which is parametric. Because a logarithmic transform is cited in Bulletin 17B (Interagency Advisory Committee on Water Data, 1982), the transform should also be applied when making the statistical test for the Pearson correlation coefficient.

The tests presented for detecting nonhomogeneity follow the six steps of hypothesis testing: (1) formulate hypotheses; (2) identify theory that specifies the test statistic and its distribution; (3) specify the level of significance; (4) collect the data and compute the sample value of the test statistic; (5) obtain the critical value of the test statistic and define the region of rejection; and (6) make a decision to reject the null hypothesis if the computed value of the test statistic lies in the region of rejection.

7.2 RUNS TEST

Statistical methods generally assume that hydrologic data measure random variables, with independence among measured values. The runs (or run) test is based on the mathematical theory of runs and can test a data sample for lack of randomness or independence (or conversely, serial correlation) (Siegel, 1956; Miller and Freund, 1965). The hypotheses follow:

H_0: The data represent a sample of a single independently distributed random variable.
H_A: The sample elements are not independent values.

If one rejects the null hypothesis, the acceptance of nonrandomness does not indicate the type of nonhomogeneity; it only indicates that the record is not homogeneous. In this sense, the runs test may detect a systematic trend or an episodic change. The test can be applied as a two-tailed or one-tailed test. It can be applied to the lower or upper tail of a one-tailed test.

The runs test is based on a sample of data for which two outcomes are possible, x_1 or x_2. These outcomes can be membership in two groups, such as exceedances or nonexceedances of a user-specified criterion such as the median. In the context of flood-record analysis, these two outcomes could be that the annual peak discharges exceed or do not exceed the median value for the flood record. A *run* is defined as a sequence of one or more of outcome x_1 or outcome x_2. In a sequence of n values, n_1 and n_2 indicate the number of outcomes x_1 and x_2, respectively, where $n_1 + n_2 = n$. The outcomes are determined by comparing each value in the data series with a user-specified criterion, such as the median, and indicating whether the data value exceeds (+) or does not exceed (–) the criterion. Values in the sequence that equal the median should be omitted from the sequences of + and – values. The solution procedure depends on sample size. If the values of n_1 and n_2 are both less than 20, the critical number of runs, n_α, can be obtained from a table. If n_1 or n_2 is greater than 20, a normal approximation is made.

The theorem that specifies the test statistic for large samples is as follows: If the ordered (in time or space) sample data, contains n_1 and n_2 values for the two possible outcomes, x_1 and x_2, respectively, in n trials, where both n_1 and n_2 are not small, the sampling distribution of the number of runs is approximately normal with mean, $\overline{U}$, and variance, S_u^2, which are approximated by:

$$\overline{U} = \frac{2n_1n_2}{n_1n_2} + 1 \tag{7.1a}$$

and

$$(S_u)^2 = \frac{2n_1n_2(2n_1n_2 - n_1 - n_2)}{(n_1 + n_2)^2(n_1 + n_2 - 1)} \tag{7.1b}$$

in which $n_1 + n_2 = n$. For a sample with U runs, the test statistic is (Draper and Smith, 1966):

$$z = \frac{U - \bar{U} - 0.5}{S_u} \tag{7.2}$$

where z is the value of a random variable that has a standard normal distribution. The 0.5 in Equation 7.2 is a continuity correction applied to help compensate for the use of a continuous (normal) distribution to approximate the discrete distribution of U. This theorem is valid for samples in which n_1 or n_2 exceeds 20.

If both n_1 and n_2 are less than 20, it is only necessary to compute the number of runs U and obtain critical values of U from appropriate tables (see Appendix Table A.5). A value of U less than or equal to the lower limit or greater than or equal to the upper limit is considered significant. The appropriate section of the table is used for a one-tailed test. The critical value depends on the number of values, n_1 and n_2. The typically available table of critical values is for a 5% level of significance when applied as a two-tailed test. When it is applied as a one-tailed test, the critical values are for a 2.5% level of significance.

The level of significance should be selected prior to analysis. For consistency and uniformity, the 5% level of significance is commonly used. Other significance levels can be justified on a case-by-case basis. Since the basis for using a 5% level of significance with hydrologic data is not documented, it is important to assess the effect of using the 5% level on the decision.

The runs test can be applied as a one-tailed or two-tailed test. If a direction is specified, that is, the test is one-tailed, then the critical value should be selected accordingly to the specification of the alternative hypothesis. After selecting the characteristic that determines whether an outcome should belong to group 1 (+) or group 2 (−), the runs should be identified and n_1, n_2, and U computed. Equations 7.1a and 7.1b should be used to compute the mean and variance of U. The computed value of the test statistic z can then be determined with Equation 7.2.

For a two-tailed test, if the absolute value of z is greater than the critical value of z, the null hypothesis of randomness should be rejected; this implies that the values of the random variable are probably not randomly distributed. For a one-tailed test where a small number of runs would be expected, the null hypothesis is rejected if the computed value of z is less (i.e., more negative) than the critical value of z. For a one-tailed test where a large number of runs would be expected, the null hypothesis is rejected if the computed value of z is greater than the critical value of z. For the case where either n_1 or n_2 is greater than 20, the critical value of z is $-z_\alpha$ or $+z_\alpha$ depending on whether the test is for the lower or upper tail, respectively.

When applying the runs test to annual maximum flood data for which watershed changes may have introduced a systematic effect into the data, a one-sided test is typically used. Urbanization of a watershed may cause an increase in the central tendency of the peaks and a decrease in the coefficient of variation. Channelization may increase both the central tendency and the coefficient of variation. Where the primary effect of watershed change is to increase the central tendency of the annual maximum floods, it is appropriate to apply the runs test as a one-tailed test with a small number of runs. Thus, the critical z value would be a negative number, and the null hypothesis would be rejected when the computed z is more negative than the critical z_α, which would be a negative value. For a small sample test, the null hypothesis would be rejected if the computed number of runs was smaller than the critical number of runs.

Example 7.1

The runs test can be used to determine whether urban development caused an increase in annual peak discharges. It was applied to the annual flood series of the rural Nolin River and the urbanized Pond Creek watersheds to test the following null (H_0) and alternative (H_A) hypotheses:

H_0: The annual peak discharges are randomly distributed from 1945 to 1968, and thus a significant trend is not present.
H_A: A significant trend in the annual peak discharges exists since the annual peaks are not randomly distributed.

The flood series is represented in Table 7.1 by a series of + and – symbols. The criterion that designates a + or – event is the median flow (i.e., the flow exceeded or not exceeded as an annual maximum in 50% of the years). For the Pond Creek and North Fork of the Nolin River watersheds, the median values are 2175 ft³/sec and 4845 ft³/sec, respectively (see Table 7.1). If urbanization caused an increase in discharge rates, then the series should have significantly more + symbols in the part of the series corresponding to greater urbanization and significantly more – symbols before urbanization. The computed number of runs would be small so a one-tailed

TABLE 7.1
Annual Flood Series for Pond Creek (q_p, median = 2175 ft³/s) and the Nolin River (Q_p, median = 4845 ft³/s)

Year	q_p	Sign	Q_p	Sign	Year	q_p	Sign	Q_p	Sign
1945	2000	–	4390	–	1957	2290	+	6510	+
1946	1740	–	3550	–	1958	2590	+	8300	+
1947	1460	–	2470	–	1959	3260	+	7310	+
1948	2060	–	6560	+	1960		+	1640	–
1949	1530	–	5170	+	1961		+	4970	+
1950	1590	–	4720	–	1962		+	2220	–
1951	1690	–	2720	–	1963		+	2100	–
1952	1420	–	5290	+	1964		+	8860	+
1953	1330	–	6580	+	1965		+	2300	–
1954	607	–	548	–	1966	4380	+	4280	–
1955	1380	–	6840	+	1967	3220	+	7900	+
1956	1660	–	3810	–	1968	4320	+	5500	+

test should be applied. While rejection of the null hypothesis does not necessarily prove that urbanization caused a trend in the annual flood series, the investigator may infer such a cause.

The Pond Creek series has only two runs (see Table 7.1). All values before 1956 are less than the median and all values after 1956 are greater than the median. Thus, $n_1 = n_2 = 12$. The critical value of 7 was obtained from Table A.5. The null hypothesis should be rejected if the number of runs in the sample is less than or equal to 7. Since a one-tailed test was used, the level of significance is 0.025. Because the sequence includes only two runs for Pond Creek, the null hypothesis should be rejected. The rejection indicates that the data are nonrandom. The increase in urbanization after 1956 may be a causal factor for this nonrandomness.

For the North Fork of the Nolin River, the flood series represents 14 runs (see Table 7.1). Because n_1 and n_2 are the same as for the Pond Creek analysis, the critical value of 7 applies here also. Since the number of runs is greater than 7, the null hypothesis of randomness cannot be rejected. Since the two watersheds are located near each other, the trend in the flood series for Pond Creek is probably not due to an increase in rainfall. (In a real-world application, rainfall data should be examined for trends as well.) Thus, it is probably safe to conclude that the flooding trend for Pond Creek is due to urban development in the mid-1950s.

7.2.1 Rational Analysis of Runs Test

Like every statistical test, the runs test is limited in its ability to detect the influence of a systematic factor such as urbanization. If the variation of the systematic effect is small relative to the variation introduced by the random processes, then the runs test may suggest randomness. In such a case, all of the variation may be attributed to the effects of the random processes.

In addition to the relative magnitudes of the variations due to random processes and the effects of watershed change, the ability of the runs test to detect the effects of watershed change will depend on its temporal variation. Two factors are important. First, change can occur abruptly over a short time or gradually over the duration of a flood record. Second, an abrupt change may occur near the center, beginning, or end of the period of record. These factors must be understood when assessing the results of a runs test of an annual maximum flood series.

Before rationally analyzing the applicability of the runs test for detecting hydrologic change, summarizing the three important factors is worthwhile.

1. Is the variation introduced by watershed change small relative to the variation due to the randomness of rainfall and watershed processes?
2. Has the watershed change occurred abruptly over a short part of the length of record or gradually over most of the record length?
3. If the watershed change occurred over a short period, was it near the center of the record or at one of the ends?

Answers to these questions will help explain the rationality of the results of a runs test and other tests discussed in this chapter.

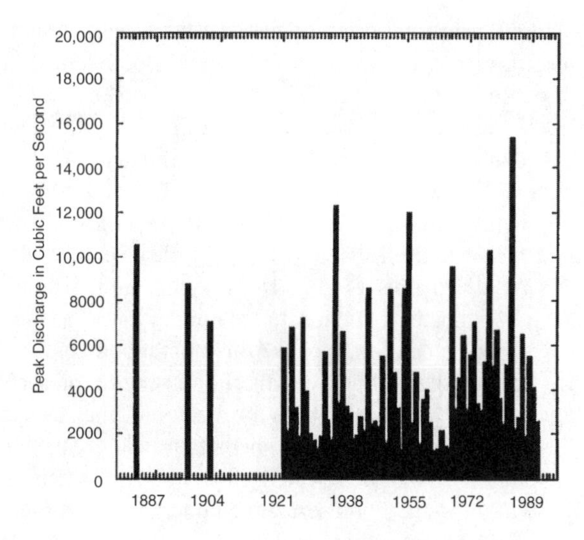

FIGURE 7.1 Annual maximum peak discharges for Ramapo River at Pompton Lakes, New Jersey.

Responses to the above three questions will include examples to demonstrate the general concepts. Studies of the effects of urbanization have shown that the more frequent events of a flood series may increase by a factor of two for large increases in imperviousness. For example, the peaks in the later part of the flood record for Pond Creek are approximately double those from the preurbanization portion of the flood record. Furthermore, variation due to the random processes of rainfall and watershed conditions appears relatively minimal, so the effects of urbanization are apparent (see Figure 2.4). The annual maximum flood record for the Ramapo River at Pompton Lakes, New Jersey (1922 through 1991) is shown in Figure 7.1. The scatter is very significant, and an urbanization trend is not immediately evident. Most urban development occurred before 1968, and the floods of record then appear smaller than floods that occurred in the late 1960s. However, the random scatter largely prevents the identification of effects of urbanization from the graph. When the runs test is applied to the series, the computed test statistic of Equation 7.2 equals zero, so the null hypothesis of randomness cannot be rejected. In contrast to the series for Pond Creek, the large random scatter in the Ramapo River series masks the variation due to urbanization.

The nature of a trend is also an important consideration in assessing the effect of urbanization on the flows of an annual maximum series. Urbanization of the Pond Creek watershed occurred over a short period of total record length; this is evident in Figure 2.4. In contrast, Figure 7.2 shows the annual flood series for the Elizabeth River, at Elizabeth, New Jersey, for a 65-year period. While the effects of the random processes are evident, the flood magnitudes show a noticeable increase. Many floods at the start of the record are below the median, while the opposite is true for later years. This causes a small number of runs, with the shorter runs near the center of record. The computed z statistic for the run test is −3.37, which is significant at the

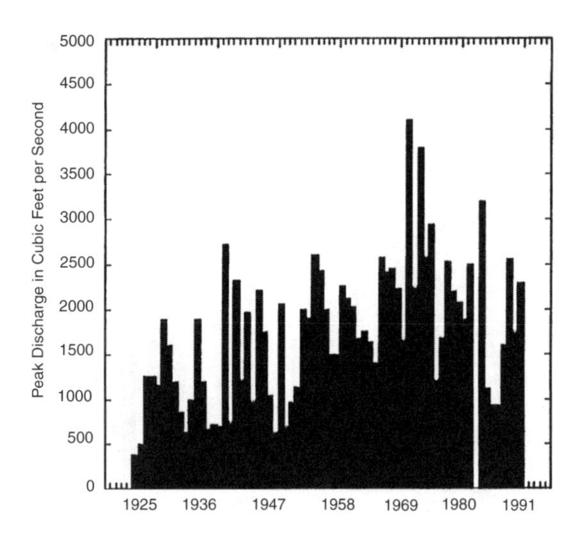

FIGURE 7.2 Annual maximum peak discharges for Elizabeth River, New Jersey.

0.0005 level. Thus, a gradual trend, especially with minimal variation due to random processes, produces a significant value for the runs test. More significant random effects may mask the hydrologic effects of gradual urban development.

Watershed change that occurs over a short period, such as that in Pond Creek, can lead to acceptance or rejection of the null hypothesis for the runs test. When the abrupt change is near the middle of the series, the two sections of the record will have similar lengths; thus, the median of the series will fall in the center of the two sections, with a characteristic appearance of two runs, but it quite possibly will be less than the critical number of runs. Thus, the null hypothesis will be rejected. Conversely, if the change due to urbanization occurs near either end of the record length, the record will have short and long sequences. The median of the flows will fall in the longer sequence; thus, if the random effects are even moderate, the flood series will have a moderate number of runs, and the results of a runs test will suggest randomness.

It is important to assess the type (gradual or abrupt) of trend and the location (middle or end) of an abrupt trend. This is evident from a comparison of the series for Pond Creek, Kentucky, and Rahway River in New Jersey. Figure 7.3 shows the annual flood series for the Rahway River. The effect of urbanization appears in the later part of the record. The computed z statistic for the runs test is -1.71, which is not significant at the 5% level, thus suggesting that randomness can be assumed.

7.3 KENDALL TEST FOR TREND

Hirsch, Slack, and Smith (1982) and Taylor and Loftis (1989) provide assessments of the Kendall nonparametric test. The test is intended to assess the randomness of a data sequence X_i; specifically, the hypotheses (Hirsch, Slack, and Smith, 1982) are:

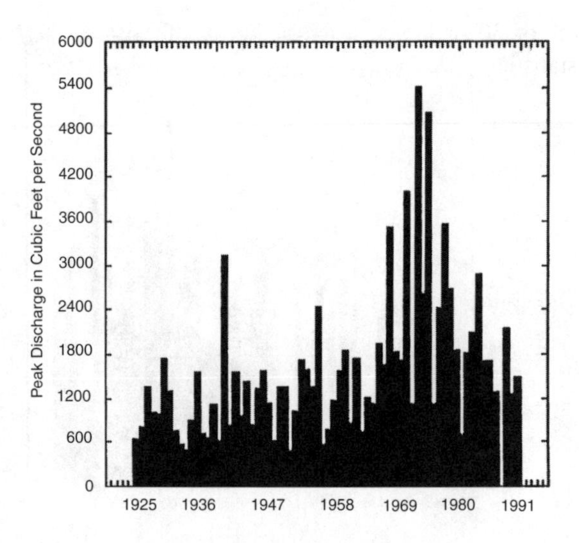

FIGURE 7.3 Annual maximum peak discharges for Rahway River, New Jersey.

H_0: The annual maximum peak discharges (x_i) are a sample of n independent and identically distributed random variables.

H_A: The distributions of x_j and x_k are not identical for all $k, j \le n$ with $k \le j$.

The test is designed to detect a monotonically increasing or decreasing trend in the data rather than an episodic or abrupt event. The above H_A alternative is two-sided, which is appropriate if a trend can be direct or inverse. If a direction is specified, then a one-tailed alternative must be specified. Gradual urbanization would cause a direct trend in the annual flood series. Conversely, afforestation can cause an inverse trend in an annual flood series. For the direct (inverse) trend in a series, the one-sided alternative hypothesis would be:

H_A: A direct (inverse) trend exists in the distribution of x_j and x_k.

The theorem defining the test statistic is as follows. If x_j and x_k are independent and identically distributed random values, the statistic S is defined as:

$$S = \sum_{k=1}^{n-1} \sum_{j=k+1}^{n} \text{sgn}(x_j - x_k) \tag{7.3}$$

where

$$z = \begin{cases} 1 & \text{if } \Theta > 0 \\ 0 & \text{if } \Theta = 0 \\ -1 & \text{if } \Theta < 0 \end{cases} \tag{7.4}$$

For sample sizes of 30 or larger, tests of the hypothesis can be made using the following test statistic:

$$z = \begin{cases} (S-1)/V^{0.5} & \text{for } S > 0 \\ 0 & \text{for } S = 0 \\ (S+1)/V^{0.5} & \text{for } S < 0 \end{cases} \tag{7.5a}$$

in which z is the value of a standard normal deviate, n is the sample size, and V is the variance of S, given by:

$$V = \frac{n(n-1)(2n+5) - \sum_{i=1}^{g} t_i(t_i - 1)(2t_i + 5)}{18} \tag{7.5b}$$

in which g is the number of groups of measurements that have equal value (i.e., ties) and t_i is the number of ties in group i. Mann (1945) provided the variance for series that did not include ties, and Kendall (1975) provided the adjustment shown as the second term of Equation 7.5b. Kendall points out that the normal approximation of Equation 7.5a should provide accurate decisions for samples as small as 10, but it is usually applied when $N \geq 30$. For sample sizes below 30, the following τ statistic can be used when the series does not include ties:

$$\tau = 2S/[n(n-1)] \tag{7.6}$$

Equation 7.6 should not be used when the series includes discharges of the same magnitude; in such cases, a correction for ties can be applied (Gibbons, 1976).

After the sample value of the test statistic z is computed with Equation 7.5 and a level of significance α selected, the null hypothesis can be tested. Critical values of Kendall's τ are given in Table A.6 for small samples. For large samples with a two-tailed test, the null hypothesis H_0 is rejected if z is greater than the standard normal deviate $z_{\alpha/2}$ or less than $-z_{\alpha/2}$. For a one-sided test, the critical values are z_α for a direct trend and $-z_\alpha$ for an inverse trend. If the computed value is greater than z_α for the direct trend, then the null hypothesis can be rejected; similarly, for an inverse trend, the null hypothesis is rejected when the computer z is less (i.e., more negative) than $-z_\alpha$.

Example 7.2

A simple hypothetical set of data is used to illustrate the computation of τ and decision making. The sample consists of 10 integer values, as shown, with the values of sgn(Θ) shown immediately below the data for each sample value.

2	5	0	3	7	1	4	9	6	8
	+	−	+	+	−	+	+	+	+
		−	−	+	−	−	+	+	+
			+	+	+	+	+	+	+
				+	−	+	+	+	+
					−	−	+	−	+
						+	+	+	+
							+	+	+
								−	−
									+

Since there are 33 + and 12 − values, S of Equation 7.3 is 21. Equation 7.6 yields the following sample value of τ:

$$\tau = \frac{2(21)}{10(9)} = 0.467$$

Since the sample size is ten, critical values are obtained from tables, with the following tabular summary of the decision for a one-tailed test:

Level of Significance	Critical τ	Decision
0.05	0.422	Reject H_0
0.025	0.511	Accept H_0
0.01	0.600	Accept H_0
0.005	0.644	Accept H_0

Thus, for a 5% level the null hypothesis is rejected, which suggests that the data contain a trend. At smaller levels of significance, the test would not suggest a trend in the sequence.

Example 7.3

The 50-year annual maximum flood record for the Northwest Branch of the Anacostia River watershed (Figure 2.1) was analyzed for trend. Since the record length is greater than 30, the normal approximation of Equation 7.5 is used:

$$z = \frac{459}{119.54} = 3.83 \tag{7.7}$$

Because the Northwest Branch of the Anacostia River has undergone urbanization, the one-sided alternative hypothesis for a direct trend is studied. Critical values of z for 5% and 0.1% levels of significance are 1.645 and 3.09, respectively. Thus, the computed value of 3.83 is significant, and the null hypothesis is rejected. The test suggests that the flood series reflects an increasing trend that we may infer resulted from urban development within the watershed.

Example 7.4

The two 24-year, annual-maximum flood series in Table 7.1 for Pond Creek and the North Fork of the Nolin River were analyzed for trend. The two adjacent watersheds have the same meteorological conditions. Since the sample sizes are below 30, Equation 7.6 will be used for the tests. S is 150 for Pond Creek and 30 for Nolin River. Therefore, the computed τ for Pond Creek is:

$$\tau = \frac{(150)2}{24(24-1)} = 0.543$$

For Nolin River, the computed τ is:

$$\tau = \frac{(30)2}{24(24-1)} = 0.109$$

For levels of significance of 5, 2.5, 1, and 0.5%, the critical values are 0.239, 0.287, 0.337, and 0.372, respectively. Thus, even at a level of significance of 0.5%, the null hypotheses would be rejected for Pond Creek. For Nolin River, the null hypothesis must be accepted at a 5% level of significance. The results show that the Pond Creek series is nonhomogeneous, which may have resulted from the trend in urbanization. Since the computed τ of 0.109 is much less than any critical values, the series for the North Fork of the Nolin River does not contain a trend.

7.3.1 Rationale of Kendall Statistic

The random variable S is used for both the Kendall τ of Equation 7.6 and the normal approximation of Equation 7.5. If a sequence consists of alternating high-flow and low-flow values, the summation of Equation 7.3 would be the sum of alternating + 1 and −1 values, such as for deforestation, which would yield a near-zero value for S. Such a sequence is considered random so the null hypothesis should be accepted for a near-zero value. Conversely, if the sequence consisted of a series of increasingly larger flows, such as for deforestation, which would indicate a direct trend, then each Θ of Equation 7.3 would be +1, so S would be a large value. If the flows showed an inverse trend, such as for afforestation, then the summation of Equation 7.3 would consist of values of −1, so S would be a large negative value. The denominator of Equation 7.6 is the maximum possible number for a sequence of n flows, so the ratio of S to $n(n - 1)$ will vary from −1, for an inverse trend to +1 for a direct trend. A value of zero indicates the absence of a trend (i.e., randomness).

For the normal approximation of Equation 7.5, the z statistic has the form of the standard normal transformation equation: $z = (x - \bar{x})/s$, where $\bar{x}$ the mean and s is the standard deviation. For Equation 7.5, S is the random variable, a mean of zero is inherent in the null hypothesis of randomness, and the denominator is the standard deviation of S. Thus, the null hypothesis of the Kendall test is accepted for values of z that are not significantly different from zero.

The Kendall test statistic depends on the difference in magnitude between every pair of values in the series, not just adjacent values. For a series in which an abrupt watershed change occurred, there will be more changes in sign of the $(x_j - x_k)$ value of Equation 7.3, which will lead to a value of S that is relatively close to zero. This is especially true if the abrupt change is near one of the ends of the flood series. For a gradual watershed change, a greater number of positive values of $(x_j - x_k)$ will occur. Thus, the test will suggest a trend. In summary, the Kendall test may detect watershed changes due to either gradual trends or abrupt events. However, it appears to be more sensitive to changes that result from gradually changing trends.

7.4 PEARSON TEST FOR SERIAL INDEPENDENCE

If a watershed change, such as urbanization, introduces a systematic variation into a flood record, the values in the series will exhibit a measure of serial correlation. For example, if the percentage of imperviousness gradually increases over all or a major part of the flood record, then the increase in the peak floods that results from the higher imperviousness will introduce a measure of correlation between adjacent flood peaks. This correlation violates the assumption of independence and stationarity that is required for frequency analysis.

The serial correlation coefficient is a measure of common variation between adjacent values in a time series. In this sense, serial correlation, or autocorrelation, is a univariate statistic, whereas a correlation coefficient is generally associated with the relationship between two variables. The computational objective of a correlation analysis is to determine the degree of correlation in adjacent values of a time or space series and to test the significance of the correlation. The nonstationarity of an annual flood series as caused by watershed changes is the most likely hydrologic reason for the testing of serial correlation. In this sense, the tests for serial correlation are used to detect nonstationarity and nonhomogeneity. Serial correlation in a data set does not necessarily imply nonhomogeneity.

The Pearson correlation coefficient (McNemar, 1969; Mendenhall and Sincich, 1992) can be used to measure the association between adjacent values in an ordered sequence of data. For example, in assessing the effect of watershed change on an annual flood series, the correlation would be between values for adjacent years in a sequential record. The correlation coefficient could be computed for either the measured flows or their logarithms but the use of logarithms is recommended when analyzing annual maximum flood records. The two values will differ, but the difference is usually not substantial except when the sample size is small. The hypotheses for the Pearson serial independence test are:

H_0: $\rho = 0$
H_A: $\rho \neq 0$

in which ρ is the serial correlation coefficient of the population. If appropriate for a particular problem, a one-tailed alternative hypothesis can be used, either $\rho > 0$ or $\rho < 0$. As an example in the application of the test to annual maximum flood data, the hypotheses would be:

H_0: The logarithms of the annual maximum peak discharges represent a sequence of n independent events.
H_A: The logarithms of the annual maximum peak discharges are not serially independent and show a positive association.

The alternative hypothesis is stated as a one-tailed test in that the direction of the serial correlation is specified. The one-tailed alternative is used almost exclusively in serial correlation analysis.

Given a sequence of measurements on the random variable x_i (for $i = 1, 2, \ldots, n$), the statistic for testing the significance of a Pearson R is:

$$t = \frac{R}{[(1 - R^2)/(n-3)]^{0.5}} \tag{7.8}$$

where n is the sample size, t (sometimes called Student's t) is the value of a statistic that has $(n - 3)$ degrees of freedom, and R is the value of a random variable computed as follows:

$$R = \frac{\sum_{i=1}^{n-1} x_i x_{i+1} - \left(\sum_{i=1}^{n-1} x_i\right)\left(\sum_{i=2}^{n} x_i\right) \Big/ (n-1)}{\left(\sum_{i=1}^{n-1} x_i^2 - \left(\sum_{i=1}^{n-1} x_i\right)^2 \Big/ (n-1)\right)^{0.5} \left(\sum_{i=2}^{n} x_i^2 - \left(\sum_{i=2}^{n} x_i\right)^2 \Big/ (n-1)\right)^{0.5}} \tag{7.9}$$

Note that for a data sequence of n values, only $n - 1$ pairs are used to compute the value of R. For a given level of significance α and a one-tailed alternative hypothesis, the null hypothesis should be rejected if the computed t is greater than $t_{v,\alpha}$ where $v = n - 3$, the degrees of freedom. Values of $t_{v,\alpha}$ can be obtained from Table A.2. For a two-tailed test, $t_{\alpha/2}$ is used rather than t_α. For serial correlation analysis, the one-tailed positive correlation is generally tested. Rejection of the null hypothesis would imply that the measurements of the random variable are not independent. The serial correlation coefficient will be positive for both an increasing trend and a decreasing trend. When the Pearson correlation coefficient is applied to bivariate data, the slope of the relationship between the two random variables determines the sign on the correlation coefficient. In serial correlation analysis of a single data sequence, only the one-sided upper test is generally meaningful.

Example 7.5

To demonstrate the computation of the Pearson R for data sequences that include dominant trends, consider the annual maximum flows for two adjacent watersheds, one undergoing deforestation (A), which introduces an increasing trend, and one undergoing afforestation (B), which introduces a decreasing trend. The two data sets are given in Table 7.2.

TABLE 7.2
Computation of Pearson *R* for Increasing Trend (A) and Decreasing Trend (B)

Year of Record	Flow A_i	Offset A_{i+1}	Product $A_i A_{i+1}$	A_i^2	A_{i+1}^2	Flow B_i	Offset B_{i+1}	Product $B_i B_{i+1}$	B_i^2	B_{i+1}^2
1	12	14	168	144	196	17	14	238	289	196
2	14	17	238	196	289	14	10	140	196	100
3	17	22	374	289	484	10	13	130	100	169
4	22	25	550	484	625	13	11	143	169	121
5	25	27	675	625	729	11	8	88	121	64
6	27	31	837	729	961	8	8	64	64	64
7	31	—	—	—	—	8	—	—	—	—
Totals	117	136	2842	2467	3284	73	64	803	939	714

The Pearson *R* for the increasing trend is:

$$R_A = \frac{2842 - 117(136)/6}{(2467 - 117^2/6)^{0.5}(3284 - 136^2/6)^{0.5}} = 0.983$$

The Pearson *R* for the decreasing trend is:

$$R_B = \frac{803 - 73(64)/6}{(939 - (73)^2/6)^{0.5}(714 - (64)^2/6)^{0.5}} = 0.610$$

Both are positive values because the sign of a serial correlation coefficient does not reflect the slope of the trend. The serial correlation for sequence *A* is higher than that for *B* because it is a continuously increasing trend, whereas the data for *B* includes a rise in the third year of the record.

Using Equation 7.8, the computed values of the test statistic are:

$$t_A = \frac{0.983}{((1 - 0.983^2)/(7-3))^{0.5}} = 10.71$$

$$t_B = \frac{0.610}{((1 - 0.610^2)/(7-3))^{0.5}} = 1.540$$

For a sample size of 7, 4 is the number of degrees of freedom for both tests. Therefore, the critical *t* for 4 degrees of freedom and a level of significance of 5% is 2.132. The trend causes a significant serial correlation in sequence *A*. The trend in series *B* is not sufficiently dominant to conclude that the trend is significant.

Example 7.6

The Pearson *R* was computed using the 24-year annual maximum series for the Pond Creek and North Fork of the Nolin River watersheds (Table 7.1). For Pond Creek, the sample correlation for the logarithms of flow is 0.72 and the computed *t* is 4.754

according to Equation 7.8. For 21 degrees of freedom and a level of significance of 0.01, the critical t value is 2.581, implying that the computed R value is statistically significantly different from zero. For the North Fork, the sample correlation is 0.065 and the computed t is 0.298 according to Equation 7.8. This t value is not statistically significantly different from zero even at a significance level of 0.60.

Example 7.7

Using the 50-year record for the Anacostia River (see Figure 2.1), the Pearson R was computed for the logarithms of the annual series ($R = 0.488$). From Equation 7.8, the computed t value is 3.833. For 47 degrees of freedom, the critical t value for a one-tailed test would be 2.41 for a level of significance of 0.01. Thus, the R value is statistically significant at the 1% level. These results indicate that the increasing upward trend in flows for the Anacostia River has caused a significant correlation between the logarithms of the annual peak discharges.

7.5 SPEARMAN TEST FOR TREND

The Spearman correlation coefficient (R_S) (Siegel, 1956) is a nonparametric alternative to the Pearson R, which is a parametric test. Unlike the Pearson R test, it is not necessary to make a log transform of the values in a sequence since the ranks of the logarithms would be the same as the ranks for the untransformed data. The hypotheses for a direct trend (one-sided) are:

H_0: The values of the series represent a sequence of n independent events.
H_A: The values show a positive correlation.

Neither the two-tailed alternative nor the one-tailed alternative for negative correlation is appropriate for watershed change.

The Spearman test for trend uses two arrays, one for the criterion variable and one for an independent variable. For example, if the problem were to assess the effect of urbanization on flood peaks, the annual flood series would be the criterion variable array and a series that represents a measure of the watershed change would be the independent variable. The latter might include the fraction of forest cover for afforestation or deforestation or the percentage of imperviousness for urbanization of a watershed. Representing the two series as x_i and y_i, the rank of each item within each series separately is determined, with a rank of 1 for the smallest value and a rank of n for the largest value. The ranks are represented by r_{xi} and r_{yi}, with the i corresponding to the ith magnitude.

Using the ranks for the paired values r_{xi} and r_{yi}, the value of the Spearman coefficient R_S is computed using:

$$R_s = 1 - \frac{6\sum_{i=1}^{N}(r_{xi} - r_{yi})^2}{n^3 - n} \tag{7.10}$$

For sample sizes greater than ten, the following statistic can be used to test the above hypotheses:

$$t = \frac{R_s}{\left[\left(1 - R_s^2\right)/(n-2)\right]^{0.5}} \tag{7.11}$$

where t follows a Student's t distribution with $n - 2$ degrees of freedom. For a one-sided test for a direct trend, the null hypothesis is rejected when the computed t is greater than the critical t_α for $n - 2$ degrees of freedom.

To test for trend, the Spearman coefficient is determined by Equation 7.10 and the test applies Equation 7.11. The Spearman coefficient and the test statistic are based on the Pearson coefficient that assumes that the values are from a circular, normal, stationary time series (Haan, 1977). The transformation from measurements on a continuous scale to ordinal scale (i.e., ranks) eliminates the sensitivity to the normality assumption. The circularity assumption will not be a factor because each flood measurement is transformed to a rank.

Example 7.8

The annual-maximum flood series (1929–1953) for the Rubio Wash is given in column 2 of Table 7.3. The percentage of impervious cover for each year is given in column 4, and the ranks of the two series are provided in columns 3 and 5. The differences in the ranks are squared and summed (column 6). The sum is used with Equation 7.10 to compute the Spearman correlation coefficient:

$$R_s = 1 - \frac{6(1424)}{25^3 - 25} = 0.4523 \tag{7.12}$$

The test statistic of Equation 7.11 is:

$$t = \frac{0.4523}{\left(\dfrac{1-(0.4523)^2}{25-2}\right)^{0.5}} = 2.432 \tag{7.13}$$

which has 23 degrees of freedom. For a one-tailed test, the critical t values (t_α) from Table A.2 and the resulting decisions are:

α %	t_α	Decision
10	1.319	Reject H_0
5	1.714	Reject H_0
2.5	2.069	Reject H_0
1	2.500	Accept H_0

Thus, for a 5% level, the trend is significant. It is not significant at the 1% level.

TABLE 7.3
Application of Spearman Test for Trend in Annual Flood Series of Rubio Wash (1929–1953)

Water Year	Discharge (cfs)	Rank of Discharge, r_q	Imperviousness (%)	Rank of Imperviousness Area, r_i	$(r_q - r_i)^2$
1929	661	2	18.0	1	1
1930	1690	12	18.5	2	100
1931	798	3	19.0	3	0
1932	1510	9	19.5	4	25
1933	2070	17	20.0	5	144
1934	1680	11	20.5	6	25
1935	1370	8	21.0	7	1
1936	1180	6	22.0	8	4
1937	2400	22	23.0	9	169
1938	1720	13	25.0	10	9
1939	1000	4	26.0	11	49
1940	1940	16	28.0	12	16
1941	1200	7	29.0	13	36
1942	2780	24	30.0	14	100
1943	1930	15	31.0	15	0
1944	1780	14	33.0	16	4
1945	1630	10	33.5	17	49
1946	2650	23	34.0	18	25
1947	2090	18	35.0	19	1
1948	530	1	36.0	20	361
1949	1060	5	37.0	21	256
1950	2290	20	37.5	22	4
1951	3020	25	38.0	23	4
1952	2200	19	38.5	24	25
1953	2310	21	39.0	25	16
					Sum = 1424

7.5.1 Rationale for Spearman Test

The Spearman test is more likely to detect a trend in a series that includes gradual variation due to watershed change than in a series that includes an abrupt change. For an abrupt change, the two partial series will likely have small differences (d_i) because it will reflect only the random variation in the series. For a gradual change, both the systematic and random variation are present throughout the series, which results in larger differences (d_i). Thus, it is more appropriate to use the Spearman serial correlation coefficient for hydrologic series where a gradual trend has been introduced by watershed change than where the change occurs over a short part of the flood record.

7.6 SPEARMAN–CONLEY TEST

Recommendations have been made to use the Spearman R_s as bivariate correlation by inserting the ordinal integer as a second variable. Thus, the x_i values would be the sequential values of the random variable and the values of i from 1 to n would be the second variable. This is incorrect because the integer values of i are not truly values of a random variable and the critical values are not appropriate for the test. The Spearman–Conley test (Conley and McCuen, 1997) enables the Spearman statistic to be used where values of the independent variable are not available.

In many cases, the record for the land-use-change variable is incomplete. Typically, records of imperviousness are sporadic, for example, aerial photographs taken on an irregular basis. They may not be available on a year-to-year basis. Where a complete record of the land use change variable is not available and interpolation will not yield accurate estimates of land use, the Spearman test cannot be used.

The Spearman–Conley test is an alternative that can be used to test for serial correlation where the values of the independent variable are incomplete. The Spearman–Conley test is univariate in that only values of the criterion variable are used. The steps for applying it are as follows:

1. State the hypotheses. For this test, the hypotheses are:
   H_0: The sequential values of the random variable are serially independent.
   H_A: Adjacent values of the random variable are serially correlated.
   As an example for the case of a temporal sequence of annual maximum discharges, the following hypotheses would be appropriate:
   H_0: The annual flood peaks are serially independent.
   H_A: Adjacent values of the annual flood series are correlated.
   For a flood series suspected of being influenced by urbanization, the alternative hypothesis could be expressed as a one-tailed test with an indication of positive correlation. Significant urbanization would cause the peaks to increase, which would produce a positive correlation coefficient. Similarly, afforestation would likely reduce the flood peaks over time, so a one-sided test for negative serial correlation would be expected.
2. Specify the test statistic. Equation 7.10 can also be used as the test statistic for the Spearman–Conley test. However, it will be denoted as R_{sc}. In applying it, the value of n is the number of pairs, which is 1 less than the number of annual maximum flood magnitudes in the record. To compute the value of R_{sc}, a second series X_t is formed, where $X_t = Y_{t-1}$. To compute the value of R_{sc}, rank the values of the two series in the same manner as for the Spearman test and use Equation 7.10 to compute the value of R_{sc}.
3. Set the level of significance. Again, this is usually set by convention, typically 5%.
4. Compute the sample value of the test statistic. The sample value of R_{sc} is computed using the following steps:
   (a) Create a second series of flood magnitudes (X_t) by offsetting the actual series (Y_{t-1}).

(b) While keeping the two series in chronological order, identify the rank of each event in each series using a rank of 1 for the smallest value and successively larger ranks for events in increasing order.
(c) Compute the difference in ranks for each pair.
(d) Compute the value of the Spearman–Conley test statistic R_{sc} using Equation 7.10.

5. Obtain the critical value of the test statistic. Unlike the Spearman test, the distribution of R_{sc} is not symmetric and is different from that of R_s. Table A.7 gives the critical values for the upper and lower tails. Enter Table A.7 with the number of pairs of values used to compute R_{sc}.
6. Make a decision. For a one-tailed test, reject the null hypothesis if the computed R_{sc} is greater than the value of Table 7.3. If the null hypothesis is rejected, one can conclude that the annual maximum floods are serially correlated. The hydrologic engineer can then conclude that the correlation reflects the effect of urbanization.

Example 7.9

To demonstrate the Spearman–Conley test, the annual flood series of Compton Creek is used. However, the test will be made with the assumption that estimates of imperviousness (column 3 of Table 7.4) are not available.

Column 7 of Table 7.4 chronologically lists the flood series, except for the 1949 event. The offset values appear in column 8. While the record includes nine floods, only eight pairs are listed in columns 7 and 8. One value is lost because the record must be offset. Thus, $n = 8$ for this test. The ranks are given in columns 9 and 10, and the difference in ranks in column 11. The sum of squares of the d_i values equals 44. Thus, the computed value of the test statistic is:

$$R_{sc} = 1 - \frac{6(44)}{8^3 - 8} = 0.476 \tag{7.14}$$

For a 5% level of significance and a one-tailed upper test, the critical value (Table A.7) is 0.464. Therefore, the null hypothesis can be rejected. The values in the flood series are serially correlated. The serial correlation is assumed to be the result of urbanization.

7.7 COX–STUART TEST FOR TREND

The Cox–Stuart test is useful for detecting positively or negatively sloping gradual trends in a sequence of independent measurements on a single random variable. The null hypothesis is that no trend exists. One of three alternative hypotheses are possible: (1) an upward or downward trend exists; (2) an upward trend exists; or (3) a downward trend exists. Alternatives (2) and (3) indicate that the direction of the trend is known *a priori*. If the null hypothesis is accepted, the result indicates that the measurements within the ordered sequence are identically distributed. The test is conducted as follows:

TABLE 7.4
Spearman and Spearman–Conley Tests of Compton Creek Flood Record

(1) Year	(2) Annual Maximum Discharge Y_i	(3) Average Imperviousness x_i (%)	(4) Rank of Y_i r_{yi}	(5) Rank of X_i r_{xi}	(6) Difference $d_i = r_{yi} - r_{xi}$
1949	425	40	1	1	0
1950	900	42	3	2	1
1951	700	44	2	3	−1
1952	1250	45	7	4	3
1953	925	47	4	5	−1
1954	1200	48	6	6	0
1955	950	49	5	7	−2
1956	1325	51	8	8	0
1957	1950	52	9	9	0
					$\Sigma d_i^2 = 16$

(7) Annual Maximum Discharge Y_t (cfs)	(8) x_t = Offset Y_t (cfs)	(9) Rank of Y_i r_{yi}	(10) Rank of X_i r_{xi}	(11) Difference $d_i = r_{yi} - r_{xi}$
900	425	2	1	1
700	900	1	3	−2
1250	700	6	2	4
925	1250	3	7	−4
1200	925	5	4	1
950	1200	4	6	−2
1325	950	7	5	2
1950	1325	8	8	0
				$\Sigma d_i^2 = 44$

1. N measurements are recorded sequentially ($i = 1, 2, \ldots, N$) in an order relevant to the intent of the test, such as with respect to the time that the measurements were made or their ordered locations along some axis.
2. Data are paired by dividing the sequence into two parts so that x_i is paired with x_j where $j = i + (N/2)$ if N is even and $j = i + 0.5\,(N + 1)$ if N is odd. This produces n pairs of values. If N is an odd integer, the middle value is not used.
3. For each pair, denote the case where $x_j > x_i$ as a +, where $x_j < x_i$ as a −, and where $x_j = x_i$ as a 0. If any pair produces a zero, n is reduced to the sum of the number of + and − values.
4. The value of the test statistic is the number of + signs.
5. If the null hypothesis is true, a sequence is expected to have the same number of + and − values. The assumptions of a binomial variate apply, so the rejection probability can be computed for the binomial distribution with $p = 0.5$. If an increasing trend is specified in the alternative hypothesis,

TABLE 7.5
Cox–Stuart Test of Trends in Baseflow

Month	Discharge at Station 1		Symbol	Discharge at Station 2		Symbol
	Year 1	Year 2		Year 1	Year 2	
January	12.2	13.3	+	9.8	9.3	–
February	13.4	15.1	+	10.1	9.4	–
March	14.2	14.8	+	10.9	9.8	–
April	13.9	14.3	+	10.8	10.2	–
May	11.8	12.1	+	10.3	9.9	–
June	10.3	9.7	–	9.4	9.5	+
July	8.9	8.6	–	8.7	9.2	+
August	8.3	7.9	–	8.5	9.1	+
September	8.5	8.6	+	9.1	9.0	–
October	9.1	9.4	+	9.4	9.2	–
November	10.2	11.6	+	9.7	9.2	–
December	11.7	12.5	+	9.8	9.3	–
			9+			3+
			3–			9–

then rejection of H_0 would occur for a large number of + values; thus, the rejection region is in the upper tail. If a decreasing trend is expected, then a large number of – values are expected, so the region of rejection is in the lower tail. If the number of + signs is small, then the region of rejection is in the lower tail. For a two-sided alternative, regions in both tails should be considered.

Example 7.10

Consider the need to detect a trend in baseflow discharges. Table 7.5 shows hypothetical data representing baseflow for two stations, one where an increasing trend is suspected and one where a decreasing trend is suspected. The data include a cyclical trend related to the annual cycle of rainfall. The intent is to examine the data independent of the periodic trend for a trend with respect to the year.

At station 1, the baseflows for year 2 are compared with those for year 1. A + indicates that flow in the second year is larger than that for the first year. For the 12 comparisons, 9 + symbols occur. Since the objective was to search for an increasing trend, a large number of + signs indicates that. Because a large number of + symbols is appropriate when testing for an increasing trend, the upper portion of the binomial distribution is used to find rejection probability. For $n = 12$ and $p = 0.5$, the cumulative function follows.

x	0	1	2	3	4	8	9	10	11	12
$F(x)$	0.0002	0.0032	0.0193	0.0730	0.1938	0.9270	0.9807	0.9968	0.9998	1.0000

The probability of nine or more + symbols is $1 - F(8) = 0.0730$. For a 5% level of significance, the null hypothesis would be accepted, with results suggesting that the data do not show an increasing trend. The decision would be to reject the null hypothesis of no trend, if a 10% level of significance was used.

For station 2, interest is in detecting a decreasing trend. The data show 3 + symbols and 9 – symbols. The computed value of the test statistic is 3. The above binomial probabilities also apply, but now the lower portion of the distribution is of interest. If all symbols were –, then it would be obvious that a decreasing trend was part of the sequence. The probability of 3 or fewer + symbols is 0.0730. For a 5% level of significance, the null hypothesis would be accepted; the trend is not strong enough to suggest significance.

Example 7.11

The annual maximum series for the Elizabeth River watershed at Elizabeth, New Jersey, for 1924 to 1988 is analyzed using the Cox–Stuart test (see Figure 7.2). The watershed experienced urban growth over a large part of the period of record. The nonstationary series is expected to include an increasing trend. The 65-year record is divided into two parts: 1924–1955 and 1957–1988. The 1956 flow is omitted in order to apply two sequences of equal length.

The trend is one of increasing discharge rates. Therefore, the test statistic is the number of + values, and the rejection probability would be from the upper part of the cumulative binomial distribution, with $p = 0.5$ and $n = 32$. The data in Table 7.6 include 26 positive symbols. Therefore, the rejection probability is $1 - F(25) = 0.0002675$. Since this is exceptionally small, the null hypothesis can be rejected, which indicates that the trend is statistically significant even at very small levels of significance.

7.8 NOETHER'S BINOMIAL TEST FOR CYCLICAL TREND

Hydrologic data often involve annual or semiannual cycles. Such systematic variation may need to be considered in modeling the processes that generate such data. Monthly temperature obviously has an underlying cyclical nature. In some climates, such as southern Florida and southern California, rainfall shows a dominant annual variation that may approximate a cycle. In other regions, such as the Middle Atlantic states, rainfall is more uniform over the course of a year. In mountainous environments, snow accumulation and depletion exhibit systematic annual variation that generally cannot be represented as a periodic function since it flattens out at zero for more than half of the year.

The detection of cyclical variation in data and the strength of any cycle detected can be an important step in formulating a model of hydrologic processes. A basic sinusoidal model could be used as the framework for modeling monthly temperature. Since monthly rainfall may show systematic variation over a year that could not

TABLE 7.6
Cox–Stuart Test of Annual Maximum Series for Elizabeth River

1924–1955	1957–1988	Symbol	No. of + Symbols	Cumulative Probability
1280	795	−		
980	1760	+	20	0.9449079
741	806	+	21	0.9749488
1630	1190	−	22	0.9899692
829	952	+	23	0.9964998
903	1670	+	24	0.9989488
418	824	+	25	0.9997325
549	702	+	26	0.9999435
686	1490	+	27	0.9999904
1320	1600	+	28	0.9999987
850	800	−	29	0.9999999
614	3330	+	30	1.0000000
1720	1540	−	31	1.0000000
1060	2130	+	32	1.0000000
1680	3770	+		
760	2240	+		
1380	3210	+		
1030	2940	+		
820	2720	+		
1020	2440	+		
998	3130	+		
3500	4500	+		
1100	2890	+		
1010	2470	+		
830	1900	+		
1030	1980	+		
452	2550	+		
2530	3350	+		
1740	2120	+		
1860	1850	−		
1270	2320	+		
2200	1630	−		

realistically be represented by a sinusoid, a composite periodic model requiring one or more additional empirical coefficients may be necessary to model such processes.

An important step in modeling of such data is the detection of the periodicity. Moving-average filtering (Section 2.3) can be applied to reveal such systematic variation, but it may be necessary to test whether cyclical or periodic variation is statistically significant. Apparent periodic variation may be suggested by graphical or moving-average filtering that may not be conclusive. In such cases, a statistical test may be warranted.

7.8.1 Background

Cyclical and periodic functions are characterized by periods of rise and fall. At the zeniths and nadirs of the cycles, directional changes occur. At all other times in the sequence, the data values will show increasing or decreasing trends, with both directions equally likely over the duration of each cycle.

Consider a sequence X_t ($t = 1, 2, \ldots, N$) in which a dominant periodic or cyclical trend may or may not be imbedded. If we divide the sequence into $N/3$ sections of three measurements, the following two runs would suggest, respectively, an increasing and a decreasing systematic trend: $(X_t < X_{t+1} < X_{t+2})$ and $(X_t > X_{t+1} > X_{t+2})$. They are referred to as *trend sequences*. If the time or space series did not include a dominant periodic or cyclical trend, then one of the following four sequences would be expected: $(X_t < X_{t+1}, X_t < X_{t+2}, X_{t+1} > X_{t+2})$; $(X_t > X_{t+1}, X_t < X_{t+2}, X_{t+1} < X_{t+2})$, $(X_t < X_{t+1}, X_t > X_{t+2}, X_{t+1} > X_{t+2})$, and $(X_t > X_{t+1}, X_t > X_{t+2}, X_{t+1} < X_{t+2})$. These four are referred to as *nontrend sequences*. A time series with a dominant periodic or cyclical trend could then be expected to have a greater number than expected of the two increasing or decreasing trend sequences. A trend can be considered significant if the number of trend sequences exceeds the number expected in a random sequence.

The above sequences are identical to sequences of a binomial variate for a sample of three. If the three values (i.e., X_t, X_{t+1}, and X_{t+2}) are converted to ranks, the six alternatives are (1, 2, 3), (1, 3, 2), (2, 1, 3), (2, 3, 1), (3, 1, 2), and (3, 2, 1). The first and last of the six are the trend sequences because they suggest a trend; the other four indicate a lack of trend. Therefore, a cyclical trend is expected to be present in a time series if the proportion of sequences of three includes significantly more than a third of the increasing or decreasing trend sequences.

7.8.2 Test Procedure

The Noether test for cyclical trend can be represented by the six steps of a hypothesis test. The hypotheses to be tested are:

H_0: The sequence does not include a periodic or cyclical trend.
H_A: The sequence includes a dominant periodic or cyclical trend.

The sample value of the test statistic is computed by separating the sequence of N values into $N/3$ sequential runs of three and counting the number of sequences in one of the two trend sequences (denoted as n_S). The total of three-value sequences in the sample is denoted as n. Since two of the six sequences indicate a cyclical trend, the probability of a trend sequence is one-third. Since only two outcomes are possible, and if the sequences are assumed independent, which is reasonable under the null hypothesis, the number of trend sequences will be binomially distributed with a probability of one-third. Since the alternative hypothesis suggests a one-tailed test, the null hypothesis is rejected if:

$$\sum_{i=n_s}^{n} \binom{n}{i} \left(\frac{1}{3}\right)^i \left(\frac{2}{3}\right)^{n-i} \le \alpha \tag{7.15}$$

where α is the level of significance. The one-sided alternative is used because the alternative hypothesis assumes that, given a direction between the first two values in the cycle sequence, the sequence tends to remain unchanged.

7.8.3 Normal Approximation

Many time series have sample sizes that are much longer than would be practical to compute the probability of Equation 7.15. For large sample sizes, the binomial probability of Equation 7.15 can be estimated via standard normal transformation. The mean (μ) and standard deviation (σ) of the binomial distribution are $\mu = np$ and $\sigma = (np(1-p))^{0.5}$. Thus, for a series with n_s trend sequences, n_s can be transformed to a z statistic:

$$z = \frac{n_s - np - 0.5}{(np(1-p))^{0.5}} \tag{7.16}$$

where z is a standard normal deviate. The subtraction of 0.5 in the numerator is a continuity correction required because the binomial distribution is discrete and the normal distribution is continuous. The rejection probability of Equation 7.16 can then be approximated by:

$$p\left(z > \frac{n_s - np - 0.5}{(np(1-p))^{0.5}}\right) \le \alpha \tag{7.17}$$

Equation 7.17 is generally a valid approximation to the exact binomial probability of Equation 7.15 if both np and $np(1-p)$ are greater than 5. Since p is equal to one-third for this test, then $np = n(1/3) = 5$ would require a sample size of at least 15. The other constraint, $np(1-p)$ gives $n(1/3)(2/3) = 5$ and requires n to be at least 22.5. Thus, the second constraint is limiting, so generally the normal approximation of Equation 7.17 is valid if $n \ge 23$.

Consider the case for a sample size of 20. In this case, the binomial solution of Equation 7.15 would yield the following binomial distribution:

$$\sum_{i=n_s}^{20} \binom{20}{i} \left(\frac{1}{3}\right)^i \left(\frac{2}{3}\right)^{20-i} \tag{7.18}$$

The normal approximation is:

$$p\left(z > \frac{n_s - 20(1/3) - 0.5}{(20(1/3)(2/3))^{0.5}}\right) = p\left(z > \frac{n_s - 7.167}{1.108}\right) \tag{7.19}$$

The following tabular summary shows the differences between the exact binomial probability and the normal approximation.

n_s	14	13	12	11	10
Equation 7.15	0.00088	0.00372	0.01297	0.03764	0.09190
z	3.241	2.767	2.293	1.818	1.344
Equation 7.19	0.0006	0.0028	0.0083	0.00346	0.0895

If n is equal to 30, then the agreement is better.

n_s	19	18	17	16	15	14
Equation 7.15	0.0007	0.0025	0.0072	0.0188	0.0435	0.0898
z	3.292	2.905	2.517	2.130	1.743	1.356
Equation 7.19	0.0005	0.0018	0.0059	0.0166	0.0407	0.0875

For both sample sizes, the normal approximation provides a good approximation. For a sample size of 30, the two yield the same decision (i.e., reject or accept) for levels of significance of 0.0001, 0.005, 0.01, 0.05, and 0.10. For the sample size of 20, the two would agree at all levels of significance except 1%.

Example 7.12

The Noether test was applied to 78 months (March 1944 to August 1950) of runoff depth (in.) for the Chestuee Creek watershed. The record was separated into 26 sequences of 3 months (see Table 7.7). For one sequence, 2 months had the same runoff depth, so it was considered as a two-value tie; that is, since the record only included the one tie, it was not used in computing the rejection probability. Thus, the probability is based on 25 sequences. Twelve of the 25 sequences suggested a trend, so the rejection probability p_α is:

$$p_\alpha = \sum_{i=12}^{25} \binom{25}{i} \left(\frac{1}{3}\right)^i \left(\frac{2}{3}\right)^{25-i} = 0.0918 \tag{7.20}$$

The probability based on a normal approximation is:

$$\hat{p}_\alpha = p\left(z > \frac{12 - 25(1/3) - 0.5}{[25(1/3)(2/3)]^{0.5}}\right) = p(z > 1.344) = 0.0895 \tag{7.21}$$

At a 5% level of significance, the null hypothesis that the cyclical trend is not significant must be accepted. However, it would be considered statistically significant at 10%. This example illustrates the importance of using a rational, physically based criterion for selecting the level of significance.

Since many of the monthly runoff sequences in Table 7.7 exhibit an annual cycle, the monthly rainfall for the Chestuee Creek watershed was subjected to Noether's test. Ties were not present, so the 78-month series included 26 sequences of three

TABLE 7.7
Application of Noether's Test to Monthly Runoff Depth (in.) for Chestuee Creek Watershed, March 1944 to August 1950

5.18	3.47	1.70	*Y*[a]
0.85	0.56	0.51	*Y*
0.63	0.71	0.47	
0.83	1.47	3.39	*Y*
2.31	1.24	1.98	
1.03	0.55	0.49	*Y*
0.30	0.48	0.90	*Y*
2.18	6.94	4.86	
2.78	1.73	2.65	
1.14	0.91	0.86	*Y*
1.31	0.83	1.35	
1.51	8.96	2.82	
2.48	1.55	1.03	*Y*
0.88	0.67	0.51	*Y*
0.50	0.39	0.54	
0.74	1.05	5.72	*Y*
3.52	2.24	0.87	*Y*
0.50	0.58	0.51	
0.32	0.32	3.65	2-TIE
3.96	6.34	3.32	
2.03	2.65	1.70	
0.99	2.70	0.78	
0.82	0.99	1.59	*Y*
1.69	4.73	3.99	
4.95	1.79	1.86	
2.54	1.33	0.95	*Y*

[a] *Y* indicates that the sequence suggests a trend.

(see Table 7.8). Ten of these suggested a cyclical trend, which is too few to suggest that the number could not be the result of randomness. The binomial rejection probability is 0.3571, which is much too high to suggest that randomness is unlikely. Since the rainfall data do not suggest an annual cycle, then the results of Noether's test of the runoff data seems correct.

7.9 DURBIN–WATSON TEST FOR AUTOCORRELATION

Analyses of hydrologic data often assume that the individual measurements are independent and identically distributed. Flood frequency analysis is only one example. The residuals of a regression analysis are assumed to be independent; however, when regression is used to fit the coefficients for modeling spatial or temporal data,

TABLE 7.8
Application of Noether's Test to Monthly Rainfall Depth (in.) for Chestuee Creek Watershed, March 1944 to August 1950

6.91	4.64	3.44	*Y*[a]
2.15	2.30	3.34	*Y*
6.62	1.15	2.46	
4.29	2.64	7.01	
3.05	4.86	5.08	*Y*
3.85	3.84	2.72	*Y*
1.56	4.73	5.17	*Y*
5.84	10.42	5.28	
4.91	4.09	6.66	
3.87	4.29	4.43	*Y*
5.75	3.79	4.03	
4.20	12.63	3.06	
3.28	2.76	4.40	
5.04	3.47	3.20	*Y*
2.92	2.10	4.18	
2.69	4.12	7.77	*Y*
6.58	2.57	2.70	
1.45	5.04	3.29	
2.50	2.88	13.28	*Y*
5.76	8.72	3.60	
4.63	4.44	2.94	*Y*
5.06	7.60	4.05	
3.96	6.87	1.18	
4.19	8.16	4.13	
7.78	1.43	6.98	
6.01	6.39	5.29	
3.84	3.70		

[a] *Y* indicates that the sequence suggests a trend.

the independence assumption of the residuals might be violated. Therefore, before reporting the work, the residuals should be checked for independence to ensure that the underlying assumption has not been violated. The Durbin–Watson test can be used to test for serial correlation in a univariate sequence of values recorded on an interval scale. It can be applied as a one-tailed or two-tailed test.

7.9.1 Test for Positive Autocorrelation

The hypotheses for the Durbin–Watson test for positive autocorrelation are:

$$H_0\text{: } \rho = 0 \tag{7.22a}$$

$$H_A\text{: } \rho > 0 \text{ (autocorrelation is positive)} \tag{7.22b}$$

where ρ is the autocorrelation coefficient for the model

$$\varepsilon_i = \rho\varepsilon_{i-1} + a_i \tag{7.23}$$

where ε_i is the residual for observation i. The test statistic is:

$$D = \frac{\sum_{i=2}^{n} (e_i - e_{i-1})^2}{\sum_{i=1}^{n} e_i^2} \tag{7.24}$$

in which e_i is the ith sample error. Critical values are a function of the sample size, the level of significance, and the number of predictor variables in the model. Because only approximate values are available for the D statistic, the test may not lead to a conclusive decision. Specifically, the decision criteria are:

$$\text{if } D \geq D_{\alpha,U}, \text{ do not reject } H_0 \tag{7.25a}$$

$$\text{if } D_{\alpha,L} \geq D, \text{ reject } H_0 \tag{7.25b}$$

$$\text{if } D_{\alpha,L} < D < D_{\alpha,U}, \text{ the test is inconclusive} \tag{7.25c}$$

7.9.2 Test for Negative Autocorrelation

To test for negative autocorrelation, the hypotheses are:

$$H_0\text{: } \rho = 0 \tag{7.26a}$$

$$H_A\text{: } \rho < 0 \text{ (autocorrelation is negative)} \tag{7.26b}$$

The computed value of the test statistic (D') is:

$$D' = 4 - D$$

where D is computed using Equation 7.24. The criteria of Equation 7.25 are then applied with D' to make a decision.

7.9.3 Two-Sided Test for Autocorrelation

For a two-sided test for autocorrelation, the hypotheses are:

$$H_0\text{: } \rho = 0 \tag{7.27a}$$

$$H_A\text{: } \rho \neq 0 \tag{7.27b}$$

To conduct the two-sided test, D and D' are computed and compared as discussed for both one-sided tests. If D and D' are greater than $D_{\alpha,U}$, accept the null hypothesis;

if one or the other is less than $D_{\alpha,L}$, reject the null hypothesis. For other possibilities, the test is inconclusive.

Example 7.13

The data matrix consists of the annual peak discharges for the Pond Creek watershed (1945–1968) and the maximum rainfall intensity during the storm that produced the annual peak. The peak discharge was regressed on the rainfall intensity, and the resulting bivariate linear regression model was used to compute the vector of predicted peak discharges (see Table 7.9). The predicted ($\hat{Y}$) and measured (Y) peak discharges were used to compute the residuals ($e = \hat{Y} = Y$), which is a vector of 24 values. Regressing e_i on e_{i-1} resulted in a serial correlation coefficient of 0.312. For the 24-year flood record, the correlation of the residuals is based on 23 pairs of values.

Assume a two-sided Durbin–Watson test is applied to the residuals, with a computed value of 1.01 for the test statistic. For a 10% level of significance, the lower and upper critical values are 1.26 and 1.44, respectively. Since the computed D

TABLE 7.9
Residuals for Regression of Pond Creek Discharge on Rainfall Intensity

	Discharge (cfs)		Residual	Intensity
Year	Predicted	Measured	(cfs)	(in./hr)
1945	2243	2000	243	0.37
1946	2132	1740	392	0.29
1947	2118	1460	658	0.28
1948	2619	2060	559	0.64
1949	2174	1530	644	0.32
1950	3634	1600	2034	1.37
1951	2313	1690	623	0.42
1952	2563	1420	1143	0.60
1953	2048	1330	718	0.23
1954	2118	607	1511	0.28
1955	1993	1380	613	0.19
1956	2243	1660	583	0.37
1957	2118	2290	−172	0.28
1958	2758	2590	168	0.74
1959	2563	3260	−697	0.60
1960	3717	2490	1227	1.43
1961	2813	2080	733	0.78
1962	2507	2520	−13	0.56
1963	2396	3360	−964	0.48
1964	2799	8020	−5221	0.77
1965	2646	4310	−1664	0.66
1966	2382	4380	−1998	0.47
1967	3342	3220	122	1.16
1968	3078	4320	−1242	0.97

is below the lower value, the null hypothesis should be rejected. The assumption of independence underlying regression is violated.

In general, we assume that the annual peak discharges are independent of each other. Thus, the rejection of the null hypothesis of independence needs an explanation. The effects of urban development introduced a trend into the flood series, especially after 1954. This has the effect of producing serially correlated discharges and, therefore, the residuals of the regression equation. The rejection of the null hypothesis of the Durbin–Watson test reflects the presence of the trend introduced by urban development.

7.10 EQUALITY OF TWO CORRELATION COEFFICIENTS

Numerous situations in hydrologic analysis require comparison of two correlation coefficients. For example, it may be of interest to compare serial correlation before and after a watershed change. In verifying a model, it may be of interest to decide whether the correlation coefficient for the verification data was significantly below the correlation coefficient achieved with the calibration data. Situations such as these can be evaluated using a statistical test for the equality of two correlation coefficients. The null and set of alternative hypotheses for making such a test follow:

H_0 : Populations 1 and 2 have the same correlation, $\rho_1 = \rho_2$.
H_{A1}: The correlation for population 1 is less than that of population 2, $\rho_1 < \rho_2$.
H_{A2}: The correlation for population 1 is greater than that of population 2, $\rho_1 > \rho_2$.
H_{A3}: The correlations of population 1 and 2 are not equal, $\rho_1 \neq \rho_2$.

The problem to be solved dictates which of the three alternative hypotheses should be tested. For the case of comparing correlations for calibration and verification data, a one-sided test showing that the correlation for the verification data is lower than that for the calibration data is the appropriate alternative. For the case of the before–after comparison, the two-sided alternative might be appropriate if it suggested that different models should be used in modeling the two periods of record.

The test statistic for performing this test is:

$$\chi^2 = \frac{(z_1 - z_2)^2 (n_1 - 3)(n_2 - 3)}{n_1 + n_2 - 6} \tag{7.28}$$

for $n_1 \geq 4$ and $n_2 \geq 4$. The statistic always uses 1 degree of freedom and the values of z_i are computed by:

$$z_i = \frac{1}{2} \ln_e \left(\frac{1 + r_i}{1 - r_i} \right) \tag{7.29}$$

for $r_i \neq 1$. The sample sizes do not have to be the same. The null hypothesis is rejected if the computed value of χ^2 is greater than the tabled value for level of significance α and 1 degree of freedom. If $n_1 = n_2$, then Equation 7.28 becomes

$$\chi^2 = 0.5(z_1 - z_2)^2(n-3) \tag{7.30}$$

Simulation tests show that the test is a good approximation if the correlations are low or if the sample sizes are greater than ten. For larger correlations, the accuracy decreases. As the sample size decreases below ten or the correlation is above 0.6, the actual level of significance decreases below the value suggested to be used in obtaining the critical value.

A one-tailed test is conducted by specifying which sample value is expected to be smallest and then using a level of significance of one-half of the tabled value. For example, using a critical value of 3.84 would provide a level of significance of 2.5%. For a 5% level of significance and a one-tailed test, the critical value would be 2.71.

Example 7.14

Assume that the calibration of a model using a sample size of 24 yielded a correlation coefficient 0.67. Based on a verification test of 17 measurements, a correlation coefficient of 0.26 was computed. The following hypotheses are tested:

$$H_0\text{: } \rho_{\text{cal}} = \rho_{\text{ver}} \tag{7.31a}$$

$$H_A\text{: } \rho_{\text{cal}} > \rho_{\text{ver}} \tag{7.31b}$$

Rejection of the null hypothesis would suggest that the model provided poorer accuracy with the implication that the model is less generally applicable than the calibration correlation would suggest. The test statistic is computed as follows:

$$z_c = 0.5 \ln_e\left(\frac{1+0.67}{1-0.67}\right) = 0.8107 \tag{7.32}$$

$$z_v = 0.5 \ln_e\left(\frac{1+0.36}{1-0.36}\right) = 0.3769 \tag{7.33}$$

$$\chi^2 = \frac{(0.8107 - 0.3769)^2(24-3)(17-3)}{24+17-3} = 1.456 \tag{7.34}$$

The critical values for a 12.5% and 5% levels of significance are 1.32 and 2.71, respectively. The rejection probability exceeds 10%, so the null hypothesis should be accepted. The difference is not sufficient to justify assuming that the model is not adequately fitted.

Example 7.15

The serial correlation coefficients for 2-week-average baseflow measurements are made for the year before and after the principal channel is cleared. The values are 0.53 and –0.09, respectively. A one-sample test on the two coefficients suggests that the smoothing in the natural channel resulted in a positive serial correlation while the cleared stream was subject to variation that did not reflect serial correlation. A two-sample test needs to be applied to assess whether they can be considered significantly different and that would indicate they should be modeled differently. The two-sample test is applied as follows:

$$z_{pre} = 0.5\ln_e\left(\frac{1+0.53}{1-0.53}\right) = 0.5901 \tag{7.35}$$

$$z_{post} = 0.5\ln_e\left(\frac{1+(-0.09)}{1-(-0.09)}\right) = -0.0902 \tag{7.36}$$

$$\chi^2 = (0.5901-(-0.0902))^2(26-3)/2 = 5.324 \tag{7.37}$$

For a 5% level of significance, the critical value is 3.84 (Table A.3). Therefore, the null hypothesis can be rejected, which indicates that the processes that influenced the serial correlation prior to the stream clearing were changed by the clearing.

7.11 PROBLEMS

7-1 Graphically show the continuity correction of Equation 7.2. Given that U is a discrete random variable, discuss the reason for the continuity correction.

7-2 If $n_1 = n_2 = 10$ and $U = 6$, what is the effect of the continuity correction in Equation 7.2 on the computed rejection probability?

7-3 The data in the table below include Manning's n, the tree density d (i.e., the average number of trees per foot for both sides within the stream reach), and the average channel width (w) for various sections of the Kankakee River in Indiana. Use the runs test to test for randomness of n, d, and w.

River Mile	n	d	w	River Mile	n	d	w
70	0.032	0.10	208	100	0.050	0.21	139
72	0.034	0.11	198	102	0.051	0.20	132
74	0.035	0.10	196	104	0.050	0.16	117
76	0.034	0.10	191	106	0.045	0.20	124
78	0.036	0.09	194	108	0.050	0.26	120
80	0.035	0.11	190	110	0.050	0.25	127
82	0.035	0.12	192	112	0.055	0.30	102
84	0.037	0.14	200	114	0.053	0.26	109

(Continued)

(Continued)

River Mile	*n*	*d*	*w*	River Mile	*n*	*d*	*w*
86	0.039	0.17	172	116	0.053	0.17	111
88	0.040	0.16	177	118	0.060	0.44	110
90	0.039	0.14	165	120	0.055	0.30	112
92	0.040	0.15	162	122	0.050	0.25	102
94	0.041	0.14	153	124	0.048	0.18	98
96	0.043	0.17	131	126	0.055	0.20	101
98	0.046	0.19	142	128	0.050	0.21	100

7-4 For each sequence shown below, use the runs test to determine whether the data are randomly distributed. Use a 5% level of significance.
(a) NFFNFNNNNFFNNFFFNFFNNNFNFNNFFFNFFFNNFNNN-FFNFNNNFFN
(b) NNNNNNNNNNNNNNNNNNNNNFFFFFFFFFFFFFFFFFFF
(c) NFNFNFNFNFNFNFNFNFNFNFNFNFNFNFNFNFNFNFNF

7-5 The data given below are the residuals (e) from a linear bivariate regression analysis and the values of the predictor variables, X. Can one reasonably conclude that the residuals are randomly distributed with respect to the predictor variable X? Use $\alpha = 0.01$.

X	1	2	3	4	5	6	7	8	9	10	11	12	13	14	15	16	17	18	19	20
e	−2	3	0	−5	2	1	3	−2	−1	−4	1	3	−6	2	3	−2	4	−1	−3	−2
X	21	22	23	24	25	26	27	28	29	30	31	32	33	34	35	36	37	38	39	10
e	−4	2	1	−1	2	1	−3	−2	4	2	−1	−2	1	3	2	4	−2	−3	−1	−1

7-6 A manufacturing process involves installation of a component on an assembly line. The number of incorrect installations per half-hour (I) is noted by the quality control section. For the 3 days of data below, can we conclude that no time bias exists in the number of incorrectly installed components? Use a 10% level of significance.

Time period	1	2	3	4	5	6	7	8	9	10	11	12	13	14	15	16
I	2	3	0	1	2	2	4	1	3	0	5	4	3	4	3	6
Time period	17	18	19	20	21	22	23	24	25	26	27	28	29	30	31	32
I	5	6	7	4	8	7	8	5	9	7	9	8	6	8	10	11
Time period	33	34	35	36	37	38	39	40	41	42	43	44	45	46	47	48
I	8	6	7	11	9	8	7	9	6	10	7	12	13	8	10	10

7-7 The following sequence of annual maximum discharges occurred over a 23-year period on a watershed undergoing urbanization. Was the urbanization sufficient to introduce a nonhomogeneous trend into the data? Use the runs test to make a decision. X = {680, 940, 870, 1230, 1050, 1160,

800, 770, 890, 540, 950, 1230, 1460, 1080, 1970, 1550, 940, 880, 1000, 1350, 1610, 1880, 1540}.

7-8 The following sequence is the daily degree-day factor for the Conejos River near Magote, Colorado, for June 1979. Using the runs test, do the values show an increasing trend? $X = \{12.75, 15.37, 17.51, 19.84, 22.04, 23.54, 20.22, 13.18, 8.37, 12.78, 20.57, 25.25, 27.04, 25.34, 24.45, 23.25, 18.83, 16.91, 15.78, 21.40, 23.99, 26.25, 23.95, 24.66, 25.52, 26.23, 29.28, 28.99, 27.81, 25.75\}$.

7-9 The following sequence is the daily mean discharge for the Conejos River near Magote, Colorado, for May 1979. Use the runs test to assess whether the discharge increases during the month. $X = \{590, 675, 520, 640, 620, 730, 820, 780, 640, 506, 448, 416, 392, 469, 980, 810, 920, 980, 1405, 1665, 1630, 1605, 1680, 1965, 1925, 2010, 2200, 2325, 2485, 2670, 2360\}$.

7-10 Develop an example ($n = 12$) that demonstrates the difference in applying the runs test to time series with an imbedded gradual trend versus an imbedded abrupt change.

7-11 Apply Kendall's test to the following data sequence for the alternative hypothesis of an increasing trend: $X = \{4, 7, 1, 2, 5, 8, 3, 6, 9\}$.

7-12 Apply Kendall's test to the following data sequence for the alternative hypothesis of a decreasing trend: $X = \{24, 16, 17, 21, 9, 14, 11, 10, 6, 18, 12, 7\}$.

7-13 Apply Kendall's test to the data of Problem 7-7 for the alternative hypothesis of an increasing trend.

7-14 Apply Kendall's test to the data of Problem 7-8 for the alternative hypothesis of an increasing trend.

7-15 The following sequence is the daily degree-day factor for the Conejos River near Magote, Colorado, for September 1979. Use Kendall's test to assess the likelihood of a decreasing trend. $D = \{25.66, 23.13, 25.07, 27.07, 25.45, 27.16, 27.37, 26.95, 25.54, 28.57, 26.57, 25.49, 18.40, 15.59, 13.12, 17.31, 19.69, 20.81, 19.11, 22.92, 18.40, 18.11, 20.87, 21.16, 22.66, 21.19, 19.99, 20.14, 20.19, 22.23\}$.

7-16 Use Kendall's test to test the values of Manning's n in Problem 7-3 for randomness.

7-17 The following data are the annual maximum discharges for the Fishkill Creek at Beacon, New York (1945 to 1968). Use Kendall's test to check the series for randomness. $Q = \{2290, 1470, 2220, 2970, 3020, 1210, 2490, 3170, 3220, 1760, 8800, 8280, 1310, 2500, 1960, 2140, 4340, 3060, 1780, 1380, 980, 1040, 1580, 3630\}$.

7-18 Develop a set of data ($n = 15$) that is dominated by an episodic event near the middle of the sequence and evaluate the data for randomness using Kendall's test.

7-19 Use Pearson's correlation coefficient to evaluate the degree association between daily degree-day factor D and daily streamflow Q for the Conejos River near Magote, Colorado, for April 1979. Test the significance of the correlation.

Day	D	Q	Day	D	Q	Day	D	Q
1	0.00	60	11	0.00	101	21	10.10	378
2	0.00	55	12	0.00	91	22	11.77	418
3	0.00	51	13	0.00	82	23	15.75	486
4	0.00	53	14	8.51	88	24	12.92	506
5	4.49	57	15	13.01	109	25	13.01	534
6	10.43	60	16	18.61	174	26	10.30	530
7	9.16	79	17	16.63	250	27	10.72	488
8	9.63	98	18	13.37	360	28	11.86	583
9	4.44	128	19	6.22	402	29	8.72	619
10	0.00	116	20	5.95	378	30	6.77	588

7-20 Use the Pearson correlation coefficient to assess the correlation between Manning's n and the tree density d for the data of Problem 7-3. Test the significance of the computed correlation.

7-21 It is acceptable to compute the correlation coefficient between river mile and Manning's n with the data of Problem 7-3? Explain.

7-22 Use the Pearson correlation coefficient to test the serial correlation of the discharges Q of Problem 7-19. Test the significance of the computed serial correlation coefficient.

7-23 Use the Pearson correlation coefficient to assess the serial correlation of the Manning's n values of Problem 7-3. Test the significance of the computed correlation.

7-24 Using the test statistic of Equation 7.8, develop a table for sample sizes from 5 to 20 for the critical value of R for a 5% level of significance and the alternative hypothesis that $p < 0$.

7-25 Discuss the applicability of the Pearson R of Equation 7.9 for time series that involve an episodic event near the middle of the series.

7-26 Use the Spearman test to assess the significance of the correlation between Manning's n and the tree density d with the data of Problem 7-3.

7-27 Using the data of Problem 7-19, test the significance of the Spearman correlation coefficient between the degree-day factor D and the discharge Q.

7-28 The following data show snow-covered area (A) for the lower zone of the Conejos River watershed for May 1979: A = {0.430, 0.411, 0.399, 0.380, 0.360, 0.343, 0.329, 0.310, 0.295, 0.280, 0.263, 0.250, 0.235, 0.215, 0.200, 0.190, 0.173, 0.160, 0.145, 0.130, 0.120, 0.105, 0.095, 0.081, 0.069, 0.059, 0.048, 0.035, 0.022, 0.011, 0.000}. Using the streamflow Q in Problem 7-9, test the significance of the Spearman correlation coefficient between A and Q.

7-29 Test the roughness coefficient (n) data of Problem 7-3 with the Spearman–Conley statistic for an increasing trend with increases in river mile.

7-30 Test the discharge data of Problem 7-9 with the Spearman–Conley statistic for an increasing trend.

7-31 Test the Fishkill Creek discharge data of Problem 7-17 for randomness using the Spearman–Conley test.

7-32 Discuss why the Cox–Stuart test may not be applicable to time or space series that involve an abrupt change.

7-33 Consider a time series with the following deterministic component: $y_d = 3.15 + 0.05t$ for $1 \leq t \leq 20$, with t being an integer. The following standardized series are used to form two random variation components: $z = \{-0.9, 1.2, -0.4, 1.2, -0.3, -0.6, 0.1, 0.3, -1.0, 1.3, -0.7, -0.2, 0.4, -0.8, -0.3, 0.4, 0.7, -0.5, -1.2, 1.1\}$. Two series are formed: $\hat{y}_1 = y_d + zN(0, 0.1)$ and $\hat{y}_2 = y_d + zN(0, 1)$ as follows: $\hat{y}_1 = \{3.11, 3.37, 3.26, 3.47, 3.37, 3.39, 3.51, 3.58, 3.50, 3.78, 3.63, 3.73, 3.84, 3.77, 3.87, 3.99, 4.07, 4.00, 3.98, 4.26\}$ and $\hat{y}_2 = \{2.3, 4.45, 2.90, 4.55, 3.10, 2.85, 3.60, 3.85, 2.60, 4.95, 3.00, 3.55, 4.20, 3.05, 3.60, 4.35, 4.70, 3.55, 2.90, 5.25\}$. Apply the Cox–Stuart test to both data sets to check for increasing trends. Based on the results, discuss the importance of the random variation relative to the systematic variation of the trend in trend detection.

7-34 Apply the Cox–Stuart test to the Manning's roughness (n) data of Problem 7-3 under the alternative hypothesis of a lack of randomness.

7-35 Apply the Cox–Stuart test to the Fishkill Creek data of Problem 7-17 under the alternative hypothesis of a lack of randomness.

7-36 Discuss the application of the Cox–Stuart test for periodic data sequences.

7-37 Apply the Cox–Stuart test to the following daily streamflow data for the Conejos River near Magote, Colorado, for June 1979. $Q = \{2210, 1890, 2095, 2240, 2230, 2485, 2875, 2940, 2395, 1690, 2540, 1785, 2205, 2650, 2845, 3080, 2765, 2440, 2070, 1630, 1670, 1860, 2080, 2175, 2055, 2090, 2185, 2255, 2210, 2260\}$. Test for randomness.

7-38 Create a table that gives the rejection probability for $5 \leq n \leq 10$ for the Noether test.

7-39 Evaluate the effects of the continuity corrections of Equations 7.16 and 7.17.

7-40 The following are the monthly total precipitation depths (P, in.) for the Wildcat Creek near Lawrenceville, Georgia, for October 1953 through September 1956. Test for a trend using the Noether test. $P = \{1.25, 1.74, 8.12, 6.43, 2.22, 4.22, 3.06, 2.50, 4.27, 2.63, 1.62, 0.51, 0.44, 3.23, 3.54, 5.01, 5.91, 2.50, 4.27, 2.55, 2.92, 3.61, 2.14, 1.05, 1.29, 3.92, 1.29, 2.13, 6.09, 7.50, 4.29, 3.51, 2.00, 4.83, 3.19, 7.11\}$.

7-41 The following are monthly total pan evaporation levels (E, in.) for the Chattooga River near Clayton, Georgia, for October 1953 through September 1956. Test for a trend using the Noether test. $E = \{3.94, 2.09, 2.79, 2.13, 3.10, 3.33, 4.91, 5.19, 7.47, 8.22, 6.36, 6.13, 4.54, 1.88, 1.29, 1.16, 2.36, 3.36, 5.95, 6.69, 6.98, 6.37, 6.24, 5.78, 3.95, 2.34, 1.40, 1.61, 2.57, 4.30, 5.80, 6.33, 7.50, 6.84, 6.51, 5.40\}$.

7-42 Use the Durbin–Watson test for negative autocorrelation with the snow-covered area data of Problem 7-28.

7-43 Use the Durbin–Watson test for positive autocorrelation for the daily mean discharge data of Problem 7-9.

7-44 Use the Durbin–Watson test for randomness with the Fishkill Creek discharge data of Problem 7-17.

7-45 Assess the usefulness of the Durbin–Watson test with periodic or cyclical data.

7-46 Assess the applicability of the Durbin–Watson test for a series that includes an episodic event marked by a noticeable increase in the mean.

8 Detection of Change in Moments

8.1 INTRODUCTION

Hydrologic variables depend on many factors associated with the meteorological and hydrologic processes that govern their behavior. The measured values are often treated as random variables, and the probability of occurrence of all random variables is described by the underlying probability distribution, which includes both the parameters and the function. Change in either the meteorological or the hydrologic processes can induce change in the underlying population through either a change in the probability function or the parameters of the function.

A sequential series of hydrologic data that has been affected by watershed change is considered nonstationary. Some statistical methods are most sensitive to changes in the moments, most noticeably the mean and variance; other statistical methods are more sensitive to change in the distribution of the hydrologic variable. Selecting a statistical test that is most sensitive to detecting a difference in means when the change in the hydrologic process primarily caused a change in distribution may lead to the conclusion that hydrologic change did not occur. It is important to hypothesize the most likely effect of the hydrologic change on the measured hydrologic data so that the most appropriate statistical test can be selected. Where the nature of the change to the hydrologic data is uncertain, it may be prudent to subject the data to tests that are sensitive to different types of change to decide whether the change in the hydrologic processes caused a noticeable change in the measured data.

In this chapter, tests that are sensitive to changes in moments are introduced. In the next chapter, tests that are appropriate for detecting change in the underlying probability function are introduced. All of the tests follow the same six steps of hypothesis testing (see Chapter 3), but because of differences in their sensitivities, the user should be discriminating in the selection of a test.

8.2 GRAPHICAL ANALYSIS

Graphical analyses are quite useful as the starting point for detecting change and the nature of any change uncovered. Changes in central tendency can be seen from a shift of a univariate histogram or the movement up or down of a frequency curve. For a periodic or cyclical hydrologic processes, a graphical analysis might reveal a shift in the mean value, without any change in the amplitude or phase angle of the periodic function.

A change in the variance of a hydrologic variable will be characterized by a change in spread of the data. For example, an increase in variance due to hydrologic change would appear in the elongation of the tails of a histogram. A change in slope of a frequency curve would also suggest a change in variance. An increase or decrease in the amplitude of a periodic function would be a graphical indication of a change in variance of the random variable.

Once a change has been detected graphically, statistical tests can be used to confirm the change. While the test does not provide a model of the effect of change, it does provide a theoretical justification for modeling the change. Detection is the first step, justification or confirmation the second step, and modeling the change is the final step. This chapter and Chapter 9 concentrate on tests that can be used to statistically confirm the existence of change.

8.3 THE SIGN TEST

The sign test can be applied in cases that involve two related samples. The name of the test implies that the random variable is quantified by signs rather than a numerical value. It is a useful test when the criterion cannot be accurately evaluated but can be accurately ranked as being above or below some standard, often implied to mean the central tendency. Measurement is on the ordinal scale, within only three possible outcomes: above the standard, below the standard, or equal to the standard. Symbols such as +, −, and 0 are often used to reflect these three possibilities.

The data consist of two random variables or a single criterion in which the criterion is evaluated for two conditions, such as before treatment versus after treatment. A classical example of the use of the sign test would be a poll in which voters would have two options: voting for or against passage. The first part of the experiment would be to record each voter's opinion on the proposed legislation, that is, for or against, + or −. The treatment would be to have those in the random sample of voters read a piece of literature that discusses the issue. Then they would be asked their opinion a second time, that is, for or against. Of interest to the pollster is whether the literature, which may possibly be biased toward one decision, can influence the opinion of voters. This is a before–after comparison. Since the before-and-after responses of the individuals in the sample are associated with the individual, they are paired. The sign test would be the appropriate test for analyzing the data.

The hypotheses can be expressed in several ways, with the appropriate alternative pair of hypotheses selected depending on the specific situation. One expression of the null hypothesis is

$$H_0\text{: } P\ (X > Y) = P\ (X < Y) = 0.5 \tag{8.1a}$$

in which X and Y are the criterion values for the two conditions. In the polling case, X would be the pre-treatment response (i.e., for or against) and Y would be the post-treatment response (i.e., for or against). If the treatment does not have a significant effect, then the changes in one direction should be balanced by changes in the other direction. If the criterion did not change, then the values of X and Y are equal, which is denoted as a tie. Other ways of expressing the null hypothesis are

$$H_0: P(+) = P(-) \tag{8.1b}$$

or

$$H_0: E(X) = E(Y) \tag{8.1c}$$

where + indicates a change in one direction while – indicates a change in the other direction, and the hypothesis expressed in terms of expectation suggests that the two conditions have the same central tendency.

Both one- and two-tailed alternative hypotheses can be formulated:

$$H_A: P(X > Y) > P(X < Y) \text{ or } P(+) > P(-) \tag{8.2a}$$

$$H_A: P(X > Y) < P(X < Y) \text{ or } P(+) < P(-) \tag{8.2b}$$

$$H_A: P(X > Y) \neq P(X < Y) \text{ or } P(+) \neq P(-) \tag{8.2c}$$

The alternative hypothesis selected would depend on the intent of the analysis.

To conduct the test, a sample of N is collected and measurements on both X and Y made, that is, pre-test and post-test. Given that two conditions are possible, X and Y, for each test, the analysis can be viewed as the following matrix:

	Post-test	
Pre-test	X	Y
X	N_{11}	N_{12}
Y	N_{21}	N_{22}

Only the responses where a change was made are of interest. The treatment would have an effect for cells (X, Y) and (Y, X) but not for cells (X, X) and (Y, Y). Thus, the number of responses N_{12} and N_{21} are pertinent while the number of responses N_{11} and N_{22} are considered "ties" and are not pertinent to the calculation of the sample test statistic. If $N = N_{11} + N_{22} + N_{12} + N_{21}$ but only the latter two are of interest, then the sample size of those affected is $n = N_{12} + N_{21}$, where $n \leq N$. The value of n, not N, is used to obtain the critical value.

The critical test statistic depends on the sample size. For small samples of about 20 to 25 or less, the critical value can be obtained directly from the binomial distribution. For large samples (i.e., 20 to 25 or more), a normal approximation is applied. The critical value also depends on the alternative hypothesis: one-tailed lower, one-tailed upper, and two-tailed.

For small samples, the critical value is obtained from the cumulative binomial distribution for a probability of $p = 0.5$. For any value T and sample size n, the

TABLE 8.1
Binomial Probabilities for Sign Test with $n = 13$; $f(T)$ = Mass Function and $F(T)$ = Cumulative Mass Function

T	$f(T)$	$F(T)$
0	0.00012	0.00012
1	0.00159	0.00171
2	0.00952	0.01123
3	0.03491	0.04614
4	0.08728	0.13342
5	0.15711	0.29053
6	0.20947	0.50000
7	0.20947	0.70947
8	0.15711	0.86658
9	0.08728	0.95386
10	0.03491	0.98877
11	0.00952	0.99829
12	0.00159	0.99988
13	0.00012	1.00000

cumulative distribution $F(T)$ is:

$$F(T) = \sum_{i=0}^{T} \binom{n}{i} p^n (1-p)^{n-T} = \sum_{i=0}^{T} \binom{n}{i} (0.5)^n \tag{8.3}$$

When the alternative hypothesis suggests a one-tailed lower test, the critical value is the largest value of T for which $\alpha \geq F(T)$. For a one-tailed upper test, the critical value is the value of T for which $\alpha \geq F(n - T)$. For a two-tailed test, the lower and upper critical values are the values of T for which $\alpha/2 \geq F(T)$ and $\alpha/2 \geq F(n - T)$.

Consider the case where $n = 13$. Table 8.1 includes the mass and cumulative mass functions as a function of T. For a 1% level of significance, the critical value for a one-tailed lower test would be 1 since $F(2)$ is greater than α. For a 5% level of significance, the critical value would be 3 since $F(4)$ is greater than α. For a one-tailed upper test, the critical values for 1% and 5% are 12 and 10. For a two-tailed test, the critical values for a 1% level of significance are 1 and 12, which means the null hypothesis is rejected if the computed value is equal to 0, 1, 12, or 13. For a 5% level of significance the critical values would be 2 and 11, which means that the null hypothesis is rejected if the computed value is equal to 0, 1, 2, 11, 12, or 13. Even though α is set at 5%, the actual rejection probability is 2(0.01123) = 0.02246, that is, 2.2%, rather than 5%. Table 8.2 gives critical values computed using the binomial distribution for sample sizes of 10 to 50.

TABLE 8.2
Critical Values for the Sign Tests[a]

n	$X_{.01}$	$X_{.05}$	n	$X_{.01}$	$X_{.05}$	n	$X_{.01}$	$X_{.05}$	n	$X_{.01}$	$X_{.05}$
10	1[b]	2[b]	20	4	6[b]	30	8	10	40	12	14
11	1	2	21	4	6	31	8	10	41	12	15[b]
12	1	2	22	5	6	32	9	11[b]	42	13	15
13	2[b]	3	23	5	7	33	9	11	43	13	15
14	2	3	24	6[b]	7	34	10[b]	11	44	14[b]	16
15	3[b]	4[b]	25	6	8[b]	35	10	12	45	14	16
16	3[b]	4	26	6	8	36	10	12	46	14	17[b]
17	3	4	27	7	8	37	11	13	47	15	17
18	3	5	28	7	9	38	11	13	48	15	18[b]
19	4	5	29	8[b]	9	39	12[b]	14[b]	49	16[b]	18
									50	16	18

[a] For the upper tail critical values for probabilities of 5% and 1%, use $X_{u.05} = n - X_{.05}$ and $X_{u.01} = n - X_{.01}$.

[b] To obtain a conservative estimate, reduce by 1. This entry has a rejection probability slightly larger than that indicated, but the probability of the lower value is much less than the indicated rejection probability. For example, for $n = 13$ and a 1% rejection probability, $X = 1$ has a probability of 0.0017 and $X = 2$ has a probability of 0.0112.

For large sample sizes, the binomial mass function can be approximated with a normal distribution. The mean and standard deviation of the binomial distribution are

$$\mu = np \tag{8.4}$$

$$\sigma = [np\ (1 - p)]^{0.5} \tag{8.5}$$

For a probability of 0.5, these reduce to $\mu = 0.5n$ and $\sigma = 0.5n^{0.5}$. Using these with the standard normal deviate gives:

$$z = \frac{x - \mu}{\sigma} = \frac{x - 0.5n}{0.5n^{0.5}} \tag{8.6}$$

Equation 8.6 gives a slightly biased estimate of the true probability. A better estimate can be obtained by applying a continuity correction to x, such that z is

$$z = \frac{(x + 0.5) - 0.5n}{0.5n^{0.5}} \tag{8.7}$$

Equation 8.7 can be rearranged to solve for the random variable x, which is the upper critical value for any level of significance:

$$x = 0.5n + 0.5zn^{0.5} - 0.5 = 0.5(n + zn^{0.5} - 1) \tag{8.8}$$

Enter Equation 8.8 with the standard normal deviate z corresponding to the rejection probability to compute the critical value. Equation 8.8 yields the upper bound. For the one-sided lower test, use $-z$. Values obtained with this normal transformation are approximate. It would be necessary to round up or round down the value obtained with Equation 8.8. The rounded value can then be used with Equation 8.7 to estimate the actual rejection probability.

Example 8.1

Mulch is applied to exposed soil surfaces at 63 sites, with each slope divided into two equal sections. The density of mulch applied to the two sections is different (i.e., high and low). The extent of the erosion was qualitatively assessed during storm events with the extent rated as high or low. The sign test is used to test the following hypotheses:

H_0: The density of mulch does not affect the amount of eroded soil.
H_A: Sites with the higher density of mulch experienced less erosion.

The statement of the alternative hypothesis dictates the use of a one-sided test.

Of the 63 sites, 46 showed a difference in eroded soil between the two sections, with 17 sites not showing a difference. Of the 46 sites, the section with higher mulch density showed the lower amount of eroded material on 41 sites. On 5 sites, the section with higher mulch density showed the higher amount of erosion. Therefore, the test statistic is 5. For a sample size of 46, the critical values for 1% and 5% levels of significance are 14 and 17, respectively (see Table 8.2). Therefore, the null hypothesis should be rejected. To use the normal approximation, the one-sided 1% and 5% values of z are −2.327 and −1.645, which yield critical values of 14.6 and 16.9, respectively. These normal approximations are close to the actual binomial values of 14 and 17.

8.4 TWO-SAMPLE *t*-TEST

In some cases, samples are obtained from two different populations, and it is of interest to determine if the population means are equal. For example, two laboratories may advertise that they evaluate water quality samples of some pollutant with an accuracy of ±0.1 mg/L; samples may be used to test whether the means of the two populations are equal. Similarly, tests could be conducted on engineering products to determine whether the means are equal. The fraction of downtime for two computer types could be tested to decide whether the mean times differ.

A number of tests can be used to test a pair of means. The method presented here should be used to test the means of two independent samples. This test is frequently of interest in engineering research when the investigator is interested in comparing an experimental group to a control group. For example, an environmental engineer might be interested in comparing the mean growth rates of microorganisms in a polluted and natural environment. The procedure presented in this section can be used to make the test.

Step 1: Formulate hypotheses. The means of two populations are denoted as μ_1 and μ_2, the null hypothesis for a test on two independent means would be:

$$H_0\text{: The means of two populations are equal.} \tag{8.9a}$$

Mathematically, this is

$$H_0\text{: } \mu_1 = \mu_2 \tag{8.9b}$$

Both one-sided and two-sided alternatives can be used:

$$H_{A1}\text{: } \mu_1 < \mu_2 \tag{8.10a}$$

$$H_{A2}\text{: } \mu_1 > \mu_2 \tag{8.10b}$$

$$H_{A3}\text{: } \mu_1 \neq \mu_2 \tag{8.10c}$$

The selection of the alternative hypotheses should depend on the statement of the problem.

Step 2: Select the appropriate model. For the case of two independent samples, the hypotheses of step 1 can be tested using the following test statistic:

$$t = \frac{\overline{X}_1 - \overline{X}_2}{S_p\left(\dfrac{1}{n_1} + \dfrac{1}{n_2}\right)^{0.5}} \tag{8.11}$$

in which $\overline{X}_1$ and $\overline{X}_2$ are the means of the samples drawn from populations 1 and 2, respectively; n_1 and n_2 are the sample sizes used to compute $\overline{X}_1$ and $\overline{X}_2$, respectively; t is the value of a random variable that has a t distribution with degrees of freedom (υ) of $\nu = n_1 + n_2 - 2$; and S_p is the square root of the pooled variance that is given by

$$S_p^2 = \frac{(n_1 - 1)S_1^2 + (n_2 - 1)S_2^2}{n_1 + n_2 - 2} \tag{8.12}$$

in which S_1^2 and S_2^2 are the variances of the samples from population 1 and 2, respectively. This test statistic assumes that the variances of the two populations are equal, but unknown.

Step 3: Select the level of significance. As usual, the level of significance should be selected on the basis of the problem. However, values of either 5% or 1% are used most frequently.

Step 4: Compute an estimate of test statistic. Samples are drawn from the two populations, and the sample means and variances computed. Equation 8.11 can be computed to test the null hypothesis of Equation 8.9.

Step 5: Define the region of rejection. The region of rejection is a function of the degrees of freedom ($\nu = n_1 + n_2 - 2$), the level of significance (α), and the statement of the alternative hypothesis. The regions of rejection for the alternative hypotheses are as follows:

If H_A is	then reject H_0 if
$\mu_1 < \mu_2$	$t < -t_\alpha$
$\mu_1 > \mu_2$	$t > t_\alpha$
$\mu_1 \neq \mu_2$	$t < -t_{\alpha/2}$ or $t > t_{\alpha/2}$

Step 6: Select the appropriate hypothesis. The sample estimate of the t statistic from step 4 can be compared with the critical value (see Appendix Table A.2), which is based on either t_α or $t_{\alpha/2}$ obtained from step 5. If the sample value lies in the region of rejection, then the null hypothesis should be rejected.

Example 8.2

A study was made to measure the effect of suburban development on total nitrogen levels in small streams. A decision was made to use the mean concentrations before and after the development as the criterion. Eleven measurements of the total nitrogen (mg/L) were taken prior to the development, with a mean of 0.78 mg/L and a standard deviation of 0.36 mg/L. Fourteen measurements were taken after the development, with a mean of 1.37 mg/L and a standard deviation of 0.87 mg/L. The data are used to test the null hypothesis that the population means are equal against the alternative hypothesis that the urban development increased total nitrogen levels, which requires the following one-tailed test:

$$H_0\text{: } \mu_b = \mu_a \tag{8.13}$$

$$H_A\text{: } \mu_b < \mu_a \tag{8.14}$$

where μ_b and μ_a are the pre- and post-development means, respectively. Rejection of H_0 would suggest that the nitrogen levels after development significantly exceed the nitrogen levels before development, with the implication that the development might have caused the increase.

Based on the sample data, the pooled variance of Equation 8.12 is

$$S_p^2 = \frac{(11-1)(0.36)^2 + (14-1)(0.87)^2}{11+14-2} = 0.4842 \tag{8.15}$$

The computed value of the test statistic is

$$t = \frac{0.78 - 1.37}{\sqrt{0.4842\left(\frac{1}{11} + \frac{1}{14}\right)}} = -2.104 \tag{8.16}$$

which has $\nu = 11 + 14 - 2 = 23$ degrees of freedom. From Table A.2, with a 5% level of significance and 23 degrees of freedom, the critical value of t is −1.714. Thus, the null hypothesis is rejected, with the implication that the development caused a significant change in nitrogen level. However, for a 1% significance level, the critical value is −2.500, which leads to the decision that the increase is not significant. This shows the importance of selecting the level of significance before analyzing the data.

8.5 MANN–WHITNEY TEST

For cases in which watershed change occurs as an episodic event within the duration of a flood series, the series can be separated into two subseries. The Mann–Whitney U-test (Mann and Whitney, 1947) is a nonparametric alternative to the t-test for two independent samples and can be used to test whether two independent samples have been taken from the same population. Therefore, when the assumptions of the parametric t-test are violated or are difficult to evaluate, such as with small samples, the Mann–Whitney U-test should be applied. This test is equivalent to the Wilcoxon–Mann–Whitney rank-sum test described in many textbooks as a t-test on the rank-transformed data (Inman and Conover, 1983).

The procedure for applying the Mann–Whitney test follows.

1. Specify the hypotheses:
   H_0: The two independent samples are drawn from the same population.
   H_A: The two independent samples are not drawn from the same population.
   The alternative hypothesis shown is presented as a two-sided hypothesis; one-sided alternative hypotheses can also be used:
   H_A: Higher (or lower) values are associated with one part of the series.
   For the one-sided alternative, it is necessary to specify either higher or lower values prior to analyzing the data.
2. The computed value (U) of the Mann–Whitney U-test is equal to the lesser of U_a and U_b where

$$U_a = n_a n_b + 0.5\, n_b\, (n_b + 1) - S_b \tag{8.17a}$$

$$U_b = n_a n_b + 0.5\, n_a\, (n_a + 1) - S_a \tag{8.17b}$$

   in which n_a and n_b are the sample sizes of subseries A and B, respectively. The values of S_a and S_b are computed as follows: the two groups are combined, with the items in the combined group ranked in order from smallest (rank = 1) to the largest (rank = $n = n_a + n_b$). On each rank, include a subscript a or b depending on whether the value is from subseries A or B. S_a and S_b are the sums of the ranks with subscript a and b, respectively. Since S_a and S_b are related by the following, only one value needs to be computed by actual summation of the ranks:

$$S_a + S_b = 0.5n(n+1) \tag{8.17c}$$

3. The level of significance (α) must be specified. As suggested earlier, α is usually 0.05.
4. Compute the test statistic value of U of step 2 and the value of Z:

$$Z = \frac{U - 0.5n_a n_b}{(n_a n_b (n_a + n_b + 1)/12)^{0.5}} \tag{8.18}$$

   in which Z is the value of a random variable that has a standard normal distribution.
5. Obtain the critical value ($Z_{\alpha/2}$ for a two-sided test or Z_α for a one-sided test) from a standard-normal table (Table A.1).
6. Reject the null hypothesis if the computed value of Z (step 4) is greater than $Z_{\alpha/2}$ or less than $-Z_{\alpha/2}$ (Z_α for an upper one-sided test or $-Z_\alpha$ for a lower, one-sided test).

In some cases, the hydrologic change may dictate the direction of change in the annual peak series; in such cases, it is appropriate to use the Mann–Whitney test as a one-tailed test. For example, if channelization takes place within a watershed, the peaks for the channelized condition should be greater than for the natural watershed. To apply the U-test as a one-tailed test, specify subseries A as the series with the smaller expected central tendency and then use U_a as the computed value of the test statistic, $U = U_a$ (rather than the lesser of U_a and U_b); the critical value of Z is $-Z_\alpha$ (rather than $-Z_{\alpha/2}$) and the null hypothesis is accepted when $Z > -Z_\alpha$.

Example 8.3

The 65-year annual maximum series for the Elizabeth River appears (Figure 7.2) to change in magnitude about 1951 when the river was channelized. To test if the channelization was accompanied by an increase in flood peaks, the flood series was divided into two series, 1924 to 1951 (28 years) and 1952 to 1988 (37 years). Statistical characteristics for the periods are given in Table 8.3.

Since the direction of change was dictated by the problem (i.e., the peak discharges increased after channelization), the alternative hypothesis is

H_A: The two independent samples are not from the same population and the logarithms of the peaks for 1924–1951 are expected to be less than those for 1952–1988.

TABLE 8.3
Annual Peak Discharge Characteristics for Elizabeth River, New Jersey, Before and After Channelization

			Characteristics of Series Logarithms		
Period	Mean (ft^3/s)	Standard Deviation (ft^3/s)	Mean	Standard Deviation	Skew
1924–1951	1133	643	3.0040	0.2035	0.57
1952–1988	2059	908	3.2691	0.2074	–0.44

Analysis of the annual peak discharges yields $U_a = 200$ and $U_b = 836$. Since a one-tailed test is made, $U = U_a$. Using Equation 8.18, the computed value of Z is –4.213. For a one-tailed test and a level of significance of 0.001, the critical value is –3.090. Since $Z < Z_\alpha$, the null hypothesis is rejected. The results of the Mann–Whitney U-test suggest that channelization significantly increased the peak discharges.

8.5.1 Rational Analysis of the Mann–Whitney Test

When applying the Mann–Whitney test, the flows of the annual maximum series are transformed to ranks. Thus, the variation within the floods of the annual series is reduced to variation of ranks. The transformation of measurements on a continuous variable (e.g., flow) to an ordinal scale (i.e., the rank of the flow) reduces the importance of the between-flow variation. However, in contrast to the runs test, which reduces the flows to a nominal variable, the importance of variation is greater for the Mann–Whitney test than for the runs test. Random variation that is large relative to the variation introduced by the watershed change into the annual flood series will more likely mask the effect of the change when using the Mann–Whitney test compared to the runs test.

This can be more clearly illustrated by examining the statistics U_a and U_b of Equations 8.17. The first two terms of the equation represent the maximum possible total of the ranks that could occur for a series with n_a and n_b elements. The third terms, S_a and S_b, represent the summations of the actual ranks in sections A and B of the series. Thus, when S_a and S_b are subtracted, the differences represent deviations. If a trend is present, then one would expect either S_a or S_b to be small and the other large, which would produce a small value of the test statistic U. If the flows exhibit random variation that is large relative to the variation due to watershed change, then it can introduce significant variation into the values of S_a and S_b, thereby making it more difficult to detect a trend resulting from watershed change. In a sense, the test is examining the significance of variation introduced by watershed change relative to the inherent natural variation of the flows.

In addition to the variation from the watershed change, the importance of the temporal location of the watershed change is important. In contrast to the runs test, the Mann–Whitney test is less sensitive to the location of the trend. It is more likely to detect a change in flows that results from watershed change near one of the ends of the series than the runs test. If watershed change occurs near an end of the series, then either N_a or N_b of Equations 8.17 would be small, but the sum of the ranks would also be small. The summation terms S_a and S_b decrease in relative proportion to the magnitudes of the first two terms. Thus, a change near either end of a series is just as likely to be detected as a change near the middle of the series.

The Mann–Whitney test is intended for the analysis of two independent samples. When applying the test to hydrologic data, a single series is separated into two parts based on a watershed-change criterion. Thus, the test would not strictly apply if the watershed change occurred over an extended period of time. Consequently, it may be inappropriate to apply the test to data for watershed change that occurred gradually with time. Relatively abrupt changes are more appropriate.

8.6 THE *t*-TEST FOR TWO RELATED SAMPLES

The *t*-test for independent samples was introduced in Section 8.4. The *t*-test described here assumes that the pairs being compared are related. The relationship can arise in two situations. First, the random sample is subjected to two different treatments at different times and administered the same test following each treatment. The two values of the paired test criterion are then compared and evaluated for a significant difference. As an example, n watersheds could be selected and the slope of the main channel computed for maps of different scales. The same computational method is used, but different map scales may lead to different estimates of the slope. The lengths may differ because the map with the coarser scale fails to show the same degree of meandering. The statistical comparison of the slopes would seek to determine if the map scale makes a significant difference in estimates of channel slope.

The second case where pairing arises is when objects that are alike (e.g., identical twins) or are similar (e.g., adjacent watersheds) need to be compared. In this case, the pairs are given different treatments and the values of the test criterion compared. For example, pairs of adjacent watersheds that are similar in slope, soils, climate, and size, but differ in land cover (forested vs. deforested) are compared on the basis of a criterion such as base flow rate, erosion rate, or evapotranspiration (ET). A significant difference in the criterion would indicate the effect of the difference in the land cover.

The hypotheses for this test are the same as those for the *t*-test for two independent samples:

H_0: $\mu_1 = \mu_2$
H_A: $\mu_1 \neq \mu_2$ (two-tailed)
$\mu_1 < \mu_2$ (one-tailed lower)
$\mu_1 > \mu_2$ (one-tailed upper)

The sampling variation for this test differs from that of the pooled variance for the test of two independent samples (Equation 8.12). For this test, the sampling variation is

$$S_m = \left(\frac{\sum d^2}{n(n-1)} \right)^{0.5} \tag{8.19}$$

in which n is the number of pairs and

$$\sum_i d^2 = \sum_i (D_i - \bar{D})^2 = \sum_i D_i^2 - \frac{\left(\sum_i D_i\right)^2}{n} \tag{8.20}$$

where D_i is the difference between the ith pair of scores and $\bar{D}$ is the average difference between scores. The test statistic is:

$$t = \bar{D}/S_m \qquad \text{with } \upsilon = n - 1 \tag{8.21}$$

This test assumes that the difference scores are normally distributed. A log transformation of the data may be used for highly skewed data. If the difference scores deviate significantly from normality, especially if they are highly asymmetric, the actual level of significance will differ considerably from the value used to find the critical test statistic.

The computed test statistic is compared with the tabled t value for level of significance α and degrees of freedom υ. The region of rejection depends on the statement of the alternative hypothesis, as follows:

If H_A is	reject H_0 if
$\mu_x < \mu_y$	$t < -t_{\alpha,\upsilon}$
$\mu_x > \mu_y$	$t > t_{\alpha,\upsilon}$
$\mu_x \neq \mu_y$	$t < -t_{\alpha/2,\upsilon}$ or $t > t_{\alpha/2,\upsilon}$

Example 8.4

Two different methods of assigning values of Manning's roughness coefficient (n) are used on eight small urban streams. The resulting values are given in columns 2 and 3 of Table 8.4. The two-tailed alternative hypothesis is used since the objective is only to decide if the methods give, on the average, similar values. The value of $\bar{D}$ is equal to –0.027/8 = –0.003375. Results of Equation 8.20 follow:

$$\sum d^2 = 0.001061 - \frac{(-0.027)^2}{8} = 0.0009699 \tag{8.22}$$

The sampling variation is computed using Equation 8.19:

$$S_m = \left(\frac{0.0009699}{8(8-1)}\right)^{0.5} = 0.004162 \tag{8.23}$$

TABLE 8.4
Comparison of Roughness Coefficients Using the Two-Sample *t*-Test for Related Samples: Example 8.5

Channel Reach	Roughness with Method 1	Roughness with Method 2	D	D^2
1	0.072	0.059	0.013	0.000169
2	0.041	0.049	–0.008	0.000064
3	0.068	0.083	–0.015	0.000225
4	0.054	0.070	–0.016	0.000256
5	0.044	0.055	–0.011	0.000121
6	0.066	0.057	0.009	0.000081
7	0.080	0.071	0.009	0.000081
8	0.037	0.045	–0.088	0.000064
			$\Sigma D = -0.027$	$\Sigma D^2 = 0.001061$

The test statistic is:

$$t = \frac{-0.003375}{0.004162} = -0.811 \tag{8.24}$$

with 7 degrees of freedom. For a two-tailed test and a level of significance of 20%, the critical value is ±1.415 (see Table A.2). Therefore, little chance exists that the null hypothesis should be rejected. In spite of differences in the methods of estimating the roughness coefficients, the methods do not give significantly different values on the average.

Example 8.5

The paired *t*-test can be applied to baseflow discharge data for two watersheds of similar size (see Table 8.5). It is generally believed that increases in the percentage of urban/suburban development cause reductions in baseflow discharges because less water infiltrates during storm events. It is not clear exactly how much land development is necessary to show a significant decrease in baseflow. The baseflows of two pairs of similarly sized watersheds will be compared. The baseflow (ft^3/sec) based on the average daily flows between 1990 and 1998 were computed for each month. Since baseflow rates depend on watershed size, both pairs have similar areas, with percentage differences in the area of 7.50% and 3.16%. Such area differences are not expected to have significant effects on the differences in baseflow.

For the first pair of watersheds (USGS gage station numbers 1593500 and 1591000) the monthly baseflows are given in columns 2 and 3 of Table 8.5. The two watersheds have drainage areas of 37.78 mi^2 and 35.05 mi^2. For the larger watershed, the percentages of high density and residential area are 1.25% and 29.27%, respectively. For the smaller watershed, the corresponding percentages are 0% and 1.08%. A similar analysis was made for a second pair of watersheds (USGS gage station

TABLE 8.5
Two-Sample *t*-Test of Baseflow Discharge Rates

	Watershed Pair 1			Watershed Pair 2		
Month (1)	X_t Gage No. 1593500 (2)	Y_t Gage No. 1591000 (3)	Difference Scores $D_t = X_t - Y_t$ (4)	X_t Gage No. 1583600 (5)	Y_t Gage No. 1493000 (6)	Difference Scores $D_t = X_t - Y_t$ (7)
Jan.	35.0	40	−5.0	20.0	25.0	−5.0
Feb.	40.0	40	0.0	25.0	25.0	0.0
Mar.	40.0	50	−10.0	30.0	30.0	0.0
Apr.	35.0	40	−5.0	30.0	30.0	0.0
May	30.0	30	0.0	25.0	20.0	5.0
June	20.0	20	0.0	17.5	12.5	5.0
July	10.0	15	−5.0	17.5	12.5	5.0
Aug.	10.0	10	0.0	15.0	10.0	5.0
Sept.	10.0	10	0.0	15.0	8.0	7.0
Oct.	12.5	10	2.5	12.5	10.0	2.5
Nov.	15.0	20	−5.0	20.0	12.0	8.0
Dec.	30.0	35	−5.0	20.0	20.0	0.0

numbers 1583600 and 1493000) that have areas of 21.09 mi^2 and 20.12 mi^2, respectively. The larger watershed has a percentage of high-density development of 1.04% and of residential development of 20.37%, while the smaller watershed has corresponding percentages of 0.04% and 1.77%, respectively. The first pair of watersheds is compared first. The mean and standard deviation of the difference scores in column 4 of Table 8.5 are −2.733 and 3.608, respectively. Using Equation 8.21 the computed value of the statistic is:

$$t = \frac{-2.733}{3.608/\sqrt{12}} = 2.624 \quad \text{with } \upsilon = 12 - 1 = 11 \tag{8.25}$$

The critical t values for levels of significance of 5%, 2.5%, and 1% are 1.796, 2.201, and 2.718, respectively. Thus, the rejection probability is approximately 1.5%. The null hypothesis of equal means would be rejected at 5% but accepted at 1%.

For the second pair of paired watersheds, the difference scores (column 7 of Table 8.5) have a mean and standard deviation of 2.708 and 3.768, respectively. The computed value of the t statistic is:

$$t = \frac{2.708}{3.768/\sqrt{12}} = 2.490 \quad \text{with } \upsilon = 12 - 1 = 11 \tag{8.26}$$

Since the critical t values are negative because this is a one-tailed lower test, the null hypothesis cannot be rejected. The mean baseflow discharges are not significantly different in spite of the differences in watershed development.

8.7 THE WALSH TEST

The Walsh test is a nonparametric alternative to the two-sample *t*-test for related samples. It is applicable to data measured on an interval scale. The test assumes that the two samples are drawn from symmetrical populations; however, the two populations do not have to be the same population, and they do not have to be normal distributions. If the populations are symmetrical, then the mean of each population will equal the median.

The null hypothesis for the Walsh test is:

$$H_0\text{: The average of the difference scores } (\mu_d) \text{ is equal to zero.} \tag{8.27a}$$

Both one-tailed and two-tailed alternative hypotheses can be tested:

$$H_{A1}\text{: } \mu_d < 0 \tag{8.27b}$$

$$H_{A2}\text{: } \mu_d > 0 \tag{8.27c}$$

$$H_{A3}\text{: } \mu_d \neq 0 \tag{8.27d}$$

The alternative hypothesis should be selected based on the physical problem being studied and prior to analyzing the data. Acceptance or rejection of the null hypothesis will imply that the two populations have the same or different central tendencies, respectively.

The test statistic for the Walsh test is computed based on the differences between the paired observations. Table 8.6 gives the test statistics, which differ with the sample size, level of significance, and tail of interest. The critical value is always zero, so for a one-tailed lower test (Equation 8.27a), one of the test statistics in column 4 of Table 8.6 should be used. For a one-tailed upper test (Equation 8.27b), one of the test statistics of column 5 should be used. For a two-tailed test (Equation 8.27c), the test statistics of both columns 4 and 5 are used.

The procedure for conducting Walsh's test is as follows:

1. Arrange the data as a set of related pairs: (x_i, y_i), $i = 1, 2, \ldots, n$.
2. For each pair, compute the difference in the scores, d_i: $d_i = x_i - y_i$ (note that the values of d_i can be positive, negative, or zero).
3. Arrange the difference scores in order of size, with the most negative value given a rank of 1 to the algebraically largest value with a rank of n. Tied values of d_i are given sequential ranks, not average ranks. Therefore, the d values are in order such that $d_1 \le d_2 \le d_3 \le \ldots \le d_n$. (Note that the subscript on d now indicates the order according to size, not the number of the pair of step 2.)
4. Compute the sample value of the test statistic indicated in Table 8.6.
5. The critical test statistic is always zero.

TABLE 8.6
Critical Values for the Walsh Test

	Significance Level of Tests		Tests Two-Tailed: Accept $\mu_\delta \neq 0$ If Either	
n	One-Tailed	Two-Tailed	One-Tailed: Accept $\mu_d < 0$ If	One-Tailed: Accept $\mu_d > 0$ If
4	.062	.125	$d_4 < 0$	$d_1 > 0$
5	.062	.125	$1/2(d_4 + d_5) < 0$	$1/2(d_1 + d_2) > 0$
	.031	.062	$d_5 < 0$	$d_1 > 0$
6	.047	.094	$\max [d_5, 1/2(d_4 + d_6)] < 0$	$\min [d_2, 1/2(d_1 + d_4)] > 0$
	.031	.062	$1/2(d_5 + d_6) < 0$	$1/2(d_1 + d_2) > 0$
	.016	.031	$d_6 < 0$	$d_1 > 0$
7	.055	.109	$\max [d_5, 1/2(d_4 + d_7)] < 0$	$\min [d_3, 1/2(d_1 + d_4)] > 0$
	.023	.047	$\max [d_6, 1/2(d_5 + d_7)] < 0$	$\min [d_2, 1/2(d_1 + d_3)] > 0$
	.016	.031	$1/2(d_6 + d_7) < 0$	$1/2(d_1 + d_2) > 0$
	.008	.016	$d_7 < 0$	$d_1 > 0$
8	.043	.086	$\max [d_6, 1/2(d_4 + d_8)] < 0$	$\min [d_3, 1/2(d_1 + d_5)] > 0$
	.027	.055	$\max [d_6, 1/2(d_5 + d_8)] < 0$	$\min [d_3, 1/2(d_1 + d_4)] > 0$
	.012	.023	$\max [d_7, 1/2(d_6 + d_8)] < 0$	$\min [d_2, 1/2(d_1 + d_3)] > 0$
	.008	.016	$1/2(d_7 + d_8) < 0$	$1/2(d_1 + d_2) > 0$
	.004	.008	$d_8 < 0$	$d_1 > 0$
9	.051	.102	$\max [d_6, 1/2(d_4 + d_9)] < 0$	$\min [d_4, 1/2(d_1 + d_6)] > 0$
	.022	.043	$\max [d_7, 1/2(d_5 + d_9)] < 0$	$\min [d_3, 1/2(d_1 + d_5)] > 0$
	.010	.020	$\max [d_8, 1/2(d_5 + d_9)] < 0$	$\min [d_2, 1/2(d_1 + d_5)] > 0$
	.006	.012	$\max [d_8, 1/2(d_7 + d_9)] < 0$	$\min [d_2, 1/2(d_1 + d_5)] > 0$
	.004	.008	$1/2(d_8 + d_9) < 0$	$1/2(d_1 + d_2) > 0$
10	.056	.111	$\max [d_6, 1/2(d_4 + d_{10})] < 0$	$\min [d_5, 1/2(d_1 + d_7)] > 0$
	.025	.051	$\max [d_7, 1/2(d_5 + d_{10})] < 0$	$\min [d_4, 1/2(d_1 + d_6)] > 0$
	.011	.021	$\max [d_8, 1/2(d_6 + d_{10})] < 0$	$\min [d_3, 1/2(d_1 + d_5)] > 0$
	.005	.010	$\max [d_9, 1/2(d_6 + d_{10})] < 0$	$\min [d_2, 1/2(d_1 + d_5)] > 0$
11	.048	.097	$\max [d_7, 1/2(d_4 + d_{11})] < 0$	$\min [d_5, 1/2(d_1 + d_8)] > 0$
	.028	.056	$\max [d_7, 1/2(d_5 + d_{11})] < 0$	$\min [d_5, 1/2(d_1 + d_7)] > 0$
	.011	.021	$\max [1/2(d_6 + d_{11}), 1/2(d_8 + d_9)] < 0$	$\min [1/2(d_1 + d_6), 1/2(d_3 + d_4) > 0$
	.005	.011	$\max [d_9, 1/2(d_7 + d_{11})] < 0$	$\min [d_3, 1/2(d_1 + d_5)] > 0$
12	.047	.094	$\max [1/2(d_4 + d_{12}), 1/2(d_5 + d_{11})] < 0$	$\min [1/2(d_1 + d_9), 1/2(d_2 + d_8)] > 0$
	.024	.048	$\max [d_8, 1/2(d_5 + d_{12})] < 0$	$\min [d_5, 1/2(d_1 + d_8)] > 0$
	.010	.020	$\max [d_9, 1/2(d_6 + d_{12})] < 0$	$\min [d_4, 1/2(d_1 + d_7)] > 0$
	.005	.011	$\max [1/2(d_7 + d_{12}), 1/2(d_9 + d_{10})] < 0$	$\min [1/2(d_1 + d_6), 1/2(d_3 + d_4)] > 0$
13	.047	.094	$\max [1/2(d_4 + d_{13}), 1/2(d_5 + d_{12})] < 0$	$\min [1/2(d_1 + d_{10}), 1/2(d_2 + d_9)] > 0$
	.023	.047	$\max [1/2(d_5 + d_{13}), 1/2(d_6 + d_{12})] < 0$	$\min [1/2(d_1 + d_9), 1/2(d_2 + d_8)] > 0$
	.010	.020	$\max [1/2(d_6 + d_{13}), 1/2(d_9 + d_{10})] < 0$	$\min [1/2(d_1 + d_8), 1/2(d_4 + d_5)] > 0$
	.005	.010	$\max [d_{10}, 1/2(d_7 + d_{13})] < 0$	$\min [d_4, 1/2(d_1 + d_7)] > 0$
14	.047	.094	$\max [1/2(d_4 + d_{14}), 1/2(d_5 + d_{13})] < 0$	$\min [1/2(d_1 + d_{11}), 1/2(d_2 + d_{10})] > 0$
	.023	.047	$\max [1/2(d_5 + d_{14}), 1/2(d_6 + d_{13})] < 0$	$\min [1/2(d_1 + d_{10}), 1/2(d_2 + d_9)] > 0$
	.010	.020	$\max [d_{10}, 1/2(d_6 + d_{14})] < 0$	$\min [d_5, 1/2(d_1 + d_9)] > 0$
	.005	.010	$\max [1/2(d_7 + d_{14}), 1/2(d_{10} + d_{11})] < 0$	$\min [1/2(d_1 + d_8), 1/2(d_4 + d_5)] > 0$
15	.047	.094	$\max [1/2(d_4 + d_{15}), 1/2(d_5 + d_{14})] < 0$	$\min [1/2(d_1 + d_{12}), 1/2(d_2 + d_{11})] > 0$
	.023	.047	$\max [1/2(d_5 + d_{15}), 1/2(d_6 + d_{14})] < 0$	$\min [1/2(d_1 + d_{11}), 1/2(d_2 + d_{10})] > 0$
	.010	.020	$\max [1/2(d_6 + d_{15}), 1/2(d_{10} + d_{11})] < 0$	$\min [1/2(d_1 + d_{10}), 1/2(d_5 + d_6)] > 0$
	.005	.010	$\max [d_{11}, 1/2(d_7 + d_{15})] < 0$	$\min [d_5, 1/2(d_1 + d_9)] > 0$

6. Compare the computed value (W_L and/or W_U) to the critical value of zero, as indicated in the table. The decision to accept or reject is indicated by:

If H_A is	reject H_0 if
$\mu_d < 0$	$W_L < 0$
$\mu_d > 0$	$W_U > 0$
$\mu_d \neq 0$	$W_L > 0$ or $W_U > 0$

Example 8.6

The data from Example 8.5 can be compared with the Walsh test. The first pair of watersheds is compared first. Since the larger watershed has the higher percentages of developed land, it would have the lower baseflows if the development has caused decreases in baseflow. Therefore, the alternative hypothesis of Equation 8.27b will be tested and the larger watershed will be denoted as x_i. The difference scores and their ranks are given in columns 4 and 5, respectively, of Table 8.7. The values of the test statistics are computed and shown in Table 8.8. The null hypothesis could be rejected for a level of significance of 4.7% but would be accepted for the smaller levels of significance. It may be of interest to note that the computed values for the three smaller levels of significance are equal to the critical value of zero. While this is not sufficient for rejection, it may suggest that the baseflows for the more highly developed watershed tend to be lower. In only 1 month (October) was the baseflow from the more developed watershed higher than that for the less developed watershed. This shows the importance of the selection of the level of significance, as well as the limitations of statistical analysis.

TABLE 8.7
Application of Walsh Test to Baseflow Discharge Data

	Watershed Pair 1				Watershed Pair 2			
Month (1)	X_i Gage No. 1593500 (2)	Y_i Gage No. 1591000 (3)	Difference Scores $d_i = x_i - y_i$ (4)	Rank of d_i (5)	X_i Gage No. 1583600 (6)	Y_i Gage No. 1493000 (7)	$d_i = x_i - y_i$ (8)	Rank of d_i (9)
Jan.	35.0	40	−5.0	2	20.0	25.0	−5.0	1
Feb.	40.0	40	0.0	7	25.0	25.0	0.0	2
Mar.	40.0	50	−10.0	1	30.0	30.0	0.0	3
Apr.	35.0	40	−5.0	3	30.0	30.0	0.0	4
May	30.0	30	0.0	8	25.0	20.0	5.0	7
June	20.0	20	0.0	9	17.5	12.5	5.0	8
July	10.0	15	−5.0	4	17.5	12.5	5.0	9
Aug.	10.0	10	0.0	10	15.0	10.0	5.0	10
Sept.	10.0	10	0.0	11	15.0	8.0	7.0	11
Oct.	12.5	10	2.5	12	12.5	10.0	2.5	6
Nov.	15.0	20	−5.0	5	20.0	12.0	8.0	12
Dec.	30.0	35	−5.0	6	20.0	20.0	0.0	5

TABLE 8.8
Computation of Sample Test Statistics for Walsh Test

(a) Watershed Pair No. 1

Level of Significance, α	Test Statistic	Sample Value	Decision
0.047	$\max\left[\frac{1}{2}(d_4 + d_{12}), \frac{1}{2}(d_5 + d_{11})\right] < 0$	max (−1.25, −2.5) < 0	Reject H_0
0.024	$\max\left[d_8, \frac{1}{2}(d_5 + d_{12})\right] < 0$	max (0, −1.25) < 0	Accept H_0
0.010	$\max\left[d_9, \frac{1}{2}(d_6 + d_{12})\right] < 0$	max (0, −1.25) < 0	Accept H_0
0.005	$\max\left[\frac{1}{2}(d_7 + d_{12}), \frac{1}{2}(d_9 + d_{10})\right] < 0$	max (1.25, 0) < 0	Accept H_0

(b) Watershed Pair No. 2

Level of Significance, α	Sample Value	Decision
0.047	max (4, 3.5) < 0	Accept H_0
0.024	max (5, 4) < 0	Accept H_0
0.010	max (5, 5.25) < 0	Accept H_0
0.005	max (6.5, 5) < 0	Accept H_0

The second pair of watersheds can be similarly compared. Again, the larger watershed is denoted as x_i, and since it has the larger percentages of development, it is expected to have lower baseflow discharges. Therefore, the alternative hypothesis of Equation 8.27b is selected. The baseflow discharges are given in columns 6 and 7 of Table 8.7. The difference scores and the ranks of the scores are given in columns 8 and 9. In this case, only one of the difference scores is negative (January). This in itself suggests that the land development did not decrease the baseflow. Application of the Walsh test yields the sample values of the test statistic shown in Table 8.8. All of these are greater than the criterion of less than zero, so the null hypothesis is accepted at all four levels of significance.

Both data sets suggest that for these two pairs of coastal watersheds, land development did not cause a significant decrease in baseflow discharge. Even though the assumptions of the two-sample t-test may be violated, t-tests of the same data support the results of the Walsh test. The critical values, whether for the untransformed data means or the logarithms of the data, are significantly within the region of acceptance.

8.8 WILCOXON MATCHED-PAIRS, SIGNED-RANKS TEST

The Wilcoxon matched-pairs, signed-ranks test is used to test whether two related groups show a difference in central tendency. The null hypothesis is that the central tendencies of two related populations are not significantly different. Thus, the test is an alternative to the parametric two-sample t-test. However, the Wilcoxon test can be used with variables that are measured on the ordinal scale. The following six-step procedure is used to test the hypotheses:

1. The null hypothesis is that two related groups do not show a difference in central tendency. Both one-sided and two-sided alternative hypotheses can be tested.
2. The test statistic (T) is the lesser of the sums of the positive and negative differences between the ranks of the values in the samples from the two groups.
3. Select a level of significance.
4. The sample value of the test statistic (T) is obtained as follows:
   a. Compute the magnitude of the difference between each pair of values.
   b. Rank the absolute value of the differences in ascending order (rank 1 for the smallest difference; rank n for the largest difference).
   c. Place the sign of the difference on the rank.
   d. Compute the sum of the ranks of the positive differences, S_p, and the sum of the ranks of the negative differences, S_n.
   e. The value of the test statistic T is the lesser of the absolute values of S_p and S_n.
5. For sample sizes of less than 50, the critical value, T_α, is obtained from Table A.9 using the sample size, n, and the level of significance as the arguments. An approximation will be given below for samples larger than 25.
6. Using μ_i to indicate the central tendency of sample i, the decision criterion is as follows:

If H_A is	Reject H_0 if
$\mu_1 < \mu_2$	$T < T_\alpha$
$\mu_1 > \mu_2$	$T < T_\alpha$
$\mu_1 \neq \mu_2$	$T < T_{\alpha/2}$

It is important to note that the decision rule for both of the one-sided alternatives is the same; this is the result of taking the absolute value in step 4e.

For sample sizes greater than 25, a normal approximation can be used. The mean ($\bar{T}$) and the standard deviation (S_T) of the random variable T can be approximated by:

$$\bar{T} = \frac{n(n+1)}{4} \tag{8.28}$$

and

$$S_T = \left[\frac{n(n+1)(2n+1)}{24}\right]^{0.5} \tag{8.29}$$

The random variable Z has a standard normal distribution, where Z is given by

$$Z = \frac{T - \bar{T}}{S_T} \tag{8.30}$$

For this test statistic, the region of rejection is

If H_A is	Reject H_0 if
$\mu_1 < \mu_2$	$Z < -Z_\alpha$
$\mu_1 > \mu_2$	$Z > Z_\alpha$
$\mu_1 \neq \mu_2$	$Z < -Z_{\alpha/2}$ or $Z > Z_{\alpha/2}$

Several additional points are important:

1. For a one-tailed test, the direction of the difference must be specified before analyzing the data. This implies that the test statistic is equal to whichever sum S_p or S_n is specified in advance, not the lesser of the two.
2. If one or more differences is zero, treat them as follows:
   a. If only 1, delete it and reduce n by 1.
   b. If an even number, split them between S_p and S_n and use an average rank.
   c. If an odd number (3 or larger), delete 1 value, reduce n by 1, and treat the remaining zeroes as an even number of zeroes.

Example 8.7

Consider the case of two 5-acre experimental watersheds, one continually maintained with a low-density brush cover and the other allowed to naturally develop a more dense brush cover over the 11-year period of record. The annual maximum discharges were recorded for each year of record on both watersheds. The results are given in Table 8.9. Expectation would be for the annual maximum discharge rates from the low density brush covered watershed (i.e., watershed 1) to be higher than those for the watershed undergoing change (i.e., watershed 2). The null hypothesis would be that the means were equal, which would have the implication that the brush density was not a dominant factor in peak discharge magnitude. A one-tailed alternative is used, as peak discharges should decrease with increasing brush density. Therefore, the sum of the negative ranks is to be used as the value of the test statistic T. As shown in Table 8.9, the sample value is 3.

For a sample size of 11 and a one-tailed test, the critical values for levels of significance of 5%, 1%, and 0.5% are 14, 7, and 5, respectively (see Table A.9). Since the computed value of T is less than the critical value even for 0.5%, the null hypothesis can be rejected, which leads to the conclusion that the mean discharge decreases as the density of brush cover increases.

TABLE 8.9
Detection of Change in Cover Density

Year	Annual Maximum for Watershed 1	Annual Maximum for Watershed 2	Difference	r_p	r_n
1983	4.7	5.1	−0.4		2
1984	6.7	6.8	−0.1		1
1985	6.3	5.6	0.7	5	
1986	3.7	3.2	0.5	3	
1987	5.2	4.6	0.6	4	
1988	6.3	5.2	1.1	6	
1989	4.3	2.4	1.9	7	
1990	4.9	2.7	2.2	8	
1991	5.9	3.4	2.5	10	
1992	3.5	1.2	2.3	9	
1993	6.7	4.0	2.7	11	
				$S_p = 63$	$S_n = 3$

8.8.1 Ties

Tied values can play an important part in the accuracy of the Wilcoxon test. Ties can arise in two ways. First, paired values of the criterion variable can be the same, thus producing a difference of zero. Second, two differences can have the same magnitude in which case they need to be assigned equal ranks. Understanding the proper way to handle ties is important to the proper use of the Wilcoxon test.

Tied differences are more easily handled than tied values of the criterion. As a general rule, if two differences have the same magnitude, then they are given the average of the two ranks that would have been assigned to them. If three differences have the same magnitude, then all three are assigned the average of the three ranks that would have been assigned to them. Consider the case where the differences of 12 pairs of scores are as follows:

$$5, 7, -3, -5, 1, -2, 4, 3, 6, 4, 5, 8$$

Note that the sequence includes a pair of threes, with one negative and one positive. These would have been assigned ranks of 3 and 4, so each is assigned a rank of 3.5. Note that the difference in the signs of the differences, +3 and −3, is irrelevant in assigning ranks. The sequence of differences also includes two values of 4, which would have been ranked 5 and 6. Therefore, each is assigned a rank of 5.5. The above sequence includes three differences that have magnitudes of 5. Since they would have been assigned ranks of 7, 8, and 9, each of them is given a rank of 8. Thus, the ranks of the 12 differences shown above are:

$$8, 11, 3.5, 8, 1, 2, 5.5, 3.5, 10, 5.5, 8, 12$$

Note that ranks of 7 and 9 are not included because of the ties.

The presence of tied criterion scores is a more important problem since the procedure assumes that the random variable is continuously distributed. However, because tied measurements occur, the occurrence of tied scores must be addressed. A pair of identical scores produces a difference of zero. The handling of a difference of zero is subject to debate. Several courses of action are possible. First, the zero value could be assigned a split rank of 0.5, with half being placed in the +rank column and 0.5 being placed in the –rank column. Second, the pair with equal scores could be discarded and the sample size reduced by 1. The problem with this alternative is that the scores were measured and even suggest that the null hypothesis of equal means is correct. Therefore, discarding the value would bias the decision toward rejection of the null hypothesis. Third, if the differences contain multiple ties, they could be assigned equal ranks and divided between the +rank and –rank columns. Fourth, the tied difference could be ranked with the rank assigned to the column that is most likely to lead to acceptance of H_0. Each of the four alternatives is flawed, but at the same time has some merit. Therefore, a fifth possibility is to complete the test using each of the four rules. If the alternatives lead to different decisions, then it may be preferable to use another test.

Example 8.8

The handling of tied scores is illustrated with a hypothetical data set of eight paired scores (columns 1 and 2 of Table 8.10). The data include one pair of ties of 3, which yields a difference of zero. Method 1 would give the value a rank of 1, as it is the smallest difference, but split the rank between the positive and negative sums (see columns 4 and 5). Method 4 assigns the rank of 1 to the negative column, as this will make the smaller sum larger, thus decreasing the likelihood of rejecting the null hypothesis. For method 2, the zero difference is omitted from the sample, which reduces the sample size to seven.

TABLE 8.10
Effect of Tied Scores with the Wilcoxon Test

			Method 1		Method 2		Method 4	
X	Y	d	+ Rank	– Rank	+ Rank	– Rank	+ Rank	– Rank
7	3	4	6.5	—	5.5	—	6.5	—
8	6	2	3.5	—	2.5	—	3.5	—
4	1	3	5.0	—	4.0	—	5.0	—
6	1	3	8.0	—	7.0	—	8.0	—
7	5	2	3.5	—	2.5	—	3.5	—
3	3	0	0.5	0.5	—	—	—	1
9	5	4	6.5	—	5.5	—	6.5	—
2	3	–1	—	2.0	—	1	—	2
		Totals	33.5	2.5	27.0	1	33.0	3

For each of the three cases, the positive and negative ranks are added, which gives the sums in Table 8.10. Using the smaller of the positive and negative ranks yields values of 2.5, 1, and 3 for the test statistic under methods 1, 2, and 4, respectively. For methods 1 and 4, the critical values are 4, 2, and 0 for a two-tailed test (see Table A.9) for levels of significance of 5%, 2%, and 1%, respectively. For these cases, H_0 would be rejected at the 5% level, but accepted at the 2% level. The tied-score pair would not influence the decision. For method 2, the critical values are 2 and 0 for 5% and 2%, respectively. In this case, the null hypothesis is rejected at the 5% level but accepted at the 2% level, which is the same decision produced using methods 1 and 4.

In this case, the method of handling ties did not make a difference in the decision. In other cases, it is possible that the method selected would make a difference, which suggests that all methods should be checked.

Example 8.9

Twenty-eight pairs of watersheds are used to examine the effect of a particular type of land development on the erosion rate. The erosion rates (tons/acre/year) are given in Table 8.11 for the unchanged watershed (X_1) and the developed watershed (X_2). Of interest is whether development increases the mean erosion rate. Therefore, a one-tailed test is applied. For testing the alternative that the mean of X_2 is greater than the mean of X_1, the summation of the positive ranks is used as the value of the test statistic. For the 28 paired watersheds, only 12 experienced greater erosion than the undeveloped watershed. Two pairs of watersheds had the same erosion rate, which yields two differences of zero. These were assigned average ranks of 1.5, with one placed in the +rank column and one in the −rank column. The sum of the positive ranks is 220. If the normal approximation of Equations 8.28 to 8.30 is applied, the mean $\bar{T}$ is 203, the standard deviation S_T is 43.91, and the z value is

$$z = \frac{220 - 203}{43.91} = 0.387$$

Critical values of z for a lower-tailed test would be −1.282 and −1.645 for 10% and 5% levels of significance, respectively. Since the computed value of z is positive, the null hypothesis of equal means is accepted.

For the standard test, critical sums of 130 and 102 are obtained from Table A.9 for 5% and 1% levels of significance, respectively. Since S_p is larger than these critical values, the null hypothesis of equal means is accepted.

Note that if the two-tailed alternative was tested, then the computed value of T would be the lesser of S_p and S_n, which for the data would have been 186. This value is still safely within the region of acceptance.

8.9 ONE-SAMPLE CHI-SQUARE TEST

Watershed change can influence the spread of data, not just the mean. Channelization may increase the central tendency of floods and either increase or decrease the spread of the data. If the channelization reduces the natural storage, then the smoothing of the flows will not take place. However, if the channelization is limited in effects, it

TABLE 8.11
Application of Wilcoxon Test to Paired Watersheds

Erosion Rate for Watershed		Absolute Difference	Rank	
1	2		+	–
35	42	7	—	8
27	18	9	11	—
62	49	13	16	—
51	57	6	—	6
43	18	25	27	—
26	54	28	—	28
30	53	23	—	25
41	41	0	1.5	—
64	57	7	8	—
46	33	13	16	—
19	32	13	—	16
55	38	17	21.5	—
26	43	17	—	21.5
42	39	3	4	—
34	45	11	—	13
72	58	14	18	—
38	23	15	19	—
23	47	24	—	26
52	62	10	—	12
41	33	8	10	—
45	26	19	23	—
29	29	0	—	1.5
12	17	5	—	5
37	30	7	8	—
66	50	16	20	—
27	49	22	—	24
55	43	12	14	—
44	43	1	3	—
			$S_p = 220$	$S_n = 186$

may cause the smaller events to show a relatively greater effect than the larger events, thereby reducing the spread. Variation in reservoir operating rules can significantly influence the spread of flood data in channels downstream of the dam. Urbanization can also change the spread of the data.

Both parametric and nonparametric tests are available for detecting change in the variance of data. Additionally, both univariate and bivariate methods are available.

The one-sample test can be used to test the variance of a single random variable against a standard of comparison σ_o^2 with the following null hypothesis:

$$H_0\colon \sigma^2 = \sigma_o^2 \tag{8.31}$$

One of the following alternative hypotheses is used depending on the problem:

$$H_A\text{: } \sigma^2 < \sigma_o^2 \text{ (one-sided upper test)} \tag{8.32a}$$

$$H_A\text{: } \sigma^2 < \sigma_o^2 \text{ (one-sided lower test)} \tag{8.32b}$$

$$H_A\text{: } \sigma^2 \neq \sigma_o^2 \text{ (two-sided test)} \tag{8.32c}$$

To test these hypotheses, the following test statistic is used:

$$\chi^2 = \frac{(n-1)S^2}{\sigma_o^2} \tag{8.33}$$

in which n is the sample size, S^2 is the sample variance, σ_o^2 is the hypothesized population variance, and χ^2 is the value of a random variable that has a chi-square distribution with $(n - 1)$ degrees of freedom. Values of χ^2 are given in Table A.3. For a given level of significance, the decision is made as follows:

If H_A is	Reject H_0 if
$\sigma^2 > \sigma_o^2$	$\chi^2 > \chi_\alpha^2$
$\sigma^2 < \sigma_o^2$	$\chi^2 < \chi_{1-\alpha}^2$
$\sigma^2 \neq \sigma_o^2$	$\chi^2 < \chi_{1-\alpha/2}^2$ or $\chi^2 > \chi_{\alpha/2}^2$

Example 8.10

Manning's roughness coefficient (n) varies considerably along a stream reach. Tables of n suggest that the standard error in an estimate of n is about 20%. If the mean roughness for a channel reach is 0.053, the standard error would then be 0.011.

The following eight values along a reach are estimated using comparisons with pictures that show typical cross sections and a representative n value:

0.036, 0.042, 0.049, 0.042, 0.057, 0.067, 0.068, 0.063

Does the variation of this sample suggest that the variation in roughness along the stream reach is too great to consider the reach as a single reach?

A one-sided upper test should be applied because a small sample variation would actually be favorable, where a larger variation would suggest that the stream should be separated on the basis of sections of unequal roughness. The sample mean and standard deviation are 0.053 and 0.01244, respectively. Equation 8.34 is used:

$$\chi^2 = \frac{(8-1)(0.01244)^2}{(0.011)^2} = 8.953 \tag{8.34}$$

For $n - 1 = 7$ degrees of freedom, the critical values (Table A.3) are 14.067 and 18.475 for 5% and 1% levels of significance, respectively. Thus, it seems safe to accept the null hypothesis that the sample variation is sufficiently small. Consequently, the sample mean can be used to represent the roughness of a single channel reach.

8.10 TWO-SAMPLE *F*-TEST

For comparing the variances of two random samples, several strategies have been recommended, with each strategy valid when the underlying assumptions hold. One of these strategies is presented here.

For a two-tailed test, an *F*-ratio is formed as the ratio of the larger sample variance (S_l^2) to the smaller sample variance (S_s^2):

$$F = S_l^2 / S_s^2 \tag{8.35}$$

with degrees of freedom of $\upsilon_1 = n_l - 1$ for the numerator and $\upsilon_2 = n_S - 1$ for the denominator, where n_l and n_S are the sample sizes for the samples used to compute S_l^2 and S_s^2, respectively. It is not necessary to specify *a priori* which sample variance will be used in the numerator. The computed *F* is compared with the tabled *F* (Table A.4), and the null hypothesis of equal variances (H_0: $\sigma_1^2 = \sigma_2^2$) is accepted if the computed *F* is less than the tabled *F*. If the computed *F* is greater than the tabled *F*, then the null hypothesis is rejected in favor of the alternative hypothesis (H_A: $\sigma_1^2 \neq \sigma_2^2$). An important note for this two-tailed test is that the level of significance is twice the value from which the tabled *F* was obtained. For example, if the 5% *F* table (Table A.4) was used to obtain the critical *F* statistic, then the decision to accept or reject the null hypothesis is being made at a 10% level of significance. This is the price paid for using knowledge from the sample that one sample has the larger variance.

For a one-tailed test, specifying which of the two samples is expected to have the larger population variance is necessary before data collection. The computed *F* statistic is the ratio of the sample variance of the group expected to have the larger population variance to the sample variance from the second group. If it turns out that the sample variance of the group expected to have the larger variance is actually smaller than that of the group expected to have the smaller variance, then the computed *F* statistic will be less than 1. The level of significance is equal to that shown on the *F* table, not twice the value as necessary for the two-tailed test. The null hypothesis is rejected if the computed *F* is greater than the critical *F*. Because the direction is specified, the null hypothesis is accepted when the computed *F* is less than the critical *F*.

Example 8.11

Two sites are being considered for an infiltration-based stormwater management system. In addition to the average infiltration capacity, it is desirable to have a relative uniform variation of infiltration capacity over the site where the facility is to be located.

Because infiltration capacity is an important factor is siting, soil tests are made at both sites. In addition to the mean infiltration capacity, the variation is also computed. Because of other siting criteria, site A is currently favored over site B and will be selected if it meets the condition that the variability of the infiltration capacity at site A is not greater than that at site B. Thus, the following hypotheses are tested:

$$H_0\text{: } \sigma_A^2 = \sigma_B^2 \tag{8.36a}$$

$$H_A\text{: } \sigma_A^2 > \sigma_B^2 \tag{8.36b}$$

If the null hypothesis is accepted, then site A will be used; if it is rejected, then site B will be used.

Measurements of infiltration capacity are made at the two sites: A = {0.9, 1.4, 1.0, 1.7} in./hr and B = {1.4, 0.9, 1.2, 1.4}. This yields standard deviations of S_A = 0.3697 and S_B = 0.2363. The computed value of the test statistic is

$$F = \frac{S_A^2}{S_B^2} = \frac{(0.3697)^2}{(0.2363)^2} = 2.448 \tag{8.37}$$

with degrees of freedom of $\upsilon_A = n_A - 1 = 3$ and $\upsilon_B = n_B - 1 = 3$. The critical F value is 9.28 for a 5% level of significance (see Table A.4). Thus, we cannot reject the null hypothesis, so site A is selected as the better site for the planned stormwater management facility.

8.11 SIEGEL–TUKEY TEST FOR SCALE

The Siegel–Tukey test is a nonparametric alternative to the F-test for two variances. The Siegel–Tukey test assumes that measurements from two continuous populations X and Y are available. The sample of data includes n measurements on X and m measurements on Y. The data must be measured in such a way that the relative magnitudes of the measurements can be used to discriminate between values; therefore, the test can be used for data on either ordinal or interval scales. The test requires that the populations from which the data are drawn have equal central tendencies. If they are known to violate this assumption, the test can still be applied by linearly transforming the data by subtracting a respective measure of central tendency from each sample. The transformation should be such that the scales of both variables remain unchanged.

The Siegel–Tukey test is for the following hypotheses:

H_0: $\sigma_x = \sigma_y$
H_A: $\sigma_x \neq \sigma_y$ (two-tailed)
$\sigma_x < \sigma_y$ (one-tailed lower)
$\sigma_x > \sigma_y$ (one-tailed upper)

If a one-tailed alternative is selected, the direction of the test must be specified before collecting the data. If the null hypothesis is true and the central tendencies are identical, then, if the two samples are pooled and ordered from smallest to largest, the pooled values should appear to be in a random order. If the null hypothesis is false, then the scales would be different, and the measurements from the population with the smallest scale should be bunched near the center

The procedure for making the Siegel–Tukey test is as follows:

1. Pool the two samples and place the values in ascending order.
2. Select the center point to use for assigning scores to the measurements. If the pooled sample size N $(= m + n)$ is odd, use the center value. If the pooled sample size is an even integer, select either the $N/2$ or $(N + 2)/2$ value as the center value.
3. The center point gets a score of 1. Integer scores are assigned in ascending order such that:
   a. When N is odd, give a score of 2 to one of the adjacent positions. Then move to the other side of the center point and assign a 3 and 4 to the two positions, with the lower score given to the position closest to the center point. Continue moving from side to side and assigning two scores.
   b. When N is even and $N/2$ was used as the center point, assign scores of 2 and 3 to the two positions to the right of the center point. Then move to the other side and assign scores of 4 and 5. Continue alternating from side to side, while assigning two scores per side.
   c. When N is even and $(N + 2)/2$ was used as the center point, assign scores of 2 and 3 to the two positions to the left of the center point. Then move to the other side and assign scores of 4 and 5. Continue alternating from side to side, while assigning two scores per side.

Note: Table 8.12 illustrates the assignment of scores for N from 4 to 11.

TABLE 8.12
Assignment of Scores for Siegel–Tukey Test

N		
4	4 $\underline{1}$ 2 3	3 2 $\underline{1}$ 4
5	4 3 $\underline{1}$ 2 5	5 2 $\underline{1}$ 3 4
6	5 4 $\underline{1}$ 2 3 6	6 3 2 $\underline{1}$ 4 5
7	7 4 3 $\underline{1}$ 2 5 6	6 5 2 $\underline{1}$ 3 4 7
8	8 5 4 $\underline{1}$ 2 3 6 7	7 6 3 2 $\underline{1}$ 4 5 8
9	8 7 4 3 $\underline{1}$ 2 5 6 9	9 6 5 2 $\underline{1}$ 3 4 7 8
10	9 8 5 4 $\underline{1}$ 2 3 6 7 10	10 7 6 3 2 $\underline{1}$ 4 5 8 9
11	11 8 7 4 3 $\underline{1}$ 2 5 6 9 10	10 9 6 5 2 $\underline{1}$ 3 4 7 8 11

4. Having assigned scores to each position, indicate the sample, X or Y, from which the value was taken. If any measured values were equal, which is referred to as a tie, the scores are averaged.
5. Compute the sum of the scores associated with sample X and the sum of scores for sample Y. The computed value of the test statistic is the lesser of the two values.
6. If n and m are both 10 or less, obtain the rejection probability from Table A.10. Otherwise, use the normal approximation:

$$z = \frac{S_x \pm 0.5 - n(N+1)/2}{(nm(N+1)/12)^{0.5}} \tag{8.38}$$

where the + is used for the upper tail and the – is used for the lower tail. If ties occur, then the denominator of Equation 8.38 should be:

$$\left(\frac{nm(N+1)}{12} - \frac{nm\left(\sum U^3 - \sum U\right)}{12N(N-1)} \right)^{0.5} \tag{8.39}$$

where U is the number of observations tied for any given position, and the sum is over all sets of ties. The probability associated with the z value of Equation 8.38 can be obtained from a table of the standard normal distribution (see Table A.1).

Example 8.12

A flood record of 7 years was collected at the inflow to a small dam; this set is denoted as X. After the dam was constructed and filled to normal depth, a flood record of 5 years was collected downstream of the outlet and at the same site where the prestructure data were collected; this set is denoted as Y. The annual maximum discharges follow:

$$X = \{11600, 18400, 16200, 10500, 17400, 12000, 22600\}$$

$$Y = \{7260, 6150, 4790, 8200, 7660\}$$

The flood storage behind the dam contributed to a reduction in central tendency. It is believed that the storage will also decrease the spread of the flows. Therefore, the following hypotheses are appropriately tested:

$$H_0\text{: } \sigma_x = \sigma_y$$

$$H_A\text{: } \sigma_x > \sigma_y$$

TABLE 8.13
Application of Siegel–Tukey Test

Pooled Series	United Score	X_a Score	Y_a Score
–4500	12	12.0	—
–3400	9	9.0	—
–3000	8	8.0	—
–2210	5	—	5.0
–850	4	—	4.0
260	1	—	1.0
660	2	—	2.0
1200	3	4.5	—
1200	6	—	4.5
2400	7	7.0	—
3400	10	10.0	—
7600	11	11.0	—
	Totals	61.5	16.5

Modeling studies indicate central tendency flows of 15,000 cfs and 7000 cfs, respectively, for the pre- and post-development periods. These flows are used to adjust the flows for the two periods by subtracting the central tendency flows from the measured flows, which yields:

$$X_a = \{-3400, 3400, 1200, -4500, 2400, -3000, 7600\}$$

$$Y_a = \{260, -850, -2210, 1200, 660\}$$

The values of X_a and Y_a are pooled and ordered in ascending order (see Table 8.13). The center point is identified as the $N/2$ value. The scores are then assigned to the remaining values, with the order following the pattern shown in Table 8.13 for even values of N. Since two of the pooled values are identical, that is, 1200 cfs, the untied scores for these two flows are averaged and the revised scores assigned to the appropriate series (see columns 3 and 4 of Table 8.13). The column sums are also given in Table 8.13. Since the alternative hypothesis is a one-tailed test that indicates series Y is expected to have a smaller variation than series X, then the sum of the scores for Y_a is used as the value of the test statistic. (Note that even if the sum of the Y_a scores was larger than the sum of the X_a scores, it would still be used as the test statistic. The smaller of the two values is used only in a two-tailed analysis.) Thus, the test statistic is 16.5.

The rejection probability is obtained from Table A.10 for $n = 7$ and $m = 5$ as 0.004. Since this is less than 5%, the results suggest that the null hypothesis can be rejected; therefore, the variation of the flows after the dam was in place is significantly less than the variation of the flows prior to construction.

Example 8.13

The large-sample assessment of differences in variation is demonstrated using the annual flood series for the Elizabeth River, New Jersey, watershed. The annual flood series for the 1924 to 1988 water years ($N = 65$) represents data for pre-urban development and post-urban development. While the development took place over an extended period of time, the series was broken into two parts, which can be referred to as the low urbanization (X) and high urbanization (Y) periods, which represent the 1924–1949 and 1950–1988 periods, respectively (see Figure 7.2). The urbanization appears to have caused an increase in both central tendency and variance. Because urbanization should increase the spread, the most appropriate alternative hypothesis would be that the variance of the first series is less than the variance of the second series.

To remove the effect of the difference in central tendency, the median of 1004 cfs was subtracted from each flow in the 1924–1949 period and the median of 1900 cfs was subtracted from each flow in the 1950–1988 period. The median-adjusted flows are given as a pooled ordered data set in columns 3 and 6 of Table 8.14. The scores for the 1924–1949 data are given in columns 4 and 7, with the 1950–1988 scores in columns 5 and 8. The sum of the X scores is used as the value of the test statistic because the one-sided alternative hypothesis indicates that the variance of the X-series discharges is the smaller of the two. The sum of 623 is used in Equation 8.31 to compute the z statistic:

$$z = \frac{623 + 0.5 - 26(65+1)/2}{(20(39)(65+1)/12)^{0.5}} = -3.14 \tag{8.40}$$

This result corresponds to a rejection probability of 0.0008, which indicates that the null hypothesis should be rejected. Therefore, the flows in the period of urban development are greater than in the pre-urban period.

8.12 PROBLEMS

8-1 Perform the following graphical analyses:
   a. On a graph of discharge versus year, graph a 30-year record of discharge for a watershed subject to urbanization that occurred over a 2-year period (i.e., episodic) near the center of the record with the effect of increasing the mean and decreasing the variance.
   b. On a single plot of frequency versus the magnitude of the discharge, show histograms for the two periods.
   c. On a single piece of frequency paper, show two frequency curves that would reflect the two conditions.

8-2 Perform the following graphical analyses:
   a. On a graph of discharge versus years, graph a 25-year record of annual maximum discharges for a watershed subject to heavy afforestation starting in about the 10th year of record.

TABLE 8.14
Siegel–Tukey Test for Equality of Variances for Pre-Urbanization (X) and Post-Urbanization (Y) of Elizabeth River Watershed, Elizabeth, New Jersey

Discharge (cfs)		Median-Adjusted Pooled Series	X Score	Y Score	Median-Adjusted Pooled Series	X Score	Y Score
X 1924–1949	Y 1950–1988						
1290	452	−1448		64	80		13
980	2530	−1198		63	96	14	
741	1740	−1105		60	220		17
1630	1860	−1100		59	230		18
829	1270	−1094		56	276	21	
903	2200	−1076		52	300		22
418	1530	−948		52	316	25	
549	795	−710		51	340		26
686	1760	−630		48	376	29	
1320	806	−586	47		420		30
850	1190	−455	44		540		33
614	952	−410		43	570		34
1720	1670	−390	40		626	37	
1060	824	−370		39	630		38
1680	702	−360		36	650		41
760	1490	−318	35		676	42	
1380	1600	−300		32	716	45	
1030	800	−270		31	820		46
820	3330	−263	28		990		49
1020	1540	−244	27		1040		50
998	2130	−230		24	1230		53
3500	3770	−184	23		1310		54
1100	2240	−175	20		1430		57
1010	3210	−174	19		1450		58
830	2940	−160		16	1870		61
1030	2720	−154	15		2496	62	
	2440	−140		12	2600		65
	3130	−101	11			623	1522
	4500	−50		8		T_x	T_y
	2890	−40		7			
	2470	−24	4				
	1900	−6	3				
	1980	0		1			
	2550	6	2				
	3350	16	5				
	2120	26	7.5				
	1850	26	7.5				
	2320	56	10				
	1630						

b. On a single plot of frequency versus flood magnitude, show the histograms for the first 10 years versus the last 15 years.
c. On a single piece of frequency paper, show frequency curves that would reflect the pre-afforestation and afforestation periods.

8-3 Show the effect of the continuity correction of Equation 8.7 for sample sizes of 25 and 50 for probabilities of less than 10%.

8-4 In Example 8.1, test the effect of mulch if only 32 of the sites showed a difference with 26 of the 32 showing lower erosion for the higher mulch rate.

8-5 As part of a stream restoration project, homeowners in the community are polled before and after the restoration effort as to the aesthetic characteristics of the stream reach. Of the 35 people polled, only 14 indicate a change, with 11 indicating a change for improved aesthetics. Is it reasonable to conclude that the restoration work improved aesthetic qualities of the reach?

8-6 Create a table of critical values for the signs test for 5% and 1% with sample sizes (n) from five to nine.

8-7 For 15 storm events, suspended solids concentrations were measured after a watershed had developed. The mean and standard deviation are 610 and 178, respectively. Before the development of the watershed, nine measurements were made with a mean and standard deviation of 380 and 115, respectively. Can we conclude that the post-development concentration is significantly greater than the predevelopment concentration?

8-8 The total phosphate concentration (mg/L) in eight samples of rainwater are {0.09, 0.62, 0.37, 0.18, 0.30, 0.44, 0.23, 0.35}. Six measurements in runoff from a local forested watershed are {0.97, 0.56, 1.22, 0.31, 0.73, 0.66}. Can we conclude that the concentration in streamflow is significantly higher than in the rainfall?

8-9 Three days after a storm event, six soil moisture measurements are made at the top of a slope and six more at the bottom. The mean and standard deviation at the top were 17 and 4.5, respectively. At the bottom, the measurements yield a mean and standard duration of 26 and 3.1, respectively. Is it reasonable to conclude that the soil moisture is higher at the bottom of the slope?

8-10 Apply the Mann–Whitney test with the data of Problem 8-8 to determine if the concentration is higher in the streamflow.

8-11 Biochemical oxygen demand (mg/L) measurements are made for drainage from a residential area and from a business district. The following measurements are from the residential area: $R = \{22, 41, 57, 36, 15, 28\}$. The measurements from the business district are $B = \{66, 73, 15, 41, 52, 38, 61, 29\}$. Use the Mann–Whitney test to decide if the populations are different.

8-12 Use the Mann–Whitney test to decide if the annual maximum discharges for the period from 1935 to 1948 (A) are from a different population than the discharges from 1949 to 1973 (B) for the Floyd River at James, Iowa. $A = \{1460, 4050, 3570, 2060, 1300, 1390, 1720, 6280, 1360, 7440, 5320,$

1400, 3240, 2710}, B = {4520, 4840, 8320, 13900, 71500, 6250, 2260, 318, 1330, 970, 1920, 15100, 2870, 20600, 3810, 726, 7500, 7170, 2000, 829, 17300, 4740, 13400, 2940, 5660}.

8-13 Trees on one watershed of each pair of eight paired watershed are removed. The annual maximum discharges for the following 8 years are recorded for all 16 watersheds, with the values for the forested (F) and deforested (D) watersheds as follows:

Pair	1	2	3	4	5	6	7	8
F	65	87	94	61	49	75	90	68
D	63	82	126	66	57	72	104	69

Use the paired t-test to decide if the deforested watersheds tend to have higher discharges.

8-14 Two methods for measuring infiltration rates are used on each of 12 watersheds. The issue is to determine whether the methods provide similar values.

a. Use the paired t-test to decide.
b. Apply the two-sample t-test (Section 8.4) and compare the results with the paired t-test.

Discuss any differences detected in the results.

A	0.221	0.314	0.265	0.166	0.128	0.272	0.334	0.296	0.187	0.097	0.183	0.207
B	0.284	0.363	0.338	0.231	0.196	0.222	0.292	0.360	0.261	0.127	0.158	0.268

8-15 Use the Walsh test to decide whether the deforested watersheds of Problem 8-13 tend to have higher discharges than the forested watershed.

8-16 Use the Walsh test to decide if the two methods of measuring infiltration in Problem 8-14 give dissimilar estimates.

8-17 Use the Wilcoxon test to decide if the deforested watersheds of Problem 8-13 tend to have higher discharges than the forested watershed.

8-18 Use the Wilcoxon test to decide if the two methods of measuring infiltration in Problem 8-14 give dissimilar estimates.

8-19 The standard deviation of 58 annual maximum discharges for the Piscataquis River is 4095 ft^3/sec. Test whether this is significantly different from 3500 ft^3/sec.

8-20 A sample of five measurements follows: x = {0.16, –0.74, –0.56, 1.07, 0.43}. Is the sample standard deviation the same as a standard normal distribution?

8-21 Using the rainfall phosphate concentration of Problem 8-8, test to determine whether the standard deviation is greater than 0.1 mg/L.

8-22 Using the BOD measurements for the business district (Problem 8-11), test to determine whether the standard deviation is significantly less than 25 mg/L.

8-23 Is it reasonable to conclude that the variance of the residential BOD measurements (Problem 8-11) is significantly smaller than the variance of the values for the business district? Use the two-sample F-test.

8-24 Is it reasonable to conclude that the discharges for the forested watersheds (Problem 8-13) have a smaller variance than the variance of the deforested watersheds? Use the two-sample F-test.

8-25 Is it reasonable to conclude that the discharges for the forested watersheds (Problem 8-13) have a smaller variance than the variance of the deforested watersheds? Use the Siegel–Tukey test.

9 Detection of Change in Distribution

9.1 INTRODUCTION

Frequency analysis (see Chapter 5) is a univariate method of identifying a likely population from which a sample was drawn. If the sample data fall near the fitted line that is used as the best estimate of the population, then it is generally safe to use the line to make predictions. However, "nearness to the line" is a subjective assessment, not a systematic statistical test of how well the data correspond to the line. That aspect of a frequency analysis is not objective, and individuals who have different standards as to what constitutes a sufficiently good agreement may be at odds on whether or not to use the fitted line to make predictions. After all, lines for other distributions may provide a degree of fit that appears to be just as good. To eliminate this element of subjectivity in the decision process, it is useful to have a systematic test for assessing the extent to which a set of sample data agree with some assumed population. Vogel (1986) provided a correlation coefficient test for normal, log-normal, and Gumbel distributions.

The goal of this chapter is to present and apply statistical analyses that can be used to test for the distribution of a random variable. For example, if a frequency analysis suggested that the data could have been sampled from a lognormal distribution, one of the one-sample tests presented in this chapter could be used to decide the statistical likelihood that this distribution characterizes the underlying population. If the test suggests that it is unlikely to have been sampled from the assumed probability distribution, then justification for testing another distribution should be sought.

One characteristic that distinguishes the statistical tests from one another is the number of samples for which a test is appropriate. Some tests are used to compare a sample to an assumed population; these are referred to as one-sample tests. Another group of tests is appropriate for comparing whether two distributions from which two samples were drawn are the same, known as two-sample tests. Other tests are appropriate for comparing samples from more than two distributions, referred to as *k*-sample tests.

9.2 CHI-SQUARE GOODNESS-OF-FIT TEST

The chi-square goodness-of-fit test is used to test for a significant difference between the distribution suggested by a data sample and a selected probability distribution. It is the most widely used one-sample analysis for testing a population distribution. Many statistical tests, such as the *t*-test for a mean, assume that the data have been drawn from a normal population, so it may be necessary to use a statistical test, such as the chi-square test, to check the validity of the assumption for a given sample of data. The chi-square test can also be used as part of the verification phase of modeling to verify the population assumed when making a frequency analysis.

9.2.1 Procedure

Data analysts are often interested in identifying the density function of a random variable so that the population can be used to make probability statements about the likelihood of occurrence of certain values of the random variable. Very often, a histogram plot of the data suggests a likely candidate for the population density function. For example, a frequency histogram with a long right tail might suggest that the data were sampled from a lognormal population. The chi-square test for goodness of fit can then be used to test whether the distribution of a random variable suggested by the histogram shape can be represented by a selected theoretical probability density function (PDF). To demonstrate the quantitative evaluation, the chi-square test will be used to evaluate hypotheses about the distribution of the number of storm events in 1 year, which is a discrete random variable.

Step 1: Formulate hypotheses. The first step is to formulate both the null (H_0) and the alternative (H_A) hypotheses that reflect the theoretical density function (PDF; continuous random variables) or probability mass function (PMF; discrete random variables). Because a function is not completely defined without the specification of its parameters, the statement of the hypotheses must also include specific values for the parameters of the function. For example, if the population is hypothesized to be normal, then μ and σ must be specified; if the hypotheses deal with the uniform distribution, values for the location α and scale β parameters must be specified. Estimates of the parameters may be obtained either empirically or from external conditions. If estimates of the parameters are obtained from the data set used in testing the hypotheses, the degrees of freedom must be modified to reflect this.

General statements of the hypotheses for the chi-square goodness-of-fit test of a continuous random variable are:

$$H_0\text{: } X \sim \text{PDF (stated values of parameters)} \tag{9.1a}$$

$$H_A\text{: } X \neq \text{PDF (stated values of parameters)} \tag{9.1b}$$

If the random variable is a discrete variable, then the PMF replaces the PDF. The following null and alternative hypotheses are typical:

H_0: The number of rainfall events that exceed 1 cm in any year at a particular location can be characterized by a uniform density function with a location parameter of zero and a scale parameter of 40.

H_A: The uniform population U (0, 40) is not appropriate for this random variable.

Mathematically, these hypotheses are

$$H_0\text{: } f(n) = \mathrm{U}(\alpha = 0,\ \beta = 40) \tag{9.2a}$$

$$H_A\text{: } f(n) \neq \mathrm{U}(\alpha = 0,\ \beta = 40) \tag{9.2b}$$

Note specifically that the null hypothesis is a statement of equality and the alternative hypothesis is an inequality. Both hypotheses are expressed in terms of population parameters, not sample statistics.

Rejection of the null hypothesis would not necessarily imply that the random variable is not uniformly distributed. It may also be rejected because one or both of the parameters, in this case 0 and 40, are incorrect. Rejection may result because the assumed distribution is incorrect, one or more of the assumed parameters is incorrect, or both.

The chi-square goodness-of-fit test is always a one-tailed test because the structure of the hypotheses are unidirectional; that is, the random variable is either distributed as specified in the null hypothesis or it is not.

Step 2: Select the appropriate model. To test the hypotheses formulated in step 1, the chi-square test is based on a comparison of the observed frequencies of values in the sample with frequencies expected with the PDF of the population, which is specified in the hypotheses. The observed data are typically used to form a histogram that shows the observed frequencies in a series of k cells. The cell bounds are often selected such that the cell width for each cell is the same; however, unequal cell widths could be selected to ensure a more even distribution of the observed and expected frequencies. Having selected the cell bounds and counted the observed frequencies for cell i (O_i), the expected frequencies E_i for each cell can be computed using the PDF of the population specified in the null hypothesis of step 1. To compute the expected frequencies, the expected probability for each cell is determined for the assumed population and multiplied by the sample size n. The expected probability for cell i, p_i, is the area under the PDF between the cell bounds for that cell. The sum of the expected frequencies must equal the total sample size n. The frequencies can be summarized in a cell structure format, such as Figure 9.1a.

The test statistic, which is a random variable, is a function of the observed and expected frequencies, which are also random variables:

$$\chi^2 = \sum_{i=1}^{k} \frac{(O_i - E_i)^2}{E_i} \tag{9.3}$$

where χ^2 is the computed value of a random variable having a chi-square distribution with ν degrees of freedom; O_i and E_i are the observed and expected frequencies in cell i, respectively; and k is the number of discrete categories (cells) into which the data are separated. The random variable χ^2 has

Cell bound	$-\infty$			....	
Cell number i	1	2	3	....	k
Observed frequency (O_i)	O_1	O_2	O_3	....	O_k
Expected frequency (E_i)	E_1	E_2	E_3	....	E_k
$(O_i - E_i)^2/E_i$	$\frac{(O_1 - E_1)^2}{E_1}$	$\frac{(O_2 - E_2)^2}{E_2}$	$\frac{(O_3 - E_3)^2}{E_3}$	....	$\frac{(O_k - E_k)^2}{E_k}$

(a)

Cell bound	0	10	20	30	40
Cell number i	1	2	3	4	
Observed frequency (O_i)	18	19	25	18	
Expected frequency (E_i)	20	20	20	20	
$(O_i - E_i)^2/E_i$	0.20	0.05	1.25	0.20	

(b)

FIGURE 9.1 Cell structure for chi-square goodness-of-fit test: (a) general structure; and (b) structure for the number of rainfall events.

a sampling distribution that can be approximated by the chi-square distribution with $k - j$ degrees of freedom, where j is the number of quantities that are obtained from the sample of data for use in calculating the expected frequencies. Specifically, since the total number of observations n is used to compute the expected frequencies, 1 degree of freedom is lost. If the mean and standard deviation of the sample are needed to compute the expected frequencies, then two additional degrees of freedom are subtracted (i.e., $\upsilon = k - 3$). However, if the mean and standard deviation are obtained from past experience or other data sources, then the degrees of freedom for the test statistic remain $\upsilon = k - 1$. It is important to note that the degrees of freedom do not directly depend on the sample size n; rather they depend on the number of cells.

Step 3: Select the level of significance. If the decision is not considered critical, a level of significance of 5% may be considered appropriate, because of convention. A more rational selection of the level of significance will be discussed later. For the test of the hypotheses of Equation 9.2, a value of 5% is used for illustration purposes.

Step 4: Compute estimate of test statistic. The value of the test statistic of Equation 9.3 is obtained from the cell frequencies of Figure 9.1b. The range of the random variable was separated into four equal intervals of ten. Thus, the expected probability for each cell is 0.25 (because the random variable is assumed to have a uniform distribution and the width of the cells is the same). For a sample size of 80, the expected frequency for each of the four cells is 20 (i.e., the expected probability times the total number of observations). Assume that the observed frequencies of 18, 19, 25, and 18 are determined from the sample, which yields the cell structure shown in Figure 9.1b.

Using Equation 9.3, the computed statistic χ^2 equals 1.70. Because the total frequency of 80 was separated into four cells for computing the expected frequencies, the number of degrees of freedom is given by $\upsilon = k - 1$, or $4 - 1 = 3$.

Step 5: Define the region of rejection. According to the underlying theorem of step 2, the test statistic has a chi-square distribution with 3 degrees of freedom. For this distribution and a level of significance of 5%, the critical value of the test statistic is 7.81 (Table A.3). Thus, the region of rejection consists of all values of the test statistic greater than 7.81. Note again, that for this test the region of rejection is always in the upper tail of the chi-square distribution.

Step 6: Select the appropriate hypothesis. The decision rule is that the null hypothesis is rejected if the chi-square value computed in step 4 is larger than the critical value of step 5. Because the computed value of the test statistic (1.70) is less than the critical value (7.81), it is not within the region of rejection; thus the statistical basis for rejecting the null hypothesis is not significant. One may then conclude that the uniform distribution with location and scale parameters of 0 and 40, respectively, may be used to represent the distribution of the number of rainfall events. Note that other distributions could be tested and found to be statistically acceptable, which suggests that the selection of the distribution to test should not be an arbitrary decision.

In summary, the chi-square test for goodness of fit provides the means for comparing the observed frequency distribution of a random variable with a population distribution based on a theoretical PDF or PMF. An additional point concerning the use of the chi-square test should be noted. The effectiveness of the test is diminished if the expected frequency in any cell is less than 5. When this condition occurs, both the expected and observed frequencies of the appropriate cell should be combined with the values of an adjacent cell; the value of k should be reduced to reflect the number of cells used in computing the test statistic. It is important to note that this rule is based on expected frequencies, not observed frequencies.

To illustrate this rule of thumb, consider the case where observed and expected frequencies for seven cells are as follows:

Cell	1	2	3	4	5	6	7
O_i	3	9	7	5	9	4	6
E_i	6	8	4	6	10	7	2

Note that cells 3 and 7 have expected frequencies less than 5, and should, therefore, be combined with adjacent cells. The frequencies of cell 7 can be combined with the frequencies of cell 6. Cell 3 could be combined with either cell 2 or cell 4. Unless physical reasons exist for selecting which of the adjacent cells to use, it is probably best to combine the cell with the adjacent cell that has the lowest expected frequency count. Based on this, cells 3 and 4 would be combined. The revised cell configuration follows:

Cell	1	2	3	4	5
O_i	3	9	12	9	10
E_i	6	8	10	10	9

The value of k is now 5, which is the value to use in computing the degrees of freedom. Even though the observed frequency in cell 1 is less than 5, that cell is not combined. Only expected frequencies are used to decide which cells need to be combined. Note that a cell count of 5 would be used to compute the degrees of freedom, rather than a cell count of 7.

9.2.2 Chi-Square Test for a Normal Distribution

The normal distribution is widely used because many data sets have shown to have a bell-shaped distribution and because many statistical tests assume the data are normally distributed. For this reason, the test procedure is illustrated for data assumed to follow a normal population distribution.

Example 9.1

To illustrate the use of the chi-square test with the normal distribution, a sample of 84 discharges is used. The histogram of the data is shown in Figure 9.2. The sample mean and standard deviation of the random variable were 10,100, and 780, respectively. A null hypothesis is proposed that the random variable is normally distributed with a mean and standard deviation of 10,100 and 780, respectively. Note that the sample moments are being used to define the population parameters in the statement of hypotheses; this will need to be considered in the computation of the degrees of freedom. Table 9.1 gives the cell bounds used to form the observed and expected frequency cells (see column 2). The cell bounds are used to compute standardized

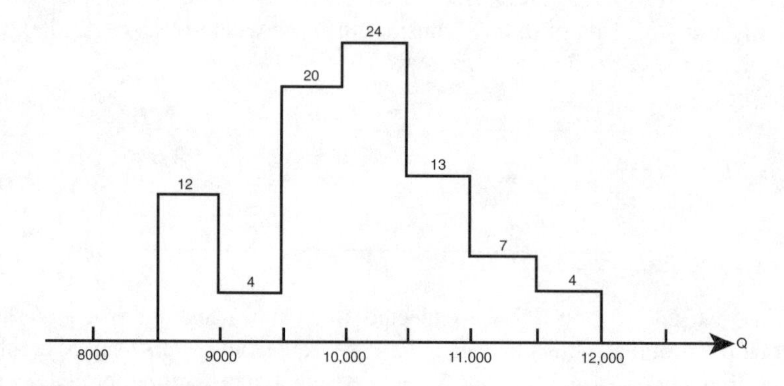

FIGURE 9.2 Histogram of discharge rate (Q, cfs).

TABLE 9.1
Computations for Example 9.1

Cell i	Cell Bound	z_i	$P(z < z_i)$	Expected Probability	Expected Frequency	Observed Frequency	$\frac{(O_i - E_i)^2}{E_i}$
1	9000	$\frac{9000 - 10,100}{780} = -1.41$	0.0793	0.0793	6.66	12	$\frac{(12 - 6.66)^2}{6.66} = 4.282$
2	9500	$\frac{9500 - 10,100}{780} = -0.77$	0.2206	0.1413	11.87	4	$\frac{(4 - 11.87)^2}{11.87} = 5.218$
3	10,000	$\frac{10,000 - 10,100}{780} = -0.13$	0.4483	0.2277	19.13	20	$\frac{(20 - 19.13)^2}{19.13} = 0.040$
4	10,500	$\frac{10,500 - 10,100}{780} = 0.52$	0.6985	0.2502	21.02	24	$\frac{(24 - 21.02)^2}{21.02} = 0.422$
5	11,000	$\frac{11,000 - 10,100}{780} = 1.15$	0.8749	0.1764	14.82	13	$\frac{(13 - 14.82)^2}{14.82} = 0.224$
6	11,500	$\frac{11,500 - 10,100}{780} = 1.79$	0.9633	0.0884	7.42 } 10.50 (7.42 + 3.08 combined)	11	$\frac{(11 - 10.50)^2}{10.50} = 0.024$
7	∞	— —	1.0000	0.0367	3.08		
				1.0000	84	84	10.209

variates z_i for the bounds of each interval (column 3), the probability that the variate z is less than z_i (column 4), the expected probabilities for each interval (column 5), the expected and observed frequencies (columns 6 and 7), and the cell values of the chi-square statistic of Equation 9.3 (column 8).

The test statistic has a computed value of 10.209. Note that because the expected frequency for the seventh interval was less than 5, both the observed and expected frequencies were combined with those of the sixth cell. Three degrees of freedom are used for the test. With a total of six cells, 1 degree of freedom was lost for n, while two were lost for the mean and standard deviation, which were obtained from the sample of 84 observations. (If past evidence had indicated a mean of 10,000 and a standard deviation of 1000, and these statistics were used in Table 9.1 for computing the expected probabilities, then 5 degrees of freedom would be used.) For a level of significance of 5% and 3 degrees of freedom, the critical chi-square value is 7.815. The null hypothesis is, therefore, rejected because the computed value is greater than the critical value. One may conclude that discharges on this watershed are not normally distributed with $\mu = 10{,}100$ and $\sigma = 780$. The reason for the rejection of the null hypothesis may be due to one or more of the following: (1) the assumption of a normal distribution is incorrect, (2) $\mu \neq 10{,}100$, or (3) $\sigma \neq 780$.

Alternative Cell Configurations

Cell boundaries are often established by the way the data were collected. If a data set is collected without specific bounds, then the cell bounds for the chi-square test cells can be established at any set of values. The decision should not be arbitrary, especially with small sample sizes, since the location of the bounds can influence the decision. For small and moderate sample sizes, multiple analyses with different cell bounds should be made to examine the sensitivity of the decision to the placement of the cell bounds.

While any cell bounds can be specified, consider the following two alternatives: equal intervals and equal probabilities. For equal-interval cell separation, the cell bounds are separated by an equal cell width. For example, test scores could be separated with an interval of ten: 100–90, 90–80, 80–70, and so on. Alternatively, the cell bounds could be set such that 25% of the underlying PDF was in each cell. For the standard normal distribution $N(0, 1)$ with four equal-probability cells, the upper bounds of the cells would have z values of -0.6745, 0.0, 0.6745, and ∞. The advantage of the equal-probability cell alternative is that the probability can be set to ensure that the expected frequencies are at least 5. For example, for a sample size of 20, 4 is the largest number of cells that will ensure expected frequencies of 5. If more than four cells are used, then at least 1 cell will have an E_i of less than 5.

Comparison of Cell Configuration Alternatives

The two-cell configuration alternatives can be used with any distribution. This will be illustrated using the normal distribution.

TABLE 9.2
Chi-Square Test of Pipe Length Data Using Equal Interval Cells

Cell	Length (ft) Range	Observed Frequency, O_i	z_i	Σp_i	p_i	$E_i = np_i$		$\frac{(O_i - E_i)^2}{E_i}$
1	0–1000	3	−1.099	0.1358	0.1358	9.506		4.453
2	1000–2000	20	−0.574	0.2829	0.1471	10.297		9.143
3	2000–3000	22	−0.050	0.4801	0.1972	13.804		4.866
4	3000–4000	8	0.474	0.6822	0.2021	14.147		2.671
5	4000–5000	7	0.998	0.8409	0.1587	11.109		1.520
6	5000–6000	2	1.523	0.9361	0.0952	6.664		0.116
7	6000–7000	3	2.047	0.9796	0.0435	3.045	11.137	0
8	7000–∞	5	∞	1.0000	0.0204	1.428		0
		70				70.000		22.769

Example 9.2

Consider the total lengths of storm-drain pipe used on 70 projects (see Table B.6). The pipe-length values have a mean of 3096 ft and a standard deviation of 1907 ft. The 70 lengths are allocated to eight cells using an interval of 1000 ft (see Table 9.2 and Figure 9.3a). The following hypotheses will be tested:

$$\text{Pipe length} \sim N(\mu = 3096, \sigma = 1907) \tag{9.4a}$$

$$\text{Pipe length} \neq N(3096, 1907) \tag{9.4b}$$

Note that the sample statistics are used to define the hypotheses and will, therefore, be used to compute the expected frequencies. Thus, 2 degrees of freedom will be subtracted because of their use. To compute the expected probabilities, the standard normal deviates z that correspond to the upper bounds X_u of each cell are computed (see column 4 of Table 9.2) using the following transformation:

$$z = \frac{X_u - \overline{X}}{S_x} = \frac{X_u - 3096}{1907} \tag{9.5}$$

The corresponding cumulative probabilities are computed from the cumulative standard normal curve (Table A.1) and are given in column 5. The probabilities associated with each cell (column 6) are taken as the differences of the cumulative probabilities of column 5. The expected frequencies (E_i) equal the product of the sample size 70 and the probability p_i (see column 7). Since the expected frequencies in the last two cells are less than 5, the last three cells are combined, which yields six cells. The cell values of the chi-square statistic of Equation 9.3 are given in column 8, with a sum of 22.769. For six cells with 3 degrees of freedom lost, the

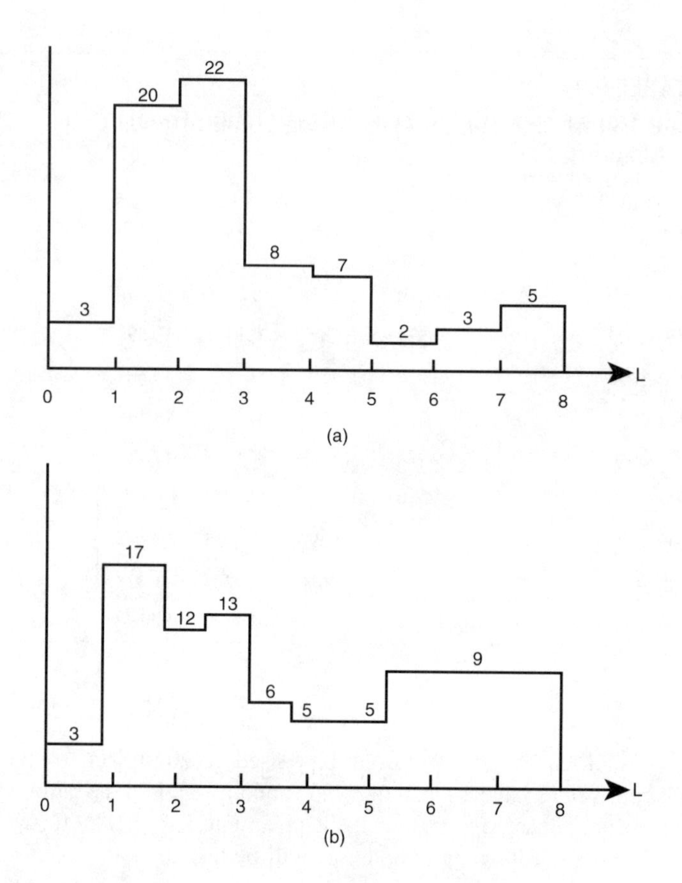

FIGURE 9.3 Frequency histogram of pipe lengths (L, ft × 10^3) using (a) equals interval and (b) equal probability cells.

critical test statistic for a 5% level of significance is 7.815. Thus, the computed value is greater than the critical value, so the null hypothesis can be rejected. The null hypothesis would be rejected even at a 0.5% level of significance ($\chi^2_{.005} = 12.84$). Therefore, the distribution specified in the null hypothesis is unlikely to characterize the underlying population.

For a chi-square analysis using the equal-probability alternative, the range is divided into eight cells, each with a probability of 1/8 (see Figure 9.3b). The cumulative probabilities are given in column 1 of Table 9.3. The z_i values (column 2) that correspond to the cumulative probabilities are obtained from the standard normal table (Table A.1). The pipe length corresponding to each z_i value is computed by (see column 3):

$$X_u = \mu + z_i\sigma = 3096 + 1907z_i \tag{9.6}$$

These upper bounds are used to count the observed frequencies (column 4) from the 70 pipe lengths. The expected frequency (E_i) is $np = 70(1/8) = 8.75$. Therefore, the computed chi-square statistic is 18.914. Since eight cells were used and 3 degrees

TABLE 9.3
Chi-Square Test of Pipe Length Data Using Equal Probability Cells

Σp	X	X	O_i	X_i	
0.125	−1.150	903	3	70/8	3.779
0.250	−0.675	1809	17	70/8	7.779
0.375	−0.319	2488	12	70/8	1.207
0.500	0.000	3096	13	70/8	2.064
0.625	0.319	3704	6	70/8	0.864
0.750	0.675	4383	5	70/8	1.607
0.875	1.150	5289	5	70/8	1.607
1.000	∞	∞	9	70/8	0.007
					18.914

of freedom were lost, the critical value for a 5% level of significance and 5 degrees of freedom is 11.070. Since the computed value exceeds the critical value, the null hypothesis is rejected, which suggests that the population specified in the null hypothesis is incorrect.

The computed value of chi-square for the equal-probability delineation of cell bounds is smaller than for the equal-cell-width method. This occurs because the equal-cell-width method causes a reduction in the number of degrees of freedom, which is generally undesirable, and the equal-probability method avoids cells with a small expected frequency. Since the denominator of Equation 9.3 acts as a weight, low expected frequencies contribute to larger values of the computed chi-square value.

9.2.3 Chi-Square Test for an Exponential Distribution

The histogram in Figure 9.4 has the general shape of an exponential decay function, which has the following PDF:

$$f(x) = \lambda e^{-\lambda x} \quad \text{for } x > 0 \tag{9.7}$$

in which λ is the scale parameter and x is the random variable. It can be shown that the method of moments estimator of λ is the reciprocal of the mean (i.e., $\lambda = 1/\bar{x}$). Probabilities can be evaluated by integrating the density function $f(x)$ between the upper and lower bounds of the interval. Intervals can be set randomly or by either the constant-probability or constant-interval method.

Example 9.3

Using the sediment yield Y data of Table B.1 and the histogram of Figure 9.4, a test was made for the following hypotheses:

TABLE 9.4
Computations for Example 9.3

Cell	Interval	$f(x)$	Expected $Nf(x)$		Observed O_i		$\lvert d_i \rvert$	d_i^2/e_i
1	$0 \le x \le 0.35125$	0.4137	15.30		22		6.70	2.93
2	$0.35125 \le x \le 1.01375$	0.3721	13.77		7		6.77	3.33
3	$1.01375 \le x \le 1.67625$	0.1359	5.03		3			
4	$1.67625 \le x \le 2.33875$	0.0497	1.84	7.93	2	8	0.07	0.00
5	$2.33875 \le x$	0.0286	1.06		3			
		1.0000	37.00		37			6.263

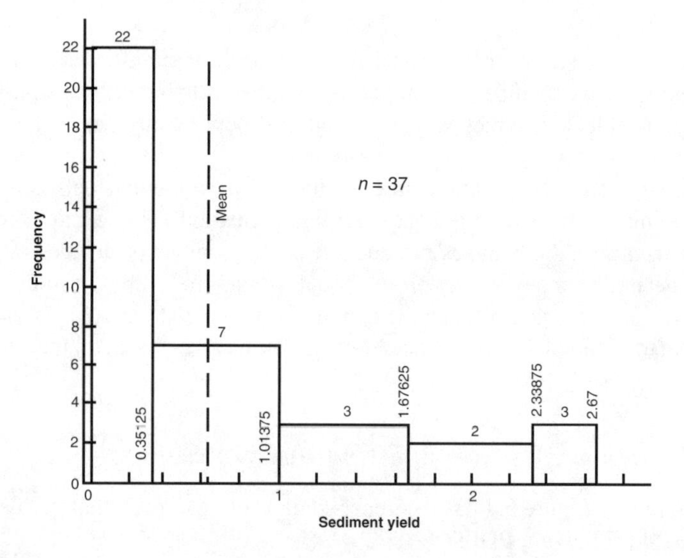

FIGURE 9.4 Histogram of sediment yield data.

H_o: Y has an exponential density function with

$$\hat{\lambda} = \frac{1}{\bar{Y}} = \frac{1}{0.6579} = 1.520 \tag{9.8}$$

H_A: Y is not exponentially distributed, with $\hat{\lambda} = 1.52$

The calculation of the computed value of chi-square is shown in Table 9.4. Although the histogram initially included five cells, the last three cells had to be combined to ensure that all cells would have an expected frequency of 5 or greater. The computed chi-square statistic is 6.263. Two degrees of freedom are lost because n and $\bar{Y}$ were used to compute the expected frequencies; therefore, with only three cells, only 1 degree of freedom remains. For levels of significance of 5% and 1% and 1 degree of freedom, the critical values are 3.841 and 6.635, respectively. Thus, the null hypothesis would be rejected for a 5% level of significance but accepted for 1%. This illustrates the importance of selecting the level of significance on the basis of a rational analysis of the importance of type I and II errors.

9.2.4 Chi-Square Test for Log-Pearson III Distribution

The log-Pearson type III distribution is used almost exclusively for the analysis of flood peaks. Whether the data points support the use of a log-Pearson III distribution is usually a subjective decision based on the closeness of the data points to the assumed population curve. To avoid this subjectivity, a statistical analysis may be a good alternative in determining whether the data points support the assumed LP3 distribution. The chi-square test is one possible analysis. Vogel's (1986) probability plot correlation coefficient is an alternative.

Two options are available for estimating probabilities. First, the LP3 density function can be integrated between cell bounds to obtain probabilities to compute the expected frequencies. Second, the tabular relationship between the exceedance probability and the LP3 deviates K can be applied. The first option would enable the use of the constant probability method for setting cell bounds; however, it would require the numerical integration of the LP3 density function. The second option has the disadvantage that getting a large number of cells may be difficult. The second option is illustrated in the following example.

Example 9.4

The 38-year record of annual maximum discharges for the Back Creek watershed (Table B.4) is used to illustrate the application of the chi-square test with the LP3 distribution. The 32-probability table of deviates (Table A.11) is used to obtain the probabilities for each cell. The sample skew is –0.731; therefore, the K values for a skew of –0.7 are used with the sample log mean of 3.722 and sample log standard deviation of 0.2804 to compute the log cell bound X (column 3 of Table 9.5) and the cell bound Y (column 4):

$$X = 3.722 + 0.2804\,K \tag{9.9a}$$

$$Y = 10^X \tag{9.9b}$$

The cell bounds Y are used to compute the observed frequencies O_i (column 5). The cumulative probability of column 1 is incremented to get the cell probabilities of column 6, which are multiplied by the sample size n to compute the expected frequencies E_i (column 7). Cells must be combined to ensure that the expected cell frequencies are at least 5. Table 9.6 gives the results of the chi-square test. The computed chi-square value is 8.05. Since the three sample moments and the sample

TABLE 9.5
Chi-Square Test for Log-Pearson III Distribution

ΣP	K	$X = \bar{X} + KS$	$Y = 10^X$	O_i	p_i	$e_i = np_i$	Sample Probability	Difference
0.0001	−5.274	2.243	175	0	0.0001	0.0038		0.0001
0.0005	−4.462	2.471	296	0	0.0004	0.0152		0.0004
0.0010	−4.100	2.572	374	0	0.0005	0.0190		0.0005
0.0020	−3.730	2.676	474	0	0.0010	0.0380		0.0010
0.0050	−3.223	2.818	658	1	0.0030	0.1140	1/38	−0.0243
0.0100	−2.824	2.930	851	0	0.0050	0.1900		−0.0213
0.0200	−2.407	3.047	1114	0	0.0100	0.3800		−0.0163
0.0250	−2.268	3.086	1219	0	0.0050	0.1900		−0.0063
0.0400	−1.967	3.170	1481	0	0.0150	0.5700		−0.0013
0.0500	−1.819	3.212	1629	1	0.0100	0.3800	2/38	−0.0126
0.1000	−1.333	3.348	2230	0	0.0500	1.9000		−0.0026
0.2000	−0.790	3.500	3166	2	0.1000	3.8000	4/38	−0.0053
0.3000	−0.429	3.602	3997	6	0.1000	3.8000	10/38	−0.0632
0.4000	−0.139	3.683	4820	8	0.1000	3.8000	18/38	−0.1739
0.4296	−0.061	3.705	5068	1	0.0296	1.1248	19/38	−0.1000
0.5000	0.116	3.755	5682	4	0.0704	2.6752	23/38	−0.1757
0.5704	0.285	3.802	6337	2	0.0704	2.6752	25/38	−0.1579
0.6000	0.356	3.822	6635	0	0.0296	1.1248	25/38	−0.0875
0.7000	0.596	3.889	7747	4	0.1000	3.8000	29/38	−0.0632
0.8000	0.857	3.962	9168	4	0.1000	3.8000	33/38	−0.0684
0.9000	1.183	4.054	11317	2	0.1000	3.8000	35/38	−0.0211
0.9500	1.423	4.121	13213	0	0.0500	1.9000		0.0289
0.9600	1.489	4.140	13788	0	0.0100	0.3800		0.0389
0.9750	1.611	4.174	14918	0	0.0150	0.5700		0.0539
0.9800	1.663	4.188	15428	0	0.0050	0.1900		0.0589
0.9900	1.806	4.228	16920	1	0.0100	0.3800	36/38	0.0426
0.9950	1.926	4.262	18283	0	0.0050	0.1900		0.0476
0.9980	2.057	4.299	19897	1	0.0030	0.1140	37/38	0.0243
0.9990	2.141	4.322	21006	0	0.0010	0.0380		0.0253
0.9995	2.213	4.343	22005	0	0.0005	0.0190		0.0258
0.9999	2.350	4.381	24040	1	0.0004	0.0152	38/38	−0.0001
1.0000	—	—	—	—	0.0001	0.0038		0.0000
				38	1.0000	38.0000		

size were used to estimate the expected probabilities, 4 degrees of freedom are lost. With 5 cells, only 1 degree of freedom is available. The critical chi-square values for 5%, 1%, and 0.5% levels of significance are 3.84, 6.63, and 7.88, respectively. Therefore, the null hypothesis of an LP3 PDF must be rejected:

$$H_O\text{: } Y \sim \text{LP3 } (\log \mu = 3.722, \log \sigma = 0.2804, \log g = -0.7) \tag{9.10}$$

TABLE 9.6
Chi-Square Test for Log-Pearson III Distribution

Cells	E_i	O_i	$(O_i - E_i)^2/E_i$
0–3166	7.6	4	1.705
3167–4820	7.6	14	5.389
4821–6635	7.6	7	0.047
6636–9168	7.6	8	0.021
9169–∞	7.6	5	0.889
	38.0	38	8.051

It appears that either the LP3 distribution is not appropriate or one or more of the sample parameters are not correct. Note that, if the LP3 deviates K are obtained from the table (Table A.11), then neither the equal probability or the equal cell width is used. In this case, the cell bounds are determined by the probabilities in the table.

9.3 KOLMOGOROV–SMIRNOV ONE-SAMPLE TEST

A frequent problem in data analysis is verifying that the population can be represented by some specified PDF. The chi-square goodness-of-fit test was introduced as one possible statistical test; however, the chi-square test requires at least a moderate sample size. It is difficult to apply the chi-square test with small samples because of the 5-or-greater expected frequency limitation. Small samples will lead to a small number of degrees of freedom. The Kolmogorov–Smirnov one-sample (KS1) test was developed for verifying a population distribution and can be used with much smaller samples than the chi-square test. It is considered a nonparametric test.

9.3.1 PROCEDURE

The KS1 tests the null hypothesis that the cumulative distribution of a variable agrees with the cumulative distribution of some specified probability function; the null hypothesis must specify the assumed population distribution function and its parameters. The alternative hypothesis is accepted if the distribution function is unlikely to be the underlying function; this may be indicated if either the density function or the specified parameters is incorrect.

The test statistic, which is denoted as D, is the maximum absolute difference between the values of the cumulative distributions of a random sample and a specified probability distribution function. Critical values of the test statistic are usually available only for limited values of the level of significance; those for 5% and 1% are given in Table A.12.

The KS1 test may be used for small samples; it is generally more efficient than the chi-square goodness-of-fit test when the sample size is small. The test requires data on at least an ordinal scale, but it is applicable for comparisons with continuous distributions. (The chi-square test may also be used with discrete distributions.)

The Kolmogorov–Smirnov one-sample test is computationally simple; the computational procedure requires the following six steps:

1. State the null and alternative hypotheses in terms of the proposed PDF and its parameters. Equations 9.1 are the two hypotheses for the KS1 test.
2. The test statistic, D, is the maximum absolute difference between the cumulative function of the sample and the cumulative function of the probability function specified in the null hypothesis.
3. The level of significance should be set; values of 0.05 and 0.01 are commonly used.
4. A random sample should be obtained and the cumulative probability function derived for the sample data. After computing the cumulative probability function for the assumed population, the value of the test statistic can be computed.
5. The critical value, D_α, of the test statistic can be obtained from tables of D_α in Table A.12. The value of D_α is a function of α and the sample size, n.
6. If the computed value D is greater than the critical value D_α, the null hypothesis should be rejected.

When applying the KS1 test, it is best to use as many cells as possible. For small and moderate sample sizes, each observation can be used to form a cell. Maximizing the number of cells increases the likelihood of finding a significant result if the null hypothesis is, in fact, incorrect. Thus, the probability of making a type I error is minimized.

Example 9.5

The following are estimated erosion rates (tons/acre/year) from 13 construction sites:

47 53 61 57 64 44 56 52 63 58 49 51 54

The values are to be compared with a study that suggested erosion rates could be represented by a normal PDF with a mean and standard deviation of 55 and 5, respectively. If a level of significance of 5% is used, would it be safe to conclude that the sample is from a normally distributed population with a mean of 55 and a standard deviation of 5?

If the data are separated on a scale with intervals of 5 tons/acre/year, the frequency distribution (column 2), sample probability function (column 3), and population probability function (column 4) are as given in Table 9.7. The cumulative function for the population uses the z transform to obtain the probability values; for example, the z value for the upper limit of the first interval is

$$z = \frac{45 - 55}{5} = -2 \tag{9.11}$$

TABLE 9.7
Example of Kolmogorov–Smirnov One-Sample Test

Range	Observed Frequency	Probability Function	Cumulative Function	Cumulative N (55, 5)	Absolute Difference
40–45	1	0.0769	0.0769	0.0228	0.0541
45–50	2	0.1538	0.2307	0.1587	0.0720
50–55	4	0.3077	0.5384	0.5000	0.0384
55–60	3	0.2308	0.7692	0.8413	0.0721
60–65	2	0.1538	0.9230	0.9772	0.0542
65–70	1	0.0770	1.0000	0.9987	0.0013
		1.0000			

TABLE 9.8
Example of Kolmogorov–Smirnov One-Sample Test

X_u	z	$F(z)$	$p(x)$	Difference
44	−2.2	0.0139	1/13	−0.063
47	−1.6	0.0548	2/13	−0.099
49	−1.2	0.1151	3/13	−0.116
51	−0.8	0.2119	4/13	−0.096
52	−0.6	0.2743	5/13	−0.110
53	−0.4	0.3446	6/13	−0.117
54	−0.2	0.4207	7/13	−0.118
56	0.2	0.5793	8/13	−0.036
57	0.4	0.6554	9/13	−0.037
58	0.6	0.7257	10/13	−0.044
61	1.2	0.8849	11/13	0.039
63	1.6	0.9452	12/13	0.022
64	1.8	0.9641	13/13	−0.036

Thus, the probability is $p(z < -2) = 0.0228$. After the cumulative functions were derived, the absolute difference was computed for each range. The value of the test statistic, which equals the largest absolute difference, is 0.0721. For a 5% level of significance, the critical value (see Table A.12) is 0.361. Since the computed value is less than D_α, the null hypothesis cannot be rejected.

With small samples, it may be preferable to create cells so that each cell contains a single observation. Such a practice will lead to the largest possible difference between the cumulative distributions of the sample and population, and thus the greatest likelihood of rejecting the null hypothesis. This is a recommended practice.

To illustrate this, the sample of 13 was separated into 13 cells and the KS1 test applied (see Table 9.8). With one value per cell, the observed cumulative probabilities would increase linearly by 1/13 per cell (see column 4). The theoretical cumulative probabilities (see column 3) based on the null hypothesis of a normal distribution ($\mu = 55$, $\sigma = 5$) are computed by $z = (x - 55)/5$ (column 2). Column 5 of Table 9.8

gives the difference between the two cumulative distributions. The largest absolute difference is 0.118. The null hypothesis cannot be rejected at the 5% level with a critical value of 0.361. While this is the same conclusion as for the analysis of Table 9.7, the computed value of 0.118 is 64% larger than the computed value of 0.0721. This is the result of the more realistic cell delineation.

Example 9.6

Consider the following sample of ten:

$$\{-2.05, -1.52, -1.10, -0.18, 0.20, 0.77, 1.39, 1.86, 2.12, 2.92\}$$

For such a small sample size, a histogram would be of little value in suggesting the underlying population. Could the sample be from a standard normal distribution, (0, 1)? Four of the ten values are negative, which is close to the 50% that would be expected for a standard normal distribution. However, approximately 68% of a standard normal distribution is within the −1 to +1 bounds. For this sample, only three out of the ten are within this range. This might suggest that the standard normal distribution would not be an appropriate population. Therefore, a statistical test is appropriate for a systematic analysis that has a theoretical basis.

The data are tested with the null hypothesis of a standard normal distribution. The standard normal distribution is divided into ten equal cells of 0.1 probability (column 1 of Table 9.9). The z value (column 2 of Table 9.9) is obtained from Table A.1 for each of the cumulative probabilities. Thus 10% of the standard normal distribution would lie between the z values of column 2 of Table 9.9, and if the null hypothesis of a standard normal distribution is true, then 10% of a sample would lie in each cell. The actual sample frequencies are given in column 3, and the cumulative frequency is shown in column 4. The cumulative frequency distribution

TABLE 9.9
Kolmogorov–Smirnov Test for a Standard Normal Distribution

(1) Cumulative Normal, *N* (0, 1)	(2) Standardized Variate, *z*	(3) Sample Frequency	(4) Cumulative Sample Frequency	(5) Cumulative Sample Probability	(6) Absolute Difference
0.1	−1.28	2	2	0.2	0.1
0.2	−0.84	1	3	0.3	0.1
0.3	−0.52	0	3	0.3	0.0
0.4	−0.25	0	3	0.3	0.1
0.5	0.00	1	4	0.4	0.1
0.6	0.25	1	5	0.5	0.1
0.7	0.52	0	5	0.5	0.2
0.8	0.84	1	6	0.6	0.2
0.9	1.28	0	6	0.6	0.3
1.0	∞	4	10	1.0	0.0

is converted to a cumulative probability distribution (column 5) by dividing each value by the sample size. The differences between the cumulative probabilities for the population specified in the null hypothesis (column 1) and the sample distribution (column 5) are given in column 6. The computed test statistic is the largest of the absolute difference shown in column 6, which is 0.3.

Critical values for the KS1 test are obtained in Table A.12 for the given sample size and the appropriate level of significance. For example, for a 5% level of significance and a sample size of 10, the critical value is 0.41. Any sample value greater than this indicates that the null hypothesis of equality should be rejected. For the data of Table 9.9, the computed value is less than the critical value, so the null hypothesis should be accepted. Even though only 30% of the sample lies between −1 and +1, this is not sufficient evidence to reject the standard normal distribution as the underlying population. When sample sizes are small, the difference between the sample and the assumed population must be considerable before a null hypothesis can be rejected. This is reasonable as long as the selection of the population distribution stated in the null hypothesis is based on reasoning that would suggest that the normal distribution is appropriate.

It is generally useful to identify the rejection probability rather than arbitrarily selecting a level of significance. If a level of significance can be selected based on some rational analysis (e.g., a benefit–cost analysis), then it is probably not necessary to compute the rejection probability. For this example, extreme tail areas of 20%, 15%, 10%, and 5% correspond to D_α values of 0.322, 0.342, 0.368, and 0.410, respectively. Therefore, the rejection probability exceeds 20% (linear interpolation gives a value of about 25.5%).

Example 9.7

The sample size of the Nolin River data (see Table B.2) for the 1945–1968 period is too small for testing with the chi-square goodness-of-fit test; only four cells would be possible, which would require a decision based on only 1 degree of freedom. The KS1 test can, therefore, be applied. If the sample log mean and log standard deviation are applied as assumed population parameters, then the following hypotheses can be tested:

$$H_0\text{: } q \sim LN\,(\log \mu = \log \bar{x},\ \log \sigma_d = \log s) = LN\,(3.6115,\ 0.27975) \tag{9.12a}$$

$$\text{H}_A\text{: } q \neq LN\,(\log \bar{x},\ \log s) \neq LN\,(3.6115,\ 0.27975) \tag{9.12b}$$

In this case, the sample statistics are used to define the parameters of the population assumed in the hypotheses, but unlike the chi-square test, this is not a factor in determining the critical test statistic.

Since the sample includes 24 values, the KS1 test will be conducted with 24 cells, one for each of the observations. By using the maximum number of cells, a significant effect is most likely to be identified if one exists. In order to leave one value per cell, the cell bounds are not equally spaced. For the 24 cells (see Table 9.10), the

TABLE 9.10
Nolin River, Kolmogorov–Smirnov One-Sample Test[a]

		Sample			Population		
Cell	Cell Bound	Observed Frequency	Cumulative Frequency	Cumulative Probability	z	Cumulative Probability	$\|D\|$
1	3.00	1	1	0.0417	−2.186	0.0144	0.0273
2	3.25	1	2	0.0833	−1.292	0.0982	0.0149
3	3.33	1	3	0.1250	−1.006	0.1572	0.0322
4	3.35	1	4	0.1667	−0.935	0.1749	0.0082
5	3.37	1	5	0.2083	−0.863	0.1941	0.0142
6	3.41	1	6	0.2500	−0.720	0.2358	0.0142
7	3.50	1	7	0.2917	−0.399	0.3450	0.0533
8	3.57	1	8	0.3333	−0.148	0.4412	0.1079
9	3.60	1	9	0.3750	−0.041	0.4836	0.1086
10	3.64	1	10	0.4167	0.102	0.5406	0.1239*
11	3.66	1	11	0.4583	0.173	0.5687	0.1104
12	3.68	1	12	0.5000	0.245	0.5968	0.0968
13	3.70	1	13	0.5417	0.316	0.6240	0.0823
14	3.72	1	14	0.5833	0.388	0.6510	0.0677
15	3.73	1	15	0.6250	0.424	0.6642	0.0392
16	3.80	1	16	0.6667	0.674	0.7498	0.0831
17	3.815	1	17	0.7083	0.727	0.7663	0.0580
18	3.8175	1	18	0.7500	0.736	0.7692	0.0192
19	3.83	1	19	0.7917	0.781	0.7826	0.0091
20	3.85	1	20	0.8333	0.853	0.8031	0.0302
21	3.88	1	21	0.8750	0.960	0.8315	0.0435
22	3.90	1	22	0.9167	1.031	0.8487	0.0680
23	3.92	1	23	0.9583	1.103	0.8650	0.0933
24	∞	1	24	1.0000	∞	1.0000	0.0000

[a] Log (x) = {3.642, 3.550, 3.393, 3.817, 3.713, 3.674, 3.435, 3.723, 3.818, 2.739, 3.835, 3.581, 3.814, 3.919, 3.864, 3.215, 3.696, 3.346, 3.322, 3.947, 3.362, 3.631, 3.898, 3.740}.

largest absolute difference between the sample and population is 0.1239. The critical values for selected levels of significance (see Table A.12) follow:

α	0.20	0.15	0.10	0.05	0.01
D_α	0.214	0.225	0.245	0.275	0.327

Since the computed value of 0.124 is much less than even the value for a 20% level of significance, the null hypothesis cannot be rejected. The differences between the sample and population distributions are not sufficient to suggest a significant difference. It is interesting that the difference was not significant given that the logarithms of the sample values have a standardized skew of −1.4, where the lognormal would

TABLE 9.11
Effect of Number of Cells on Test Statistic

Number of Cells	Cumulative Probability		$\|D\|$
	Sample	Population	
24	0.4167	0.5406	0.1239
12	0.4167	0.5406	0.1239
8	0.3750	0.4836	0.1086
6	0.3337	0.4412	0.1079
4	0.5000	0.5968	0.0968
2	0.5000	0.5968	0.0968

have a value of zero. For the sample size given, which is considered small or moderate at best, the difference is not sufficient to reject H_0 in spite of the skew. Of course, the test might accept other assumed populations, which would suggest that a basis other than statistical hypothesis testing should be used to select the distribution to be tested, with the test serving only as a means of verification.

When applying the KS1 test, as well as other tests, it is generally recommended to use as many cells as is practical. If less than 24 cells were used for the Nolin River series, the computed test statistic would generally be smaller than that for the 24 cells. Using the cell bounds shown in Table 9.10, the absolute difference can be computed for other numbers of cells. Table 9.11 gives the results for the number of cells of 24, 12, 8, 6, 4, and 2. In general, the computed value of the test statistic decreases as the number of cells decreases, which means that a significant finding would be less likely to be detected. In the case of Table 9.11, the decision to accept H_0 remains the same, but if the critical test statistic were 0.11, then the number of cells used to compute the test statistic would influence the decision.

Example 9.8

In Example 9.4, the chi-square test was applied with the Back Creek annual maximum series to test the likelihood of the log-Pearson III distribution. The same flood series is applied to the LP3 distribution with the KS1 test. The cumulative probability functions for the LP3 and observed floods are given in columns 1 and 8 of Table 9.5, respectively. The differences between the two cumulative functions are given in column 9. The largest absolute difference is 0.1757. For levels of significance of 10%, 5%, and 1%, the critical values from Table A.12 are 0.198, 0.221, and 0.264, respectively. Thus, the null hypothesis should not be rejected, which is a different decision than indicated by the chi-square test. While the chi-square test is more powerful for very large samples, the KS1 test is probably a more reliable test for small samples because information is not lost by the clumping of many observations into one cell. Therefore, it seems more likely that it is legitimate to assume that the Back Creek series is LP3 distributed with the specified moments.

9.4 THE WALD–WOLFOWITZ RUNS TEST

Tests are available to compare the moments of two independent samples. For example, the two-sample parametric *t*-test is used to compare two means. In some cases, the interest is only whether or not the two samples have been drawn from identical populations regardless of whether they differ in central tendency, variability, or skewness. The Wald–Wolfowitz runs test can be used to test two independent samples to determine if they have been drawn from the same continuous distribution. The test is applied to data on at least an ordinal scale and is sensitive to any type of difference (i.e., central tendency, variability, skewness). The test assumes the following hypotheses:

H_0: The two independent samples are drawn from the same population.
H_A: The two independent samples are drawn from different populations.

The hypotheses can be tested at a specified level of significance. If a type I error is made, it would imply that the test showed a difference when none existed. Note that the hypotheses do not require the specification of the distribution from which the samples were drawn, only that the distributions are not the same.

The test statistic r is the number of runs in a sequence composed of all $2n$ values ranked in order from smallest to largest. Using the two samples, the values would be pooled into one sequence and ranked from smallest to largest regardless of the group from which the values came. The group origin of each value is then denoted and forms a second sequence of the same length. A run is a sequence of one or more values from the same group. For example, consider the following data:

Group A	7	6	2	12	9	7
Group B	8	4	3	4		

These data are pooled into one sequence, with the group membership shown below the number as follows:

2	3 4 4	6 7 7	8	9 12	
A	B B B	A A A	B	A A	
1	2	3	4	5	run number

The pooled, ranked sequence includes five runs. Thus, the computed value of the test statistic is 5.

To make a decision concerning the null hypothesis, the sample value of the test statistic is compared with a critical value obtained from a table (Table A.13), and if the sample value is less than or equal to the critical value, the null hypothesis is rejected. The critical value is a function of the two sample sizes, n_A and n_B, and the level of significance. It is rational to reject H_0 if the computed value is less than the critical value because a small number of runs results when the two samples show a lack of randomness when pooled. When the two samples are from the same population, random variation should control the order of the values, with sampling variation producting a relatively large number of runs.

As previously indicated, the Wald–Wolfowitz runs test is sensitive to differences in any moment of a distribution. If two samples were drawn from distributions with different means, but the same variances and skews (e.g., $N[10, 2]$ and $N[16, 2]$), then the values from the first group would be consistently below the values of the second group, which would yield a small number of runs. If two samples were drawn from distributions with different variances but the same means and skews (e.g., $N[10, 1]$ and $N[10, 6]$), then the values from the group with the smaller variance would probably cluster near the center of the combined sequence, again producing a small number of runs. If two samples were drawn from distributions with the same mean and standard deviation (e.g., $Exp[b]$ and $N[b, b]$) but different skews, the distribution with the more negative skew would tend to give lower values than the other, which would lead to a small number of runs. In each of these three cases, the samples would produce a small number of runs, which is why the critical region corresponds to a small number of runs.

The test statistic, which is the number of runs, is an integer. Therefore, critical values for a 5% or 1% level of significance are not entirely relevant. Instead, the PMF or cumulative mass function for integer values can be computed, and a decision based on the proportion of the mass function in the lower tail. The cumulative PMF is given by:

$$p(r < R) = \begin{cases} \dfrac{1}{\dbinom{n_1+n_2}{n_1}} \displaystyle\sum_{r=2}^{R} (2)\binom{n_1-1}{\frac{r}{2}-1}\binom{n_2-1}{\frac{r}{2}-1} & \text{for even values of } r \quad (9.13a) \\[2em] \dfrac{1}{\dbinom{n_1+n_2}{n_1}} \displaystyle\sum_{r=1}^{R} \left[\binom{n_1-1}{k-1}\binom{n_2-1}{k-2}+\binom{n_1-1}{k-2}\binom{n_2-1}{k-1}\right] & \text{for odd values of } r \end{cases}$$

(9.13b)

in which k is an integer and is equal to $0.5\,(r + 1)$. A table of values of R and its corresponding probability can be developed for small values of n_1 and n_2 (see Table 9.12). Consider the case where n_1 and n_2 are both equal to 5. The cumulative probabilities for values of R of 2, 3, 4, and 5 are 0.008, 0.040, 0.167, and 0.357, respectively. This means that, with random sampling under the conditions specified by the null hypothesis, the likelihood of getting exactly two runs is 0.008; three runs, 0.032; four runs, 0.127; and five runs, 0.190. Note that it is not possible to get a value for an exact level of significance of 1% or 5%, as is possible when the test statistic has a continuous distribution such as the normal distribution. For this case, rejecting the null hypothesis if the sample produces two runs has a type I error probability of 0.8%, which is close to 1%, but strictly speaking, the decision is slightly conservative. Similarly, for a 5% level of significance, the decision to reject should be made when the sample produces three runs, but three runs actually has a type I error probability of 4%. In making decisions with the values of Table 9.12, the critical number of runs should be selected such that the probability shown in

TABLE 9.12
Critical Values for Wald–Wolfowitz Test

n_1	n_2	R	p	n_1	n_2	R	p
2	2	2	0.3333	4	4	2	0.0286
2	3	2	0.2000	4	4	3	0.1143
2	4	2	0.1333	4	5	2	0.0159
2	5	2	0.0952	4	5	3	0.0714
2	5	3	0.3333	4	5	4	0.2619
3	3	2	0.1000	4	5	2	0.0079
3	4	2	0.0571	5	5	2	0.0079
3	5	2	0.0357	5	5	3	0.0397
3	5	3	0.1429	5	5	4	0.1667

Note: See Table A.13 for a more complete table.

Table 9.12 is less than or equal to the tabled value of R. This decision rule may lead to some troublesome decisions. For example, if $n_1 = 5$ and $n_2 = 8$ and a 5% level of significance is of interest, then the critical number of runs is actually three, which has a type I error probability of 1%. If four is used for the critical value, then the level of significance will actually be slightly larger than 5%, specifically 5.4%. If the 0.4% difference is not of concern, then a critical number of four runs could be used.

The Wald–Wolfowitz runs test is not especially powerful because it is sensitive to differences in all moments, including central tendency, variance, and skew. This sensitivity is its advantage as well as its disadvantage. It is a useful test for making a preliminary analysis of data samples. If the null hypothesis is rejected by the Wald–Wolfowitz test, then other tests that are sensitive to specific factors could be applied to find the specific reason for the rejection.

9.4.1 Large Sample Testing

Table A.13 applies for cases where both n_1 and n_2 are less than or equal to 20. For larger samples, the test statistic r is approximately normal under the assumptions of the test. Thus, decisions can be made using a standard normal transformation of the test statistic r:

$$z = \frac{r - \mu_r}{\sigma_r} \tag{9.14}$$

or, if a continuity correction is applied,

$$z = \frac{|r - \mu_r| - 0.5}{\sigma_r} \tag{9.15}$$

where the mean μ_r and standard deviation σ_r are

$$\mu_r = \frac{2n_1 n_2}{n_1 + n_2} + 1 \quad (9.16a)$$

$$\sigma_r = \left[\frac{2n_1 n_2 (2n_1 n_2 - n_1 - n_2)}{(n_1 + n_2)^2 (n_1 + n_2 - 1)} \right]^{0.5} \quad (9.16b)$$

The null hypothesis is rejected when the probability associated with the left tail of the normal distribution is less than the assumed level of significance.

9.4.2 Ties

While the Wald–Wolfowitz test is not a powerful test, the computed test statistic can be adversely influenced by tied values, especially for small sample sizes. Consider the following test scores of two groups, *X* and *Y*:

X	2	3	5	7		
Y	7	8	8	9	11	12

Two possible pooled sequences are

2	3	5	7	7	8	8	9	11	12	
X	*X*	*X*	*X*	*Y*	*Y*	*Y*	*Y*	*Y*	*Y*	sequence #1
X	*X*	*X*	*Y*	*X*	*Y*	*Y*	*Y*	Y	*Y*	sequence #2

The two sequences differ in the way that the tied value of 7 is inserted into the pooled sequences. For the first sequence, the score of 7 from group *X* is included in the pooled sequence prior to the score of 7 from group *Y*; this leads to two runs. When the score of 7 from group *Y* is inserted between the scores of 5 and 7 from group *X*, then the pooled sequence has four runs (see sequence 2). The location of the tied values now becomes important because the probability of two sequences is 1% while the probability of four sequences is 19%. At a 5% level of significance, the location of the tied values would make a difference in the decision to accept or reject the null hypothesis.

While tied values can make a difference, a systematic, theoretically justified way of dealing with ties is not possible. While ways of handling ties have been proposed, it seems reasonable that when ties exist in data sets, all possible combinations of the data should be tried, as done above. Then the analyst can determine whether the ties make a difference in the decision. If the positioning of ties in the pooled sequence does make a difference, then it may be best to conclude that the data are indecisive. Other tests should then be used to determine the possible cause of the difference in the two samples.

Example 9.9

The annual maximum peak discharges for the 10 years following clear cutting on watershed A follow:

47 38 42 33 34 30 20 26 21 20

A nearby watershed (B) that did not experience clear cutting but is essentially the same in other watershed characteristics experienced the following annual maximums for the same period:

16 12 19 18 25 23 17 24 19 16

The Wald–Wolfowitz test is applied to test the significance of the observed differences in the distributions of the discharges.

The pooled sample is as follows, with the runs indicated:

12	16	16	17	18	19	19	20	20	21	23	24	25	26	30	33	34	38	42	47
B	B	B	B	B	B	B	A	A	A	B	B	B	A	A	A	A	A	A	A

The pooled sample consists of four runs, that is, $r = 4$. The critical value of R is 6, which is obtained from Table A.13 for $n_A = n_B = 10$ and a 5% level of significance. Since the number of runs is not greater than the critical value, it is reasonable to reject the null hypothesis that the samples are from identical populations. Evidently, the deforestation caused a temporary increase in the discharges, with the subsequent deforestation causing a reduction in the annual maximum discharges. Thus, during the period of deforestation the clear-cut watershed has a generally higher mean discharge and discharges with a larger variance, thus appearing to have a different underlying distribution.

Example 9.10

The ratios of average monthly runoff depth to the average monthly rainfall depth are computed for an urbanized (U) and a natural (R) watershed of similar size for each month.

Month	U	R
Jan.	0.41	0.76
Feb.	0.38	0.68
Mar.	0.36	0.65
Apr.	0.35	0.53
May	0.34	0.48
June	0.33	0.40
July	0.31	0.37
Aug.	0.31	0.32
Sept.	0.33	0.37
Oct.	0.34	0.42
Nov.	0.35	0.56
Dec.	0.39	0.62

The goal is to decide whether the distributions of the ratios are significantly different. A plot of these ratios versus the month of occurrence shows obvious differences, but the two distributions are of similar magnitude during the summer months.

The Wald–Wolfowitz test is applied by forming a single sequence from the smallest to largest ratio, with the source of the values identified below the value.

0.31	0.31	0.32	0.33	0.33	0.34	0.34	0.35	0.35	0.36	0.37	0.37
U	*U*	*R*	*U*	*U*	*U*	*U*	*U*	*U*	*U*	*R*	*R*
0.38	0.39	0.40	0.41	0.42	0.48	0.53	0.56	0.62	0.65	0.68	0.76
U	*U*	*R*	*U*	*R*	*R*	*R*	*R*	*R*	*R*	*R*	*R*

This produces eight runs. The critical value can be obtained from Table A.13 for $n_U = n_R = 12$. The critical value for a 5% level of significance is 7. Therefore, the large number of runs indicates that the null hypothesis cannot be rejected with the conclusion that the two distributions are the same.

This result shows a problem with small samples. While a difference obviously exists and can be related to the physical processes of the two hydrologic regimes, the very small samples of 12 would require a larger difference in order to be considered significant. The level of significance is also a deciding factor in this case. If a larger level of significance were used, the critical value would be higher, and then the difference would be statistically significant. Therefore, the example illustrates that, when the critical and computed number of runs are nearly the same, the use of an arbitrary level of significance should not be considered in making a decision.

Example 9.11

Numerous methods have been proposed for estimating watershed and channel slopes. Some are simple to compute, such as where the slope is the elevation difference divided by the length, while others are more tedious, such as those involving division of the flow length into subsections and computing a weighted slope based on the proportion of the length or area represented by the slope. If the two methods provide values that could be represented by the same probability distribution, then it seems reasonable that the less tedious method could be used to compute slopes.

Two methods of computing channel slope are compared. The simple method divided the difference in elevation of the endpoints by the channel length; this is referred to as the simple slope method. The second method sections the channel into n segments where the slope is relatively constant within each segment and computes the slope by:

$$S_c = \left[\frac{n}{\sum_{i=1}^{n} (1/S_i)^{0.5}} \right]^2 \tag{9.17}$$

in which S_i is the slope of segment i; this is referred to as the complex slope method. Sample data are used to test the following hypotheses:

H_0: The two samples of channel slope are from the same population.
H_A: The two samples of channel slope are drawn from different populations.

Acceptance of the null hypothesis would suggest that the two methods do not provide different estimates of slope; thus, the simple method could be adopted without a loss of accuracy. If the null hypothesis is rejected, then the data would suggest that the two methods produce different estimates and further investigation is warranted before adopting a method.

Table 9.13 contains estimates of the channel slope from 16 watersheds computed with the simple and complex methods. The pooled sequence (column 4) is used to create the run sequence (column 5). The run sequence consists of 21 runs. Based on the critical values in Table A.13, 21 runs are highly likely for sample sizes of $n_1 = n_2 = 16$. Eleven or fewer runs would be required to reject the null hypothesis at a level of significance of 5%. Therefore, it seems reasonable to assume that the two methods of computing channel slope give comparable estimates.

Example 9.12

The snow cover on a watershed can be quantified by expressing it as a percentage or fraction of the total drainage area. The fraction of snow cover is an input to numerous hydrologic models used in regions where snow contributes a significant portion of the flood- or water-yield runoff. The rate at which snow melts varies with elevation zone, the amount of precipitation that occurs during the melt season, and the cumulative degree-days during the melt season.

The Conejos River watershed near Magote, Colorado, drains an area of 282 mi^2. Because the topography varies over the watershed, it can be divided into three elevation zones. Snow on the higher elevation zone, zone 3, melts much later in the season than snow on the two lower zones, in part because of the colder temperatures at the higher elevation. However, the cumulative degree-days during the period when the snow-covered area decreases from 100% to 90% appears to be less than that at the lower elevation zones even though the lengths of time required are similar. Table 9.14 gives the number of days and the cumulative degree-days required for the snow-covered area to drop from 100% to 90% for zones 2 and 3. The two samples of six measurements of the cumulative degree-days, with the zone indicated, is pooled into the following sequence:

1.81	6.67	11.67	18.57	18.72	30.20	35.58	58.89	64.47	78.72	86.73	116.55
2	2	2	2	2	3	3	3	2	3	3	3

This pooled sequence is characterized by four runs. Table A.13 gives the following critical values r and their corresponding probabilities p for $n_1 = n_2 = 6$:

r	2	3	4	5	6
p	0.002	0.013	0.067	0.175	0.392

TABLE 9.13
Application of Wald–Wolfowitz Runs Test to Channel Slope Estimation

Watershed	Simple Slope	Complex Slope	Pooled Sequence	Run Sequence
1	0.1640	0.1867	0.0060	C
2	0.0213	0.0226	0.0094	S
3	0.0258	0.0259	0.0103	S
4	0.0470	0.0530	0.0104	C
5	0.0103	0.0104	0.0119	S
6	0.0119	0.0147	0.0147	C
7	0.0388	0.1246	0.0213	S
8	0.0664	0.0670	0.0226	C
9	0.0649	0.0735	0.0258	S
10	0.0387	0.0452	0.0259	C
11	0.0094	0.0060	0.0262	C
12	0.0484	0.0262	0.0275	S
13	0.1320	0.1212	0.0279	C
14	0.0275	0.0280	0.0280	C
15	0.0287	0.0295	0.0287	S
16	0.0287	0.0279	0.0287	S
			0.0295	C
			0.0387	S
			0.0388	S
			0.0452	C
			0.0470	S
			0.0484	S
			0.0530	C
			0.0649	S
			0.0664	S
			0.0670	C
			0.0735	C
			0.1212	C
			0.1246	C
			0.1320	S
			0.1640	S
			0.1867	C

Thus, the sample value of four runs has only a 6.7% chance of occurring if the two samples of the cumulative degree-days are from the same population. While the null hypothesis could not be rejected at the 5% level of significance, the probability is small enough to suggest looking for a cause of the disparity.

The means shown in Table 9.14 indicate that the mean cumulative degree-day requirement for the 10% depletion is much less in zone 2. Thus, a Mann–Whitney U test was made on the two samples. The computed U is 3 while the critical U values for levels of significance of 10%, 5%, 2% and 1% are 7, 5, 3, and 2, respectively.

TABLE 9.14
Application of Wald–Wolfowitz Runs Test to Snow Depletion Data for Conejos River Watershed

	Zone 2		Zone 3	
Year[a]	N_d [b]	CDD [b]	N_d [b]	CDD [b]
1973	9	18.57	8	116.55
1974	9	1.81	8	78.72
1975	9	64.47	9	86.73
1976	6	6.67	7	30.20
1978	11	11.67	10	35.58
1979	9	18.72	9	58.89
		20.32	Mean	67.78
		22.62	S_d	32.52

[a] Complete data for 1977 are not available because the melt began before data collection was initiated.
[b] (CDD = cumulative degree-days during the number of days, N_d, required for the snow-covered area to drop from 100% to 90%).

Thus, the null hypotheses of an equality of means could be rejected at the 2% level. The results of the Wald–Wolfowitz test was, therefore, the result of a disparity in the central tendency of the cumulative degree-days between two zones.

9.5 KOLMOGOROV–SMIRNOV TWO-SAMPLE TEST

The Kolmogorov–Smirnov two-sample (KS2) test is used to test the null hypothesis that two independent samples are not different in distribution characteristics. The test can be used for either of the following sampling programs: (1) one population with two independently drawn samples subjected to different treatments; or (2) two samples drawn from different populations and then subjected to the same treatment. The test is sensitive to differences in any distributional characteristic: central tendency or location, dispersion or scale, and shape.

The procedure for conducting the test depends on the sample sizes. The following two cases are presented here:

Case A: Small samples ($n \le 40$) of equal size
Case B: Large samples ($n \ge 20$)

For each case, both one-tailed and two-tailed alternatives can be used. Case A cannot be applied for sample sizes less than four, and the test cannot be used when the sample sizes are unequal and less than 20. For the case where the samples are between 20 and 40 and equal in size, the two alternatives should yield the same decision.

The rationale of the KS2 test is as follows: If the cumulative frequencies (for equal sample sizes) or cumulative probabilities (for unequal sample sizes) of two independent samples from the same population are plotted on a single graph, the differences between the two graphs should be independent of the distribution of the underlying population; if the two sample distributions are from the same population, the differences should not be significant beyond what is expected from sampling variation.

For the one-tailed tests, it is necessary to specify before the analysis (i.e., step 1) that sample 1 is the group with the cumulative distribution expected to have the more rapid rise. If it turns out that the other distribution has the more rapid rise, then the computed value of the test statistic will be negative and the null hypothesis will be accepted.

9.5.1 Procedure: Case A

For small samples of equal size, histograms are tabulated for both samples and then placed in cumulative form. The cumulative histogram for samples 1 and 2 are denoted as $F_1(X)$ and $F_2(X)$, respectively. For the two-tailed alternative, the value of the test statistic D is the largest absolute difference D between corresponding ordinates of the cumulative frequency histograms:

$$D = \text{maximum } |F_1(X) - F_2(X)| \quad \text{for any } X \tag{9.18}$$

To conduct a one-tailed test, it is necessary to specify the sample that is expected to have the fastest rise, which is then specified as sample 1. The value of D is then the maximum difference (but not the absolute value):

$$D = \text{maximum } (F_l(X) - F_2(X)) \quad \text{for any } X \tag{9.19}$$

The test statistic of Equation 9.18 cannot be negative; the test statistic of Equation 9.19 for the one-tailed test can be negative.

The critical value of the test statistic depends on the sample size n, the level of significance, and on whether the test is one-tailed or two-tailed. The critical value of the test statistic D_α is obtained from a table (see Table A.14), and the null hypothesis is rejected if the computed D of either Equation 9.18 or Equation 9.19 is greater than the critical value.

Example 9.13

Samples of a water quality indicator are taken from the effluents of two distilleries, one of which produces a bourbon-type product while the second produces a spirit-type product. The purpose of the study is to determine whether the pollutant levels of the discharges are of the same population. Since the problem statement does not specify that one should be larger than the other, a two-tailed test is assumed. If the null hypothesis is accepted, then similar water treatment facility designs may be used. The following values were obtained from the effluent of the bourbon distillery:

25,500	27,200	29,300	26,400	31,200
27,800	28,600	26,200	30,700	32,400
28,200	27,400	29,900	26,200	30,200

The following values were obtained from the effluent of the distillery that produces the spirit product:

32,600	31,400	35,100	33,300	29,600
30,200	33,700	28,800	30,100	34,600
29,800	32,100	27,400	31,500	27,500

Cumulative frequencies can be computed for the two samples of 15; the cumulative frequencies and the absolute differences are given in Table 9.15. The largest absolute difference, which equals 6, occurred for the third interval. For a level of significance of 5%, the critical value, D_α, is 8. Therefore, since $D < D_\alpha$, the null hypothesis cannot be rejected. One can conclude that the distillery wastes are similar in the water quality indicator.

To show that the histogram interval is not a trivial decision, the same data are distributed into histogram intervals defined by the individual values. Table 9.16 gives the values for both series ranked from smallest to largest. If the value in column 1 is from the bourbon distillery effluent (X), the rank is increased in column 2. If the value in column 1 is from the spirits distillery effluent (Y), the rank is increased in column 3. Note that for values of 27,400 and 30,200, both ranks are increased. The difference between the ranks is shown in column 4. The largest difference is 7, which occurs for three different histogram intervals. The critical value of 8 is, therefore, not exceeded, so the null hypothesis is not rejected.

It is important to note that the interval width influences the results. For the interval used in Table 9.15, the computed test statistic is 6. For the histograms of Table 9.16, the computed value is 7. While the decision to accept H_0 is the same in this case, it may not always be. For small samples, it is generally better to use the

TABLE 9.15
Example of Kolmogorov–Smirnov Two-Sample Test

	Frequencies		Cumulative Functions		
Range	Bourbon	Spirit	Bourbon	Spirit	Absolute Difference
24,000–26,000	1	0	1	0	1
26,000–28,000	6	2	7	2	5
28,000–30,000	4	3	11	5	6
30,000–32,000	3	4	14	9	5
32,000–34,000	1	4	15	13	2
34,000–36,000	0	2	15	15	0

TABLE 9.16
Kolmogorov–Smirnov Two-Sample Test for Example 9.13

X or *Y*	*X* Rank	*Y* Rank	Absolute Difference	*X* or *Y*	*X* Rank	*Y* Rank	Absolute Difference
25,500	1	0	1	30,100	11	6	5
26,200	3	0	3	30,200	12	7	5
26,400	4	0	4	30,700	13	7	6
27,200	5	0	5	31,200	14	7	7
27,400	6	1	5	31,400	14	8	6
27,500	6	2	4	31,500	14	9	5
27,800	7	2	5	32,100	14	10	4
28,200	8	2	6	32,400	15	10	5
28,600	9	2	7	32,600	15	11	4
28,800	9	3	6	33,300	15	12	3
29,300	10	3	7	33,700	15	13	2
29,600	10	4	6	34,600	15	14	1
29,800	10	5	5	35,100	15	15	0
29,900	11	5	6				

ranking system shown in Table 9.16. This will ensure that a significant finding will be found if one exists.

9.5.2 Procedure: Case B

With Case B, the two sample sizes can be different so the test statistics are standardized by forming cumulative probability distributions, rather than the cumulative frequency distributions of Case A. The cumulative probability distribution $P(X)$ is formed by dividing each cumulative frequency histogram by the respective sample size. For a two-tailed test, the computed value of the test statistic is the maximum absolute difference between the sample cumulative probability distributions:

$$D = \text{maximum } |P_1(X) - P_2(X)| \quad \text{for any } X \tag{9.20}$$

The critical value of the test statistic D_α is computed by

$$D_\alpha = K\left(\frac{n_1 + n_2}{n_1 n_2}\right)^{0.5} \tag{9.21}$$

where K is a function of the level of significance. The null hypothesis is rejected if D computed from Equation 9.20 is greater than D_α of Equation 9.21.

To conduct a one-tailed test, it is necessary to specify which sample is expected to have the fastest rise, which is specified as sample 1. The value of D, which is computed by

$$D = \text{maximum } (P_1(X) - P_2(X)) \quad \text{for any } X \tag{9.22}$$

is then used to compute the value of the test statistic, χ^2:

$$\chi^2 = 4D^2\left(\frac{n_1 n_2}{n_1 + n_2}\right) \tag{9.23}$$

where χ^2 is the value of a random variable having a chi-square distribution with 2 degrees of freedom. It is important to note that Equation 9.23 is used only with Equation 9.22, but not with Equation 9.20. The null hypothesis is rejected if the computed χ^2 is greater than the critical chi-square value, χ^2_α.

Example 9.14

Case B is illustrated here using measurements of a groundwater contaminant from up-gradient and down-gradient wells adjacent to a landfill. Over a period of time, 36 measurements were made at the up-gradient well and 20 measurements at the down-gradient well. The data are used to test the following hypotheses:

H_0: The distributional characteristics of the contaminant in the two wells are from the same population.
H_A: The distribution of the contaminant in the downgradient well has a higher proportion of large concentrations.

Since the alternative hypothesis indicates a direction, a one-tailed test is used. The cumulative probability distribution for the up-gradient well is expected to rise earlier if the landfill contributed to groundwater contamination, so it will be designated as sample 1. The results are given in Table 9.17. The largest difference is 0.867. Since

TABLE 9.17
Kolmogorov–Smirnov Two-Sample Test

Concentration	Frequency		Cumulative Frequency		Cumulative Probability		
(ppb)	$f_1(x)$	$f_2(x)$	$f_1(x)$	$f_2(x)$	$p_1(x)$	$p_2(x)$	Difference
0–10	6	0	6	0	0.167	0	0.167
10–20	10	0	16	0	0.444	0	0.444
20–30	11	0	27	0	0.750	0	0.750
30–40	4	1	31	1	0.861	0.05	0.811
40–50	2	0	33	1	0.917	0.05	0.867[a]
50–60	1	2	34	3	0.944	0.15	0.794
60–70	1	5	35	8	0.942	0.40	0.572
70–80	0	7	35	15	0.942	0.75	0.222
80–90	1	4	36	19	1.000	0.95	0.050
90–100	0	1	36	20	1.000	1.00	0

[a] Computed value of test statistic.

this is a one-tailed test for Case B analysis, the chi-square statistic of Equation 9.23 is computed using $D = 0.867$:

$$\chi^2 = 4(0.867)^2\left(\frac{20(36)}{20+36}\right) = 38.66 \tag{9.24}$$

Even for a 0.005 level of significance, the computed chi-square statistic is much greater than the critical value (38.66 vs. 10.60). Thus, the null hypothesis is rejected and the distributional characteristics of the down-gradient well are assumed to have a larger proportion of high concentrations.

9.6 PROBLEMS

9-1 Use the chi-square goodness-of-fit test to decide if the following data are from a uniform distribution $U(0, 100)$: U_i = {94, 67, 44, 70, 64, 10, 01, 86, 31, 40, 70,74, 44, 79, 30, 13, 70, 22, 55, 45, 68, 12, 58, 50, 13, 28, 77, 42, 29, 54, 69, 86, 54, 11, 57, 01, 29, 72, 61, 36, 11, 62, 65, 44, 98, 22, 48, 11, 49, 37}.

9-2 Use the chi-square goodness-of-fit test to decide if the following data are from a uniform distribution, $U(0, 100)$: U_i = {62, 16, 18, 23, 64, 50, 57, 50, 54, 04, 09, 08, 17, 14, 63, 17, 56, 10, 17, 11, 57, 21, 40, 41, 45, 41, 46, 18, 55, 32, 57, 44, 12, 64, 12, 01, 13, 68, 13, 48, 60, 30, 57, 13, 14, 28, 46, 51, 02, 49}.

9-3 Use the chi-square goodness-of-fit test to decide if the following data are from a standard normal distribution, $N(0, 1)$: z = {0.20, −1.84, −0.75, 0.59, −0.33, 1.81, 1.26, −0.62, 0.52, 0.91, 0.60, 1.20, −0.34, −1.07, −1.02, 0.50, 0.17, 1.67, −0.98, −0.89, −0.98, −1.49, −0.71, −1.62, −1.12, −1.30, −0.45, 0.17, 0.70, −0.37, −0.70, −0.01, 1.22, −1.14, 0.55, 0.78, −0.40, −0.70, 1.45, −0.96, −0.23, −0.09, 0.17, 0.28, −0.05, −0.14, −1.64, 0.44, −0.48, −2.21}.

9-4 Use the chi-square goodness-of-fit test to decide if the following data are from a standard normal distribution $N(0, 1)$: z = {−0.03, 2.37, −0.37, 1.16, −0.28, 0.07, 0.23, 2.16, 1.32, 2.33, 2.76, −0.23, 2.27, 2.83, 0.13, 0.78, 0.17, 1.30, 0.26, 0.48, 0.80, 0.58, −0.11, 0.68, 1.11, −0.40, 0.33, −0.33, −0.18, 1.72, 1.26, −0.27, 0.24, 0.10, 0.52, 1.44, 0.77, 0.44, 0.13, 0.31, 1.22, 2.61, 0.26, 0.05, 1.88, 1.80, 0.61, 0.16, 0.38, 0.79}.

9-5 Use the chi-square goodness-of-fit test to decide if the following data are from an exponential distribution: X = {0.39, 1.34, 0.06, 0.01, 0.43, 1.61, 0.79, 0.11, 0.98, 1.19, 0.03, 0.02, 0.70, 0.68, 0.31, 0.58, 2.96, 1.44, 1.97, 1.46, 0.92, 0.08, 0.67, 1.24, 1.18, 0.18, 1.70, 0.14, 0.49, 0.78, 1.24, 1.16, 1.02, 0.34, 1.37, 2.70, 1.65, 0.50, 0.02, 2.05, 1.24, 0.26, 0.01, 2.11, 0.70, 0.08, 2.26, 0.55, 1.87, 1.66, 1.49, 1.52, 0.46, 0.48, 0.87, 0.13, 1.15, 1.17, 0.34, 0.20}.

9-6 Use the chi-square goodness-of-fit test to decide if the Floyd River data (1935–1973) are from a log-Pearson type III distribution: X = {1460,

4050, 3570, 2060, 1300, 1390, 1720, 6280, 1360, 7440, 5320, 1400, 3240, 2710, 4520, 4840, 8320, 13900, 71500, 6250, 2260, 318, 1330, 970, 1920, 15100, 2870, 20600, 3810, 726, 7500, 7170, 2000, 829, 17300, 4740, 13400, 2940, 5660}.

9-7 Use the chi-square goodness-of-fit test to decide if the Shoal Creek, Tennessee, annual-minimum 7-day low flows are from a log-Pearson type III distribution: $X = \{99, 90, 116, 142, 99, 63, 128, 126, 93, 83, 83, 111, 112, 127, 97, 71, 84, 56, 108, 123, 120, 116, 98, 145, 202, 133, 111, 101, 72, 80, 80, 97, 125, 124\}$.

9-8 Using the data from Problem 9-1 and the Kolmogorov–Smirnov one-sample test, decide if the data are from a uniform distribution $U(0, 100)$.

9-9 Using the data from Problem 9-2 and the Kolmogorov–Smirnov one-sample test, decide if the data are from a uniform distribution $U(0, 100)$.

9-10 Using the data from Problem 9-3 and the Kolmogorov–Smirnov one-sample test, decide if the data are from a standard normal distribution.

9-11 Using the data from Problem 9-4 and the Kolmogorov–Smirnov one-sample test, decide if the data are from a standard normal distribution.

9-12 Using the data from Problem 9-5 and the Kolmogorov–Smirnov one-sample test, decide if the data are from an exponential distribution.

9-13 Using the data from Problem 9-6 and the Kolmogorov–Smirnov one-sample test to decide if the data are from a log-Pearson type III distribution.

9-14 Using the data from Problem 9-7 and the Kolmogorov–Smirnov one-sample test to decide if the data are from a log-Pearson type III distribution.

9-15 Use the Wald–Wolfowitz runs test to decide if the following two samples are from the same population. Discuss the likely cause of any difference detected.

A	17	6	8	4	10	7
B	14	13	3	1		

9-16 Use the Wald–Wolfowitz runs test to decide if the following two samples are from the same population. Discuss the likely cause of any difference detected.

C	37	21	58	41	47	62		
D	20	33	16	6	26	12	51	28

9-17 Use the Wald–Wolfowitz runs test to decide if the first nine discharges of Problem 9-6 are from the same population as the remainder of the sample. Discuss the cause of any difference detected.

9-18 Use the Kolmogorov–Smirnov two-sample test to decide if the first 19 years of the record in Problem 9-7 are from the same population as the remainder of the sample.

9-19 The following are monthly low flows for a wooded (W) and an agricultural (A) watershed of similar size. Use the Kolmogorov–Smirnov two-sample test to decide if the two samples are from the same population. Discuss any differences detected.

W	16.4	15.7	14.9	14.6	13.9	13.3	13.0	12.4	12.1	13.6	14.7	15.1
A	14.8	14.2	13.5	13.0	12.7	12.1	12.0	10.7	10.6	11.2	11.9	13.1

9-20 The following data are soil moisture percentages for two fields, one with a clay soil and one with a sandy soil. Use the Kolmogorov–Smirnov two-sample test to decide if the percentages are from the same distribution.

Field 1	8.9	10.6	7.3	11.1	12.2	8.3	9.5	7.9	10.3	9.2
Field 2	12.3	11.7	13.1	10.4	13.2	11.6	10.3	9.5	12.2	8.1

10 Modeling Change

10.1 INTRODUCTION

The analysis of hydrologic data is assumed to be capable of detecting the effects of watershed change. Graphical methods and statistical tests can be used to support hypotheses developed on the basis of independent information. However, detection is rarely the ultimate goal. Instead, the purpose of detection is generally to model the effect of the watershed change so that the effect can be extrapolated to some future state. For example, if the study design requires the specification of ultimate watershed development, knowing the hydrologic effect of partial development in a watershed may enable more accurate forecasts of the effects that ultimate development will have.

Detecting hydrologic effects of watershed change is the first step. Modeling the effect is necessary in order to forecast the effects of future change. Therefore, one graph or one statistical test will probably be inadequate to detect hydrologic change. Several statistical tests may be required in order to understand the true nature of the effect. Is the change in the distribution of the hydrologic variable due to a change in the function itself or to a change in one or more of the moments? Does a statistically significant serial correlation reflect a lack of independence or is it due to an unidentified secular trend in the data? Questions such as these emphasize the need for a thorough effort in detecting the true effect. An inadequate effort may ultimately lead to an incorrect assessment of the hydrologic effect and, therefore, an incorrect forecast.

Assuming that the correct hydrologic effect has been detected, the next step is to properly model the effect. Given the substantial amount of natural variation in hydrologic data, it is important to use the best process and tools to model the hydrologic effect of change. Using a simple model is best only when it is accurate. With the availability of computer-aided modeling tools, the ability to model nonstationary hydrologic processes has greatly improved. This chapter provides a discussion of the fundamentals of a few basic modeling tools. A complete introduction to all modeling tools would require several volumes.

10.2 CONCEPTUALIZATION

The first phase of modeling, conceptualization, is often lumped with the second phase, formulation. While the two phases have some similarities and commonalities, they also differ in important ways. The conceptualization phase is more general and involves assembling resources, including underlying theory and available data measurements. It involves important decisions about the purpose of the model, approximate complexity of the model, physical processes to be modeled, and criteria that will be used to judge model accuracy.

The first step is to decide on the purpose of the model, that is, the task for which the model is being designed. For example, a model may be designed to adjust measured annual maximum discharges in a nonstationary series for the effects of urbanization. It may be intended to provide a prediction of the effect but not the magnitude of the annual maximum discharge. By limiting the purpose of a model, the required complexity can usually be reduced, but the flexibility of the model is also limited. When identifying the purpose of a model, it is important to try to identify the range of potential model users to ensure that it will be adequate for their purposes. This may prevent its misuse.

As part of the use of the model, the type of model must be identified. Models can be empirical, theoretical, or some combination of the two. Models can be real-time forecast models or design storm models. Continuous, multi-event, and single-event models are possible. Models can be deterministic or stochastic. A model may provide a single point estimate or the entire distribution of the output variable. Decisions about the type of model will influence its flexibility and quite likely its accuracy.

The next aspect of model conceptualization is to identify the variables that will be involved, including both the criterion or output variables and the predictor or independent variables. The fewer variables involved, the less data required, both for calibration and for use. However, the fewer variables involved, the less flexible the model and the fewer situations where the model will be applicable. It is generally unwise to require inputs that are not readily available. For example, the lack of spatial land-use data from past decades makes it very difficult to accurately model the effects of land use on the annual flood series for a locale.

Model variables largely reflect the physical processes to be modeled. Decisions about the hydrologic, meteorologic, and geomorphic processes to be represented in the model affect the required variables. For some very simple empirical models, decisions about the processes to be modeled are unnecessary. The available data are largely known and the primary decision concerns the functional form to be used.

The expected availability of data is another consideration in model conceptualization. It makes little sense to propose a model for which data for the necessary inputs are rarely available. The quality, or accuracy, and quantity of data generally available are both factors to be considered. For each input variable, the range of measured data commonly available must be compared to the range over which the model is expected to function. For example, if data from watersheds with a range of impervious area from 0% to 20% are available, a model developed from such data may not be accurate for predictions on watersheds at 50% to 75% imperviousness. This would require extrapolation well beyond the bounds of the calibrated data.

In the conceptualization stage, it is also important to decide the level of accuracy required and the criteria that will be used to judge the adequacy of the model. Accuracy criteria such as correlation coefficients, model biases, and standard errors of estimate are typically used in empirical modeling, but factors such as the failure of correlation coefficients for small databases must be addressed. The selection of the level of significance to be used in hypothesis testing should be made at this stage before data are assembled and used to calibrate the model. Some thought must be given to the potential problem that the model will not meet the accuracy criteria established for its use. The feedback loop to revise the model or expend the resources necessary to collect more data should be considered.

It may seem that this somewhat nonquantitative phase of modeling is not important, while in reality it is well known that an improperly conceived model may lead to poor decisions.

Example 10.1

Consider the problem of revising the TR-55 (Soil Conservation Service, 1986) peak-discharge, model-adjustment factor for ponded storage. Currently, the adjustment uses only a multiplication factor F_p, where the value of F_p depends only on the percentage of the watershed area in ponds or swampy lands. Thus, the peak discharge is the product of four factors: F_p; the drainage area, A; the depth of runoff, Q; and the unit peak discharge, q_u:

$$q_p = F_p q_u A Q \tag{10.1}$$

This adjustment factor does not account for the depth of the pond d_p or the surface area of the watershed that drains to pond A_c, both of which are important.

In this case, the type of model is set by the type of model to which the new model will be attached. Since the discharge model of Equation 10.1 is a deterministic, single-event, design-storm model, the new method of adjusting for ponds will be classified as the same type of model. The adjustment should also be relatively simple, such as the multiplication factor F_p of Equation 10.1.

The statement of the problem indicates that the model would be conceptually more realistic if both d_p and A_c were used to determine the volume of required storage. If the objective is to provide a storage volume either at one location within the watershed or distributed throughout the watershed that can contain all of the runoff from the contributed area A_c for design depth Q, then the volume that needs to be stored is $A_c Q$. If the pond or swamp lands have a total area of A_p, then the depth of storage would be:

$$d_p = (A_c / A_p) Q$$

If all runoff during the design storm is contained within the pond, then the area that drains to the watershed outlet is $A - A_c$. Thus, the adjustment factor F_p of Equation 10.1 is not needed, and the peak discharge can be computed by

$$q_p = q_u (A - A_c) Q \tag{10.2}$$

The conceptual development of this model (Equation 10.2) assumes that the surface area that drains to storage does not produce any runoff. The conceptual development of Equation 10.1 does not require that assumption. Instead, Equation 10.1 assumes that the entire watershed contributes flow to the watershed outlet, that the storage areas are effective in storing water, and that the depth of storage is not relevant. However, the depth of storage is relevant. Equation 10.1 assumes that the same value of F_p is used regardless of the available storage, which is conceptually faulty. The volume of storage available and the location of the storage are conceptually realistic, and the model should be formulated to incorporate such concepts.

10.3 MODEL FORMULATION

In the conceptualization phase of modeling, specific forms of model components are not selected. The physical processes involved and the related variables are identified. In the formulation stage, the algorithm to be calibrated is formalized. Specific functional forms are selected, often after considering graphical summaries of the available data. Interactions between model components must be specified as well as ways that data will be summarized in order to compute the accuracy and rationality of the calibrated model considered. The main objective in this phase of modeling is the assembling of the model that will be calibrated, or fitted, to the measured data and then assessed.

10.3.1 TYPES OF PARAMETERS

Functional forms used in models can vary in position, spread, and form. Specifically, the characteristics of a function depend on parameters of the function. Three general types of parameters are available: location, scale, and shape. Location parameters can be used to position a function on the ordinate, the abscissa, or both. Scale parameters control the spread or variation of the function. Shape parameters control the form of a function. Not all models include all types of parameters. Some models include just one parameter, while other models require one or more of each type of parameter to represent data.

In the bivariate linear model $\hat{Y} = a + bX$, the intercept a acts as a location parameter and the slope coefficient b acts as a scale parameter. The intercept positions the line along the y-axis, and it can thus be classed as a y-axis location parameter. The slope coefficient scales the relationship between Y and X.

A more general model is

$$\hat{Y} = C_1 + C_3(X - C_2)^{C_4} \tag{10.3}$$

in which C_1 acts as a y-axis location parameter, C_2 acts as an x-axis location parameter, C_3 acts as a scale parameter, and C_4 acts as a shape parameter. The two location parameters enable the function to shift location along the axes. The shape parameter C_4 enables the function to take on a linear form ($C_4 = 1$), a zero-sloped line ($C_4 = 0$), an increasing function with a decreasing slope ($0 < C_4 < 1$), or an increasing function with an increasing slope ($C_4 > 1$). C_4 could also be negative, which yields corresponding shapes below the axis created with C_4 greater than 0.

The point of Equation 10.3 is that greater flexibility in form can be achieved through more complex functions. The price that must be paid is the potential decrease in the degrees of freedom due to the greater number of coefficients that must be fit and the requirement of a fitting method that can be used to fit the more complex model structures. Increasing the flexibility of a model by adding additional parameters may create problems of irrationality. A more flexible structure may be problematic if nonsystematic variation in the data causes the function to produce irrational effects. Polynomial functions are one example. Flexibility is both an advantage and disadvantage of complex functions. Polynomials can fit many data sets and produce good correlation coefficients, but they often suffer from polynomial swing.

Example 10.2

Equation 10.3 includes two location parameters, a scale parameter, and a shape parameter. The effect of variation in each of these is shown in Figure 10.1. In Figure 10.1(a), the *y*-axis location parameter C_1 is varied, with the line moving vertically upward as C_1 is increased. In Figure 10.1(b), the *x*-axis location parameter C_2 is varied, with the line moving horizontally to the right as C_2 increases. In Figure 10.1(c), the value of the scale parameter C_3 is varied, with the line increasing in slope as C_3 increases; note that the intercept remains fixed as C_1 and C_2 are constant. In Figure 10.1(d), the shape parameter C_4 is varied. For C_4 equal to 1, the function is linear. For C_4 greater than 1, the function appears with an increasing slope (concave upward). For C_4 less than 1, the function has a decreasing slope (concave downward). In Figure 10.1(e), both C_3 and C_4 are varied, with the shape and scale changing.

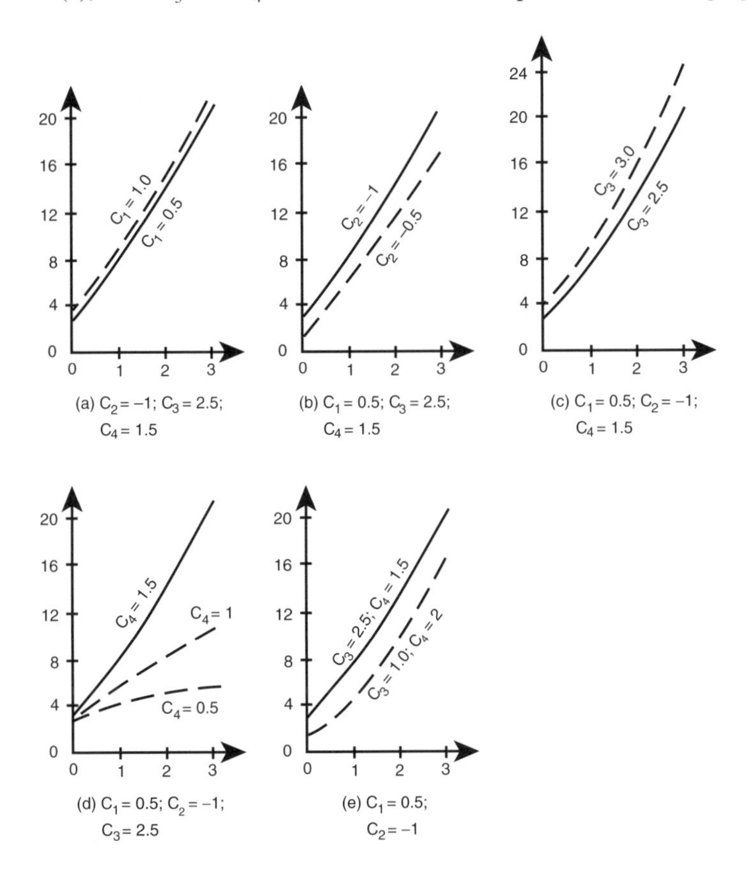

FIGURE 10.1 Effect of parameters of Equation 10.3.

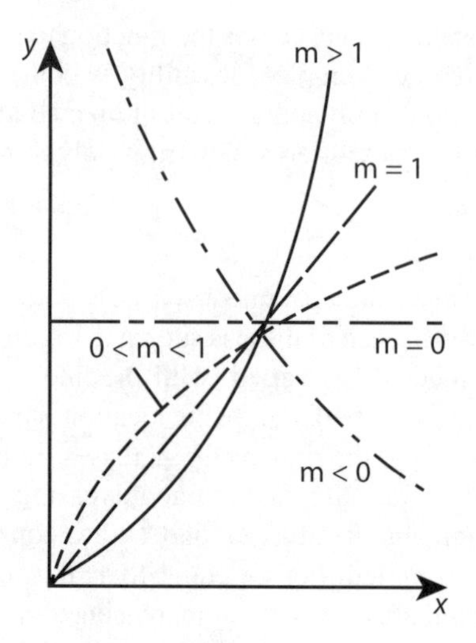

FIGURE 10.2 Characteristic shapes for $Y = KX^m$.

10.3.2 Alternative Model Forms

Nonlinear functions can take on a variety of forms. Those to be introduced here will be separated into five groups: the power model, logarithmic models, exponential models, trigonometric functions, and composite models. In some cases, theory may lead to one of these forms. In other cases, a form may be selected solely because it takes on the trend exhibited by the data.

The one-predictor power model has the general form

$$\hat{Y} = b_0 X^{b_1} \tag{10.4}$$

where b_0 and b_1 are fitting coefficients. This is used widely in hydrologic analysis and in engineering in general. Figure 10.2 shows the shape that can be assumed by the constant-intercept power model. Equation 10.4 is an example of this type of model with one predictor variable.

Logarithmic models can take on a variety of forms. Figure 10.3 shows a few possibilities. Additional location and scale parameters can be included in the function when fitting data.

Exponential models are shown in Figure 10.4. They can also be modified by including location and scale parameters. The logistic curve is also based on the exponential function; both growth and decay forms of the logistic function are shown in Figure 10.5.

Figure 10.6 shows some of the forms that hyperbolic trigonometric functions can assume. Sine, cosine, and tangent functions can also be used. Location and scale parameters should be applied to the functions.

Power, exponential, logarithmic, and trigonometric functions can be combined to form flexible functions. A wide variety of shapes can be fit by combining the power model with an exponential function (see Figure 10.7). Combining a trigonometric

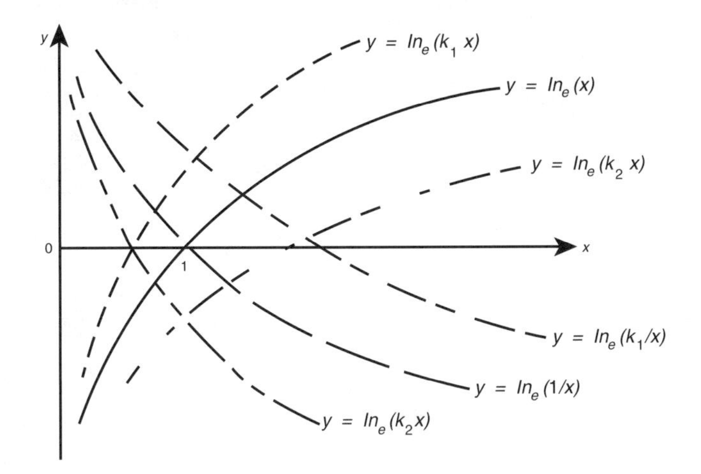

FIGURE 10.3 Characteristic shapes of logarithmic models ($k_1 > k_2$).

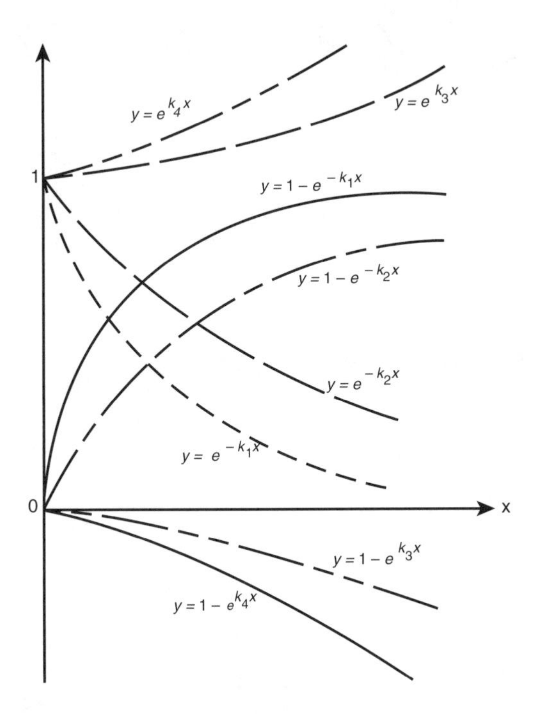

FIGURE 10.4 Characteristic shapes of exponential models ($k_1 > k_2$; $k_4 > k_3$).

function with an exponential function produces the decaying sinusoid of Figure 10.8. Other combinations are possible.

It should be obvious that almost all of these functions cannot be fit with ordinary least squares. However, they can be fit using a numerical least-squares algorithm, that is, numerical optimization (McCuen, 1993).

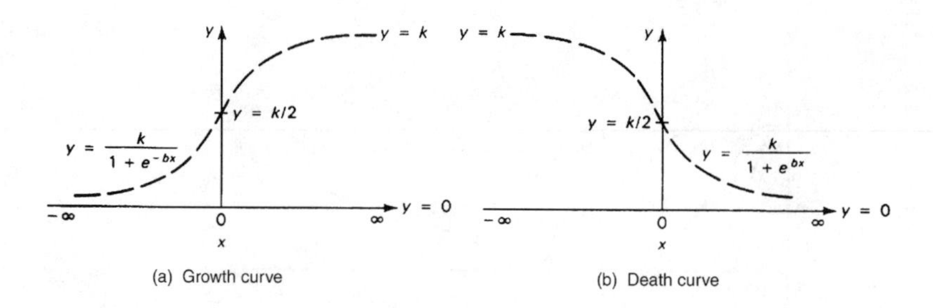

FIGURE 10.5 Characteristic shapes for logistic models.

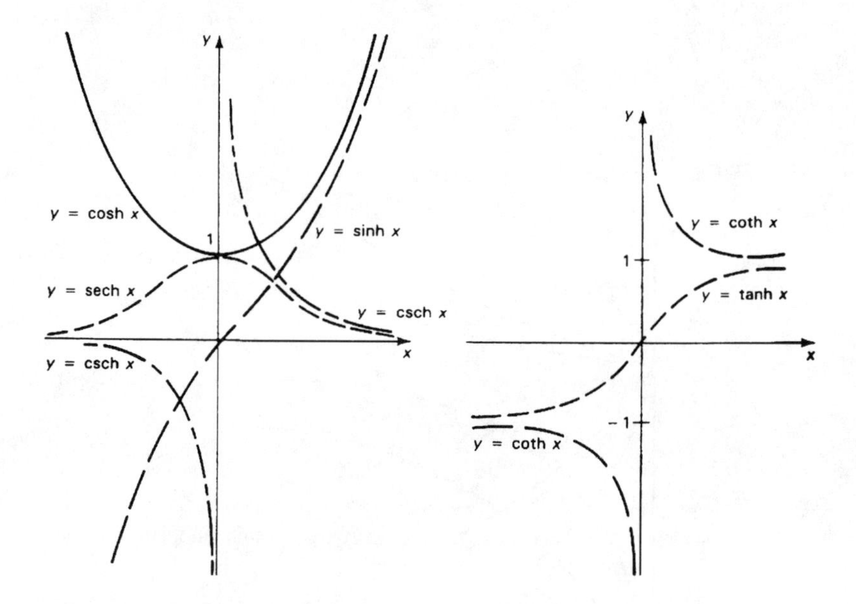

FIGURE 10.6 Characteristic shapes for hyperbolic trigonometric functions.

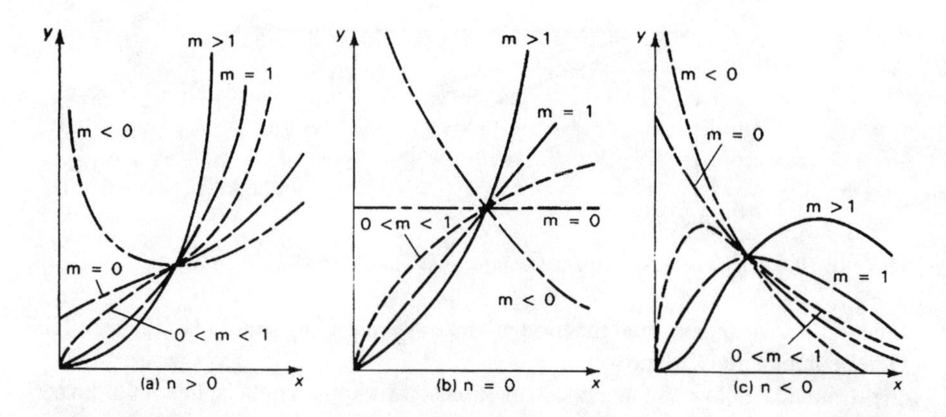

FIGURE 10.7 Characteristic shapes for $Y = KX^m e^{nx}$.

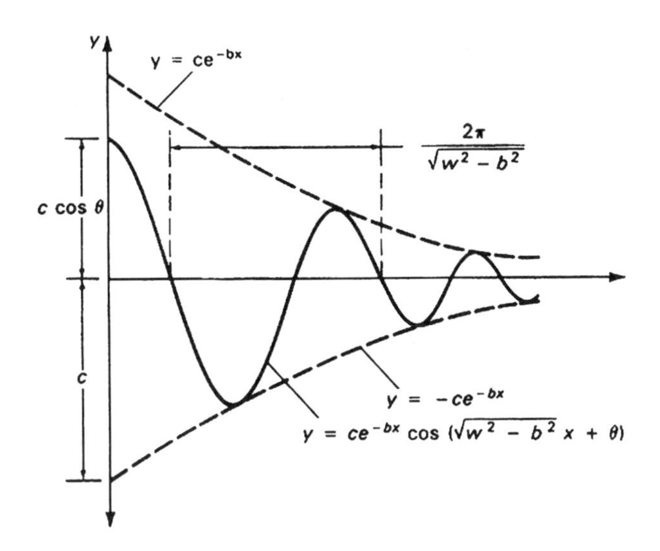

FIGURE 10.8 Characteristic shape of decaying sinusoid.

10.3.3 Composite Models

Computers have increased our ability to fit model coefficients, even with least squares as the fitting criterion, as well as facilitating the use of a wider array of functional forms. When measured data reflect different physical processes, it is likely that one functional form will not accurately reflect the data variation. In such cases, composite models (McCuen, 1993) provide a good alternative to single-valued functions.

A composite model consists of two or more functions, with the functions applicable to the regions representing the physical processes that underlie the measured data. The functions should be made to match at the boundaries of the regions so that the model is not discontinuous at the boundary. To ensure continuity of effect at a boundary, it may also be necessary to impose a constraint that ensures continuity of slope at a boundary. For each constraint imposed on the model, a fitting coefficient must be sacrificed, which reduces the flexibility of the model to fit the data but also ensures some measure of rationality.

Consider the case in which the graph of a data set indicates an initial steeply sloped rise from the origin of the graph, but with a decreasing slope. At some point, the curvature ends and the data appear to follow a linear trend. The first part of the graph might suggest a power model from $x = 0$ to the boundary between the two regions, which is denoted as x_c. Thus, the composite model has the form

$$\hat{Y} = \begin{cases} ax^b & \text{for } x \le x_c \quad (10.5a) \\ c + d(x - x_c) & \text{for } x > x_c \quad (10.5b) \end{cases}$$

where a, b, c, and d are empirical coefficients. In order for the functions to match at the boundary, one of the four coefficients would need to be set to ensure continuity of magnitude. For the slopes of the two functions to match at the boundary, a second coefficient is dedicated to the continuity of slope constraint. Thus, only two of the four coefficients would be available for adjustment to fit the data, with the other two coefficients dedicated to satisfying the constraints. The continuity of magnitude and continuity of slope constraints follow:

$$\hat{y}_1 = \hat{y}_2 \ @ \ x = x_c \tag{10.6a}$$

and

$$\frac{d\hat{y}_1}{dx} = \frac{d\hat{y}_2}{dx} \ @ \ x = x_c \tag{10.6b}$$

It is not always necessary to apply the continuity of slope constraint if the data suggest that the constraint will be achieved without applying it. For simple functions, analytical solutions are possible, but as the functions used for the individual components of the model become more complex, numerical fitting becomes necessary.

Example 10.3

Consider the case of a set of x versus y with a decreasing slope, such as a variable-intercept power model (see Figure 10.2), for low values of x and a linear trend for larger values of x. Thus, the model would be:

$$\hat{y} = \begin{cases} ae^{bx}x^c & \text{for } x \le x_c \quad (10.7a) \\ d + f(x - x_c) & \text{for } x > x_c \quad (10.7b) \end{cases}$$

This model has five coefficients to be fitted with the data. Assume that the data show considerable scatter so that continuity-of-magnitude and continuity-of-slope constraints are necessary. For the continuity of magnitude constraint, the functions would have equal magnitudes at x_c:

$$ae^{bx_c}x_c^c = d + f(x_c - x_c) = d \tag{10.8}$$

Therefore, the left side of Equation 10.8 can be substituted into Equation 10.7b. For the continuity of slope, the derivatives of Equations 10.7 are set equal:

$$ae^{bx_c}x_c^c[(c/x_c) + b] = f \tag{10.9}$$

Equation 10.9 can also be substituted into Equation 10.7b, which yields:

$$\begin{aligned} \hat{y} &= ae^{bx_c}x_c^e + ae^{bx_c}x_c^c[(c/x_c) + b](x - x_c) \\ &= ae^{bx_c}x_c^c[1 + (c/x_c + b)(x - x_c)] \end{aligned} \tag{10.10}$$

Note that Equation 10.10 is still a linear function. Equation 10.10 is used in place of Equation 10.7b as part of the two-function model. The model now has three coefficients that are fitted to the data. Values of d and f can then be computed using Equations 10.8 and 10.9.

10.4 MODEL CALIBRATION

The method of fitting empirical coefficients to a set of data depends primarily on model complexity. As complexity increases, the required effort, knowledge requirements, and required experience in modeling increase. With the availability of software that fits models to data, the software user should not calibrate models without having a complete understanding of the calibration process.

Simple models can be fitted using analytical least squares. In such cases, the model structure is generally linear or can be transformed to a linear form, the model or coefficients will not be subjected to constraints, and minimizing the sum of squares of the errors is the only criterion to use in deciding the values of the coefficients that are considered "best." Understanding the fitting process requires little background knowledge and application of a program requires little experience.

In comparison to simple models, fitting more complex models, often to more involved data sets, requires (1) a better understanding of the fitting process, (2) proper handling of constraints, (3) balancing of competing fitting criteria, and (4) the ability to assess the greater amount of output. While the fitting of simple models only involves the solution of a set of simultaneous equations, the fitting of complex models is generally iterative and requires both initial estimates of the unknowns and values of the step sizes needed for the underlying "trial-and-error" search process. The unknown coefficients of simple models are most often fitted without constraints; however, complex model fitting allows constraints on the coefficients to help ensure rationality. Simple models generally are fit with one criterion function, specifically minimizing the sum of the squares. Complex models may also use least squares but attempts are also made to provide an unbiased solution. Complex models have numerous advantages, but if the fitting is not properly executed, the complex model may actually be a poorer representation of the real system than a much simpler model.

10.4.1 Least-Squares Analysis of a Linear Model

The components of an unconstrained regression analysis include the model, objective function, and data set. The data set consists of a set of n observations on p predictor variables and one criterion variable, where n should be, if possible, at least four times greater than p. The data set can be viewed as a matrix having dimensions of n by $(p + 1)$. The principle of least squares is used as the objective function. The model, in raw-score form, is

$$\hat{y} = b_0 + b_1 X_1 + b_2 X_2 + \cdots + b_p X_p \tag{10.11}$$

in which X_j ($j = 1, 2,\ldots, p$) are the predictor variables, b_j ($j = 1, 2,\ldots, p$) are the partial regression coefficients, b_0 is the intercept coefficient, and $\hat{y}$ is the criterion variable. Using the least squares principle and the model of Equation 10.11, the objective function becomes

$$F = \min \sum_{i=1}^{n} e_i^2 = \min \sum_{i=1}^{n} \left(b_0 + \sum_{j=1}^{p} b_j X_{ij} - y_i \right)^2 \tag{10.12}$$

in which F is the value of the objective function. It should be noted that the predictor variables include two subscripts, with i indicating the observation and j the specific predictor variable.

The method of solution is to take the $(p + 1)$ derivatives of the objective function, Equation 10.12, with respect to the unknowns b_j ($j = 0, 1, \ldots, p$), setting the derivatives equal to zero, and solving for the unknowns. A set of $(p + 1)$ normal equations is an intermediate result of this process.

As an example, consider the case where $p = 2$; thus Equation 10.11 reduces to

$$\hat{y} = b_0 + b_1 X_1 + b_2 X_2 \tag{10.13}$$

Also, the objective function, Equation 10.12, is given by

$$F = \min \sum_{i=1}^{n} (b_0 + b_1 X_{i1} + b_2 X_{i2} - y_i)^2 \tag{10.14}$$

The resulting derivatives follow:

$$\frac{\partial F}{\partial b_0} = 2\sum_{i=1}^{n} (b_0 + b_1 X_{i1} + b_2 X_{i2} - y_i)(1) = 0 \tag{10.15a}$$

$$\frac{\partial F}{\partial b_1} = 2\sum_{i=1}^{n} (b_0 + b_1 X_{i1} + b_2 X_{i2} - y_i)(X_{i1}) = 0 \tag{10.15b}$$

$$\frac{\partial F}{\partial b_2} = 2\sum_{i=1}^{n} (b_0 + b_1 X_{i1} + b_2 X_{i2} - y_i)(X_{i2}) = 0 \tag{10.15c}$$

Rearranging Equations 10.15 yields a set of normal equations:

$$nb_0 + b_1 \sum_{i=1}^{n} X_{i1} + b_2 \sum_{i=1}^{n} X_{i2} = \sum_{i=1}^{n} y_i \tag{10.16a}$$

$$b_0 \sum_{i=1}^{n} X_{i1} + b_1 \sum_{i=1}^{n} X_{i1}^2 + b_2 \sum_{i=1}^{n} X_{i1} X_{i2} = \sum_{i=1}^{n} X_{i1} y_i \tag{10.16b}$$

$$b_0 \sum_{i=1}^{n} X_{i2} + b_1 \sum_{i=1}^{n} X_{i1} X_{i2} + b_2 \sum_{i=1}^{n} X_{i2}^2 = \sum_{i=1}^{n} X_{i2} y_i \tag{10.16c}$$

The solution of the three simultaneous equations would yield values of b_0, b_1, and b_2.

10.4.2 Standardized Model

When the means and standard deviations of the predictor and criterion variables are significantly different, round-off error, which results from the inability to maintain a sufficient number of significant digits in the computations, may cause the partial regression coefficient to be erroneous. Thus, most multiple regression analyses are computed using a standardized model,

$$Z_y = t_1 Z_1 + t_2 Z_2 + \cdots + t_p Z_p \tag{10.17}$$

in which t_j ($j = 1, 2, \ldots, p$) are called standardized partial regression coefficients, and Z_y and Z_j ($j = 1, 2, \ldots, p$) are the criterion variable and the predictor variables, respectively, expressed in standardized form; specifically, for $i = 1, 2, \ldots, n$, they are computed by

$$Z_{iy} = \frac{y_i - \bar{y}}{S_y} \tag{10.18}$$

and

$$Z_{ij} = \frac{X_{ij} - \overline{X_j}}{S_j} \tag{10.19}$$

in which S_y is the standard deviation of the criterion variable and S_j ($j = 1, 2, \ldots, p$) are standard deviations of the predictor variables. It can be shown that the standardized partial regression coefficients (i.e., the t_j's) and the partial regression coefficients (i.e., the b_j's of Equation 10.11) are related by

$$b_j = \frac{t_i S_y}{S_j} \tag{10.20}$$

The intercept coefficient can be computed by

$$b_0 = \bar{y} - \sum_{j=1}^{p} b_j \overline{X_j} \tag{10.21}$$

Thus, the raw score model of Equation 10.11 can be computed directly from the standardized model, Equation 10.17, and Equations 10.20 and 10.21.

10.4.3 Matrix Solution of the Standardized Model

The correlation matrix can be used in solving for the standardized partial regression coefficients. The solution is represented by

$$\boldsymbol{R}_{11} t = \boldsymbol{R}_{12} \tag{10.22}$$

in which $\boldsymbol{R}_{11}$ is the $p \times p$ matrix of intercorrelations between the predictor variables, $\boldsymbol{t}$ is a $p \times 1$ vector of standardized partial regression coefficients, and $\boldsymbol{R}_{12}$ is the $p \times 1$ vector of predictor-criterion correlation coefficients. Since $\boldsymbol{R}_{11}$ and $\boldsymbol{R}_{12}$ are known while $\boldsymbol{t}$ is unknown, it is necessary to solve the matrix equation, Equation 10.22, for the $\boldsymbol{t}$ vector; this involves premultiplying both sides of Equation 10.22 by $\boldsymbol{R}_{11}^{-1}$ (i.e., the inverse of $\boldsymbol{R}_{11}$) and using the matrix identities, $\boldsymbol{R}_{11}^{-1}\boldsymbol{R}_{11} = \boldsymbol{I}$ and $\boldsymbol{I}\boldsymbol{t} = \boldsymbol{t}$, where $\boldsymbol{I}$ is the unit matrix:

$$\boldsymbol{R}_{11}^{-1}\boldsymbol{R}_{11}\boldsymbol{t} = \boldsymbol{R}_{11}^{-1}\boldsymbol{R}_{12} \tag{10.23a}$$

$$\boldsymbol{I}\boldsymbol{t} = \boldsymbol{R}_{11}^{-1}\boldsymbol{R}_{12} \tag{10.23b}$$

$$\boldsymbol{t} = \boldsymbol{R}_{11}^{-1}\boldsymbol{R}_{12} \tag{10.23c}$$

It is important to recognize from Equation 10.23c that the elements of the $\boldsymbol{t}$ vector are a function of both the intercorrelations and the predictor-criterion correlation coefficients. If $\boldsymbol{R}_{11} = \boldsymbol{I}$, then $\boldsymbol{R}_{11}^{-1} = \boldsymbol{I}$, and $\boldsymbol{t} = \boldsymbol{R}_{12}$. This suggests that since the t_j values serve as weights on the standardized predictor variables, the predictor-criterion correlations also reflect the importance, or "weight," that should be given to a predictor variable. However, when $\boldsymbol{R}_{11}$ is quite different from $\boldsymbol{I}$ (i.e., when the intercorrelations are significantly different from zero), the t_j values will provide considerably different estimates of the importance of the predictors than would be indicated by the elements of $\boldsymbol{R}_{12}$.

10.4.4 Intercorrelation

It is evident from Equation 10.23c that intercorrelation can have a significant effect on the t_j values and thus on the b_j values. In fact, if the intercorrelations are significant, the t_j values can be irrational because of the inversion of the $\boldsymbol{R}_{11}$ matrix. It should be evident that irrational regression coefficients would lead to irrational predictions. Thus, it is important to assess the rationality of the coefficients.

The irrationality results from the difficulty of taking the inverse of the $\boldsymbol{R}_{11}$ matrix, which corresponds to the round-off-error problem associated with the solution of the normal equations. A matrix in which the inverse cannot be evaluated is called a *singular matrix*. A *near-singular matrix* is one in which one or more pairs of the standardized normal equations are nearly identical.

The determinant of a square matrix, such as a correlation matrix, can be used as an indication of the degree of intercorrelation. The determinant is a unique scalar value that characterizes the intercorrelation of the $\boldsymbol{R}_{11}$ matrix; that is, the determinant is a good single-valued representation of the degree of linear association between the normal equations.

General rules for interpreting the value of a determinant follow:

1. The determinant of a correlation matrix will lie between 0.0 and 1.0.
2. If the intercorrelations are zero, the determinant of a correlation matrix equals 1.0.

3. As the intercorrelations become more significant, the determinant approaches zero.
4. When any two rows of a correlation matrix are nearly equal, the determinant approaches zero.

As a rule of thumb, a determinant of $\boldsymbol{R}_{11}$ greater than 0.5 suggests that the intercorrelations are not sufficiently significant to cause irrational partial regression coefficients. A value less than 0.2 will indicate that irrationality is very likely. A value between 0.2 and 0.5 would indicate the distinct possibility of irrational coefficients. This is only a rule of thumb, and the regression coefficients should be checked for rationality even when the value of the determinant is greater than 0.5. This rule of thumb is not applicable for small samples.

It should be remembered that a regression coefficient may be irrational in magnitude as well as sign. The sign of the coefficient indicates the type of relationship; a negative sign indicates that as the predictor variable increases (decreases), the criterion variable will decrease (increase). The sign of the standardized partial regression coefficient should be the same as the sign of the predictor-criterion correlation coefficient, even when the sign of the predictor-criterion correlation coefficient is irrational. Irrational signs on predictor-criterion correlation coefficients often occur in small samples when the predictor variable is not overly important with respect to the criterion variable. One should also remember that a regression coefficient may be irrational in magnitude even when it is rational in sign; however, irrationality in magnitude is much more difficult to assess than irrationality in sign.

10.4.5 Stepwise Regression Analysis

In exploratory studies of hydrologic modeling, we often wish to use the basic linear form of Equation 10.11, but are uncertain as to which hydrologic variables are important. Obviously, in a cause-and-effect model such questions do not arise. But in practical hydrologic modeling they do arise. For example, consider monthly streamflow at some station. We could predict this streamflow to some level of accuracy using the total rainfall depth on the watershed during the month. But watershed storage will cause runoff in the current month to depend on rainfall that occurred in one or more previous months. How many previous months of rainfall should be included in the model? Consider another situation. Suppose that we have two sets of hydrologic data from a watershed. One set is the record before some change in land use in the area. The other set is a record after land use. Can we show a difference in the two sets by including some variable representing change?

The objective of stepwise regression is to develop a prediction equation relating a criterion (dependent) variable to one or more predictor variables. Although it is a type of multiple regression analysis, it differs from the commonly used multiple regression technique in that stepwise regression, in addition to calibrating a prediction equation, uses statistical criteria for selecting which of the available predictor variables will be included in the final regression equation; the multiple regression technique includes all available predictor variables in the equation and often is plagued by irrational regression coefficients. Stepwise regression usually avoids the irrational coefficients because the statistical criteria that are used in selecting the predictor variables usually eliminate predictor variables that have high intercorrelation.

Two hypothesis tests are commonly used in making stepwise regression analyses: the total F-test and the partial F-test. While they provide statistical assessments of the models at each step, they should never be used blindly. Experience suggests that they are not good indicators of the best model, at least when the level of significance is selected arbitrarily, such as 5%. It is much better to select a model based on traditional goodness-of-fit concepts. Model rationality is the best criterion, but the standard error ratio S_e/S_y, the coefficient of multiple determination R^2, and the accuracy of the coefficients are better criteria. Since F-tests are often the primary basis for stepwise decisions, they should be fully understood.

The objective of a total F-test is to determine whether the criterion variable is significantly related to the predictor variables that have been included in the equation. It is a test of significance of the following null (H_0) and alternative hypotheses (H_A):

$$H_0\text{: } \beta_1 = \beta_2 = \cdots = \beta_q = 0 \tag{10.24a}$$

$$H_A\text{: At least one regression coefficient is significantly different from zero.} \tag{10.24b}$$

In these hypotheses, q is the number of predictor variables included in the equation at step q, and β_i ($i = 1, 2, \ldots, q$) are the population regression coefficients. The null hypothesis is tested using the total F statistic,

$$F = \frac{R_q^2 / q}{\left(1 - R_q^2\right)/(n - q - 1)} \tag{10.25}$$

in which R_q is the multiple correlation coefficient for the equation containing q predictor variables, and n is the number of observations on the criterion variable (i.e., the sample size.) The null hypothesis is accepted if F is less than or equal to the critical F value, F_α, which is defined by the selected level of significance α and the degrees of freedom $(q, n - q - 1)$ for the numerator and denominator, respectively. If the null hypothesis is accepted, one must conclude that the criterion variable is not related to any of the predictor variables that are included in the equation. If the null hypothesis is rejected, one or more of the q predictor variables are statistically related to the criterion variable; this does not imply that all of the predictor variables are necessary, but this is usually assumed.

The partial F-test is used to test the significance of one predictor variable. It can be used to test the significance of either the last variable added to the equation or for deleting any one of the variables that is already in the equation; the second case is required to check whether the reliability of prediction will be improved if a variable is deleted from the equation. The null and alternative hypotheses are

$$H_0\text{: } \beta_k = 0 \tag{10.26a}$$

$$H_A\text{: } \beta_k \neq 0 \tag{10.26b}$$

where β_k is the regression coefficient for the predictor variable under consideration. The hypothesis is tested using the following test statistic:

$$F = \frac{\text{fraction increase in explained variation due to subject variable}/\nu_1}{\text{fraction of unexplained variation of the prediction equation}/\nu_2} \tag{10.27}$$

in which ν_1 and ν_2 are the degrees of freedom associated with the quantities in the numerator and denominator, respectively. In general, $\nu_1 = 1$ and $\nu_2 = n - q - 1$. For example, when selecting the first predictor variable, the test statistic is

$$F = \frac{(R_1^2 - R_0^2)/1}{(1 - R_1^2)/(n-2)} = \frac{R_1^2}{(1 - R_1^2)/(n-2)} \tag{10.28}$$

in which R_1 and R_0 are the correlation coefficients between the criterion variable and the first predictor variable and no predictor variables, respectively, and n is the sample size. The test statistic, which can be used for either the case when a second predictor variable is being added to the equation or the case when the two variables already in an equation are being tested for deletion, is

$$F = \frac{(R_2^2 - R_1^2)/1}{(1 - R_2^2)/(n-3)} \tag{10.29}$$

in which R_1 and R_2 are the correlation coefficients between the criterion variable and a prediction equation having one and two predictor variables, respectively. The null hypothesis, Equation 10.26a, is accepted when the test statistic F is less than or equal to the critical F value, F_α. If $F < F_\alpha$, the predictor variable that corresponds to $\beta_k(X_k)$ is not significantly related to the criterion variable when the other $(k - 1)$ predictor variables are in the equation; in other words, adding X_k to the prediction equation does not result in a significant increase in the explained variation. If the null hypothesis is rejected (i.e., $F > F_\alpha$), then the variable X_k makes a contribution toward the prediction accuracy that is significant even beyond the contribution made by the other $(k - 1)$ predictor variables. When the partial F-test is used to check for deletion of a predictor variable, a significant F value (i.e., $F > F_\alpha$) indicates that the variable should not be dropped from the equation. Partial F statistics are computed for every variable at each step of a stepwise regression.

Example 10.4

The data matrix (Table B.5) includes six predictor variables and the criterion variable, which is the snowmelt runoff volume for the 4-month period of April 1 through July 31. Annual values are available for a 25-year period. The correlation matrix is characterized by high intercorrelation. This suggests that inclusion of all predictors will lead to an irrational model. Thus, stepwise regression analysis is a reasonable approach to the problem of developing a prediction model. A summary of the stepwise regression analysis is given in Table 10.1.

TABLE 10.1
Stepwise Regression Summary for Snowmelt Runoff Data

Variable Entered	R	R^2	ΔR^2	S_e	S_e/S_y	Total F	Partial Regression Coefficient b_0	b_1	b_2	b_3	b_4	b_5	b_6
X_5	0.887	0.782	0.782	12.22	0.471	85.1	−16.86	—	—	—	—	2.518	—
X_3	0.895	0.801	0.014	12.09	0.466	44.3	−8.01	—	—	1.352	—	1.314	—
X_1	0.903	0.815	0.014	11.91	0.459	30.9	−17.31	2.619	—	1.504	—	1.206	—
X_4	0.904	0.817	0.002	12.15	0.469	22.3	−16.41	2.658	—	1.136	0.470	1.219	—
X_2	0.906	0.820	0.003	12.37	0.477	17.3	−8.84	2.306	−1.325	1.265	0.971	1.429	—
X_6	0.906	0.820	0.000	12.71	0.490	13.7	−9.31	2.321	−1.283	1.226	1.010	1.448	−0.093

Variable Entered	Standardized Partial Regression Coefficient t_1	t_2	t_3	t_4	t_5	t_6	Standardized Error Ratio for Coefficient b_1	b_2	b_3	b_4	b_5	b_6
X_5	—	—	—	—	0.89	—	—	—	—	—	0.11	—
X_3	—	—	0.44	—	0.46	—	—	—	0.81	—	0.77	—
X_1	0.12	—	0.49	—	0.43	—	0.78	—	0.72	—	0.83	—
X_4	0.12	—	0.37	0.12	0.43	—	0.79	—	1.29	2.60	0.84	—
X_2	0.11	−0.25	0.41	0.25	0.50	—	0.96	1.77	1.19	1.57	0.77	—
X_6	0.11	−0.24	0.40	0.26	0.51	−0.01	1.00	2.10	1.55	1.90	0.87	28.4

Variable Entered	Partial F to Enter Variable X_1	X_2	X_3	X_4	X_5	X_6
X_5	0.0	56.1	84.4	61.3	85.1	39.8
X_3	1.2	0.0	1.5	1.1	—	0.0
X_1	1.6	0.3	—	0.1	—	0.0
X_4	—	0.1	—	0.1	—	0.0
X_2	—	0.3	—	—	—	0.1
X_6	—	—	—	—	—	0.001

Using the classical approach to model selection, partial F statistics were evaluated for each step using a 5% level of significance. Only the first step is necessary. The predictor X_5 enters the model since it has the largest partial F. For step 2, the largest partial F is 1.53, which is not statistically significant so the bivariate model (Y vs. X_5) would be selected when the partial F-test is used with a 5% level of significance as the decision criterion. Since the total F, which equals the partial F for step 1, is significant, the first model is accepted. The slope coefficient has a rational sign. The intercept coefficient is negative and about 43% of the mean value of Y. If X_5 equaled 6.68 in. or less, then the predicted runoff volume would be negative. Since the two smallest measured values of X_5 were 8.8 in. and 13.1 in., the lower limit of rationality (6.68 in.) does not appear to be a major problem; however, it is a concern.

In terms of goodness of fit, the first model decreases the standard error from 25.93 in. (S_y) to 12.22 in. (47% of S_y). Adding additional variables can produce a S_e of 11.91 in., but this is not a significant improvement over the S_e for the first model. The correlation coefficient is 0.887. The multiple regression model that includes all of the predictor variables has an R of 0.906, which is only a 3% increase in the explained variance (ΔR^2) over the one-predictor model. In this case, the partial F-test would provide the same model identified using the goodness-of-fit statistics.

Evaluating the accuracy of the regression coefficients with the $S_e(b)/b$ ratios also suggests that the first model is best. For step 1, the ratio is 0.108. For step 2 both ratios are quite large, that is, above 0.76. This suggests that the regression coefficients for the second model are inaccurate.

Note that the S_e increases at step 4, even though the correlation coefficient continues to increase with each step. This occurs because the increase in ΔR^2 is marginal but the model loses an additional degree of freedom with each step. Thus, even though R^2 continues to increase, the model reliability actually declines.

Example 10.5

Project cost data (Table B.6) were analyzed to develop a simple model for predicting the cost of drainage systems. The standard deviation of the criterion variable is \$115,656. The ten predictor variables include physical characteristics of the site and size components of the system. A summary of the results is given in Table 10.2.

The first two steps decrease the S_e to \$64,000, which is 55% of S_y. Thereafter, the decrease is minimal, with a minimum of about \$59,000 which is 51% of S_y. This would suggest the two-predictor model, although it does not discount the use of more than two.

The increase in explained variance (ΔR^2) is large for the first two steps, but only 1.8% at step 3. The multiple regression model with all ten predictors explains about 7% more variation than the two-predictor model. While 7% improvement in the explained variance is usually considered worthwhile, it is probably not in this case as it would require adding eight additional variables and would yield an irrational model.

In terms of rationality, none of the intercepts are highly rational, although the intercept for the one-predictor model (X_{10}) is at least positive. All of the other models have negative intercepts. The slope coefficients are rational until step 6. At step 6,

TABLE 10.2
Stepwise Regression Summary for Project Cost Data

Step	Variable Entered	$\|R\|$	R	R^2	ΔR^2	S_e	S_e/S_y	Total F	Intercept
1	X_{10}	1.00	0.71	0.51	0.51	81958	0.71	69.4	31756
2	X_8	0.85	0.84	0.70	0.20	63999	0.55	79.2	–43894
3	X_5	0.84	0.84	0.72	0.02	62556	0.54	56.6	–84771
4	X_4	0.38	0.86	0.74	0.02	60571	0.52	46.6	–89083
5	X_6	0.16	0.87	0.75	0.01	59865	0.52	38.7	–107718
6	X_7	0.03	0.87	0.76	0.01	59164	0.51	33.4	–114088
7	X_3	0.03	0.88	0.77	0.01	58743	0.51	29.4	–170607
10	All	0.00	0.88	0.77	—	59359	0.51	20.3	–178598

Standard Error Ratio $S_e(b)/b$ for Coefficient

Step	b_1	b_2	b_3	b_4	b_5	b_6	b_7	b_8	b_9	b_{10}
1	—	—	—	—	—	—	—	—	—	0.12
2	—	—	—	—	—	—	—	0.15	—	0.14
3	—	—	—	—	0.49	—	—	0.14	—	0.14
4	—	—	—	0.43	0.42	—	—	0.29	—	0.13
5	—	—	—	0.40	0.55	0.63	—	0.33	—	0.22
6	—	—	—	0.40	0.59	0.47	0.63	0.36	—	0.23

Partial F to Enter for Variable

Step	X_1	X_2	X_3	X_4	X_5	X_6	X_7	X_8	X_9	X_{10}
1	0.4	1.2	0.0	32.1	0.8	45.0	28.2	60.5	36.8	69.4
2	0.6	0.5	0.9	33.3	1.3	4.1	2.5	44.5	3.0	—
3	0.0	0.3	2.4	3.7	4.1	3.3	0.5	—	0.3	—
4	0.0	0.1	1.3	5.4	—	1.5	0.8	—	0.3	—
5	0.0	0.8	0.9	—	—	2.5	0.5	—	0.1	—
6	0.1	0.5	1.2	—	—	—	2.5	—	0.7	—

Standardized Partial Regression Coefficient

Step	t_1	t_2	t_3	t_4	t_5	t_6	t_7	t_8	t_9	t_{10}
1	—	—	—	—	—	—	—	—	—	0.71
2	—	—	—	—	—	—	—	0.48	—	0.52
3	—	—	—	—	0.13	—	—	0.49	—	0.52
4	—	—	—	0.22	0.15	—	—	0.33	—	0.53
5	—	—	—	0.24	0.12	0.15	—	0.30	—	0.42
6	—	—	—	0.23	0.11	0.23	−0.22	0.27	—	0.57

the coefficient b_7 has an irrational sign. Therefore, rationality does not appear to be a critical decision criterion for this data set.

The standard error ratios for the regression coefficients, $S_e(b)/b$, are small for the first two steps. For step 3, the largest ratio is 0.49, which suggests the corresponding regression coefficient, b_5, is not sufficiently accurate. A value less than 0.5

is generally acceptable, while a ratio above 0.5 suggests that the coefficient is inaccurate.

The goodness-of-fit statistics indicate that at least two variables will be needed. The third variable to enter (X_5) is the smallest pipe diameter. This would require some computations unless it is set by the drainage policy. The variable at step 4 (X_4) is the number of inlets, which would require a preliminary layout of the drainage layout. Thus, inclusion of more than two predictor variables will require more effort than would be required for the first two predictors, the area to be developed and pipe length.

These criteria provide convincing evidence that only two predictor variables are required. When using a 5% level of significance, the partial F-test would indicate that four predictor variables should be included. The computed partial F at step 5 is 2.54, with the critical value being 3.99. In this example, the decision based on the statistical hypothesis test with a level of significance of 5% does not agree with the decision based on the other more important criteria.

10.4.6 Numerical Optimization

Models used in hydrologic analysis and design are often fit using analytical least squares. This approach is certainly appropriate for linear models and for some very simple nonlinear models such as Equation 10.4. The fitting procedure requires an objective function, which is to minimize the sum of the squares of the errors. Analytical optimization involves taking derivatives of the objective function, which is expressed as a function of the model, and setting the derivatives equal to zero. The analytical evaluation of the derivatives is possible only because the model structure is simple.

When a model structure is too complex such that the unknowns cannot be fit analytically, numerical procedures can be used. The numerical procedure is similar to the analytical procedure in that we are still interested in derivatives and the values of the unknowns where the derivatives are zero. But because of the complexity of the model and the inability to analytically derive explicit expressions of the derivatives, the derivatives must be assessed numerically, with an iterative approach to the point where they are zero. Specific examples of numerical optimization methods include the methods of steepest descent, pattern search, and conjugate gradient.

Response Surfaces

Although analytical optimization requires the evaluation of first derivatives, the fact that they are set equal to zero does not ensure optimality. The vanishing of the first derivative is a necessity, but by no means a sufficient condition for a point to be a local minimum, for the tangent is also horizontal at a maximum and at an inflection point, or a flat place. This observation leads to the concept of a stationary point. By definition, a stationary point is any point at which the gradient vanishes. The five types of stationary points are minima, maxima, valleys, ridges, and saddle points. Figure 10.9 shows these five types of stationary points for systems involving one and two unknowns. In the case of the minima and maxima, both global (G) and local (L) optimal values are shown. The usual objective is to find the global optimum

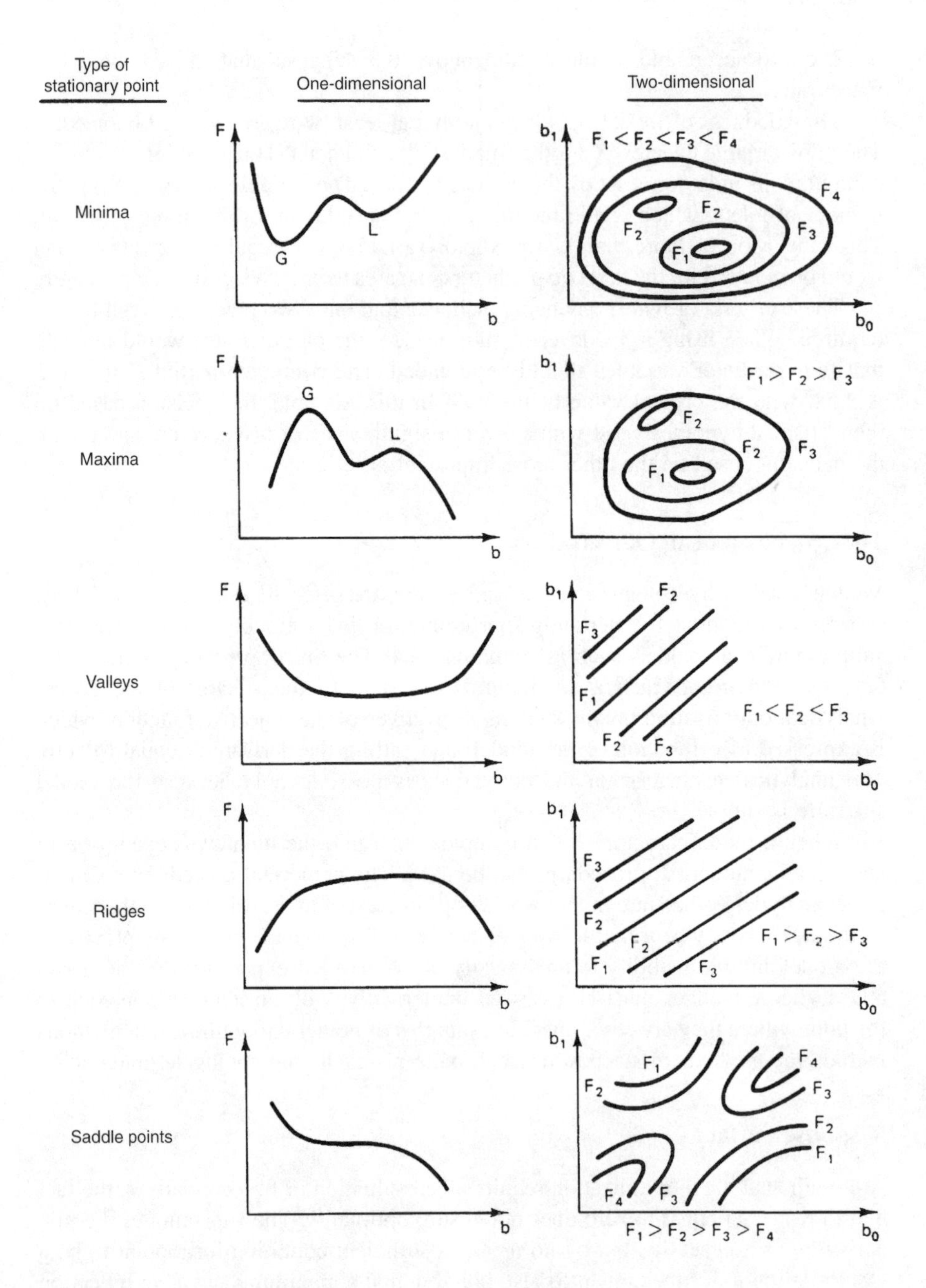

FIGURE 10.9 Stationary points. F is the volume of the objective function, with F_1, F_2, F_3, and F_4 being specific values of F; b, b_0, and b_1 are coefficients for which optimum values are sought.

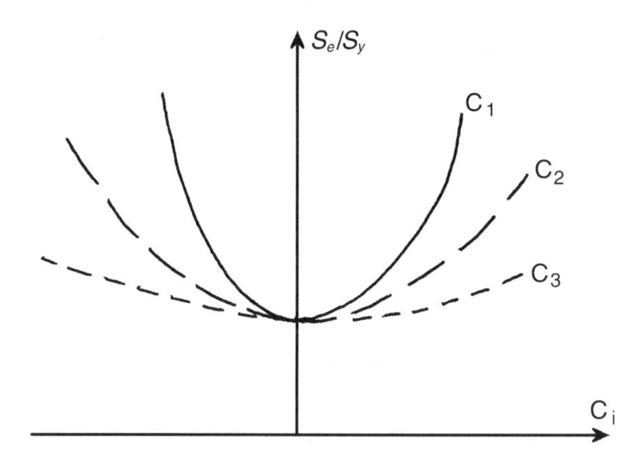

FIGURE 10.10 Use of one-dimensional response surfaces to evaluate coefficient importance.

value, although the local optimum may be a better alternative if the global optimum is associated with irrational coefficients.

Response surfaces represent the relationship between the fitting coefficients and the objective function. One- and two-dimensional response surfaces can be presented graphically, as in Figure 10.9. Like graphs of variables, that is, *x* versus *y* plots, graphs of response surfaces convey very important information. One-coefficient response surfaces can be used to identify the relative importance of the fitted coefficients, the stability of the optimum solution, and the degree to which error in a corresponding predictor variable introduces error in the criterion variable. Two-coefficient response surfaces indicate the extent of interaction between the coefficients, as well as providing the same information given by one-coefficient response surfaces.

The one-dimensional response surfaces in Figure 10.10 show the variation of the S_e/S_y, which is a measure of the relative accuracy, for three coefficients of a model. Assuming that the model only involves three unknowns, then the base values are optimum since S_e/S_y is a minimum for each coefficient. Assuming that the changes in the coefficients are comparable, C_1 is the most important coefficient while C_3 is the least important. A small change in C_1 will cause a large change in relative accuracy. The insensitivity of C_3 suggests that almost any value of C_3 will produce the same level of accuracy.

Figure 10.11 shows the one-dimensional response surfaces for two coefficients of a model. While C_4 is more important than C_5, the optimum solution has not been found since C_5 is not at its minimum. Further decrease in the least-squares objective function could be achieved by decreasing the value of C_5 from its current value. Figure 10.11 gives a hint of a problem with numerical methods of calibration. Specifically, the search process may stop when the important coefficients are near to their optimum values but the less important coefficients have not approached their best values. This results in an approximately optimum solution. Some search procedures are more efficient than others and are able to move closer to the optimum. Figure 10.11 also suggest the general rule that the values of insensitive coefficients are unlikely to reflect the underlying physical process because they do not always

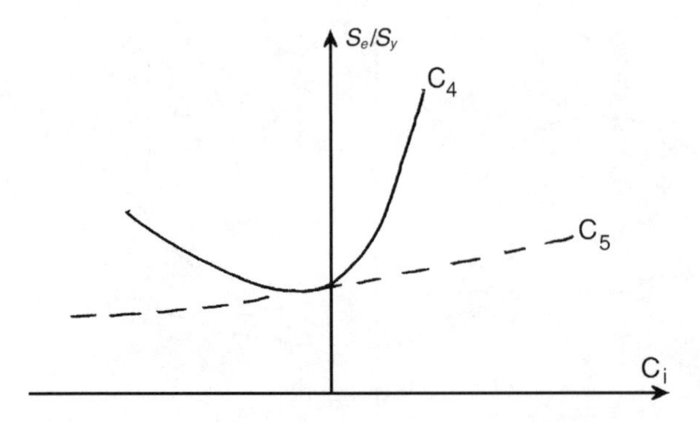

FIGURE 10.11 One-dimensional response surfaces for a nonoptimum solution.

approach their true values. This indicates the value of plotting and interpreting response surfaces after the algorithm indicates that convergence has taken place. Figure 10.11 also shows that a coefficient, C_4, is not necessarily equally sensitive in both directions. The response surface for C_4 shows that the accuracy criterion is much more sensitive to increases in value than it is to decreases in value.

In addition to the benefits of graphing one-dimensional response surfaces, two-dimensional surfaces also provide a clear picture of the optimization process. Figure 10.12 a shows a two-dimensional response surface for two coefficients that do not interact and that are of similar importance. The uniform spacing of the isolines indicates that the coefficients are of similar importance.

Figure 10.12(b) shows two coefficients of unequal importance that do not interact. Coefficient C_3 is insensitive since the criterion S_e/S_y changes very little for a change in C_3. Conversely, a small change in C_4 causes a large change in the criterion, so it is a more important coefficient in the optimization process.

The two coefficients of Figure 10.12(c) are highly interactive, as evidenced by the slope of the principal diagonal of the isolines. The same value of the objective function can be achieved for two quite different sets of values: (C_{5a}, C_{6a}) and (C_{5b}, C_{6b}). The objective of fitting the coefficients is to descend over the response surface to the point of minimum S_e/S_y. Interacting coefficients often reduce the efficiency of numerical optimization because it is more time consuming to move over the response surface of Figure 10.12(c) than the response surface of Figure 10.12(a). Interacting coefficients are much like highly intercorrelated variables in multiple regression in that they produce irrationalities. Interactive coefficients often suggest that the model components which they represent convey similar information or that the model structure forces the coefficients to be interactive. Interaction may also suggest that the model can be simplified with little loss of accuracy.

Response Surfaces of Model Bias

Least-squares regression analysis of a bivariate model, $\hat{y} = a + bx$, or a multiple regression model yields an unbiased model. Nonlinear models, even the bivariate power model, can be highly biased. While a model that minimizes the error variation, that is, min Σe^2, is a desirable criterion, it is also of interest to have an unbiased

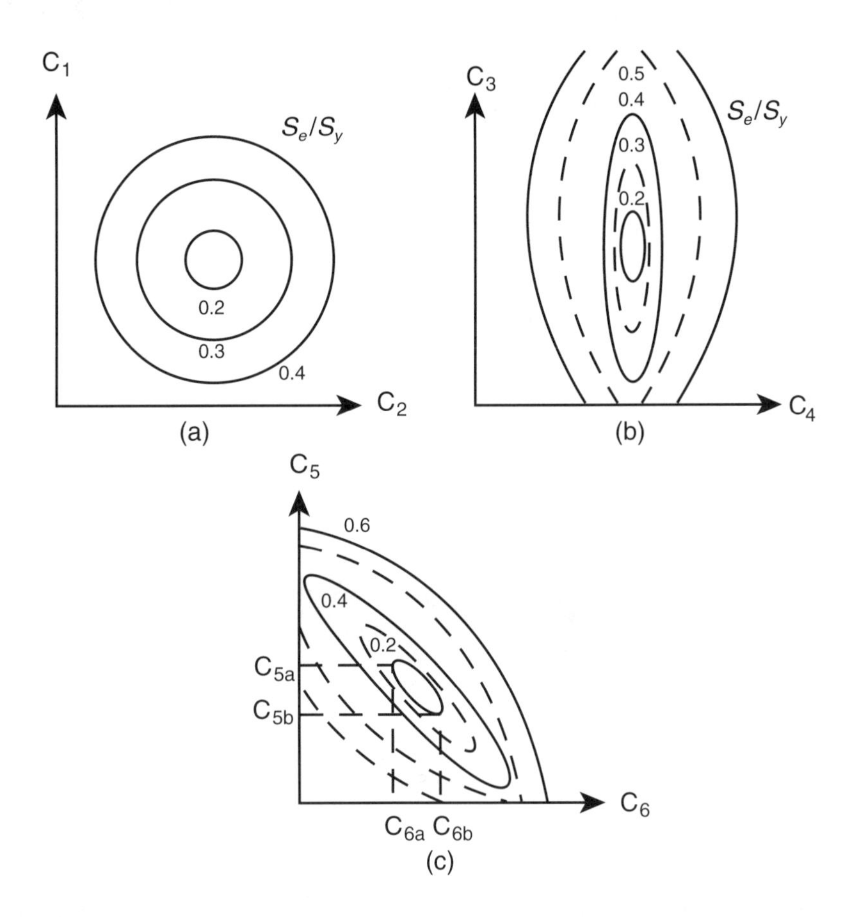

FIGURE 10.12 Two-dimensional response surfaces: (a) no interaction and equal coefficient importance; (b) no interaction, with C_4 more important than C_3; and (c) interaction, but equal coefficient importance.

model. In many cases, these two goals, that is, min S_e and zero bias, conflict. It may be better to have a set of coefficients that compromise on these two criteria.

Figure 10.13 shows the response surfaces for the relative accuracy (S_e/S_y) and the relative bias ($\bar{e}/\bar{y}$) for a two-coefficient (C_1 and C_2) model. The application of least squares would yield optimum coefficient values of C_{10} and C_{20} (see Figure 10.13). At this point, the relative accuracy is about 0.40. However, the relative-bias response surface shows a large relative bias of about 18% at the point of minimum sum of squares. Any combination of values of C_1 and C_2 that falls on the line of zero bias would meet the goal of an unbiased model. However, it is obvious that to obtain an unbiased model, the goal of minimum S_e/S_y would have to be sacrificed. The values of C_{11} and C_{21} yield a bias of zero but increase the standard error ratio by about 0.08 (from 0.40 to 0.48). This is generally considered detrimental to model accuracy just as the 18% under prediction is considered unacceptable. One possible compromise is to use C_{12} and C_{22}, which gives unbiased estimates but has an S_e/S_y of about 0.465. If this is not considered acceptable, then the model will

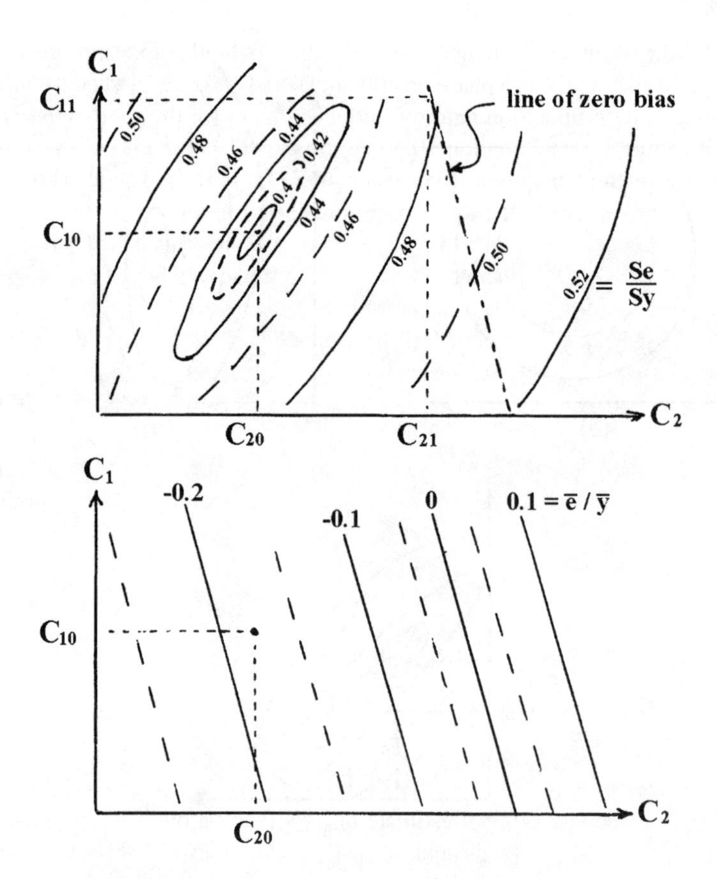

FIGURE 10.13 Two-parameter response surfaces of the standard error ratio (S_e/S_y) and the relative bias ($\bar{e}/\bar{Y}$) for two fitted coefficients, C_1 and C_2.

cause under-prediction of about 10% and have a standard error ratio of about 0.44. These values of C_1 and C_2 may yield a model that is a better representation of the underlying physical processes than either the zero-bias model or the minimum sum-of-squares model. The need to compromise on the two criterion adds an element of subjectivity into the modeling process that is not desirable, but it is probably necessary. As models and fitting techniques become more complex, it is often necessary for the modeler to be more involved in the decision process.

Example 10.6

Many hydrometeorological variables are characterized by an annual cycle. Temperature reaches a maximum in summer and a minimum in winter, with the variation having a periodic form,

$$\hat{T} = M + A\sin(2\pi ft + \phi) \tag{10.30a}$$

where M is the mean annual temperature, A is the amplitude of the sine curve, f is the frequency, t is the time, ϕ is a phase or shift angle, and $\hat{T}$ is the predicted temperature.

A numerical optimization requires initial estimates for the coefficients. The data used to illustrate fitting Equation 10.30 are 5 years (1954–1958) of mean monthly temperature for the Chattooga River watershed (Table B.7). The average temperature appears to be about 60°F, so a value of 60 will be used as an initial estimate of M. The larger values tend to be about 75°F and smaller values about 35°F to 40°F, so a value of 15 will be used as the amplitude A. The time t is assigned a value of 0 for January and 11 for December; it cycles from 0 to 11 for each year. Thus, the frequency coefficient f will have a value of 1/12 as the initial estimate. Since the sine curve crosses the mean about April 1, when $t = 3$, then the phase angle ϕ should be approximately $3/12 \times 2\pi$, which gives an estimate of 1.57 as the initial value for ϕ.

Numerical optimization also requires incrementors, or step sizes, for each of the coefficients. Since the step sizes are automatically decreased as the optimum is approached, the initial step sizes can be relatively large. A value of 0.5°F will be used for both M and A. Since f should be 0.083, a small value of 0.001 will be used for the step size for f. The initial value ϕ may be in error by half a month, or $\Delta\phi =$ 0.3, so a step size of 0.1 will be used for ϕ.

Based on these inputs, the first trial yielded the model

$$\hat{T} = 57.73 + 17.24[2\pi(0.0817)t + 1.561] \tag{10.30b}$$

which yields a standard error of estimate of 2.98°F and a bias of 0.00; thus, $S_e/S_y =$ 0.238 and $\bar{e}/\bar{Y} = 0$. The model is unbiased and provides reasonably accurate estimates. The residuals do not show a trend nor a consistent over- or under-prediction for any one season. The one-dimensional response surfaces indicates that variation in M is the most important, followed by f, with variation A and ϕ being much less important.

Example 10.7

Fifteen duplicate measurements (n = 30) of biochemical oxygen demand (BOD) from the Pigeon River (mile 63) taken during the summer of 1965. The data were fit to a logistic curve:

$$B\hat{O}D = \frac{K}{1 + e^{r(t - t_0)}} \tag{10.31a}$$

A graphical analysis suggested the following initial estimates: K = 3.5 mg/L, r = –0.25, and t_0 = 5 days. The numerically calibrated model is

$$B\hat{O}D = \frac{6.865}{1 + e^{-0.226(t - 4.366)}} \tag{10.31b}$$

and has the following goodness-of-fit statistics: $S_e = 0.309$ mg/L, $S_e/S_y = 0.181$, and bias = 0.009 mg/L. The bias is essentially zero.

Example 10.8

Twenty-five measurements of the dissolved oxygen concentration were made with Spree River water (Pleissner, 1910). The predictor variable is the time in days. Plotting the data suggested an exponential model,

$$\hat{y} = 8.47 - K(1 - e^{-rt}) \tag{10.32a}$$

where the constant 8.47 is the value measured at time $t = 0$. A graphical analysis indicated that reasonable initial estimates are as follows: $K = 4.4$ ppm, $r = 0.6$ days^{-1}. Numerical optimization yielded the following model:

$$\hat{y} = 8.47 - 3.834(1 - e^{-0.45t}) \tag{10.32b}$$

with the following goodness-of-fit statistics: $S_e = 0.055$ ppm, $S_e/S_y = 0.055$, bias = −0.004 ppm.

Example 10.9

The diversity of problems for which semivariogram modeling has been used in hydrologic analysis suggests that it is an important tool for statistical estimation. To evaluate the weights for the kriging model, a semivariogram model must be selected. The spherical model appears to be the most generally applicable model.

Karlinger and Skrivan (1981) computed the sample semivariogram points using the mean annual precipitation data for stations in the Powder River Basin in Montana and Wyoming. Mean annual precipitation is an important variable in regions where restoration of coal strip-mining lands is necessary. The sample points (see Table 10.3) are based on the mean annual precipitation at 60 stations in the region. The points show the trend that is typical of many semivariogram plots. The first three points have values of the semivariogram $\gamma(h)$ less than 4.1, while five of the last six values are greater than 4.5. Except for the value of $\gamma(h)$ at a separation distance of 187.5 miles, the sample values show little scatter about the trend. Additionally, it appears that the data would be best fit using a semivariogram model that includes an intercept (i.e., the nugget effect). While Karlinger and Skrivan used a linear semivariogram model, a spherical model will be used here:

$$\hat{\gamma}(h) = \begin{cases} \gamma_n + (\gamma_r - \gamma_n)\left[1.5\left(\frac{h}{r}\right) - 0.5\left(\frac{h}{r}\right)^3\right] & \text{for } h \le r \quad (10.33a) \\ \gamma_r & \text{for } h > r \quad (10.33b) \end{cases}$$

TABLE 10.3
Database for Fitting Semivariogram Model and Distribution of Errors

Separation Distance (mi)	$\gamma(h)$ (in.^{2})	$\hat{\gamma}(h)$ (in.^{2})	$\hat{\gamma}(h)-\gamma(h)$ (in.^{2})
17.2	3.94	3.60	–0.34
42.5	3.38	3.95	0.57
59.4	4.06	4.17	0.11
76.6	5.16	4.38	–0.78
87.5	4.16	4.51	0.35
109.4	5.03	4.74	–0.29
128.1	4.34	4.92	0.58
145.3	4.94	5.06	0.12
160.9	4.53	5.17	0.64
187.5	7.05	5.29	–1.76
237.5	4.53	5.33	0.80
			0.00 Sum

TABLE 10.4
Summary of Fitting of Semivariogram Model

	Value of Coefficients			Incrementors					
Trial	$C(1)$	$C(2)$	$C(3)$	$\Delta C(1)$	$\Delta C(2)$	$\Delta C(3)$	S_e/S_y	S_e	Bias
1	3.0	5.5	200.0	0.1	0.1	5.0	0.8774	0.836	0
2	3.212	5.291	186.9	0.01	0.01	1.0	0.8774	0.836	0
3	3.212	5.292	186.9	0.001	0.001	0.1	0.8774	0.836	0

in which γ_n is the nugget effect, γ_r the sill, r the radius of influence, and h the separation distance. Values for γ_n, γ_r, and r must be estimated from data.

The numerical optimization strategy requires initial estimates of the unknowns. Reasonable estimates of the three coefficients can be obtained from a plot of the sample points. Except for the point at a separation distance of 187.5 miles, most of the points at large separation distances fall near a value of 5.5; thus, a value of 5.5 in.^{2} will be used as the initial estimate of the sill, γ_r. If a line is drawn through the first few points and projected back to a separation distance of 0.0, a nugget effect of 3.0 is not unreasonable as an initial estimate. The same straight line intersects the sill at a separation distance of about 200 miles, which is used as an initial estimate of r.

The initial estimates of γ_r, γ_n, and r were used with the data of Table 10.3, which were estimated from a graph supplied by Karlinger and Skrivan, to calibrate Equations 10.33. The numerical optimization analysis converged, with the values of γ_n, γ_r, and r shown in Table 10.4. In comparison with the accuracy of the initial parameter estimates, the numerical fitting reduced the error sum of squares by about 13%.

The nugget effect increased from 3.00 in.^{2} to 3.21 in.^{2}, while the estimate of the radius of influence decreased from 200 miles to 187 miles. The value of the sill changed from 5.5 in.^{2} to 5.29 in.^{2}.

The resulting parameters are given in Table 10.4. The final spherical semivariogram model is

$$\hat{\gamma}(h) = \begin{cases} 3.21 + (5.29 - 3.21)\left[1.5\left(\dfrac{h}{187}\right) - 0.5\left(\dfrac{h}{187}\right)^3\right] & \text{for } h \le 187 \quad (10.34a) \\ 5.29 & \text{for } h > 187 \quad (10.34b) \end{cases}$$

The predicted values and errors are given in Table 10.3. Although the model is unbiased, the error variation is still significant. The standard error of estimate equals 0.836 in.^{2}, while the standard deviation of the $\gamma(h)$ values equals 0.953 in.^{2}; thus, the spherical model of Equations 10.34 reduced the error variation by only 13%. The large error variation may reflect either the small samples that were available to estimate the 11 sample points of Table 10.3 or the absence of a significant spatial structure of mean annual precipitation in the Powder River basin. Certainly, the observation at a separation distance of 187.5 miles is responsible for a large part of the error variation.

10.4.7 Subjective Optimization

Subjective optimization is an alternative to analytical and numerical optimization and is necessary when one of the following conditions exist:

1. The model structure is sufficiently complex such that analytical derivatives cannot be computed and numerically evaluating the derivatives is inefficient.
2. The database is sufficiently massive that the computer storage requirements for numerical optimization would be exceeded.
3. Fitting criteria beyond least squares and bias are of interest.
4. Complex constraints on model parameters and predicted values of the criterion variable(s) need to be applied.

The Stanford watershed model and its offspring, including HSPF, are models that require subjective optimization to fit the model parameters.

The term *subjective* may actually be a misnomer. It is an accurate descriptor of the calibration process in that it allows the user to use personal knowledge of both the model and the data in fitting the unknowns. It is an inaccurate descriptor if one believes that a systematic process of fitting the unknowns is not followed. It is identical to both analytical and numerical optimization in that correct use follows a systematic process and unbiased, precise estimates of the unknowns are desired. It is similar to numerical optimization in that multiple trials are made and that the process leads to movement across multiple response surfaces in search of the global optimum solution.

The multiple criteria aspect of subjective optimization is a significant departure from both analytical and numerical optimization. Least-squares minimization is the sole criterion involved with analytical optimization. In numerical optimization, movement across the response surface is made using least-squares minimization, but the bias response surface is an important factor in selecting the final parameter values. Generally, subjective optimization involves fitting numerous objective functions. For example, with a continuous simulation watershed model, the individual calibrating the model may wish to consider the following criteria in obtaining the optimum set of coefficients:

1. Over the course of a year of daily discharge rates, 10 to 15 larger-than-normal discharges may be considered peaks and coefficients that affect the peaks will need to be fitted to provide good precision and little bias.
2. Other model coefficients will need adjustment so that the model predicts low-flow discharges with good precision and little bias.
3. Model coefficients that control the recession slopes of the larger storm events will need to be adjusted so that they are accurately fitted.
4. The runoff volumes of individual events, as well as the total annual runoff volume, should closely match the actual runoff volumes.
5. Coefficients that control model storage components (e.g., groundwater storage) need to be adjusted so that end-of-year storages remain relatively stable from year to year unless hydrologic conditions for that year, such as an extreme drought, would suggest that storage would likely increase or decrease.
6. Coefficients and storages that control the initial part of the total hydrograph may need adjustment to ensure a good fit during the first few weeks of the simulation.

Balancing the demands of optimizing each of these criteria is essential to providing a solution that is truly the global optimum in both a statistical sense and in terms of the model accurately representing the hydrologic processes of the watershed being modeled.

Another aspect of subjective optimization is the change or changes made to the coefficients that are being calibrated. This corresponds to the step sizes in numerical optimization. If the changes made are too small, then the time to optimize will be lengthened. If the change to a coefficient is too large, then the solution may jump beyond its optimum value and actually suggest that improvement in the previous value cannot be made. Knowledge of the sensitivity of the model output to the coefficients is essential to accurate optimization. To be successful in optimizing a model to a set of data requires that the user have a good working knowledge of the model and its sensitivity, the optimization process, and the watershed to which the model is being calibrated.

Some general guidelines for calibrating a complex model using subjective optimization follow:

1. Make reasonable initial estimates of the fitting coefficients and assess the goodness of fit based on criteria such as the six discussed above.

2. Identify the major sources of imprecision (e.g., recessions that are too flat) and bias (e.g., underestimation of peaks or overestimation of baseflow).
3. Based on the most significant errors identified in step 2 and knowledge of the relationship between these errors and the fitting coefficients, adjust the values of two to four of the coefficients and re-run the model. Given that many fitting coefficients of complex models are highly correlated (e.g., decreasing the baseflow will also decrease the total flow and therefore the fitting of the peaks), it is best to try to change only coefficients that have a low intercorrelation (i.e., are not related to the same model components).
4. Repeat steps 2 and 3 until changes make little or no difference in the goodness of fit criteria.

Derivatives are a central element of both analytical and numerical optimization. The former uses $d\text{F}/d\text{b}$ while the latter uses the numerical equivalent, $\Delta\text{F}/\Delta\text{b}$. While derivatives are not explicitly computed when subjectively optimizing a complex model, they are implicit in the use of the user's knowledge of model sensitivity to decide on size of the change to make for each coefficient. Sensitivity functions are derivatives, and thus, since subjective optimization depends on knowledge of model sensitivity, it is based on derivatives just like analytical and numerical optimization.

10.5 MODEL VERIFICATION

The verification of a fitted model is an important, but often neglected, step. The failure to properly verify an empirical model can lead to very inaccurate estimates. Assuming that others will try to apply a fitted model, the omission of the verification stage, and especially reporting the verification results, can lead to gross misapplications of the model.

The verification stage of the modeling process is intended to be an independent check on the model. The calibration phase resulted in a set of fitted coefficients, an assessment of the rationality of trends predicted by the model, and an assessment of the model reliability, that is, goodness of fit. The verification phase has essentially the same set of objectives but with the important distinction that the analyses are made independently of the data used in the calibration phase. In the verification phase, it is important to ensure that the model will function accurately within the bounds for which the model was intended, which may be well beyond the bounds of the calibration data.

10.5.1 SPLIT-SAMPLE TESTING

Split-sample analysis is a common modeling approach to verification of a model. In general, the process is to randomly divide the data available for calibration into two parts, with one part used for calibration and the remaining data used for verification or testing of the model. Random separation is important to provide some sense of independence. While split-sample testing has the advantage of ensuring some resemblance of verification, it has the disadvantage of reducing the sample size used for calibration and, therefore, the accuracy of the coefficients. Goodness-of-fit statistics for the calibration and verification subsets are compared. If the verification statistics are significantly poorer than those obtained from the calibration phase, then the model is generally considered to be unverified. A two-coefficient test of correlation could be used to assess the statistical significance of calibration and verification correlations.

10.5.2 Jackknife Testing

Recognizing the problem of using a sample size comprised of 50% of the available data, which in hydrologic analysis is often small, an alternative from of split-sample testing has been used. The alternative is referred to as the jackknife method (Mosteller and Tukey, 1977). Several variations of the jackknife concept have been proposed. One of these alternatives is presented here. The method makes n calibrations, where n is the sample size, but each calibration is based on $n - 1$ pieces of data, with the piece not used in calibration used as the verification sample point. The error for the jackknifed point is computed, with the procedure yielding n estimates of the error. These n values are then used to make the standard goodness-of-fit calculations, that is, $\bar{e}$, $\bar{e}/\bar{Y}$, S_e, S_e/S_y, and R^2.

The jackknife method of verification is illustrated with measurements on two random variables X and Y (see Figure 10.14). The Y variable is regressed on X for the two-coefficient (a and b) linear model. Four calibrations are performed, with the intercept a and slope coefficient b computed for each combination of three observations. For each (a, b) pair, the omitted value of X is used to predict Y from which the error is computed as the difference between the predicted and measured Y. Once all four regressions are made and the errors e computed, verification statistics such as S_e, S_e/S_y, $\bar{e}$, $\bar{e}/\bar{y}$, R, and R^2 can be computed. These values can then be evaluated for model accuracy. They can also be compared to the values of the same statistics for a calibration based on all $n = 4$ values.

	Set 1	Set 2	Set 3	Set 4
Data sets	X_1 Y_1 X_2 Y_2 X_3 Y_3	X_1 Y_1 X_2 Y_2 X_4 Y_4	X_1 Y_1 X_3 Y_3 X_4 Y_4	X_2 Y_2 X_3 Y_3 X_4 Y_4
Calibrate model	$\hat{y} = a_4 + b_4X$	$\hat{y} = a_3 + b_3X$	$\hat{y} = a_2 + b_2X$	$\hat{y} = a_1 + b_1X$
Predicted value	$\hat{y}_4 = a_4 + b_4X_4$	$\hat{y}_3 = a_3 + b_3X_3$	$\hat{y}_2 = a_2 + b_2X_2$	$\hat{y}_1 = a_1 + b_1X_1$
Error	$e_4 = \hat{y}_4 - y_4$	$e_3 = \hat{y}_3 - y_3$	$e_2 = \hat{y}_2 - y_2$	$e_1 = \hat{y}_1 - y_1$
Bias	$\bar{e} = \frac{1}{4}\Sigma_{i=1}^{4} e_i$			
Standard error	$S_e = \left[\frac{1}{4-2}\Sigma_{i=1}^{4} e_i^2\right]^{0.5}$			

FIGURE 10.14 Schematic of the jackknife method of model verification applied to data with a record length of four.

Sensitivity studies are an important part of model verification. A complete sensitivity analysis of a model can show the effect of inaccuracies of fitted model coefficients, the effect of extrapolation beyond the range of the calibration data, the effect of errors in the inputs, the importance of components of a model, and the relative importance of the predictor variables. These analyses verify the rationality of the model but also provide some measure of the risk of using the model.

Simulation analyses are also useful in evaluating a model. Based on the moments of the sample data, a very large number of samples can be generated and used to evaluate the distributions of the coefficients and thus the likelihood that the model will provide accurate estimates. Confidence intervals for the model can also be approximated using simulation.

Example 10.10

The jackknife and split-sample testing are applied to rainfall-runoff models developed from data for the Muddy Creek watershed near High Point, North Carolina. Rainfall (P) and runoff (R) depths were available for 18 storm events. Since the emphasis is on the application of the verification techniques, linear bivariate models are used for all analyses.

For split-sample testing, the measured data were divided randomly into two groups of nine storm events. Three regressions were calibrated: all data and two sets of nine events. In practice, only the data randomly selected as the calibration data would be calibrated. The calibration results are given in Table 10.5(a).

If the first set of nine was the calibration set, the accuracy would be good, with a correlation coefficient of 0.872, which is highly significant (rejection probability of 0.0006). The calibrated model is

$$\hat{R} = -0.6063 + 0.8209P \qquad (10.35)$$

The calibrated model was used to predict the runoff for the measured rainfalls of the second set (see column 3 of Table 10.5(b)). The goodness of fit statistics for model 1 are given in Table 10.5(d), with a correlation of 0, a bias of 1.366 in., a relative bias of 52% of the mean, and a standard error ratio of 1.02. All of the goodness-of-fit statistics suggest that Equation 10.35 is a poor model. The verification statistics suggest that the mean runoff would be a more accurate predictor than the equation. This occurs because the verification statistics do not support the accuracy suggested by the calibration statistics, which indicate an accurate model.

If data set 2 is used to calibrate a model, the following linear equation is the result:

$$\hat{R} = 0.3435 + 0.1291P \qquad (10.36)$$

The goodness-of-fit statistics indicate that Equation 10.36 is a poor model, with a correlation of zero and a standard error ratio greater than 1 (see Table 10.5(a)).

TABLE 10.5
Split-Sample Testing of Storm Event Rainfall-Runoff Data for the Muddy Creek at High Point, North Carolina, January–February 1935

(a) Calibration Results

Model	Intercept	Slope	R	S_e	S_e/S_y	n
All data	−0.68102	0.65449	0.6771	1.442	0.7585	18
1	−0.60631	0.82087	0.8722	1.137	0.5229	9
2	0.34346	0.12907	0.0000	1.030	1.0444	9

(b) Prediction of Set 2 Using Model 1

i	X_2	Y_2	Y_{2p}	Error
1	0.13	0.13	−0.500	−0.630
2	1.52	0.04	0.641	0.601
3	2.92	1.45	1.791	0.341
4	3.14	0.28	1.971	1.691
5	3.90	3.05	2.595	−0.455
6	4.13	1.19	2.784	1.594
7	4.50	0.12	3.088	2.968
8	4.81	0.30	3.342	3.042
9	5.08	0.42	3.564	3.144

(c) Prediction of Set 1 Using Model 2

i	X_1	Y_1	Y_{1p}	Error
1	1.30	1.78	0.511	−1.269
2	1.50	0.22	0.537	0.317
3	1.69	0.03	0.562	0.532
4	2.62	0.26	0.682	0.422
5	3.76	4.00	0.829	−3.171
6	5.13	3.10	1.006	−2.094
7	5.84	4.92	1.097	−3.823
8	6.67	3.65	1.204	−2.446
9	7.18	5.88	1.270	−4.610

(d) Goodness-of-Fit Statistics of Split-Sample Models

Model	Bias (in.)	Relative Bias	S_e (in.)	S_e/S_y	R
1	1.366	0.310	2.218	1.020	0.000
2	−1.794	0.166	2.890	2.929	0.000

The model is used to predict the runoff for the data of set 1 (see Table 10.5(c)), with the goodness-of-fit statistics shown in Table 10.5(d). The verification statistics are much poorer than the calibration statistics, with an underprediction bias of −1.79 inches (relative bias of −231%), a correlation coefficient of 0, and a standard error ratio of 2.93.

The jackknife method was applied to the same data. Eighteen regressions were made, each based on 17 sample points. For example, when the first data point ($P = R = 0.13$ in.) was omitted, the other 17 points yielded the following model (see Table 10.6):

$$R = -0.9229 + 0.7060P \tag{10.37}$$

which has a correlation coefficient of 0.668, a standard error of 1.47 in., and a standard error ratio of 0.7687, all of which suggest a good model. Equation 10.37 is then used to predict the runoff for the rainfall of 0.13 in., with a predicted value of –0.83 in. (see column 9 of Table 10.6). The other jackknifed regressions yielded the predictions in column 9 of Table 10.6 and the errors in column 10. The errors yielded a standard error of 1.62 in., a standard error ratio of 0.85, a correlation coefficient of 0.564, and a bias of –0.04 in. The standard error ratio of 0.85 is not good but it suggests that the model based on all of the data ($n = 18$) is reasonably accurate.

The split-sample and jackknife analyses suggest different results. The jackknife analysis suggests that the data can provide reasonable results. While the sample of 18 yields a correlation coefficient of 0.6771 (see Table 10.5(a)), the verification analysis yields a correlation of 0.56, which is not a major decrease ($\Delta R^2 = 0.14$). The split-sample testing suggests that the data are not adequate to develop on accurate model, with the verification goodness-of-fit statistics being very poor. The different assessment by the two methods of the verification accuracy points out the problems with the two methods. The jackknife method would be expected to suggest a model that is only slightly less accurate than the calibration results because $n - 1$ values are used in the subcalibrations. The split-sample analysis would most likely suggest relatively poor results because the sample is one-half of the total sample size. The standard error is known to become poorer for such a drop is sample size, n to $n/2$.

10.6 ASSESSING MODEL RELIABILITY

Assessment of a model's prediction accuracy is important. Since sufficient data to perform a split-sample test are rare in hydrologic analysis, it is often necessary to use the calibration data as the basis for assessing the goodness of fit. That is, reliability is evaluated with the statistics derived from the calibration results. Just as the model coefficients are used as sample estimates of the population coefficients, the sample goodness-of-fit statistics are used as estimates of the reliability of future predictions.

Several criteria can be used to assess the reliability of a fitted model, including the following:

1. Rationality of the coefficients and, for a multiple-predictor variable model, the relative importance of the predictor variables, both of which can be assessed using the standardized partial regression coefficients.
2. Degree to which the underlying assumptions of the model are met.

TABLE 10.6
Results of Jackknife Assessment of Storm Event Rainfall-Runoff Analysis for Muddy Creek at High Point, North Carolina, January–February 1935

i	P (in.)	RO (in.)	Intercept	Slope	R	S_e	S_e/S_y	Prediction RO (in.)	Error (in.)
1	0.13	0.13	−0.92292	0.70604	0.66781	1.47	0.7687	−0.831	−0.961
2	1.30	1.78	−1.03043	0.72159	0.71243	1.42	0.7248	−0.092	−1.872
3	1.50	0.22	−0.66481	0.65146	0.66109	1.49	0.7749	0.312	0.092
4	1.52	0.04	−0.62645	0.64432	0.65770	1.49	0.7780	0.353	0.313
5	1.69	0.03	−0.60745	0.64114	0.65861	1.49	0.7772	0.476	0.446
6	2.62	0.26	−0.58664	0.64134	0.67007	1.47	0.7666	1.094	0.834
7	2.92	1.45	−0.70369	0.65712	0.67729	1.49	0.7598	1.215	−0.235
8	3.14	0.28	−0.58295	0.64534	0.67835	1.46	0.7588	1.443	1.163
9	3.76	4.00	−0.79813	0.65079	0.70585	1.37	0.7316	1.649	−2.351
10	3.90	3.05	−0.73351	0.64986	0.68265	1.46	0.7547	1.801	−1.249
11	4.13	1.19	−0.65517	0.66085	0.68410	1.47	0.7533	2.074	0.884
12	4.50	0.12	−0.66114	0.68394	0.71945	1.37	0.7173	2.417	2.297
13	4.81	0.30	−0.70115	0.69561	0.72451	1.37	0.7119	2.645	2.345
14	5.08	0.42	−0.73846	0.70717	0.73021	1.36	0.7056	2.854	2.434
15	5.13	3.10	−0.66876	0.64408	0.66576	1.48	0.7706	2.635	−0.465
16	5.84	4.92	−0.54665	0.58675	0.64311	1.41	0.7909	2.880	−2.040
17	6.67	3.65	−0.68580	0.65644	0.64890	1.49	0.7858	3.693	0.043
18	7.18	5.88	−0.33513	0.52246	0.57766	1.38	0.8430	3.416	−2.464
Calibration			−0.68102	0.65449	0.67714	1.44	0.7585		
Jackknife testing					0.56446	1.62	0.8509		
						Bias = −0.0437			

3. Standard error of estimate (S_e).
4. Relative standard error (S_e/S_y).
5. Correlation coefficient (R).
6. Model bias ($\bar{e}$) and relative bias ($\bar{e}/\bar{Y}$).
7. Accuracy of the fitted coefficients as measured by the standard errors of the coefficients.
8. F statistic for the analysis of variance.

All of these criteria may not apply to every model, but they are important criteria commonly computed in model development.

10.6.1 Model Rationality

Rationality is probably the single most important criterion in assessing a model. Two methods are commonly used for assessing rationality. First, a model should provide rational predictions over all possible values of the predictor variable(s). Second, when a model coefficient is directly related to a predictor variable, such as in a linear multiple regression model, the coefficient should accurately reflect the effect of X on Y. A model that provides irrational predictions or indicates an irrational relationship between the criterion variable and a predictor variable should be used very cautiously, if at all.

The intercept coefficient of a linear model should be rational. As the values of the predictor variables approach specific target values, the intercept coefficient should approach a physically rational value. For example, if the evaporation (E) is related to the air temperature (T), air speed (W), and relative humidity (H), the following model form might be used:

$$\hat{E} = b_0 + b_1 T + b_2 W + b_3 H \tag{10.38}$$

As the air temperature approaches 32°F, $\hat{E}$ should approach zero because the water will freeze. As W approaches zero, the air mass overlying the water body will remain stagnant and, eventually, will become saturated. Thus, $\hat{E}$ will approach zero. As H approaches 100%, the air mass overlying the water body will be saturated and the evaporation will approach zero. The intercept coefficient should be such that for these conditions, the estimated value of $\hat{E}$ is zero. If the resulting predictions at these target conditions are not rational, the intercept coefficient could be irrational.

The slope coefficients (b_1, b_2, and b_3 of Equation 10.38) should also be rational. Since the slope coefficients represent the change in E for a given change in the value of the predictor variables (T, W, and H), it is easy to check for rationality in sign. In Equation 10.38, b_1 and b_2 should be positive, while b_3 should be negative; as relative humidity increases, the evaporation rate should decrease, and thus b_3 will be negative. It is more difficult to check the rationality of the magnitude of a slope coefficient. Since the slope coefficients are a function of the units of both E and the corresponding predictor variable, it may be easier to interpret the rationality of the coefficient by converting a slope coefficient b_i to the corresponding value of the standardized partial regression coefficient, t_i,

$$t_i = \frac{b_i S_i}{S_y} \tag{10.39}$$

in which S_i is the standard deviation of predictor variable i and S_y is the standard deviation of the criterion variable. The t_i values are dimensionless and independent of the units of Y and X_i; thus, they are easier to interpret than the b_i values. A t_i value should be between 0 and 1 in absolute value. A value near 1 indicates an important predictor; a value near 0 indicates that the corresponding predictor variable is not important. Thus, the larger the value, the more important a predictor variable is considered. The t_i values are similar to correlation coefficients. In fact, for the bivariate linear model, the t_i value is equal to the correlation coefficient. Therefore, a value should not be greater than 1 in absolute value. Where the intercorrelations between the predictor variables are significant, the absolute value of a t_i may exceed 1. This is the result of the matrix inversion involved in Equation 10.23c. The model should be considered irrational because the t_i value is irrational, and the corresponding b_i values should be considered irrational. An irrational model should be used with caution, if at all; this is especially true when the model is used outside the range of the data that is used for calibrating the model. After all, if the b_i value is not rational, it implies that the model does not give a true indication of the rate of charge of Y with respect to change in X_i.

10.6.2 Bias in Estimation

While models are sometimes developed to make a single prediction, a model will most often be used to make several predictions, either in time or in space. For example, a snowmelt runoff model developed for a watershed can be used to make runoff projections for several successive years before the model is recalibrated with the data from the additional years. Peak discharge models can be developed for a region and used to make design discharge estimates at many locations in the region. Whether the multiple estimates are made in time or space, and recognizing that models cannot be perfectly accurate on every estimate, one hope is that, on the average, the model does not overestimate or underestimate the true value. A model that would consistently over- or under-estimate is said to be biased. The bias is the difference between the long-term predicted value and the true value. Unfortunately, the true value is not known, but the average of the differences between the predicted values ($\hat{Y}_i$) and the measured values (Y_i) in a sample is used as an estimate of the bias,

$$\text{bias} = \bar{e} = \frac{1}{n}\sum_{i=1}^{n}(\hat{Y}_i - Y_i) = \frac{1}{n}\sum_{i=1}^{n} e_i \tag{10.40}$$

A model that consistently overestimates the measured value is said to have a positive bias, while consistent underestimation is a negative bias. Ideally, a model should be unbiased, but bias is only one criterion for assessing a model. Some bias may be acceptable when using an unbiased model would lead to poorer reliability with respect to other criteria (see Section 10.4.6).

10.6.3 Standard Error of Estimate

Given measurements on a single random variable Y, the mean of the sample is the most accurate estimate when a prediction is needed. The mean is an unbiased estimate in that the deviations from the mean sum to zero. In comparison to any other estimate made with the sample, the sum of the squares of the deviations from the mean will be smaller than the sum of squares for any other value. Thus, the standard deviation (S_y) of the values is a minimum (since S_y is just the square root of the sum of squares divided by the degrees of freedom). In summary, for a single random variable the mean is the best estimate in that it is unbiased and has a minimum error variance, and the standard deviation of the values is a measure of the accuracy of the mean.

One way to reduce the error variance of the estimated value is to relate Y to one or more predictor variables X. Then the objective is to provide a relationship between Y and the X variables such that the estimated values $\hat{Y}$ are unbiased and have a minimum sum of squares of the deviations. Dividing the sum of squares by the degrees of freedom yields the error variance S_e^2, the square root of which is the standard error of estimate, S_e. If the equation between Y and X is worthwhile, then S_e will be significantly smaller than S_y. That is, the equation reduces the error deviation from S_y to S_e. Thus, the ratio S_e/S_y is a measure of the improvement in the accuracy of prediction due to the prediction equation. When S_e/S_y is near 0.0, the model significantly improves the accuracy of prediction over predictions made with the mean. When S_e/S_y is near 1.0, the model provides little improvement in prediction accuracy when compared to the mean. The S_e/S_y ratio can actually be larger than the mean for small sample sizes; this is the result of the small number of degrees of freedom.

If two prediction models with different S_e/S_y ratios are available, then it is necessary to decide if the smaller ratio is significantly lower than the larger ratio. This is usually easiest to decide by examining the competing values of S_e. If the smaller S_e is physically better, then the corresponding model is preferred. This concept is best illustrated using an example. Assume project cost data has a standard deviation of \$116,000, which means that, assuming the errors are normally distributed about the mean, 68% of the projects will have an error less than \$116,000 if the mean is used as an estimate. Assume also that a prediction model is developed between project cost and a predictor variable. If the S_e for the model is \$75,000, then S_e/S_y is 0.647. In comparison to the S_y of \$116,000, the S_e is a physically meaningful reduction in the error variance, and so the model is an improvement over the mean. Now assume that a second model is developed that uses one additional predictor variable from data matrix. The second model yields a S_e of \$68,000 which yields $S_e/S_y = 0.586$. Compared to S_y, the reduction in S_e of \$6,000 may or may not be highly meaningful. If it is not meaningful in project planning, then the second model does not provide a meaningful improvement in accuracy. Other criteria, such as the ease of data collection, could be the deciding factor in selecting one of the two models. For example, if it were time consuming to develop a value for the second variable, then the ease of data collection as a decision criterion would be more important than the additional $6,000. If the S_e of the second model was $40,000 ($S_e/S_y = 0.345$) instead of $68,000, this would be considered a significant improvement in accuracy when compared to a S_e of $74,000, so the S_e would probably be the deciding criterion.

Mathematically, the standard error of estimate (S_e) is

$$S_e = \left(\sum e^2/\upsilon\right)^{0.5} = \left(\sum (\hat{Y} - Y)^2/\upsilon\right)^{0.5} \tag{10.41}$$

in which $\hat{Y}$ is the predicted value and Y is the measured value of the random variable; υ is the degrees of freedom, which depends on the model and sample size. If all of the predicted values equal the corresponding measured values, then the S_e is zero, which means that the model provides perfect predictions. If all predicted values equal the mean, then, depending on the degrees of freedom, the S_e will equal the standard deviation of Y. In this case, the model does not improve the accuracy of predictions.

It is of interest of note that S_e can actually be greater than S_y. S_y is based on $n - 1$ degrees of freedom. If the model has fewer than $n - 1$ degrees of freedom and the $\sum e^2$ for the mean, the S_e is greater than S_y. This implies that the mean provides better predictions of Y than the model does.

10.6.4 Correlation Coefficient

The correlation coefficient (R) is an index of the degree of linear association between two random variables. The magnitude of R indicates whether the model will provide accurate predictions of the criterion variable. Thus, R is often computed before the regression analysis is performed in order to determine whether it is worth the effort to perform the regression; however, R is always computed after the regression analysis because it is an index of the goodness of fit. The correlation coefficient measures the degree to which the measured and predicted values agree and is used as a measure of the accuracy of future predictions. It must be recognized that, if the measured data are not representative of the population (i.e., data that will be observed in the future), the correlation coefficient will not be indicative of the accuracy of future predictions.

The square of the correlation coefficient (R^2) equals the percentage of the variance in the criterion variable that is explained by the predictor variable. Because of this physical interpretation, R^2 is a meaningful indicator of the accuracy of predictions.

Strictly speaking, the correlation coefficient applies only to the linear model with an intercept. For example, it can be used for the linear model of Equation 10.11. It also applies to curvilinear models, such as polynomial models, that are based on a linear separation of variation ($TV = EV + UV$). It does not apply to the power model of Equation 10.4, except when the power model is expressed in the log-transform structure; but then the numerical value applies to the prediction of the logarithms and not the predicted values of Y for which estimates are required.

The standard error of estimate is a much better measure of goodness of fit than the correlation coefficient. The standard error has the advantages of having the same units (dimensions) as Y; properly accounting for the degrees of freedom; and being valid for nonlinear models as well as linear models. The correlation coefficient is a dimensionless index, which makes it easy to assess goodness of fit, but its value may be high only because the total variation is small such that the explained variation seems large, that is, *EV*/*TV* is large because *TV* is small.

A major problem with the correlation coefficient is that the degrees of freedom equals the sample size (n). This is acceptable only for large samples and models with very few coefficients that need fitting. In modeling, 1 degree of freedom should be subtracted from n for each coefficient that is quantified with the data. For small samples and models with a large number of fitted coefficients, the number of degrees of freedom may be small; in such cases, the correlation coefficient may be large, but the reliability of the model is still poor. This is best illustrated with an example. Assume that we have six pairs of Y and X measurements, with a systematic trend not evident from a graph of Y versus X. If a linear model is fitted, the fit would be poor, with a near-zero correlation coefficient resulting. If terms of a polynomial are successively added, then the equation would provide a better fit but the structure would not show a rational trend. We know that a fifth-order polynomial with six coefficients would fit the data points exactly, and so the explained variance would equal the total variance ($EV = TV$) and the correlation coefficient would equal 1. But the polynomial does not reflect the true relationship between Y and X, and thus, the correlation coefficient is an inaccurate measure of the reliability of prediction. In the case of a fifth-order polynomial, there would be no degrees of freedom, so the standard error is infinitely large, that is, undefined; in this case, the S_e is a more rational measure of reliability than the correlation coefficient of 1.

Another problem with the correlation coefficient is that it assumes the model is unbiased. For a biased model, a correlation coefficient may be very erroneous. It is better to compute the standard error of estimate and use it as a measure of model reliability. A better index of correlation can be computed using

$$R = \left[1 - \left(\frac{n-\upsilon}{n-1}\right)\left(\frac{S_e}{S_y}\right)^2\right]^{0.5} \tag{10.42}$$

This will not yield the same numerical value as the correlation coefficient computed with the standard equation, but it is a more correct measure of goodness of fit because it is based on the correct number of degrees of freedom. It is sometimes referred to as the adjusted correlation coefficient. Critical values of the Pearson correlation coefficient are given in Table A.15. Values in absolute value that are greater than the tabled value are statistically significant.

10.7 PROBLEMS

10-1 Outline the development of a unit hydrograph model that changes scale as time changes. A dynamic unit hydrograph model can use the same functional form as the storm time varies, but the parameters of the function change with storm time.

10-2 The power model form, $\hat{y} = ax^b$, is widely used in hydrologic modeling. Discuss whether it can be a conceptual model of a hydrologic process or only an empirical prediction equation.

10-3 Conceptualize a time-of-concentration model that accounts for the effects of urbanization with time.

10-4 Conceptualize a spatially varying model of the dispersion of a concentrated input, such as a pipe or small stream, that discharges into a open area such as a wetland or small pond.

10-5 For the function of Equation 10.3, show the numerical variation in $\hat{y}$ for variation in each of the four coefficient as its value varies.

10-6 Develop a composite model, y versus x, that is linear for low values of x and a power model for high values of x such that the constraints of Equation 10.6 are embedded into the coefficients.

10-7 Develop a composite model, y versus t, that is an exponential decay function for low values of t and linear for high values of t, with constraints of equality of magnitude and slope at the intersection point t_c.

10-8 Develop the normal equations for the following linear model: $\hat{y} = b_1 x_1 + b_2 x_2$.

10-9 Transform the standardized multiple regression model of Equation 10.17 to the form of Equation 10.11.

10-10 Determine the standardized partial regression coefficients t_1 and t_2 for the following correlation matrix:

$$R = \begin{bmatrix} 1.00 & 0.38 & 0.66 \\ 0.38 & 1.00 & 0.47 \\ 0.66 & 0.47 & 1.00 \end{bmatrix}$$

10-11 Determine the standardized partial regression coefficients t_1 and t_2 for the following correlation matrix:

$$R = \begin{bmatrix} 1.00 & -0.43 & 0.74 \\ -0.43 & 1.00 & -0.36 \\ 0.74 & -0.36 & 1.00 \end{bmatrix}$$

10-12 Calculate the determinant of the following matrices:

$$\text{(a)} \begin{vmatrix} 1.00 & 0.56 & 0.82 \\ 0.56 & 1.00 & -0.49 \\ 0.82 & -0.49 & 1.00 \end{vmatrix} \qquad \text{(b)} \begin{vmatrix} 1.00 & 0.92 & 0.87 \\ 0.92 & 1.00 & -0.12 \\ 0.87 & -0.12 & 1.00 \end{vmatrix}$$

Discuss the implications for irrationality of the three-predictor multiple regression model.

10-13 Calculate the determinant of the following matrices and discuss the implications for irrationality of three-predictor multiple regression models:

$$\text{(a)}\ R = \begin{vmatrix} 1.00 & 0.12 & 0.27 \\ 0.12 & 1.00 & -0.31 \\ 0.27 & -0.31 & 1.00 \end{vmatrix} \qquad \text{(b)} \begin{vmatrix} 1.00 & -0.78 & -0.94 \\ -0.78 & 1.00 & 0.62 \\ -0.94 & 0.62 & 1.00 \end{vmatrix}$$

10-14 The following are the correlation coefficients for each step of a stepwise regression analysis involving 62 measurements of five predictor variables. Compute the total and partial F statistics at each step and test their statistical significance.

Step	1	2	3	4	5
R	0.12	0.36	0.43	0.67	0.71

10-15 Show how the value of the total F statistic that is statistically significant at 5% varies with sample sizes from 10 to 50 (in increments of 10).

10-16 If the correlation coefficient at the second step of a stepwise regression analysis is 0.12 for a sample size of 15, what change in R^2 will be necessary at the next step to be significant addition at a 5% level of significance? How does the change in R^2 vary with sample size (e.g., $n = 25, 50, 100$)? What are the implications of this? How would the changes differ if the correlation was 0.82, rather than 0.12?

For exercises 17 to 21, do the following:

a. Plot each pair of the variables and discuss (1) the degree of correlation between each pair (low, moderate, high); (2) the possible effect of intercorrelation between predictor variables; and (3) the apparent relative importance of the predictor variables.
b. Compute the correlation matrix.
c. Compute the determinant of the matrix of intercorrelations between the predictor variables.
d. Compute the partial and standardized partial regression coefficients.
e. Compute the goodness-of-fit statistics (R, S_e) and discuss the implications.
f. Discuss the relative importance of the predictor variables.
g. Compute and analyze the residuals.

10-17 The peak discharge (Q) of a river is an important parameter for many engineering problems, such as the design of dam spillways and levees. Accurate estimates can be obtained by relating Q to the watershed area (A) and the watershed slope (S). Perform a regression analysis using the following data.

A (mi²)	S (ft/ft)	Mean Annual Discharge Q (ft³/sec)
36	0.005	50
37	0.040	40
45	0.004	45
87	0.002	110
450	0.004	490
550	0.001	400
1200	0.002	650
4000	0.0005	1550

10-18 Sediment is a major water quality problem in many rivers. Estimates are necessary for the design of sediment control programs. The sediment load (Y) in a stream is a function of the size of the contributing drainage area (A) and the average stream discharge (Q). Perform a regression analysis using the following data.

A ($\times 10^3$ mi^2)	Discharge Q (ft^3/sec)	Sediment yield Y (millions of tons/yr)
8	65	1.6
19	625	6.4
31	1450	3.0
16	2400	1.6
41	6700	19.0
24	8500	15.0
3	1500	1.2
3	3500	0.4
3	4300	0.3
7	12000	1.0

10-19 The cost of producing power (P) in mils per kilowatt hour is a function of the load factor (L) in percent and the cost of coal (C) in cents per million Btu. Perform a regression analysis to develop a regression equation for predicting P.

L	C	P
85	15	4.1
80	17	4.5
70	27	5.6
74	23	5.1
67	20	5.0
87	29	5.2
78	25	5.3
73	14	4.3
72	26	5.8
69	29	5.7
82	24	4.9
89	23	4.8

10-20 Visitation to lakes for recreation purposes varies directly as the population of the nearest city or town and inversely as the distance between the lake and the city. Develop a regression equation to predict visitation (V) in person-days per year, using as predictors the population (P) of the nearby town and the distance (D) in miles between the lake and the city.

D	P	V
10	10500	22000
22	27500	98000
31	18000	24000
18	9000	33000
28	31000	41000
12	34000	140000
21	22500	78000
13	19000	110000
33	12000	13000

10-21 The stream re-aeration coefficient (r) is a necessary parameter for computing the oxygen-deficit sag curve for a stream reach. The coefficient is a function of the mean stream velocity in feet per second (X_1), the water depth in feet (X_2), and the water temperature in degrees Celsius (X_3). Use the following data for the steps outlined above.

X_1	X_2	X_3	r
1.4	1.3	14	2.89
1.7	1.3	23	4.20
2.3	1.5	17	4.17
2.5	1.6	11	3.69
2.5	1.8	20	3.78
2.8	1.9	15	3.56
3.2	2.3	18	3.35
3.3	2.5	11	2.69
3.5	2.6	25	3.58
3.9	2.9	19	3.06
4.1	3.3	13	2.41
4.7	3.4	24	3.30
5.3	3.5	16	3.05
6.8	3.7	23	4.18
6.9	3.7	21	4.08

10-22 The results of a stepwise regression analysis are given in Tables 10.4 and 10.5, including the correlation coefficient matrix, the means and standard deviations of the six predictor variables, the partial F statistics for entering and deletion, the multiple correlation coefficient, and the regression coefficients. Perform a complete regression analysis to select a model for prediction. State the reasons for selecting the model and make all additional computations that are necessary to support your decision. The analysis is based on a sample size of 71.

11 Hydrologic Simulation

11.1 INTRODUCTION

Evaluating the effects of future watershed change is even more difficult than evaluating the effects of change that has already taken place. In the latter case, some data, albeit nonstationary, are available. Where data are available at the site, some measure of the effect of change is possible, with the accuracy of the effect dependent or the quantity and quality of the data. Where on-site data are not available, which is obviously true for the case where watershed change has not yet occurred, it is sometimes necessary to model changes that have taken place on similar watersheds and then *a priori* project the effects of the proposed change onto the watershed of interest. Modeling for the purpose of making an *a priori* evaluation of a proposed watershed change would be a task simulation task.

Simulation is a category of modeling that has been widely used for decades. One of the most notable early uses of simulation that involved hydrology was with the Harvard water program (Maass et al., 1962; Hufschmidt and Fiering, 1966). Continuous streamflow models, such as the Stanford watershed model (Crawford and Linsley, 1964) and its numerous offspring, are widely used in the simulation mode. Simulation has also been used effectively in the area of flood frequency studies (e.g., Wallis, Matalas, and Slack, 1974).

Because of recent advances in the speed and capacity of computers, simulation has become a more practical tool for use in spatial and temporal hydrologic modeling. Effects of spatial changes to the watershed on the hydrologic processes throughout the watershed can be simulated. Similarly, temporal effects of watershed changes can be assessed via simulation. For example, gradual urbanization will change the frequency characteristics of peak discharge series, and simulation can be used to project the possible effects on, for example, the 100-year flood. It is also possible to develop confidence limits on the simulated flood estimates, which can be useful in making decisions.

While simulation is a powerful tool, it is important to keep in mind that the simulated data are not real. They are projected values obtained from a model, and their accuracy largely depends on the quality of the model, including both its formulation and calibration. An expertly developed simulation program cannot overcome the inaccuracy introduced by a poorly conceived model; however, with a rational model simulation can significantly improve decision making.

11.1.1 Definitions

Before defining what is meant by hydrologic simulation, it is necessary to provide a few definitions. First, a *system* is defined herein as a set of processes or components that are interdependent with each other. This could be a natural system such as a watershed, a geographic system such as a road network, or a structural system such as a high-rise building. Because an accident on one road can lead to traffic congestion on a nearby road, the roads are interdependent.

Second, a *model* is defined herein as a representation of a real system. The model could be either a physical model, such as those used in laboratories, or a mathematical model. Models can be developed from either theoretical laws or empirical analyses. The model includes components that reflect the processes that govern the functioning of the system and provides for interaction between components.

Third, an *experiment* is defined for our purposes here as the process of observing the system or the model. Where possible, it is usually preferable to observe the real system. However, the lack of control may make this option impossible or unrealistic. Thus, instead of observing the real system, simulation enables experiments with a model to replace experiments on the real system when it is not possible to control the real system.

Given these three definitions, a preliminary definition can now be provided for simulation. Specifically, *simulation* is the process of conducting experiments on a model when we cannot experiment directly on the system. The uncertainty or randomness inherent in model elements is incorporated into the model and the experiments are designed to account for this uncertainty. The term *simulation run* or *cycle* is defined as an execution of the model through all operations for a length of simulated time. Some additional terms that need to be defined are as follows:

1. A *model parameter* is a value that is held constant over a simulation run but can be changed from run to run.
2. A *variable* is a model element whose value can vary during a simulation run.
3. *Input variables* require values to be input prior to the simulation run.
4. *Output variables* reflect the end state of the system and can consist of single values or a vector of values.
5. *Initial conditions* are values of model variables and parameters that establish the initial state of the model at the beginning of a simulation run.

11.1.2 Benefits of Simulation

Simulation is widely used in our everyday lives, such as flight simulators in the space and aircraft industries. Activities at the leading amusement parks simulate exciting space travel. Even video games use simulation to mimic life-threatening activities.

Simulation is widely used in engineering decision making. It is a popular modeling tool because it enables a representation of the system to be manipulated when manipulating the real system is impossible or too costly. Simulation allows the time or space framework of the problem to be changed to a more convenient framework. That is, the length of time or the spatial extent of the system can be expanded or compressed. Simulation enables the representation of the system to be changed in order to better understand the real system; of course, this requires the model to be a realistic representation of the system. Simulation enables the analyst to control any or all model parameters, variables, or initial conditions, which that is not possible for conditions that have not occurred in the past.

While simulation is extremely useful, it is not without problems. First, it is quite possible to develop several different, but realistic, models of the same system. The different models could lead to different decisions. Second, the data that are used to calibrate the model may be limited, so extrapolations beyond the range of the measured data may be especially inaccurate. Sensitivity analyses are often used to assess how a decision based on simulation may change if other data had been used to calibrate the model.

11.1.3 Monte Carlo Simulation

The interest in simulation methods started in the early 1940s for the purpose of developing inexpensive techniques for testing engineering systems by imitating their real-world behavior. These methods are commonly called Monte Carlo simulation techniques. The principle behind the methods is to develop an analytical model, which is usually computer based, that predicts the behavior of a system. Then parameters of the model are calibrated using data measured from a system. The model can then be used to predict the response of the system for a variety of conditions. Next, the analytical model is modified by incorporating stochastic components into the structure. Each input parameter is assumed to follow a probability function and the computed output depends on the value of the respective probability distribution. As a result, an array of predictions of the behavior are obtained. Then statistical methods are used to evaluate the moments and distribution types for the system's behavior.

The analytical and computational steps of a Monte Carlo simulation follow:

1. Define the system using a model.
2. Calibrate the model.
3. Modify the model to allow for random variation and the generation of random numbers to quantify the values of random variables.
4. Run a statistical analysis of the resulting model output.
5. Perform a study of the simulation efficiency and convergence.
6. Use the model in decision making.

The definition of the system should include its boundaries, input parameters, output (or behavior) measures, architecture, and models that specify the relationships of input and output parameters. The accuracy of the results of simulation are highly dependent on an accurate definition for the system. All critical parameters and variables should be included in the model. If an important variable is omitted from the model, then the calibration accuracy will be less than potentially possible, which will compromise the accuracy of the results. The definition of the input parameters should include their statistical or probabilistic characteristics, that is, knowledge of their moments and distribution types. It is common to assume in Monte Carlo simulation that the architecture of the system is deterministic, that is, nonrandom. However, model uncertainty is easily incorporated into the analysis by including bias factors and measures of sampling variation of the random variables. The results of these generations are values for the input parameters. These values should then be substituted into the model to obtain an output measure. By repeating the procedure *N* times (for *N* simulation cycles), *N* response measures are obtained. Statistical methods can now be used to obtain, for example, the mean value, variance, or distribution type for each of the output variables. The accuracy of the resulting measures for the behavior are expected to increase by increasing the number of simulation cycles. The convergence of the simulation methods can be investigated by studying their limiting behavior as *N* is increased.

Example 11.1

To illustrate a few of the steps of the simulation process, assume that theory suggests the relationship between two variables, *Y* and *X*, is linear, $Y = a + bX$. The calibration data are collected, with the followings four pairs of values:

X	2	4	6	8
Y	3	7	9	8

The mean and standard deviation of the two variables follow: $\bar{X} = 5.0$, $S_x = 2.582$, $\bar{Y} = 6.75$, and $S_y = 2.630$. Using least squares, fitting yields $a = 2.5$, $b = 0.85$, and a standard error of estimate $S_e = 1.7748$. The goal is to be able to simulate random pairs of *X* and *Y*. Values of *X* can be generated by assuming that *X* is normally distributed with $\mu_x = \bar{X}$, $\sigma_x = S_y$, and the following linear model:

$$\hat{X} = \mu_x + zS_x \tag{11.1a}$$

in which z is a standard normal deviate $N(0, 1)$. The generated values of *X* are then used in the generation of values of *Y* by

$$\hat{Y} = 2.5 + 0.85X + zS_e \tag{11.1b}$$

in which z is $N(0, 1)$. The last term represents the stochastic element of the model, whereas the first two terms represent the deterministic portion of the model in that it yields the same value of $2.5 + 0.85X$ whenever the same value of *X* is used.

Eight values of z are required to generate four pairs of (*X*, *Y*). Four values of z are used to generated the values of *X* with Equation 11.1a. The generated values of *X* are then inserted into Equation 11.1b to generate values of *Y*. Consider the following example:

z	$\hat{X}$	z	$\hat{Y}$
−0.37	4.04	0.42	6.68
0.82	7.12	−0.60	7.48
0.12	5.31	1.03	8.84
−0.58	6.50	−0.54	7.06

The sample statistics for the generated values of X and Y are $\bar{X} = 5.74$, $S_x = 1.36$, $\bar{Y} = 7.515$, and $S_y = 0.94$. These deviate considerably from the calibration data, but are within the bounds of sampling variation for a sample size of four.

The above analyses demonstrate the first four of the six steps outlined. The linear model was obtained from theory, the model was calibrated using a set of data, a data set was calibrated, and the moments of the generated data were computed and compared to those of the calibration data. To demonstrate the last two steps would require the generation of numerous data sets such that the average characteristics of all generated samples approached the expected values. The number of generated samples would be an indication of the size of the simulation experiment.

11.1.4 Illustration of Simulation

The sampling distribution of the mean is analytically expressed by the following theorem:

> *If a random sample of size n is obtained from a population that has the mean μ and variance σ^2, then the sample mean $\bar{X}$ is a value of a random variable whose distribution has the mean μ and the variance σ^2/n.*

If this theorem were not known from theory, it could be uncovered with simulation. The following procedure illustrates the process of simulating the sampling distribution of the mean:

1. From a known population with mean μ and variance σ^2, generate a random sample of size n.
2. Compute the mean $\bar{X}$ and variance S^2 of the sample.
3. Repeat steps 1 and 2 a total of N_s times, which yields N_s values of $\bar{X}$ and S^2.
4. Repeat steps 1 to 3 for different values of μ, σ^2, and n.
5. For each simulation run (i.e., steps 1 to 3) plot the N_s values of $\bar{X}$, examine the shape of the distribution of the $\bar{X}$ values, compute the central tendency and spread, and relate these to the values of μ, σ^2, and n.

The analysis of the data would show that the theorem stated above is valid.

For this example, the model is quite simple, computing the means and variances of samples. The input parameters are μ, σ^2, and n. The number of samples generated, N_s, is the length of a simulation run. The number of executions of step 4 would be the number of simulation runs. The output variables are $\bar{X}$ and S^2.

Example 11.1 and the five steps described for identifying the sampling distribution of the mean illustrate the first four steps of the simulation process. In the fifth step, the efficiency and convergence of the process was studied. In Example 11.1, only one sample was generated. The third step of the above description indicates that N_s samples should be generated. How large does N_s need to be in order to identify the sampling distribution? The process of answering this question would be a measure of the convergence of the process to a reliable answer. Assuming that the above five steps constitute a valid algorithm for identifying the sampling distribution of the mean, the number of simulations needed to develop the data would indicate that the algorithm has converged to a solution.

If the effect of the assumed population from which the n values in step 1 were sampled was of interest, the experiment could be repeated using different underlying populations for generating the sample values of X. The outputs of step 5 could be compared to assess whether the distribution is important. This would qualitatively evaluate the sensitivity of the result to the underlying distribution. A sensitivity analysis performed to assess the correctness of the theorem when the population is finite of size N rather than infinite would show that the variance is $\sigma^2(N - n)/[n(N - 1)]$ rather than σ^2/n.

11.1.5 Random Numbers

The above simulation of the distribution of the mean would require a random-number generator for step 1. Random numbers are real values that are usually developed by a deterministic algorithm, with the resulting numbers having a uniform distribution in the range (0, 1). A sequence of random numbers should also satisfy the condition of being uncorrelated, that is, the correlation between adjacent values equals zero. The importance of uniform random numbers is that they can be transformed into real values that follow any other probability distribution of interest. Therefore, they are the initial form of random variables for most engineering simulations.

In the early years of simulation, mechanical random-number generators were used, such as, drawing numbered balls, throwing dice, or dealing out cards. Many lotteries are still operated this way. After several stages of development, computer-based, arithmetic random-number generators were developed that use some analytical generating algorithm. In these generators, a random number is obtained based on a previous value (or values) and fixed mathematical equations. Therefore, a seed is needed to start the process of generating a sequence of random numbers. The main advantages of arithmetic random-number generators over mechanical generators are speed, that they do not require memory for storage of numbers, and repeatability. The conditions of a uniform distribution and the absence of serial correlation should also be satisfied. Due to the nature of the arithmetic generation of random numbers, a given seed should result in the same stream of random values every time the algorithm is executed. This property of repeatability is important for debugging purposes of the simulation algorithm and comparative studies of design alternatives for a system.

11.2 COMPUTER GENERATION OF RANDOM NUMBERS

A central element in simulation is a random-number generator. In practice, computer packages are commonly used to generate the random numbers used in simulation; however, it is important to understand that these random numbers are generated from a deterministic process and thus are more correctly called *pseudo-random numbers*. Because the random numbers are derived from a deterministic process, it is important to understand the limitations of these generators.

Random-number generators produce numbers with specific statistical characteristics. Obviously, if the generated numbers are truly random, an underlying population exists that can be represented by a known probability function. A single die is the most obvious example of a random-number generator. If a single die was rolled many times, a frequency histogram could be tabulated. If the die were a fair die, the sample histogram for the generated *population* would consist of six bars of equal height. Rolling the die produces values of a random variable that has a discrete mass function. Other random-number generators would produce random numbers having different distributions, including continuously distributed random variables. When a computerized random-number generator is used, it is important to know the underlying population.

11.2.1 Midsquare Method

The midsquare method is one of the simplest but least reliable methods of generating random numbers. However, it illustrates problems associated with deterministic procedures. The general procedure follows:

1. Select at random a four-digit number; this is referred to as the *seed.*
2. Square the number and write the square as an eight-digit number using preceding (lead) zeros if necessary.
3. Use the four digits in the middle as the new random number.
4. Repeat steps 2 and 3 to generate as many numbers as necessary.

As an example, consider the seed number of 2189. Squaring this yields the eight-digit number 04791721, which gives the first random number of 7917. The following sequence of 7 four-digit numbers results from using 2189 as the seed:

04791721
62678889
46076944
00591361
34963569
92833225
69422224

Note that a leading 0 was included in the first number, and that two leading zeros were required for the fourth number. At some point one of these numbers must recur. At that point the same sequence that occurred on the first pass will repeat itself. The sequence of generated numbers is no longer a sequence of random numbers. For example, if the four-digit number of 3500 occurred, the following sequence would result:

12250000
06250000
06250000
06250000

Such a sequence is obviously not random and would not pass statistical tests for randomness. While the procedure could be used for very small samples, it is limited in that a number will recur after more than a few values are generated. While the use of five-digit numbers could be used to produce ten-digit squares, the midsquare method has serious flaws that limit its usefulness. However, it is useful for introducing the concept of random-number generation.

11.2.2 Arithmetic Generators

Many arithmetic random-number generators are available, including the midsquare method, linear congruential generators, mixed generators, and multiplicative generators. All of these generators are based on the same principle of starting with a seed and having fixed mathematical equations for obtaining the random value. The resulting values are used in the same equations to obtain additional values. By repeating this recursive process N times, N random number in the range (0, 1) are obtained. However, these methods differ according to the algorithms used as the recursive model. In all recursive models, the period for the generator is of concern. The period is defined as the number of generated random values before the stream of values starts to repeat itself. It is always desirable to have random-number generators with large periods, such as much larger than the number of simulation cycles needed in a simulation study of a system.

11.2.3 Testing of Generators

Before using a random-number generator, the following two tests should be performed on the generator: a test for uniformity and a test of serial correlation. These tests can be performed either theoretically or empirically. A theoretical test is defined as an evaluation of the recursive model itself of a random-number generator. The theoretical tests include an assessment of the suitability of the parameters of the model without performing any generation of random numbers. An empirical test is a statistical evaluation of streams of random numbers resulting from a random-number generator. The empirical tests start by generating a stream of random numbers, that is, N random values in the range (0, 1). Then, statistical tests for distribution types, that is, goodness-of-fit tests such as the chi-square test, are used to assess the uniformity of the random values. Therefore, the objective in the uniformity test is to make sure that the resulting random numbers follow a uniform continuous probability distribution.

To test for serial correlation, the Spearman–Conley test (Conley and McCuen, 1997) could be used. The runs test for randomness is an alternative. Either test can be applied to a sequence of generated values to assess the serial correlation of the resulting random vector, where each value in the stream is considered to come from a different but identical uniform distribution.

11.2.4 Distribution Transformation

In simulation exercises, it is necessary to generate random numbers from the population that underlies the physical processes being simulated. For example, if annual floods at a site follow a log-Pearson type III distribution, then random numbers having a uniform distribution would be inappropriate for generating random sequences of flood flows. The problem can be circumvented by transforming the generated uniform variates to log-Pearson type III variates.

Distribution transformation refers to the act of transforming variates x from distribution $f(x)$ to variates y that have distribution $f(y)$. Both x and y can be either discrete or continuous. Most commonly, an algorithm is used to generate uniform variates, which are continuously distributed, and then the uniform variates are transformed to a second distribution using the cumulative probability distribution for the desired distribution.

The task of distribution transformation is best demonstrated graphically. Assume that values of the random variate x with the cumulative distribution $F(x)$ are generated and values of a second random variate y with the cumulative distribution $F(y)$ are needed. Figure 11.1(a) shows the process for the case where both x and y are discrete random variables. After graphing the cumulative distributions $F(x)$ and $F(y)$, the value of x

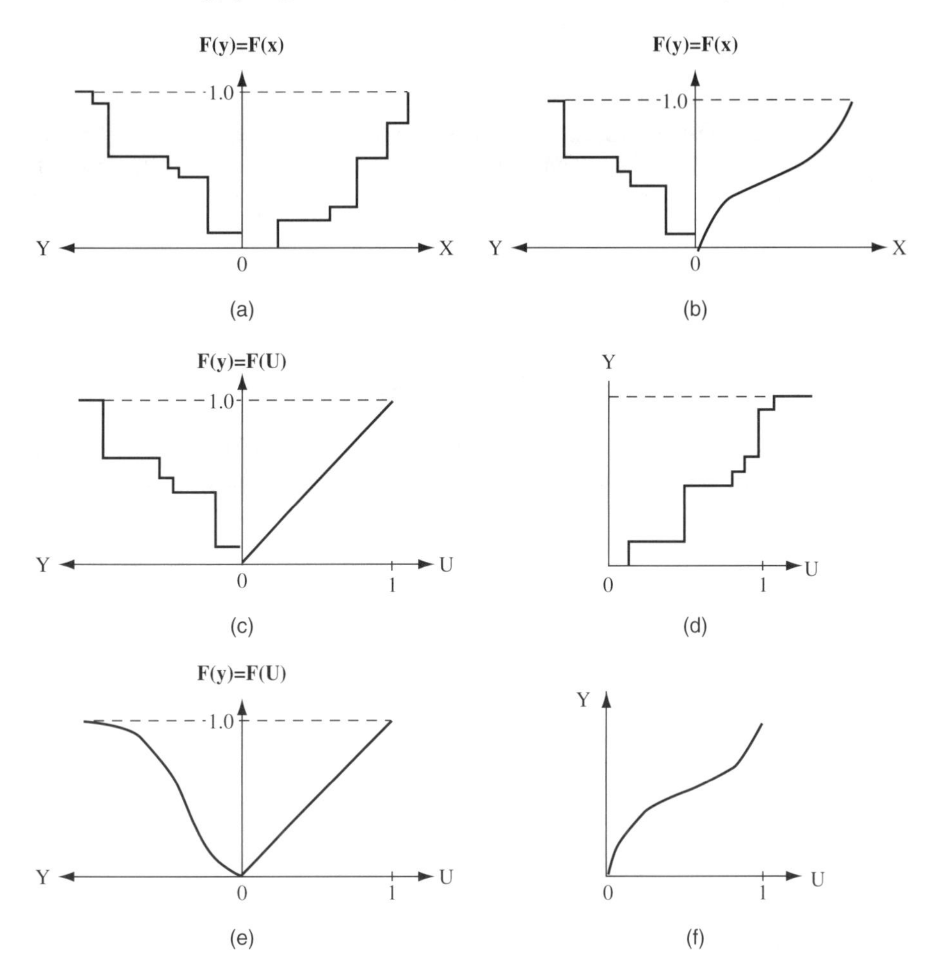

FIGURE 11.1 Transformation curves: (a) X and Y are discrete random variables; (b) X continuous, Y discrete; (c, d) X is $U(0, 1)$, Y is discrete; and (e, f) Y is continuous, U is $U(0, 1)$.

is entered on the x-axis and the value of its cumulative probability found. The cumulative value for y is assumed to equal the cumulative value of x. Therefore, the value of y is found by moving horizontally from $F(x)$ to $F(y)$ and then down to the y-axis, where the value of y_i is obtained. Given a sample of n values of x_i, a sample of n values of y_i is generated by repeating this process.

The same transformation process can be used when x is a continuously distributed random variable. Figure 11.1(b) shows this case. Because many random number generators generate uniformly distributed random numbers, $F(x)$ is most often the cumulative uniform distribution. This is illustrated in Figure 11.1(c). Since the cumulative distribution for a uniform variate is a constant-sloped line, Figure 11.1(c) can be simplified to Figure 11.1(d). Both Figure 11.1(c) and 11.1(d) show y as a discretely distributed random variable. Figures 11.1(e) and 11.1(f) show the corresponding graphs when y is a continuously distributed random variable.

Example 11.2

Assume that the number of runoff producing storms per year (x in column 1 of Table 11.1) has the mass function given in column 2 of Table 11.1. Using $f(x)$ in column 2, the cumulative mass function $F(x)$ is formed (see column 3 of Table 11.1). The rule for transforming the uniform variate u to a value of the discrete random variable x follows:

$$0 \text{ if } u_i \leq F(0) \tag{11.2a}$$

$$i \text{ if } F(i-1) < u_i \leq F(i) \tag{11.2b}$$

Assume that ten simulated values of an annual maximum flood record are needed. Then ten uniform variates u_i would be generated (see column 5). Using the transformation algorithm of Equations 11.2, the values of u_i are used to obtain generated values of x_i (see column 6). For example, u_1 is 0.62. Entering column 3, u_1 lies between $F(2)$ of 0.5 and $F(3)$ of 0.70; therefore, x_1 equals 3. The valuc of x_7 of 0.06

TABLE 11.1
Continuous to Discrete Transformation

(1) x	(2) $f(x)$	(3) $F(x)$	(4) Simulation	(5) Uniform Variate	(6) Simulated
0	0.10	0.10	1	0.62	3
1	0.15	0.25	2	0.17	1
2	0.25	0.50	3	0.43	2
3	0.20	0.70	4	0.96	6
4	0.15	0.85	5	0.22	1
5	0.10	0.95	6	0.86	5
6	0.05	1.00	7	0.06	0
			8	0.34	2
			9	0.57	3
			10	0.40	2

is the case where Equation 11.2a would be applied. Since u_7 is less than $F(0)$ of 0.1 (column 3), then x_7 is equal to 0. The ten simulated values of x_i are given in column 6.

The number of runoff-producing storms for the ten simulated years of record represent a sample and should reflect the distribution from which they were generated. Histograms of a sample and a population can, in general, be compared to assess the representativeness of the sample. For this example, a sample size of ten is too small to construct a meaningful histogram. For the purpose of comparison of this simulation, the sample ($\bar{x}$) and population (μ) means will be computed:

$$\bar{x} = \frac{1}{n}\sum x_i = \frac{25}{10} = 2.5 \tag{11.3a}$$

and

$$\begin{aligned} \mu &= \sum x_i f(x_i) \\ &= 0(0.1) + 1(0.15) + 2(0.25) + 3(0.20) + 4(0.15) + 5(0.1) + 6(0.05) = 2.65 \end{aligned} \tag{11.3b}$$

The difference between the two values only reflects the sampling variation of the mean for the small sample size. As the number of simulated values is increased, the value of the sample mean would approach the population mean of 2.65. The sample and population variances could also be computed and compared.

Example 11.3

Example 11.2 illustrates the transformation of a continuous variate to a discrete variate. The process is similar when the transformed variate is continuously distributed. Consider the case where it is necessary to obtain a simulated sequence of continuously distributed variates that have the following density function $f(x)$:

$$f(x) = \begin{cases} 6.4x & \text{for } 0.50 \le x \le 0.75 \quad (11.4a) \\ 0 & \text{otherwise} \quad (11.4b) \end{cases}$$

The cumulative distribution of x is

$$F(x) = \begin{cases} 3.2(x^2 - 0.25) & \text{for } 0.50 \le x \le 0.75 \quad (11.5) \\ 0 & \text{otherwise} \quad (11.6) \end{cases}$$

Equation 11.5 can be used to transform uniform variates u_i to variates x_i that have the density function of Equation 11.4.

If the uniform variates have location and scale parameters of 0 and 1, respectively, then the values of u_i can be set equal to $F(x)$ of Equation 11.5a and the value of x_i computed by

$$x_i = \left(\frac{F(x)}{3.2} + 0.25\right)^{0.5} \tag{11.7}$$

The transformation of six uniform variates to x_i values appears in Table 11.2.

TABLE 11.2
Continuous to Continuous Transformation

(1) i	(2) u_i	(3) x_i
1	0.41	0.615
2	0.80	0.707
3	0.16	0.548
4	0.63	0.668
5	0.78	0.703
6	0.28	0.581

11.3 SIMULATION OF DISCRETE RANDOM VARIABLES

11.3.1 Types of Experiments

The following four types of experiments are introduced in this section as examples of discrete variable experiments: binomial, multinomial, Markov chain, and Poisson. A binomial experiment is one with n independent trials, with each trial limited to two possible outcomes and the probabilities of each outcome remaining constant from trial to trial. Examples of binomial experiments include the flip of a coin, even versus odd on the roll of a die, or failure versus nonfailure. In a multinomial experiment, the n trials are independent and more than two outcomes are possible for each trial. The roll of a die that has six possible outcomes is one example. In a Markov chain experiment, the value of the discrete random variable in one time period depends on the value experienced in the previous time period. Thus, it is a sequential, nonindependent experimental approach. In a Poisson experiment, the discrete random variable is the number of occurrences in an interval (spatial or temporal). For example, the number of hurricanes hitting the U.S. coast in a year could be represented as a Poisson distribution.

11.3.2 Binomial Distribution

A discrete random variable that has two possible outcomes can follow a binomial mass function. Four assumptions underlie a binomial process:

- Only two possible outcomes are possible for each trial.
- Each experiment consists of n trials.
- The n trials are independent.
- The probability of both outcomes remains constant from trial to trial.

The probability of exactly x occurrences in n trials is given by

$$p(x) = \binom{n}{x} p^x (1-p)^{n-x} \quad \text{for } x = 0, 1, 2, \ldots, n \tag{11.7}$$

where $\binom{n}{x}$ is the binomial coefficient and equals $\frac{n!}{x!(n-x)!}$. In this section, the notation $b(x; n, p)$ is used for the binomial distribution and indicates the probability computed with Equation 11.8.

Binomial experiments can be simulated using uniform variates. The procedure is as follows for N experiments, with each having n trials and an outcome A with an occurrence probability of p and an outcome B with an occurrence probability of $1 - p$:

1. For each trial in the experiment, generate a uniform variate, u_i.
2. If $u_i < p$, then assume outcome A has occurred; otherwise, assume outcome B has occurred.
3. Determine the number of outcomes A, x, that occurred in the n trials.
4. After completing the N experiments, with each having n trials, compute the relative frequency of x outcomes.

Example 11.4

Assume three trials per experiment is of interest where the probability (p) of outcome A is 0.3. For example, if we have an unbalanced coin with a probability of 0.3 of getting a head, then we would be interested in the probabilities of getting 0, 1, 2, or 3 heads in three flips of the coin. From the binomial mass function of Equation 11.8, we can compute the following population probabilities:

$$b(0; 3, 0.3) = \binom{3}{0} 0.3^0 0.7^3 = 0.343 \tag{11.9a}$$

$$b(1; 3, 0.3) = \binom{3}{1} 0.3^1 0.7^2 = 0.441 \tag{11.9b}$$

$$b(2; 3, 0.3) = \binom{3}{2} 0.3^2 0.7^1 = 0.189 \tag{11.9c}$$

$$b(3; 3, 0.3) = \binom{3}{3} 0.3^3 0.7^0 = 0.027 \tag{11.9d}$$

The sum of these probabilities is 1.

To illustrate the simulation of binomial probabilities, let $N = 15$. For each of the 15 experiments, three uniform variates are generated, and the number of u_i values less than 0.3 counted, which is used as the sample estimate of x. One possible simulation of 15 experiments is shown in Table 11.3.

From the 15 experimental estimates of x, seven 0s, six 1s, one 2, and one 3 were generated. This yields sample estimates of the probabilities of $p(x = 0) = 7/15 = 0.467$, $p(x = 1) = 6/15 = 0.400$, $p(x = 2) = 1/15 = 0.067$, and $p(x = 3) = 1/15 = 0.066$, respectively. These sample values can be compared with the population values of Equations 11.9. The differences are obvious. Specifically, the sample probability

TABLE 11.3
Generation of a Binomial Distribution

Experiment	u_i Variates	x
1	0.78, 0.51, 0.43	0
2	0.85, 0.95, 0.22	1
3	0.88, 0.36, 0.20	1
4	0.79, 0.55, 0.71	0
5	0.11, 0.09, 0.70	2
6	0.11, 0.70, 0.98	1
7	0.79, 0.45, 0.86	0
8	0.62, 0.11, 0.77	1
9	0.45, 0.40, 0.13	1
10	0.37, 0.31, 0.12	1
11	0.54, 0.99, 0.95	0
12	0.79, 0.36, 0.37	0
13	0.73, 0.67, 0.60	0
14	0.56, 0.40, 0.71	0
15	0.24, 0.21, 0.23	3

of $x = 0$ is greater than that for $x = 1$, while the population probability of $x = 0$ is less than the sample value for $x = 1$. As N becomes larger and larger, the sample estimates of the probabilities would more closely approximate the population probabilities of Equations 11.9.

Example 11.5

Consider the case of a project that has a design life of 5 years, with the probability of failure in any one year equal to 0.1. A binomial process such as this case can be simulated using $n = 5$ and $p = 0.1$, with values of $X = 0, 1, 2, 3, 4$, and 5. To estimate the probability of no failures, the case of $X = 0$ would be of interest. Ten experiments were made and the sample estimate of the probability of a project not failing were computed as shown in Table 11.4(a). The sample ($\hat{p}$) and population (p) probabilities are shown in Table 11.4(b). For a sample of 10, the sample estimates are in surprisingly good agreement with the population values. The sample estimate of the probability of a project not failing in 5 years is 0.60.

11.3.3 Multinomial Experimentation

Multinomial experimentation is a generalization of binomial experimentation in which each trial can have more than two outcomes. A multinomial experiment has n independent trials with k mutually exclusive outcomes. Outcome i has a probability p_i of occurring on any one trial, and in an experiment x_i is used to denote the number of outcome i. The multinomial mass function is

$$f(x_1, x_2, \ldots, x_k) = \frac{n!}{x_1! x_2! \ldots x_k!} p_1^{x_1} p_2^{x_2} \ldots p_k^{x_k} \tag{11.10a}$$

TABLE 11.4
Generation of Design Life

(a) Simulation Results for a Binomial Distribution

Project	u_i Variates	x
1	0.05, 0.37, 0.23, 0.42, 0.05	2
2	0.60, 0.32, 0.79, 0.45, 0.20	0
3	0.67, 0.86, 0.66, 0.55, 0.23	0
4	0.41, 0.03, 0.90, 0.98, 0.74	1
5	0.94, 0.58, 0.31, 0.45, 0.31	0
6	0.69, 0.93, 0.77, 0.37, 0.72	0
7	0.55, 0.69, 0.01, 0.51, 0.58	1
8	0.33, 0.58, 0.72, 0.22, 0.97	0
9	0.09, 0.29, 0.81, 0.44, 0.68	1
10	0.29, 0.54, 0.75, 0.36, 0.29	0

(b) Sample ($\hat{p}$) and Population (p) Probabilities for Random Variable X

X	Sample Probability ($\hat{p}$)	Population Probability (p)
0	0.60	0.5905
1	0.30	0.3280
2	0.10	0.0729
3	0.00	0.0081
4	0.00	0.0005
5	0.00	0.0000
Column summations	1.00	1.0000

subject to the constraints that

$$\sum_{i=1}^{k} x_i = n \tag{11.10b}$$

$$\sum_{i=1}^{k} p_i = 1 \tag{11.10c}$$

Equations 11.10 represent the population probabilities.

Example 11.6

Consider the case where flood damage is divided into three mutually exclusive groups: no damage, minor damage, and major damage, with probabilities of 0.80, 0.15, and 0.05, respectively. The probability that in five floods, four will produce no damage and minor damage will occur in the fifth flood is:

$$p(x_1 = 4,\ x_2 = 1,\ x_3 = 0) = \frac{5!}{4!\ 1!\ 0!}(0.8)^4(0.15)^1(0.05)^1 = 0.3072 \tag{11.11}$$

Similarly, the probability that major damage would not result during five floods is

$$\sum_{\substack{\text{all } i \\ \text{such} \\ i+j=5}} \sum_{\substack{\text{all } j \\ \text{that}}} p(i, j, 0) \tag{11.12}$$

11.3.4 Generation of Multinomial Variates

Given a sequence of uniformly distributed random numbers u_i, values of the multinomial outcome x_i can be simulated by the following algorithm:

$$\text{if } u_i \le F(1) \quad \text{then } x_i = \text{outcome } 1 \tag{11.13a}$$

$$\text{if } F(j-1) < u_i \le F(j) \quad \text{then } x_i = \text{outcome } j \tag{11.13b}$$

in which $F(\,)$ is the cumulative mass function based on the outcome probabilities p_i.

Example 11.7

Assume that during any month a wetland can be in any one of the following four states: flooded (F), the permanent pool can be at normal level (N), no pool but saturated ground (S), or dry (D). Assume that the probabilities of these conditions are 0.1, 0.7, 0.15, and 0.05, respectively. The cumulative mass function is, therefore, $F(j) = \{0.1, 0.8, 0.95, 1.0\}$. In generating 1 year of record, the pool is at normal level if the uniform variate is between 0.1 and 0.8, while it is dry if u_i is greater than 0.95.

Consider the following 12 uniform variates, with each one used to generate a wetland condition for the month:

i	1	2	3	4	5	6	7	8	9	10	11	12
u_i	0.70	0.81	0.89	0.62	0.38	0.31	0.19	0.94	0.08	0.43	0.14	0.56
x_i	N	S	S	N	N	N	N	S	F	N	N	N

The wetland experienced 8 normal months, which gives a sample probability of 0.67 that is quite close to the population probability of 0.70. The wetland was never dry but in a period of 12 months, the chance of a dry condition is small since in any month the probability of the dry condition is only 5%. The true probability of $1F$, $8N$, and $3S$ in a 12-month period is computed with Equation 11.10a:

$$p(1F,\ 8N,\ 3S,\ 0D) = \frac{12!}{1!\ 8!\ 3!\ 0!}(0.1)^1(0.7)^8(0.15)^3(0.05)^0 = 0.0385$$

While this might seem like a small probability, it is quite reasonable given that number of combinations of F, N, S, and D in a 12-month period is large. Note that in this case, it does not specify the order of the ($1F$, $8N$, $3S$, $0D$) occurrences.

The generation of the probabilities of F, N, S, and D may seem useless since they can be computed from Equation 11.10. But generated sequences may be necessary to examine probabilities such as that of four dry (D) months in succession or the average length of time between dry periods. These other probabilities cannot be computed with Equation 11.10.

11.3.5 Poisson Distribution

Discrete random variables that can take on integer values may be represented by a Poisson mass function, which has the following form:

$$p_X(x) = \frac{e^{-\lambda}(\lambda)^x}{x!} \quad \text{for } x = 0, 1, 2, \ldots \tag{11.14}$$

where x is the number of occurrences in a reference period t, λ is a parameter of the function, and 0! is defined as 1. The cumulative function is

$$F_X(x) = \sum_{i=0}^{x} \frac{e^{-\lambda}(\lambda)^i}{i!} \quad \text{for } x = 0, 1, 2, \ldots \tag{11.15}$$

The mean and variance of the population both equal λ, so the sample mean can be used as an estimator of the population value of λ.

The Poisson distribution can be generated using the inverse transformation method. Uniform random variates can be used to generate random variates that have a Poisson distribution for any λ. The procedure for any given λ is as follows:

1. Calculate the cumulative function $F_{Xi}(x_i)$ for $x_i = 0, 1, 2, \ldots$
2. Generate a unit variate u_i in the range (0, 1).
3. If $F_{Xi}(x_{i-1}) \leq u_i < F_{Xi}(x_i)$, then the Poisson variate is x_i.

Example 11.8

The number of floods per year (n) on a river in northern Idaho was obtained from streamflow records over a 38-year period. The annual number of floods is given in Table 11.5 for each year. The number of years N_x of x floods per year is given in column 2 of Table 11.6. The sample probability of x is given in column 3. A total of 120 floods occurred during the 38 years, which yields a mean annual flood count of 3.158; thus, the sample estimate of the Poisson parameter λ is 3.158. The assumed population mass function is

$$p_{X_1}(x) = \frac{e^{-3.158}(3.158)^x}{x!} \quad \text{for } x = 0, 1, 2, \ldots \tag{11.16}$$

The population probabilities are given in column 4 of Table 11.6. The cumulative distribution of the population is given in column 5.

TABLE 11.5
Annual Flow Frequency above 28,000 cfs on Clearwater River at Kamiah, Idaho

Year	*n*	Year	*n*	Year	*n*	Year	*n*	Year	*n*
1911	4	1919	2	1927	5	1935	3	1943	6
1912	3	1920	2	1928	2	1936	5	1944	1
1913	4	1921	2	1929	3	1937	2	1945	2
1914	3	1922	3	1930	1	1938	4	1946	7
1915	1	1923	3	1931	2	1939	2	1947	3
1916	5	1924	2	1932	4	1940	2	1948	6
1917	5	1925	3	1933	4	1941	1		
1918	3	1926	3	1934	4	1942	3		

TABLE 11.6
Poisson Approximation of Flood Counts

	Sample		Population	
X	N_X	N_X/N	Mass Function *P*(*x*)	Cumulative Function *F*(*x*)
0	0	0.000	0.043	0.043
1	4	0.105	0.134	0.177
2	10	0.263	0.212	0.389
3	11	0.290	0.223	0.612
4	6	0.158	0.176	0.788
5	4	0.105	0.111	0.899
6	2	0.053	0.059	0.958
7	1	0.026	0.026	0.984
≥8	0	0.000	0.016	1.000

Assume that it is necessary to simulate the number of floods in a 10-year period. The assumed cumulative function is given in column 5 of Table 11.6. For a given vector of uniform variates u_i (column 2 of Table 11.7), the generated number of floods per year is shown in column 3 of Table 11.7. A total of 26 floods occurred in the 10-year period. Based on this sample of ten simulated values, the mean number of floods is 26/10 = 2.6. As the number of years for which the number of floods generated increases, the sample mean would be expected to move closer to the assumed population mean of 3.158.

11.3.6 Markov Process Simulation

Markov process modeling differs from the previous analyses in that independence of sequential events is not assumed. In a *Markov process*, the value of the random variable x_t depends only on its value x_{t-1} for the previous time period $t - 1$, but not the values of x that occurred for time periods before $t - 1$ (i.e., $t - 2$, $t - 3$, etc.). Assume that the random variable can take on any one of a set of states A, where A

TABLE 11.7
Synthetic Series of Flood Frequency

Year	u_i	x_i	$F_{Xi}(x_{i-1})$	$F_{Xi}(x_i)$
1	0.13	1	0.043	0.177
2	0.04	0	0.000	0.043
3	0.66	4	0.612	0.788
4	0.80	5	0.788	0.899
5	0.77	4	0.612	0.788
6	0.52	3	0.389	0.612
7	0.30	2	0.177	0.389
8	0.14	1	0.043	0.177
9	0.37	2	0.177	0.389
10	0.65	4	0.612	0.788
Column summation		26		

consists of n states a_i ($i = 1, 2, \ldots, n$). Then by definition of a Markov process, the probability that x exists in state j if it was in state i in the previous time period is denoted by $p(x_t = a_j | x_{t-1} = a_i)$. For example, assume that (1) a watershed can be classified as being wet (W) or dry (D), which would depend on the value of one or more criteria; and (2) the probabilities of being wet or dry are stationary. Then the conditional probabilities of interest are as follows:

$$P(x_t = W | x_{t-1} = W) \tag{11.17a}$$

$$P(x_t = W | x_{t-1} = D) \tag{11.17b}$$

$$P(x_t = D | x_{t-1} = W) \tag{11.17c}$$

$$P(x_t = W | x_{t-1} = D) \tag{11.17d}$$

Alternatively, if we assumed that three states (wet, moderate, and dry) were possible, then nine conditional probabilities would be of interest. The probabilities of Equations 11.17 are called *one-step transition probabilities* because they define the probabilities of making the transition from one state to another state in sequential time periods. The cumbersome notation of Equations 11.17 can be simplified to p_{ij} where the first subscript refers to the previous time period $t - 1$ and the second subscript refers to the present time period t. For example, P_{26} would indicate the probability of transitioning from state 2 in time period $t - 1$ to state 6 in time period t. In a *homogeneous* Markov process, the transition probabilities are independent of time.

As the number of states increases, the number of probabilities increases proportionally. For this reason, the transition probabilities are often presented in a square matrix:

$$\underline{P} = [p_{ij}] = \begin{bmatrix} p_{11} & p_{12} & \cdots & p_{1n} \\ p_{12} & p_{22} & \cdots & p_{2n} \\ \vdots & \vdots & & \vdots \\ p_{n1} & p_{n2} & \cdots & p_{nn} \end{bmatrix} \tag{11.18}$$

The one-step transition probability matrix is subject to the constraint that the sum of the probabilities in any row must equal 1:

$$\sum_{j=1}^{n} p_{ij} = 1 \tag{11.19}$$

To illustrate the transformation of the probabilities of Equation 11.17 to the matrix of Equation 11.18, assume that the probability of a dry day following a dry day is 0.9, while the probability of a wet day following a wet day is 0.4, then the one-step transition matrix is:

$$\begin{array}{c c} & \begin{array}{cc} D & W \end{array} \\ \begin{array}{c} D \\ W \end{array} & \begin{bmatrix} 0.9 & 0.1 \\ 0.6 & 0.4 \end{bmatrix} \end{array}$$

The remaining two probabilities are known because of the constraint of Equation 11.19. Thus, the probability of a dry day following a wet day is 0.6.

The Markov process is a stochastic process, and therefore, it must begin randomly. This requires an initial state probability vector that defines the likelihood that the random variable x will be in state i at time 0; this vector will be denoted as $\underline{P}^{(0)}$, and it has n elements.

Two other vectors are of interest: the nth-step state probability vector and the steady-state probability vector, which are denoted as $P^{(n)}$ and $P^{(\infty)}$, respectively. The nth-step vector gives the likelihood that the random variable x is in state i at time n. The steady-state vector is independent of the initial conditions.

Given the one-step transition probability matrix P and the initial-state vector, the nth-step vector can be computed, as follows:

$$p^{(1)} = p^{(0)} \underline{P} \tag{11.20a}$$

$$p^{(2)} = p^{(1)} \underline{P} \tag{11.20b}$$

$$\vdots \qquad \qquad \vdots$$

$$p^{(n)} = p^{(n-1)} \underline{P} \tag{11.20d}$$

The matrix $p^{(n)}$ gives the probability of being in state j in n steps given $p^{(0)}$ and $\underline{P}$.

Generation of Markovian Random Variates

Since Markovian processes involve sequential correlation, the generation of variates requires the random selection of an initial state, as well as the random generation of transition probabilities. Modeling a Markovian process also requires the initial-state probability vector and the one-step-transition probability matrix. These are usually estimated empirically from observations of the system. The steady-state probability vector can be used with a uniform random variate u_i to obtain the initial state or alternatively, the initial state can be obtained under the assumption that each state is equally likely.

If the steady-state probability vector is used, then the initial state of x is obtained with a uniform variate u_0 by

$$x_0 = \begin{cases} \text{state } 1 & \text{if } u_0 \leq p_1^{(0)} \quad (11.21a) \\ \text{state } i & \text{if } p_{i-1}^{(0)} < u_0 \leq p_i^{(0)} \quad (11.21b) \end{cases}$$

If all of the n initial states are assumed to be equally likely, then

$$x_0 = \begin{cases} \text{state } 1 & \text{if } u_0 \leq 1/n \quad (11.22a) \\ \text{state } i & \text{if } \frac{(i-1)}{n} < u_0 \leq \frac{i}{n} \quad (11.22b) \end{cases}$$

To generate states for all subsequent steps, it is necessary to form the cumulative transition probability matrix $\sum \underline{P}$ where the jth element of row k of $\sum \underline{P}$ is denoted as $\sum P_k$ and is computed by

$$\sum P_k = \sum_{i=1}^{j} P_{ki} \quad (11.23)$$

To obtain the state for any time t given a uniform random variate u_i, the following decision rule is used:

$$x_t = \begin{cases} \text{state } 1 & \text{if } u_i \leq \sum P_{k1} \quad (11.24a) \\ \text{state } j & \text{if } \sum P_{k(j-1)} < u_i \leq \sum P_{kj} \quad (11.24b) \end{cases}$$

The sequence of x_i is computed using a sequence of uniform variates and the decision rule of Equations 11.24.

Example 11.9

Assume that the one-step transition matrix for three states is

$$\underline{P} = \begin{bmatrix} 0.4 & 0.5 & 0.1 \\ 0.3 & 0.7 & 0.0 \\ 0.0 & 0.8 & 0.2 \end{bmatrix}$$

Also assume that initially each state is equally likely; thus, Equation 11.22 is used to define the initial state. A sequence of uniform variates follows:

$$u_i = \{0.22, 0.64, 0.37, 0.92, 0.26, 0.03, 0.98, 0.17\}$$

The goal is to generate a sequence of seven values for the three discrete states, 1 to 3. The first uniform variate, 0.22, is used to obtain the initial state. Since the problem involves three states, then Equation 11.22a indicates that state 1 is the initial state (i.e., since $u_0 = 0.22 < 1/n = 0.33$, then $x_0 = 1$. The cumulative transition matrix is

$$\sum \underline{P} = \begin{bmatrix} 0.4 & 0.9 & 1.0 \\ 0.3 & 1.0 & 1.0 \\ 0.0 & 0.8 & 1.0 \end{bmatrix} \tag{11.25}$$

Since state 1 is the initial state, row 1 of Equation 11.25 is used to find the first value of x. Since $u_1 = 0.64$, which is greater than 0.4 and less than 0.9, then the system transitions from state 1 to state 2. To find x_2, row 2 of Equation 11.25 is used. With $u_2 = 0.37$, which is greater than 0.3 and less than 1.0, row 2 indicates that the value of x at state 2 is $x_2 = 2$; thus, the system remains in state 2 and row 2 is used again. Using $u_3 = 0.92$, the system remains in state 2. For $u_4 = 0.26$, the cumulative probabilities in row 2 of Equation 11.25 would suggest that the state of the system transitions back to state 1, that is, $x_4 = 1$. Now row 1 of Equation 11.25 is used. The value $u_5 = 0.03$ is less than 0.4, so $u_5 = 1$. The value $u_6 = 0.98$ indicates that the state will transition itself to state 3, that is, $x_6 = 3$. The value $u_7 = 0.17$ is greater then 0.0 and less than 0.8, which causes the system to move to state 2. In summary, the sequence of randomly generated Markov variates is

$$\underline{X} = \{2,\ 2,\ 2,\ 1,\ 1,\ 3,\ 2\}$$

The initial-state variable x_0 is not included in the sequence. Given a longer sequence of random variates u_i, a longer sequence of values of x_i could be generated.

11.4 GENERATION OF CONTINUOUSLY DISTRIBUTED RANDOM VARIATES

The inverse transformation method can be used to generate continuously distributed random variates if the cumulative distribution function $F(x)$ is known and a uniform variate $U(0, 1)$ generator is available. If the density function $f(x)$ is known, then $F(x)$ can be derived analytically or numerically. Since it would be time consuming to compute $F(x)$ for each value of $U(0, 1)$, $F(x)$ is generally computed for a reasonable set of x and then for a given value $U(0, 1)$. The corresponding value of x is interpolated, generally by linear interpolation, to obtain an approximate value of the random variable. The finer the scale used to describe $F(x)$, the better the accuracy obtained for x.

11.4.1 UNIFORM DISTRIBUTION, $U(\alpha, \beta)$

If values of a uniform distribution with location and scale parameters of α and β are needed, the following transformation is made:

$$U_i = \alpha + (\beta - \alpha)u_i \quad \text{for } i = 1, 2, \ldots \tag{11.26}$$

where u_i is a variate for $U(0, 1)$.

11.4.2 Triangular Distribution

While the triangular distribution would have little practical use, it is very useful to demonstrate the development of a transformation function for a continuously distributed variable. As with all such transformations, values of the uniform variate u_i are known, and the goal is to generate a random sequence of values x_i that have the probability density function $f(x)$. The function $f(x)$ can be integrated to obtain the cumulative function $F(x)$, which is an expression that contains x. Since both u_i and $F(x)$ would be known, the cumulative distribution can be solved for x as a function of u_i.

Assume that the random variable x has a density function that is the slope of a right triangle with the lower end at $x = \alpha$ and the vertical side at $x = \beta$, where $\alpha < \beta$. This has the density function

$$f(x) = \frac{2x}{\beta^2 - \alpha^2} \tag{11.27}$$

The cumulative function is

$$F(x) = \frac{x^2 - \alpha^2}{\beta^2 - \alpha^2} \tag{11.28}$$

Solving for the random variable x and letting $u_i = F(x)$ yields

$$x_i = \sqrt{\alpha^2 + u_i(\beta^2 - \alpha^2)} \tag{11.29}$$

For given values of α and β and a uniform variate $U(0, 1)$, a value of x that has a triangular distribution can be computed.

Example 11.10

Given $\alpha = 0$ and $\beta = 1$, Equation 11.29 is

$$x_i = u_i^{0.5} \tag{11.30a}$$

For the following array of uniform variates,

$$u_i = \{0.22, 0.74, 0.57, 0.08, 0.40, 0.93, 0.61, 0.16, 0.89, 0.35\}$$

the corresponding triangular variates are

$$x_i = \{0.47, 0.86, 0.75, 0.28, 0.63, 0.96, 0.78, 0.40, 0.94, 0.59\}$$

Note that while the u_i values are evenly distributed over the range from 0 to 1, the values of x_i are bunched more to the right, as would be expected for a random variable with a triangular distribution.

Instead of $\alpha = 0$ and $\beta = 1$, assume that $\alpha = 2$ and $\beta = 4$, Equation 11.29 becomes

$$x_i = [4 + u_i(4^2 - 2^2)]^{0.5} = [4 + 12u_i]^{0.5} \tag{11.30b}$$

Using the same uniform variates, the sequence of x_i values is

$$x_i = \{2.58, 3.59, 3.29, 2.23, 2.97, 3.89, 3.36, 2.43, 3.83, 2.86\}$$

Again, the values of x_i are bunched more toward the right boundary of 4 than are the u_i values to the right boundary of 1.

11.4.3 Normal Distribution

Assuming that the mean μ and standard deviation σ are known, values of the random variate x can be generated using the standard normal transform equation:

$$x_i = \mu + z_i\sigma \tag{11.31}$$

in which z_i is a standard normal deviate computed with a uniform variate u_i from $U(0, 1)$. Sample estimators $\bar{X}$ and S can be used in place of the population values μ and σ. The value of the cumulative standard normal distribution $F(x)$ is assumed to equal u_i, and thus a z_i value can be generated. Again, approximate values can be obtained by interpolation from a tabular summary of z versus $F(z)$.

An alternative method of generating normal variates uses a sequence of uniform variates. Approximate values of standard normal deviates $N(0, 1)$ can be obtained using a series of k uniform variates $U(0, 1)$ by the following:

$$z_i = \frac{\left(\sum_{j=1}^{k} u_j\right) - k/2}{(k/12)^{0.5}} \tag{11.32}$$

For many cases, k can be set to 12, which yields the following $N(0, 1)$ generator:

$$z_i = \left(\sum_{j=1}^{12} u_j\right) - 6 \tag{11.33}$$

Example 11.11

Generate ten random normal numbers with a mean of 6 and a standard deviation of 1.5 using the following ten uniform variates: 0.22, 0.74, 0.57, 0.08, 0.40, 0.93, 0.61, 0.16, 0.89, and 0.35. Equation 11.31 becomes

$$x_i = 6 + 1.5z_i \tag{11.34}$$

The values of z_i corresponding to the u_i values are obtained by interpolation in a table of cumulative standard normal deviates. For example, the first uniform variate is 0.22. Using this as the value of $F(z)$, the corresponding value of z from Table A.1 is –0.772. The sequence of standard normal deviates is

$$z_i = \{-0.772, 0.643, 0.176, -1.405, -0.253, 1.476, 0.279, -0.995, 1.226, -0.386\}$$

Therefore, the values of the random variate X is obtained from Equation 11.34 are

$$x_i = \{4.842, 6.965, 6.264, 3.892, 5.620, 8.214, 6.418, 4.508, 7.839, 5.421\}$$

This sequence of x_i values has a mean of 5.998 and a standard deviation of 1.412, both of which are close to the assumed population values.

Example 11.12

The procedure of the previous example has the advantage that each value of u_i provides one value of x_i with a normal probability density function. The method has two disadvantages. First, it requires linear interpolation from entries of a cumulative normal table. Second, it would require some rule for extrapolating values at the two ends of the table. Equation 11.33 is a valid alternative, but it requires 12 values of u_i for each generated value of x_i.

Consider the following 12 uniform variates: u_i = {0.22, 0.74, 0.57, 0.08, 0.40, 0.93, 0.61, 0.16, 0.89, 0.35, 0.26, 0.75}. Using Equation 11.33, the x value is 5.96 – 6 = –0.04.

Given that computers can rapidly generate uniform variates and that a rule for handling values u_i values near 0 and 1 is not needed, this approach is useful for generating normal deviates.

11.4.4 Lognormal Distribution

If the $\log_{10}$ mean μ_y and $\log_{10}$ standard deviation σ_y are known, then values of random variable X that have a lognormal distribution can be generated using standard normal deviates z_i with

$$X_i = 10^{\mu_y + z_i \sigma_y} \tag{11.35}$$

It is important to note the following:

1. For natural logarithms, use e rather than 10 in Equation 11.35.
2. The following inequality relationships apply: $\log_{10} \mu \neq \mu_y$ and $\log \sigma \neq \sigma_y$.
3. The z_i values of Equation 11.35 must be generated from values of u_i.
4. If the population mean and standard deviation are not known, then sample estimators can be used.

Example 11.13

Assume that the log mean and log standard deviation are 1.35 and 0.22, respectively, and that the z_i values of Example 11.9 are the array of standard normal deviates. Values of X that have a lognormal distribution can be computed by

$$X_i = 10^{1.35+0.22z_i} \tag{11.36}$$

The first uniform variate of 0.22 was transformed to a z value of –0.772. Therefore, Equation 11.36 gives

$$X_1 = 10^{1.35+0.22(-0.772)} = 15.1$$

The ten values of the lognormally distributed random variable X follow:

$$X_i = \{15.1, 31.0, 24.5, 11.0, 19.7, 47.3, 25.8, 13.5, 41.7, 18.4\}$$

The sample log mean and log standard deviation for the ten generated values are 1.350 and 0.207, respectively. The larger the sample of generated values, the closer the moments of the computed values will approximate the assumed population values. Values generated by $y_i = 1.35 + 0.22z_i$ would be normally distributed. The relationship $X_i = 10^{y_i}$ transforms the normally distributed values to lognormal distributed values.

Example 11.14

Table 11.8 provides the 7-day, low-flow discharge series for a 27-year period (1939–1965), including the statistics for both the discharges and discharge logarithms. The 2-year, 10-year, and 100-year 7-day, low flows are

$$\log Q_2 = 1.957 - 0(0.134) = 1.957 \qquad \therefore Q_2 = 90.6 \text{ ft}^3/\text{s}$$

$$\log Q_{10} = 1.957 - 1.2817(0.134) = 1.785 \quad \therefore Q_{10} = 61.0 \text{ ft}^3/\text{s}$$

$$\log Q_{100} = 1.957 - 2.3267(0.134) = 1.645 \quad \therefore Q_{100} = 44.2 \text{ ft}^3/\text{s}$$

The sample statistics were used with 20 random normal deviates (column 5 of Table 11.8) to generate 20 logarithmic flows that have a normal distribution (column 6), which are then transformed to discharges (column 7). The mean and standard deviation of the generated values are also given in Table 11.8, which can be compared with the corresponding values of the measured data. The difference between the moments of the actual and generated series are a result of the small sample. If the record length of the generated series was longer, the differences would be less.

TABLE 11.8
Seven-Day Low Flows (ft³/sec) for Maury River near Buena Vista, Virginia

Year	Flow, Q	Log Q	Simulated Year	Variate $N(0, 1)$	Simulated Log Q	Simulated Q
1939	103	2.013	1	−0.646	2.044	111
1940	171	2.233	2	−0.490	2.023	105
1941	59	1.771	3	−0.067	1.966	92
1942	127	2.104	4	−1.283	2.129	135
1943	94	1.973	5	0.473	1.894	78
1944	66	1.820	6	−0.203	1.984	96
1945	81	1.908	7	0.291	1.918	83
1946	79	1.898	8	−1.840	2.204	160
1947	102	2.009	9	1.359	1.775	60
1948	176	2.246	10	−0.687	2.049	112
1949	138	2.140	11	−0.811	2.066	116
1950	125	2.097	12	0.738	1.858	72
1951	95	1.978	13	−0.301	1.997	99
1952	116	2.064	14	0.762	1.855	72
1953	86	1.935	15	−0.905	2.078	120
1954	70	1.845	16	0.435	1.899	79
1955	96	1.982	17	0.741	1.858	72
1956	66	1.820	18	−1.593	2.171	148
1957	64	1.806	19	0.038	1.952	90
1958	93	1.968	20	0.797	1.850	71
1959	81	1.908		Mean	1.978	98.5
1960	92	1.964		Standard deviation	0.117	27.4
1961	103	2.013				
1962	95	1.978				
1963	61	1.785				
1964	54	1.732				
1965	69	1.839				
Mean	94.9	1.957				
Standard deviation	31.3	0.134				

The moments of the generated series can be used to compute the 2-, 10-, and 100-year low flows:

$$\log Q_2 = 1.975 - 0(0.117) = 1.978 \qquad \therefore Q_2 = 95.1 \text{ ft}^3/\text{s}$$

$$\log Q_{10} = 1.978 - 1.2817(0.117) = 1.828 \quad \therefore Q_{10} = 67.3 \text{ ft}^3/\text{s}$$

$$\log Q_{100} = 1.978 - 2.3267(0.117) = 1.706 \quad \therefore Q_{100} = 50.8 \text{ ft}^3/\text{s}$$

For the small sample size, the simulated discharges are in good agreement with the measured values.

11.4.5 Log-Pearson Type III Distribution

Approximate deviates for the log-Pearson III distribution can be generated by integrating the density function for given values of its three parameters and then interpolating within the x versus $F(x)$ relationship. The three parameters are the log mean (μ_y), log standard deviation (σ_y), and log skew coefficient (g_S). Using a uniform variate $U(0, 1)$, enter the cumulative function and obtain the deviate K. Then compute the variable using the transform

$$X = 10^{\mu_y + KS_y} \tag{11.37}$$

where μ_y and S_y are the log mean and log standard deviation, respectively.

Values of log-Pearson III deviates can be obtained by interpolation from a table of deviates, such as the table in Bulletin 17B (Table A.11, in Interagency Advisory Committee on Water Data, 1982). For the given value of the standardized skew, the uniform variate is entered as the cumulative probability and, using linear interpolation, the value of the deviate is read from the appropriate column. For example, if $u_i = 0.72$ and the standardized skew is 0.3, the LP3 deviate is –0.61728.

Example 11.15

Assume that the values for the three parameters are $\mu_y = 3.41$, $\sigma_y = 0.63$, and $g_S =$ 0.6. The uniform variates follow:

$$U_i = \{0.62, 0.27, 0.46, 0.81\}$$

The values of the LP3 deviate K are obtained from Table A.11 for the g_S value:

$$K = \{-0.39109, 0.55031, 0.002686, -0.89149\}$$

Using these values of K with Equation 11.37 yields values of X with the following log-Pearson III distribution:

$$X_i = \{1457, 5711, 2580, 705\}$$

11.4.6 Chi-Square Distribution

A chi-square deviate with υ degrees of freedom can be generated using standard normal deviates z_i and uniform $U(0, 1)$ deviates u_i as follows:

$$\chi_i = -0.5 \ln_e \left[\prod_{j=1}^{\upsilon/2} u_j \right] \quad \text{for } \upsilon \text{ even} \tag{11.38a}$$

$$\chi_i = -0.5 \ln_e \left[\prod_{j=1}^{(\upsilon-1)/2} u_j \right] + z_i^2 \quad \text{for } \upsilon \text{ odd} \tag{11.38b}$$

Example 11.16

If $\upsilon = 4$, then two uniform deviates are needed for each chi-square deviate. Given

$$U_i = \{0.62, 0.09, 0.38, 0.79\}$$

the two chi-square deviates are

$$x_1 = -0.5 \ln_e[0.62(0.09)] = 1.443$$

$$x_2 = -0.5 \ln_e[0.38(0.79)] = 0.602$$

Example 11.17

If $\upsilon = 5$, then each chi-square deviate generated will require two uniform deviates and one standard normal deviate. Given

$$u_i = \{0.62, 0.09, 0.38, 0.79\}$$

$$z_i = \{-0.53, 0.82\}$$

the two chi-square deviates are

$$x_1 = -0.5 \ln_e[0.62(0.09)] + (-0.53)^2 = 1.742$$

$$x_2 = -0.5 \ln_e[0.38(0.79)] + (0.82)^2 = 1.274$$

11.4.7 Exponential Distribution

The inverse transform for generating exponential deviates x_i with mean b is

$$x_i = -b \ln_e u_i \tag{11.39}$$

where u_i is a uniform variate $U(0, 1)$. The cumulative function for the exponential distribution is $F(x) = 1 - e^{-x/b}$ where both the mean and standard deviation are equal to its scale parameter b.

Example 11.18

Assume that four exponential deviates are needed for a mean of 5.3. Given the following uniform deviates,

$$u_i = \{0.91, 0.46, 0.58, 0.23\}$$

the exponential deviates are derived from

$$\begin{aligned} x_i &= -5.3 \ln_e u_i \\ &= \{0.4998, 4.1156, 2.8871, 7.7893\} \end{aligned}$$

11.4.8 Extreme Value Distribution

Rainfall extremes are often represented by an extreme value distribution, which has the cumulative function

$$F(x) = 1 - \exp\{-\exp[(x - a)/b]\} \tag{11.40}$$

in which a is a location parameter, the mode; and b is a scale parameter. The extreme value has a mean of $[a + b\Gamma'(1)] = a - 0.57721b$ and a standard deviation of $b\pi/6^{0.5}$. Using the method of moments, the parameters can be estimated by

$$a = \bar{x} + 0.45S \tag{11.41a}$$

$$b = 0.7797S \tag{11.41b}$$

Extreme value variates can be generated by

$$x_i = a + b\ln_e(\ln_e U_i) \tag{11.42}$$

in which U_i is a uniform variate, $U(0, 1)$.

Example 11.19

Assume that the annual maximum 24-hour rainfall intensity (in./hr) follows an extreme value distribution with a mean of 0.13 in./hr and a standard deviation of 0.07 in./hr. To simulate a record of annual maximum 24-hour rainfall intensities, the parameters are computed from Equations 11.41:

$$a = 0.13 + 0.45(0.07) = 0.1615$$

$$b = 0.7797(0.07) = 0.0546$$

Using Equation 11.42, extreme value variates are computed by

$$x_i = 0.1615 + 0.0546\ln_e(-\ln U_i)$$

For example, for a uniform variate of 0.27, the corresponding extreme value variate is

$$x = 0.1615 + 0.0546\ \ln_e(-\ln_e(0.27))$$

$$= 0.1615 + 0.0546\ \ln_e(1.30933)$$

$$= 0.1615 + 0.0546(0.2695) = 0.1762$$

A 10-year generated series of extreme-value, distributed, annual-maximum rainfall intensities for given uniform variates follows:

U_i	0.91	0.46	0.63	0.70	0.09	0.36	0.21	0.84	0.58	0.22
X_i	0.033	0.148	0.119	0.105	0.209	0.163	0.186	0.066	0.128	0.184

The mean and standard deviation of the generated series are 0.134 in./hr and 0.0556 in./hr, respectively.

11.5 APPLICATIONS OF SIMULATION

In Section 11.1.4 the process of simulation was used to generate samples to show that a theorem was correct. Simulation has its primary use to make analyses when theory cannot provide solutions. Numerous problems have arisen in hydrology where theory was totally inadequate to provide solutions. For example, Bulletin 17B recommends an outlier detection method that assumes an underlying normal distribution, even though the log-Pearson type III distribution is used to represent the peak discharge data. Spencer and McCuen (1996) used simulation to develop outlier tests for LP3 distributed variables. Bulletin 17B also includes a method for handling historic flows. McCuen (2000) demonstrated the origin of the Bulletin 17B historic adjustment method and used simulation to extend the method for use with the LP3 distribution. Vogel (1986) used simulation to extend the probability plot correlation coefficient to test the applicability of the Gumbel distribution. Theoretical solutions are often available for random variables that follow a normal distribution. These solution may be inaccurate if applied to a skewed distribution such as the LP3 distribution. Simulation enables the importance of the normality assumption to be tested, and if the theory is not applicable to a situation, simulation can be used to develop an applicable solution. A few examples will be used to outline solution procedures for using simulation in problem solving.

Example 11.20

The runoff curve number (*CN*) is a widely used input in hydrologic design. While it is treated as a constant when it is obtained from a *CN* table, in reality the *CN* is a random variable that varies from storm to storm. In model studies for evaluating the effects of land use change, it may be of interest to examine the sensitivity of the effects to variation in the *CN*. Simulation could be used to develop confidence intervals on the *CN*. The following procedure could be applied:

1. Identify a probability distribution function that will accurately reflect the spread of the *CN*s. The function must be a bounded distribution because *CN*s have both upper and lower limits.
2. Assume that the mean value is equal to the tabled *CN* value, or some transformation of it. As an empirical alternative, the mean *CN* could be estimated from sample data for a particular set of conditions.
3. Obtain an estimate of the storm-to-storm variation of *CN*s for the hydrologic conditions for which the *CN* is to be used.
4. Use an estimation method to obtain values for the distribution parameters assumed in step 1. For example, either maximum likelihood estimation or method of moments estimation could be applied.
5. Using the distribution function of step 1 and the parameters of step 4, find the confidence bounds of the distribution for selected levels of confidence.

While the gamma distribution is bounded at one end and unbounded at the other end, it can be applied to solve this problem by substituting (100 – *CN*) as the variable and applying it for curve numbers above 40. For a land use condition for which the

CN table does not include a value, rainfall depth and runoff depth values could be used to estimate the mean and standard deviation of the *CN*. These could be used to obtain the moment estimators of the gamma distribution. Then simulation is applied to generate the distribution of the *CN* from which bounds of the confidence interval for selected levels of confidence are obtained.

Example 11.21

Simulation is widely used with continuous hydrograph models. Typically, such models have numerous storage cells with several layers to reflect different zones in the soil profile. The number of fitting coefficients often exceeds 20, with some coefficients having an option to be varied from month to month. Proper fitting of such models requires considerable knowledge and experience.

To illustrate the introduction of randomness into a continuous hydrograph model, a simple one-storage-unit model is developed here. The model uses the daily rainfall depth as an input, and the depths of runoff and evapotranspiration losses are the outputs. The daily rainfall is randomly generated such that it has serial correlation defined by a Markov process. The occurrence of rainfall is defined by two states, wet (*W*) and dry (*D*). The probability of rainfall or lack of it on any day depends probabilistically on the rainfall state of the previous day. The probabilities are input for the following matrix:

	day *t*+1	
day *t*	***D***	***W***
D	P_{DD}	P_{DW}
W	P_{WD}	P_{WW}

with $P_{DD} + P_{DW} = 1$ and $P_{WD} + P_{WW} = 1$. A uniform random number u_i determines the rainfall condition of the next day. For example, if it is dry on day t and u_i is less than P_{DD}, then it is dry on day $t + 1$. If it is wet on day t and u_i is less than P_{WD}, then it is dry on day $t + 1$. On wet days, rain depth is assumed to be the extreme value distributed with parameters a and b.

All of the daily rainfall is assumed to enter the storage cell. Continuity of mass yields the representative relationship

$$\text{storage}_t = \text{storage}_{t-1} + P_t \tag{11.43}$$

in which P_t is the depth of rainfall on day t.

Evapotranspiration (E_t) is computed on a daily basis as a depth (in.). The value is treated as a random variable. Also, it varies on an annual cycle. The model for computing the daily E_t consists of a constant, a sine curve offset in time to yield maximum rates in the summer, and a random component to reflect the daily variation that results from variation in cloud cover, temperature, and soil moisture:

$$E_t = p_1 + p_2 \sin(2\pi d/365 + \phi) + \epsilon \tag{11.44}$$

in which p_1 is the annual average daily E_t, p_2 is the amplitude, d is the day of the year, that is, 1 to 365, ϕ is the phase angle used to control the date of the start of the sine curve, and $\in$ is the random value. The random component was designed to have a lower bound but no upper bound so it was modeled as an exponential variate, with the mean value on the sine curve.

The continuity of mass was used to adjust the E_T loss:

$$\text{storage}_t = \text{storage}_t - E_t \tag{11.45}$$

The E_t depth is subtracted after the daily rainfall P_t was added.

Q_t, the depth of runoff on day t, was assumed to be proportional to the amount of water in storage:

$$Q_t = b * \text{storage}_t \tag{11.46}$$

where b is a constant. The loss due to runoff is used to adjust the storage:

$$\text{storage}_t = \text{storage}_t - Q_t \tag{11.47}$$

The discharge Q_t is based on the storage after the daily E_t has been removed.

To model the effect of watershed change, the E_t was assumed to vary with time, as if afforestation was taking place. A constant secular trend was assumed, such that the average E_t was the same for the case of no watershed change. Thus, instead of being a constant, p_1 of Equation 11.44 was replaced with a linear function that started at a lower rate and increased as the density of the forest cover increased. The same 11-year sequence of daily rainfall was used for both analyses.

Table 11.9 summarizes the two analyses. For the case of no change in forest cover, the E_t changed very little from year to year. However, as the rainfall changed, the runoff changed. Runoff was approximately 60% of rainfall for both simulations; however, in the case of afforestation, the ratio of runoff to rainfall tended to decrease as the E_t increased.

Simulation studies based on rainfall-runoff models are generally conducted to study an effect. For example, the effect of afforestation on the skew coefficient of the annual maximum flood series could be evaluated by simulating 10,000 sequences of 50-year records, computing the log skew coefficient for each sequence, and evaluating the resulting distributions of the 10,000 skews; a comparison of the two distributions would suggest the effect of afforestation on flood skew.

Another possible simulation study might be to assess the effect of deforestation on the 7-day, 10-year low flow. The rainfall-runoff model would have to be capable of simulating daily flows and the effects of changes in forest cover. For each simulated year of record, the 7-day low flow would be computed. The series of annual low flows would then be subjected to a frequency analysis, and the 10-year discharge computed. By simulating a large number of n-year sequences, the distribution of the 7-day, 10-year low flow could be determined. By performing this analysis for different scenarios of deforestation, the effect of deforestation on the 7-day, 10-year low flow could be assessed.

TABLE 11.9
Summary of 11-Year Simulations of Annual Rainfall Depth, Evapotranspiration, Discharge, and Change in Storage for (a) No Change in Forest Cover and (b) Afforestation

Year	N_d	N_w	P (in.)	E_t (in.)	Q (in.)	ΔS (in.)
			(a) No Change in Forest Cover			
1	265	100	42.80	15.78	24.58	2.43
2	285	80	32.34	15.64	22.77	–6.07
3	280	85	35.11	15.71	20.45	–1.05
4	272	93	38.60	15.73	19.80	3.08
5	274	91	38.26	15.64	22.87	–0.25
6	267	98	41.13	15.68	24.51	0.93
7	263	102	44.44	15.71	26.49	2.25
8	270	95	39.62	15.56	25.13	–1.07
9	279	86	36.91	15.63	24.57	–3.29
10	269	96	39.64	15.55	22.68	1.41
11	267	98	39.62	15.71	23.60	0.31
			428.47	172.34	257.45	–1.32
			(b) Afforestation			
1	265	100	42.80	9.15	27.56	6.09
2	285	80	32.34	10.34	27.81	–5.80
3	280	85	35.11	11.73	25.10	–1.72
4	272	93	38.60	13.07	23.39	2.14
5	274	91	38.26	14.31	25.21	–1.26
6	267	98	41.13	15.68	25.54	–0.09
7	263	102	44.44	17.04	26.19	1.21
8	270	95	39.62	18.22	23.51	–2.11
9	279	86	36.91	19.62	21.62	–4.33
10	269	96	39.64	20.86	18.41	0.38
11	267	98	39.62	22.35	17.99	-0.72
			428.47	172.37	262.33	–6.21

Note: N_d = number of dry days; N_w = number of days with rain; P = annual rainfall depth; E_t = evapotranspiration; Q = discharge; and ΔS = change in storage area.

In this example, the simulation involved two stochastic variates, one for the daily rainfall and one for the daily evapotranspiration. In formulating a simulation model, each input variable may be treated as a stochastic variate. For example, if E_t were computed from measurements of daily temperature, humidity, and air speed, then each of those variables may include both deterministic and stochastic components. It is important to check the rationality of the random variation in the computed E_t

that results from the functions that define the randomness of temperature, humidity, and air speed. The interaction among the three inputs may suppress or inflate the computed variation in E_t such that it is biased.

Example 11.22

Simulation can be used in the verification phase of modeling to evaluate the accuracy of assumptions. One assumption commonly made in flood frequency analysis is that the Weibull plotting position formula is appropriate for use with the log-Pearson type III analysis. The plotting position is used in frequency analyses to plot the sample data. Then the graph is visually assessed as to whether the data can be represented by the assumed distribution. If this is a valid assumption, then one would expect the probability computed with the Weibull formula to closely match the true probability. This is easily tested by simulation.

The following procedure could be used to check the reliability of the Weibull plotting position formula as a tool in flood-frequency analysis where the Bulletin 17B (Interagency Advisory Committee on Water Data, 1982) log-Pearson III distribution is assumed.

1. Select the criteria that will be used to evaluate the quality of the assumption. For this case, the bias in and accuracy of the computed probability will be the criteria.
2. The assumed population statistics must be set, in this case the mean μ, the standard deviation σ, and the standardized skew γ. These can be varied from one simulation analysis to another, but they must be held constant for all simulated samples within a single simulation experiment.
3. The sample size N must be set, although this can be changed from one simulation experiment to another to test whether the bias and accuracy criteria vary with N.
4. The number of simulated samples N_S should be set. This should always be tested to ensure that the conclusion is stable.
5. The decision variable should be identified. For example, the largest event in the sample might be used, or separate analyses may be made for each of the three largest events. For illustration purposes, assume that the largest event Y_L will be used as the decision variable.
6. Compute the Weibull probability P_W for the sample size being tested. For the largest event, this would be equal to $1/(N + 1)$.
7. Make the N_S simulations and perform the following steps:
   (a) Generate a random sample of size N using μ, σ, and γ with the generation method of Section 11.4.5.
   (b) Compute the sample log mean $\bar{Y}$, log standard deviation S_y, and log standardized skew g.
   (c) For the sample of N, find the largest sample value Y_L.
   (d) Computed the LP3 deviate K, where $K = (Y_L - \bar{Y})/S_y$.
   (e) Using K and the skew g, find the sample probability P_S.

Each of the N_S simulations will yield a single estimate of P_S. The bias (B_p) and the precision (P) can be computed from the N_s estimates of P_s and the Weibull probability P_W:

$$B_p = \frac{1}{N_s}\sum_{i=1}^{N_s}(P_{si} - P_w) \tag{11.48}$$

$$P = \left[\frac{1}{N_s - 1}\sum_{i=1}^{N_s}(P_{si} - P_w)^2\right] \tag{11.49}$$

The physical significance of the statistics will then need to be assessed by evaluating the effects of such errors, such as in the assumed error in the return period that results from using the sample values.

In this case, the population statistics μ, σ, and γ are also known. So, the population probability P_p could also be computed by $K_p = (Y_L - \mu)/\sigma$ and the population skew γ. In this case, the sample values of P_s could be compared to the population probability P_p to assess the true bias and accuracy.

11.6 PROBLEMS

11-1 Identify a system and its components for each of the following:
   a. A 200-square-mile watershed.
   b. A 1-acre residential lot.
   c. Groundwater aquifer.
   d. A 2-mile-long section of a river.

11-2 Justify the use of the word *model* to describe the peak-discharge rational method, $q_p = CiA$.

11-3 Describe a laboratory experiment to show that the Reynolds number 2000 is a reasonable upper limit for laminar flow.

11-4 Describe a statistical experiment that would determine if a regression model is not biased.

11-5 The following theorem defines the sampling distribution of the variance: If the variance S^2 computed from a sample of size n taken from a normally distributed population with the variance σ^2, then

$$\chi^2 = \frac{(n-1)S^2}{\sigma^2}$$

is the value of a random variable that has a chi-square distribution with parameter $\upsilon = n - 1$.

Develop the steps for simulating the sampling distribution of the variance identified in the theorem.

11-6 The following test statistic is used to test the statistical significance of a sample correlation coefficient R computed from a bivariate sample of size n:

$$t = \frac{R}{\left(\frac{1-R^2}{n-2}\right)^{0.5}}$$

in which test statistic t has a t distribution with parameter $\upsilon = n - 2$. Develop the steps for simulating the sampling distribution of t and its accuracy for testing the following hypotheses: H_0: $\rho = 0$ and H_A: $\rho \neq 0$.

11-7 Simulate the probability distribution of the roll of a single dice using a sample of 50 rolls. Compare the sample and population distributions. Use a chi-square goodness-of-fit test or a Kolmogorov–Smirnov one-sample test to evaluate the reasonableness of your sample.

11-8 From a deck of playing cards, create a new deck with the following cards: spades—2, 3, 4, 5; hearts—2, 3, 4; clubs—2, 3; and diamonds—2. Using this new population, simulate the probability of drawing a red card by sampling 50 times with replacement (shuffle the cards after each draw). Compare the population and sample probabilities.

11-9 Generate ten uniform variates, 0 to 1, using the midsquare method with a seed of 1234.

11-10 Generate 25 uniform variates, 0 to 1, using the midsquare method with a seed of 7359. Compare the sample and population means and standard deviations.

11-11 Assume that a discrete random variable has the following mass function: $p(x) = \{0.1, 0.3, 0.25, 0.15, 0.1, 0.06, 0.04\}$ for $x = 1, 2, \ldots, 7$. Use the following uniform variates to generate values of x: $U_i = \{0.72, 0.04, 0.38, 0.93, 0.46, 0.83, 0.98, 0.12\}$.

11-12 Assume that a discrete random variable x can take on values of 1, 2, 3. The uniform variates $U_i = \{0.18, 0.53, 0.12, 0.38, 0.92, 0.49, 0.76, 0.21\}$ are transformed to the following values of $x_i = \{1, 3, 1, 2, 3, 2, 3, 2\}$. Approximate the mass function of x.

11-13 Assume that the density function of a continuous random variable is $f(x) = 1.6 - 0.4x$ for $1 \le x \le 2$. Determine the transformation function that transforms uniform variates to variates of x. Compute the values of x for the following uniform variates: $U_i = \{0.27, 0.66, 0.40, 0.92, 0.08\}$.

11-14 Let a discrete random variable x represent the overtopping or not overtopping of any one of three culverts at an industrial park.

a. If the probability of overtopping is 0.15 determine the population probabilities of 0, 1, 2, and 3 over toppings in a storm event.
b. Determine the long-term mean.
c. Simulate the number of over toppings in two storm events using the following uniform variates: $U_i = \{0.42, 0.91, 0.62, 0.11, 0.37, 0.53\}$.

11-15 The annual probability of a cofferdam being overtopped by a flood is historically 17%. Simulate overtopping at 14 independent sites where the dams are in place for 2 years, with flooding in any one year being independent of flooding in the next year. Compare the sample and assumed population probabilities. U_i = {01, 07, 11, 56, 75, 80, 12, 18, 87, 51, 85, 72, 96, 39, 12, 35, 27, 41, 63, 06, 00, 84, 43, 38, 23, 31, 90, 04}. Compare the population and sample means.

11-16 The mean number of computer viruses received per week on a computer system is 2.3. Use the following uniform variates to generate a 20-week record. Compare the sample and assumed population means. U_i = {06, 54, 80, 48, 52, 17, 27, 21, 64, 69, 75, 48, 67, 65, 75, 43, 79, 24, 54, 99}.

11-17 The mean numbers of annual out-of-bank flows of a small stream before and after urbanization are 0.6 and 2.3, respectively. Use the following uniform variates to simulate 20 years of flows before and after development. Compare the sample and population means. U_i = {10, 80, 17, 67, 83, 05, 31, 23, 08, 07, 40, 00, 60, 44, 65, 70, 16, 31, 73, 05}.

11-18 Assume that a random variable can be in any one of four states (*A*, *B*, *C*, or *D*). If in state *A*, *B*, *C*, or *D*, the probability of remaining in that state is 0.35, 0.65, 0.50, and 0.45, respectively. If in state *A*, the probabilities of moving to states *B*, *C*, and *D* are 0.30, 0.25, and 0.10, respectively. When in state *B*, the probabilities of moving to states *A*, *C*, and *D* are 0.10, 0.20, and 0.05, respectively. When in state *C*, the probabilities of moving to states *A*, *B*, and *D* are 0.05, 0.20, and 0.25, respectively. When in state *D*, the probabilities of moving to states *A*, *B*, and *C* are 0.05, 0.15, and 0.35, respectively. Develop the one-step transition probability matrix. Use the following sequence of uniform variates to simulate 20 steps of the process, with the first variate used to select the initial state. Compute the sample probabilities of being in each state. U_i = {46, 97, 27, 15, 94, 68, 68, 04, 28, 84, 21, 47, 41, 24, 32, 98, 02, 54, 70, 79, 73}.

11-19 For the following set of transition probabilities, simulate 15 state-to-state transitions with the uniform variates shown:

$$P = \begin{bmatrix} 0.6 & 0.3 & 0.1 \\ 0.2 & 0.5 & 0.3 \\ 0.1 & 0.4 & 0.5 \end{bmatrix}$$

U_i = {31, 49, 87, 12, 27, 41, 07, 91, 72, 64, 63, 42, 06, 66, 82, 71}.

11-20 Transform the following uniform variates $U(0, 1)$ to $U(4, 8)$. U_i = {0.76, 0.14, 0.53, 0.88, 0.27}.

11-21 Develop the function to transform uniform variates to values of the random variable x if the density function of x is a right triangle with the vertical side at $x = \alpha$ and the downward sloping side ending at $x = \beta$, where $\alpha < \beta$.

11-22 The exponential density function is $f(x) = \lambda e^{-\lambda x}$, for $0 \le x \le \infty$. Develop the transformation function to transform a uniform variate to a variate x that would have an exponential distribution with scale parameter λ.

11-23 The Pareto density function for shape parameter c is $f(x) = cx^{-c-1}$ for the range $1 \le x \le \infty$. Develop the transformation function to transform a uniform variate U to a variate from a Pareto distribution with shape parameter c.

11-24 Use Equation 11.33 to transform the following 12 uniform variates to a single standard normal deviate. U_i = {01, 86, 77, 18, 21, 91, 66, 11, 84, 65, 48, 75}.

11-25 What are the lower and upper limits of standard normal deviates obtained with Equation 11.33? If k in Equation 11.32 is set at 24, are the upper and lower bounds the same as for Equation 11.33 where k is 12?

11-26 Using a table of the cumulative standard normal distribution (Table A.1) generate a series of ten normal deviates ($\mu = 3$, $\sigma = 0.5$) with the following uniform deviates: U_i = {14, 68, 13, 04, 88, 44, 73, 24, 82, 25}.

11-27 If the log mean μ_y and log standard deviation σ_y are 2.7 and 0.4, respectively, generate 14 lognormal deviates using the following uniform deviates: U_i = {72, 11, 79, 75, 36, 07, 12, 92, 61, 89, 93, 08, 23, 52}.

11-28 If the log mean, log standard deviation, and log skew are 3.4, 0.65, and 0.4, respectively, generate ten log-Pearson III deviates using the following uniform variates: U_i = {34, 88, 14, 13, 85, 56, 25, 62, 57, 42}.

11-29 Using the moments given in Problem 11-28, determine the uniform variates $U(0, 1)$ that would give the following log-Pearson III variates: x = {5940, 365, 2650}.

11-30 Using the following uniform variates, generate six variates that are from a chi-square distribution with $\upsilon = 2$. U_i = {12, 74, 38, 92, 51, 27}.

11-31 Using the following uniform variates, generate three variates that are from a chi-square distribution with $\upsilon = 3$. U_i = {66, 08, 43, 79, 52, 28}.

11-32 Using the following uniform variates, generate three variates that are from a chi-square distribution with $\upsilon = 4$. U_i = {81, 16, 47, 93, 25, 36}.

11-33 Generate six variates that have an exponential distribution with a scale parameter of 2.5 using the following uniform variates: U_i = {76, 23, 47, 51, 09, 82}.

11-34 Assume the long-term mean and standard deviation of a process are 6.4 and 1.1, respectively. Generate variates that have an extreme value distribution. Use the following uniform variates: U_i = {11, 30, 73, 61, 39, 88, 06}.

12 Sensitivity Analysis

12.1 INTRODUCTION

Modeling is central to many hydrologic decisions, both in analysis and in design. Decisions in hydrologic engineering often require the development of a new model or the use of an existing model. When developing a new model, the response of the model should closely mimic the response of the real-world system being modeled. Before using an existing model, the modeler should verify that the model is appropriate for solving the type of problem being addressed. When a modeler uses an existing model developed by someone else, the modeler should be sure that he or she completely understands the way that the model responds to the input. In all of these cases, the modeler should understand the model being used. Sensitivity analysis is a modeling tool that is essential to the proper use of a model because it enables the model user to understand the importance of variables and the effects of errors in inputs on computed outputs.

Questions such as the following can be answered by performing a sensitivity analysis:

- Which are the most important variables of a model?
- What is the effect of error in inputs on the output predicted with the model?
- What physical hydrologic variables are most likely related to fitted model coefficients?
- In calibrating a model, are all of the model coefficients equally important to the accuracy of the model?

Each of these questions suggests a relationship between two or more variables or the effect of one factor on another. In modeling terms, they are concerned with sensitivity, whether it is the sensitivity to errors, the sensitivity of goodness-of-fit criteria, or the sensitivity to watershed characteristics. Questions such as those presented above can be answered by applying an important modeling tool, specifically sensitivity analysis.

Of special interest in hydrologic analysis is the detection of effects of land use change. Watershed change can cause abrupt or gradual change in measured flood data or in characteristics of the data, such as the 100-year peak discharge. Just as the amount of watershed change is important, the spatial pattern of land use also can significantly influence measured hydrologic data. Specific questions that may be of interest include:

- How sensitive are annual maximum discharges to gradual or abrupt land use change?
- How sensitive are the moments of an annual flood series to historic information?
- How sensitive is the 100-year peak discharge to the existence of an outlier in the sample data?
- How sensitive is the 100-year peak discharge to changes in channel roughness?

Modeling studies are often used to answer such questions. The clarity of the answers to such questions depends, in part, on the way that the results are presented. Better decisions will be made when the results are organized and systematically presented. An understanding of the foundations of sensitivity analysis should ensure that the results are presented in ways that will ensure that the modeling efforts lead to the best possible decisions.

12.2 MATHEMATICAL FOUNDATIONS OF SENSITIVITY ANALYSIS

12.2.1 DEFINITION

Sensitivity is the rate of change in one factor with respect to change in another. Although such a definition is vague in terms of the factors involved, nevertheless it implies a quotient of two differentials. Stressing the nebulosity of the definition is important because, in practice, the sensitivity of model parameters is rarely recognized as a special case of the concept of sensitivity. The failure to recognize the generality of sensitivity has been partially responsible for the limited use of sensitivity as a tool for the design and analysis of hydrologic models.

12.2.2 THE SENSITIVITY EQUATION

The general definition of sensitivity can be expressed in mathematical form using a Taylor series expansion of the explicit function:

$$O = f(F_1, F_2, \ldots, F_n) \tag{12.1}$$

where O is often a model output or the output of one component of a model and the F_i are factors that influence O. The change in factor O resulting from change in a factor F_i is given by the Taylor series:

$$f(F_i + \Delta F_i, F_{j|j\neq i}) = O_0 + \frac{\partial O_0}{\partial F_i}\Delta F_i + \frac{1}{2!}\frac{\partial^2 O_0}{\partial F_i^2}\Delta F_i^2 + \cdots \tag{12.2}$$

in which O_0 is the value of O at some specified level of each F_i. If the nonlinear terms of Equation 12.2 are small in comparison with the linear terms, then

Equation 12.2 reduces to

$$f(F_i + \Delta F_i,\ F_{j|j\neq i}) \ = \ O_0 + \frac{\partial O_0}{\partial F_i}\Delta F_i \tag{12.3}$$

Thus, the incremental change in O is

$$\Delta O_0 = f(F_i + \Delta F_i,\ F_{j|j\neq 1}) \ - \ O_0 = \left(\frac{\partial O_0}{\partial F_i}\right)\Delta F_i \tag{12.4}$$

Since it is based only on the linear terms of the Taylor series expansion, Equation 12.4 is referred to herein as the *linearized sensitivity equation*. It measures the change in factor O that results from change in factor F_i. The linearized sensitivity equation can be extended to the case where more than one parameter is changed simultaneously. The general definition of sensitivity is derived from Equations 12.1 and 12.4:

$$S = \frac{\partial O_0}{\partial F_i} = \frac{f(F_i + \Delta F_i,\ F_{j|j\neq i}) \ - f(F_1,\ \mathrm{F}_2, \ldots,\ \mathrm{F}_\mathrm{n})}{\Delta F_i} \tag{12.5}$$

12.2.3 Computational Methods

The general definition of sensitivity, which is expressed in mathematical from by Equation 12.5, suggests two methods of computation. The left-hand side of Equation 12.5 suggests that the sensitivity of O to changes in factor F_i can be estimated by differentiating the explicit relationship of Equation 12.1 with respect to factor F_i:

$$S = \frac{\partial O_0}{\partial F_i} \tag{12.6}$$

Analytical differentiation is not used extensively for evaluating the sensitivity of hydrologic models because the complexity of most hydrologic models precludes analytical differentiation.

The method of *factor perturbation*, which is the computational method suggested by the right side of Equation 12.5, is the more commonly used method in hydrologic analysis. The right-hand side of Equation 12.5 indicates that the sensitivity of O to change in F_i can be derived by incrementing F_i by an amount ΔF_i and computing the resulting change in the solution O. The sensitivity is the ratio of the two changes and can be expressed in finite difference form as follows:

$$S = \frac{\Delta O_0}{\Delta F_i} = \frac{f(F_i + \Delta F_i,\ \mathrm{F}_{j|j\neq i}) - f(F_1,\ F_2, \ldots,\ F_n)}{\Delta F_i} \tag{12.7}$$

12.2.4 Parametric and Component Sensitivity

A simplified system or a component of a more complex system is described by three functions: the input function, the output function, and the system response, or transfer, function. The transfer function is the component(s) of the system that transforms the input function into the output function. In a simple form, it could be a probability distribution function that depends on one or more parameters. As another example, the transfer function could be an empirical time-of-concentration formula that depends on inputs for the length, slope, and roughness. In a more complex form, the transfer function could consist of all components of a continuous simulation model.

Sensitivity analyses of models can be used to measure the effect of parametric variations on the output. Such analyses focus on the output and response functions. Using the form of Equation 12.5 parametric sensitivity can be mathematically expressed as

$$S_{pi} = \frac{\partial O}{\partial P_i} = \frac{f(P_i + \Delta P_i;\ P_{j|j \neq i}) - f(P_1,\ P_2, \ldots,\ P_n)}{\Delta P_i} \tag{12.8}$$

where O represents the output function and P_i is the parameter of the system response function under consideration.

Unfortunately, the general concept of sensitivity has been overshadowed by parametric sensitivity. As hydrologic models have become more complex, the derivation of parametric sensitivity estimates has become increasingly more difficult and often impossible to compute. However, by considering the input and output functions, the general definition of sensitivity (Equation 12.5) can be used to define another form of sensitivity. Specifically, *component sensitivity* measures the effect of variation in the input function I on the output function:

$$S_c = \frac{\partial O}{\partial I} = \frac{\Delta O}{\Delta I} \tag{12.9}$$

Combining component and parameter sensitivity functions makes it feasible to estimate the sensitivity of parameters of complex hydrologic models. For example, in the simplified two-component model of Figure 12.1, the sensitivity of Y to variation in P_1 and the sensitivity of Z to variation in P_2 are readily computed using sensitivity as defined by Equation 12.6:

$$S_1 = \frac{\partial Y}{\partial P_1} \quad \text{and} \quad S_2 = \frac{\partial Z}{\partial P_2} \tag{12.10}$$

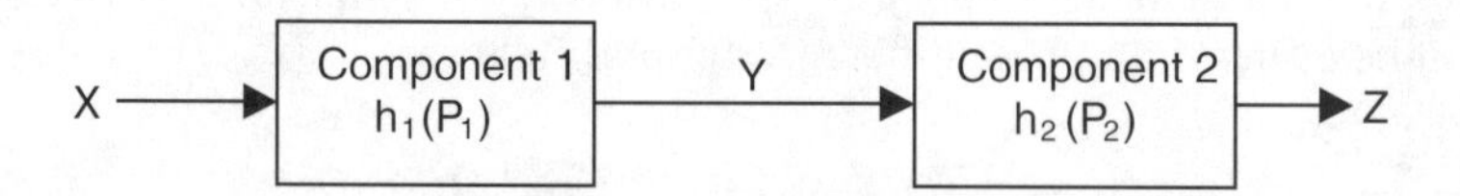

FIGURE 12.1 Two-component model.

However, the sensitivity of the output from component 2 to change in the parameter of component 1 cannot always be estimated directly from the differential $\partial Z/\partial P_1$. In such case, the component sensitivity function of component 2 can be used with the parametric sensitivity function S_1 to estimate the sensitivity of Z to change in P_1. Specifically, the sensitivity of $\partial Z/\partial P_1$ equals the product of the component sensitivity function $\partial Z/\partial Y$ and the parametric sensitivity function $\partial Y/\partial P_1$:

$$\frac{\partial Z}{\partial P_1} = \frac{\partial Z}{\partial Y}\frac{\partial Y}{\partial P_1} \tag{12.11}$$

Whereas the differentials $\partial Z/\partial Y$ and $\partial Y/\partial P_1$ are often easily derived, an explicit sensitivity function $\partial Z/\partial P_1$ can be computed only for very simple models. When a solution cannot be obtained analytically, the numerical method of Equation 12.7 must be used.

Example 12.1

Consider the simple two-component model of Figure 12.1 as a hydrologic model with X being the rainfall input, $h_1(P_1)$ being a unit hydrograph representation of the watershed processes, Y being the flow from the watershed into the channel system, $h_2(P_2)$ being the channel system, and Z being the discharge at the watershed outlet. The unit hydrograph component is represented by Zoch's (1934) simple, linear storage (or single linear reservoir) unit hydrograph:

$$Y_t = Ke^{-KX_t} \tag{12.12}$$

where X_t and Y_t are the input (rainfall) and output (surface runoff) at time t, and K is the unit hydrograph parameter. The transfer of Y_t into Z_t, the discharge at the watershed outlet, uses the Convex routing (U.S. Soil Conservation Service, 1974) method:

$$Z_t = wY_t + (1 - w)Y_{t-1} \tag{12.13}$$

in which w is the parameter of the component and corresponds to the routing coefficient of the Convex routing procedure.

In performing a complete sensitivity analysis of this model, two component sensitivities ($\partial Z_t/\partial Y_t$ and $\partial Y_t/\partial X_t$), three parametric sensitivities ($\partial Z_t/\partial K$, $\partial Z_t/\partial w$, and $\partial Y_t/\partial K$), and one model sensitivity ($\partial Z_t/\partial X_t$) would be of interest. For this simple model, they can be computed either analytically or numerically. Analytically, the two component sensitivities are

$$\frac{\partial Y_t}{\partial X_t} = -K^2e^{-KX_t} \tag{12.14}$$

$$\frac{\partial Z_t}{\partial Y_t} = w \tag{12.15}$$

The three parametric sensitivities are

$$\frac{\partial Y_t}{\partial K} = e^{-KX_t}(1 - KX_t) \tag{12.16}$$

$$\frac{\partial Z_t}{\partial w} = Y_t - Y_{t-1} \tag{12.17}$$

$$\frac{\partial Z_t}{\partial K} = \frac{\partial Z_t}{\partial Y_t} \cdot \frac{\partial Y_t}{\partial K} = we^{-KX_t}(1 - KX_t) \tag{12.18}$$

The model sensitivity function is

$$\frac{\partial Z_t}{\partial X_t} = \frac{\partial Z_t}{\partial Y_t} \cdot \frac{\partial Y_t}{\partial X_t} = -K^2 we^{-KX_t} \tag{12.19}$$

The parametric sensitivity of Equation 12.18 and the model sensitivity of Equation 12.19 are both functions of a component's sensitivity function.

The analytical sensitivity functions of Equations 12.14 to 12.19 can be computed because the model is relatively simple in structure. When the model structure becomes more complex, analytical functions cannot be computed. For complex models, the sensitivities are computed numerically. The above sensitivity functions could be approximated by $\Delta Y_t/X_t$, $\Delta Z_t/\Delta Y_t$, $\Delta Y_t/\Delta K_t$, $\Delta Z_t/\Delta w_t$, $\Delta Z_t/\Delta K_t$, and $\Delta Z_t/\Delta X_t$. Depending on the incremental changes ΔK, Δw, ΔX_t, and ΔY_t, the numerical approximations should closely match the analytical assessments of the sensitivities.

12.2.5 Forms of Sensitivity

Sensitivity can be expressed in three forms: absolute, relative, and deviation. The form in which sensitivity values are presented depends on the intended use. Sensitivity values computed with the definition of Equation 12.5 are in absolute form. Such values cannot be used for the comparison of parametric sensitivities because values computed using Equation 12.5 are not invariant to the dimensions of either factor O or F_i. If F_1 and F_2 have different units, then $\partial O/\partial F_1$ and $\partial O/\partial F_2$ would have different units and could, therefore, not be legitimately compared.

Dividing the numerator of Equation 12.5 by O_0 and the denominator by F_i provides an estimate of the relative change in O with respect to a relative change in F_i:

$$R_s = \frac{\partial O/O_0}{\partial F_i/F_i} = \left(\frac{\partial O}{\partial F_i}\right)\frac{F_i}{O_0} \tag{12.20}$$

Equation 12.20 is the mathematical definition of relative sensitivity. It could also computed as a numerical approximation:

$$R_s = \frac{\Delta O/O_0}{\Delta F_i/F_i} \tag{12.21}$$

Relative sensitivity values are invariant to the dimensions of O and F_i and thus provide a valid means for comparing factor sensitivities.

Deviation sensitivity is a third form of sensitivity. It is presented differently than either absolute or relative sensitivity. Deviation sensitivity uses Equation 12.4 as its basis. Specifically, the sensitivity is quantified as the change in the output ΔO:

$$\Delta O = \left(\frac{\partial O}{\partial F_i}\right)\Delta F_i \sim \frac{\Delta O}{\Delta F_i} \cdot \Delta F_i \tag{12.22}$$

in which the two forms on the right-hand side of Equation 12.22 differ only in that the absolute sensitivity is computed analytically or numerically. The advantage of deviation sensitivity is that it has the same units as the variable O. Thus, values of deviation sensitivity are comparable between factors F_i and F_j, and the values have a physical interpretation. For example, if O was the project cost, then differences (ΔO) to errors in two factors, such as the area of land development F_1 and the building floor area F_2, could be compared in terms of their effects on total project cost.

Example 12.2

Continuing with the model presented in Example 12.1, the relative sensitivity functions can be computed for the parameters, the components, and the model:

$$\frac{\partial Y / Y}{\partial K / K} = \left(\frac{\partial Y}{\partial K}\right)\left(\frac{K}{Y}\right) = 1 - KX \tag{12.23}$$

$$\frac{\partial Z / Z}{\partial w / w} = \left(\frac{\partial Z}{\partial w}\right)\left(\frac{w}{Z}\right) = \frac{Y_t - Y_{t-1}}{Y_t + \left(\frac{1}{w} - 1\right)Y_{t-1}} \tag{12.24}$$

$$\frac{\partial Z / Z}{\partial K / K} = \left(\frac{\partial Z}{\partial K}\right)\left(\frac{K}{Z}\right) = \frac{K(1 - KX)e^{-KX}}{Y_t + \left(\frac{1}{w} - 1\right)Y_{t-1}} \tag{12.25}$$

$$\frac{\partial Y / Y}{\partial X / X} = \left(\frac{\partial Y}{\partial X}\right)\left(\frac{X}{Y}\right) = -KX \tag{12.26}$$

$$\frac{\partial Z / Z}{\partial Y / Y} = \left(\frac{\partial Z}{\partial Y}\right)\left(\frac{Y}{Z}\right) = \frac{1}{1 + \left(\frac{1}{w} - 1\right)\frac{Y_{t-1}}{Y_t}} \tag{12.27}$$

$$\frac{\partial Z / Z}{\partial X / X} = \left(\frac{\partial Z}{\partial X}\right)\left(\frac{X}{Z}\right) = \frac{-K^2 X e^{-KX}}{Y_t + \left(\frac{1}{w} - 1\right)Y_{t-1}} \tag{12.28}$$

Each of the above functions is dimensionless, with all of the relative sensitivities being a function of time since X_t, Y_t, and Z_t are time functions.

To illustrate the relative sensitivities, assume that for a particular watershed $K = 0.25$, $w = 0.35$, and for two time periods $X_1 = 1.1$ and $X_2 = 0.46$. Using Equation 12.12, the values of Y are $Y_1 = 0.1899$ and $Y_2 = 0.2228$. If Y_0 is assumed to be 0, then $Z_1 = 0.0665$ and $Z_2 = 0.2014$. Therefore, the relative sensitivities are

$$\frac{\partial Y/Y}{\partial K/K} = \{0.725,\ 0.885\}$$

$$\frac{\partial Z/Z}{\partial w/w} = \{1.0,\ 0.0591\}$$

$$\frac{\partial Z/Z}{\partial K/K} = \{0.7288,\ 0.3438\}$$

$$\frac{\partial Y/Y}{\partial X/X} = \{-0.275,\ -0.115\}$$

$$\frac{\partial Z/Z}{\partial Y/Y} = \{1.0,\ 0.3884\}$$

$$\frac{\partial Z/Z}{\partial X/X} = \{-0.2764,\ -0.0447\}$$

Each of these shows the values for the two time periods. Some show considerable temporal variation of the relative sensitivities. In the first time period, w is more important than K, while in the second time period K is the more important of the two parameters. In both time periods, the second component is more important than the first component. Rainfall is more important to the runoff in the first time period compared with the second time period.

Example 12.3

Consider the case of a two-component model similar to Figure 12.1. The first component will represent the magnitude of the storm that produces a tidal wave, which has a magnitude defined by the second component. The transfer function of the first component is an exponential decay:

$$h_1(a,b,X) = e^{a+bX} \tag{12.29a}$$

and the transfer function of the second component is

$$h_2(c,Y) = \sin cY \tag{12.29b}$$

in which a, b, and c are parameters. The model could be used in a simulation mode in which the magnitude of the tidal wave varied with the storm magnitude X, which would be the stochastic input. Parameter c might also be a stochastic variate, such that the same tidal wave would not be the same for the same storm magnitude.

Two component sensitivity functions could be computed:

$$\frac{\partial Z}{\partial Y} = c \times \cos(cY) \tag{12.30a}$$

$$\frac{\partial Y}{\partial X} = e^{a+bX} \times \frac{\partial(a+bX)}{\partial X} = be^{a+bX} \tag{12.30b}$$

Parametric sensitivity functions can be computed for each of the three parameters:

$$\frac{\partial Y}{\partial a} = e^{a+bX} \times \frac{\partial(a+bX)}{\partial a} = e^{a+bX} \tag{12.31a}$$

$$\frac{\partial Y}{\partial b} = e^{a+bX} \times \frac{\partial(a+bX)}{\partial b} = Xe^{a+bX} \tag{12.31b}$$

$$\frac{\partial Z}{\partial c} = \cos(cY) \times \frac{\partial(cY)}{\partial c} = Y\cos(cY) \tag{12.31c}$$

$$\frac{\partial Z}{\partial a} = \frac{\partial Z}{\partial Y} \times \frac{\partial Y}{\partial a} = c \times \cos(cY) \times e^{a+bX} \tag{12.31d}$$

$$\frac{\partial Z}{\partial b} = \frac{\partial Z}{\partial Y} \times \frac{\partial Y}{\partial b} = c \times \cos(cY) \times Xe^{a+bX} \tag{12.31e}$$

The above sensitivity functions are in absolute sensitivity form. In order to compare values for a particular case, the relative sensitivity functions can be formed as follows:

$$\frac{\partial Z}{\partial Y}\left(\frac{Y}{Z}\right) = c \times \cos(cY)\left(\frac{Y}{\sin(cY)}\right) = \frac{cY}{\tan(cY)} \tag{12.32a}$$

$$\frac{\partial Y}{\partial X}\left(\frac{X}{Y}\right) = be^{a+bX}\left(\frac{X}{e^{a+bX}}\right) = bX \tag{12.32b}$$

$$\frac{\partial Y}{\partial a}\left(\frac{a}{Y}\right) = e^{a+bX}\left(\frac{a}{e^{a+bX}}\right) = a \tag{12.32c}$$

$$\frac{\partial Y}{\partial b}\left(\frac{b}{Y}\right) = Xe^{a+bX}\left(\frac{b}{e^{a+bX}}\right) = bX \tag{12.32d}$$

$$\frac{\partial Z}{\partial c}\left(\frac{c}{Z}\right) = Y\cos(cY)\left(\frac{c}{\sin cY}\right) = \frac{cY}{\tan(cY)} \tag{12.32e}$$

$$\frac{\partial Z}{\partial a}\left(\frac{a}{Z}\right) = c \times \cos(cY)e^{a+bX}\left(\frac{a}{\sin cY}\right) = \frac{ace^{a+bX}}{\tan(cY)} \tag{12.32f}$$

$$\frac{\partial Z}{\partial b}\left(\frac{b}{Z}\right) = cXe^{a+bX}\cos(cY)\left(\frac{b}{\sin cY}\right) = \frac{bcXe^{a+bX}}{\tan cY} \tag{12.32g}$$

Some interesting comparisons of these relative sensitivities can be made. Relative sensitivities can be a constant, such as that of Y with respect to a (Equation 12.32c). Relative sensitivities can vary with the magnitude of the input, such as that of Y with respect to b in Equation 12.32d. Relative sensitivities can be equal, such as $(\partial Y/Y)/(\partial X/X) = (\partial Y/Y)/(\partial b/b)$ and $(\partial Z/Z)/(\partial Y/Y) = (\partial Z/Z)/(\partial c/c)$. All of the relative sensitivities depend on one or more of the parameters (a, b, and c). As the parameters and the input values change, the importance of the parameters and inputs changes. Thus, comparisons should only be made for conditions that are meaningful with respect to the states of the system.

12.2.6 A Correspondence between Sensitivity and Correlation

The Pearson product-moment correlation coefficient R is used as a measure of the linear association between two random variables:

$$R = \frac{\sum xy}{\sqrt{\sum x^2 \sum y^2}} \tag{12.33}$$

where x is the difference between the random variable X and the mean value $\bar{X}$, and y is the difference between Y and $\bar{Y}$. The square of the correlation coefficient represents the proportion of the variance in Y that can be attributed to its linear regression on X.

Least-squares regression analysis is commonly used to derive a linear relationship between two random variables. A strong structural similarity exists between correlation analysis and regression analysis. The linear regression coefficient b can be determined from the correlation coefficient and the standard deviations of X and Y:

$$b = \frac{\sum xy}{\sum x^2} = \frac{RS_y}{S_x} \tag{12.34}$$

Equation 12.34 suggests that for values of R near 1, changes in X will produce comparatively large changes in Y. As the value of R decreases, the change in Y

decreases because b is relatively smaller. Equation 12.34 applies to the bivariate case, y versus x. In the multiple regression case with p predictors, the standardized partial regression coefficient t_i can be used as a measure of the importance or relative sensitivity of the corresponding predictor variable:

$$t_i = b_i \left(\frac{S_{xi}}{S_y} \right) \tag{12.35}$$

in which b_i is the partial regression coefficient for predictor X_i, S_{xi} is the standard deviation of predictor variable X_i, and S_y is the standard deviation of the criterion variable y. Since t_i is dimensionless, it corresponds to the relative sensitivity.

For a linear regression equation that relates two random variables X and Y, the sensitivity of the dependent variate Y to variation in the independent variate X can be determined by differentiating the regression equation with respect to X:

$$\frac{\partial Y}{\partial X} = b \tag{12.36}$$

Equation 12.36 indicates that the regression coefficient represents the rate of change in with respect to change in X. Furthermore, Equations 12.34 and 12.36 suggest that a direct correspondence exists between correlation and sensitivity. Substituting Equation 12.34 into Equation 12.36 shows that the sensitivity of Y with respect to X varies directly with the correlation coefficient. This implies that the sensitivity of a variable is a measure of its importance, as the correlation coefficient is generally considered a measure of importance. The correspondence between sensitivity and importance is supported by the results of empirical investigations with hydrologic simulation models. For example, Dawdy and O'Donnell (1965) found that the more sensitive parameters of the model were in better agreement with the true values than were the less sensitive parameters.

We generally assume that highly correlated variables are important to each other. For example, in stepwise regression analysis, the predictor variables that are finally selected are typically highly correlated with the criterion variable but independent of each other. Thus, the correlation coefficient is associated with importance. Combining Equations 12.34 and 12.36 shows that the absolute sensitivity equals the product of the correlation coefficient and the ratio of two constants. Thus, if correlation is a measure of importance, then sensitivity is a measure of importance. The advantage of the general concept of sensitivity is that it applies to all models whereas the correlation coefficient assumes a linear model.

Example 12.4

The data of Table 12.1 provide values of the runoff in the month of March (RO_M), with values of the rainfall for both March (P_M) and February (P_F) as predictor variables. A linear regression analysis of the data yields (in.)

$$RO_M = -0.0346 + 0.5880 P_M + 0.1238 P_F \tag{12.37}$$

TABLE 12.1
Selected Monthly Rainfall and Runoff, White Hollow Watershed

Year	P_M (in.)	P_F (in.)	RO_M (in.)
1935	9.74	4.11	6.15
1936	6.01	3.33	4.93
1937	1.30	5.08	1.42
1938	4.80	2.41	3.60
1939	4.15	9.64	3.54
1940	5.94	4.04	2.26
1941	2.99	0.73	0.81
1942	5.11	3.41	2.68
1943	7.06	3.89	4.68
1944	6.38	8.68	5.18
1945	1.92	6.83	2.91
1946	2.82	5.21	2.84
1947	2.51	1.78	2.02
1948	5.07	8.39	3.27
1949	4.63	3.25	3.05
1950	4.24	5.62	2.59
1951	6.38	8.56	4.66
1952	7.01	1.96	5.40
1953	4.15	5.57	2.60
1954	4.91	2.48	2.52
1955	8.18	5.72	6.09
1956	5.85	10.19	4.58
1957	2.14	5.66	2.02
1958	3.06	3.04	2.59
ΣX	116.35	119.58	82.39
$\Sigma X_1 X_j$	663.1355	589.8177	458.9312
$\Sigma X_2 X_j$		753.1048	435.9246
ΣY^2			331.4729

The model explained 78.4% (R^2) of the variation in RO_M, which suggests that estimates of RO_M should be reasonably accurate. Taking the derivatives of the prediction equation results in the following values of absolute sensitivity:

$$\frac{\partial RO_M}{\partial P_M} = 0.5880 \tag{12.38}$$

and

$$\frac{\partial RO_M}{\partial P_F} = 0.1238 \tag{12.39}$$

The mean values of P_M and P_F are 4.85 in. and 4.98 in., respectively, and the standard deviations are 2.076 in. and 2.615 in., respectively.

The deviation sensitivities can be used to evaluate the effect of errors in the rainfall measurements. Several sources of errors should be considered. For example, errors in estimates of RO_M can occur because of either instrument errors in measurement of P_M and P_F or the inability of the gage to reflect the rainfall over the drainage area. If either of these sources of error amounts to 5% of the estimate of P_M or P_F, the error in RO_M that results from the error in the rainfall estimate can be computed from the deviation sensitivity and the linear sensitivity equation (Equation 12.4). For example, if the rainfall is 5 in. in any one month, the error in the resulting runoff due to a 5% measurement error would be

$$\Delta RO_M = \frac{\partial RO_M}{\partial P_M}\Delta P_M = 0.5880[(0.05)(5)] = 0.147 \text{ in.} \tag{12.40a}$$

$$\Delta RO_M = \frac{\partial RO_M}{\partial P_F}\Delta P_F = 0.1238[0.05(5)] = 0.031 \text{ in.} \tag{12.40b}$$

If 5 in. of rain fell in both February and March, the predicted March runoff would be 3.52 in. Thus, the error in the March runoff resulting from the 5% error in the March rainfall would be about 4% of the runoff but only 0.9% of the March runoff for the 5% error in February rainfall. The error for February rainfall is relatively insignificant compared to the error for the March rainfall.

The relative sensitivities of the inputs can also be computed. At the mean values, the relative sensitivity of RO_M to P_M and P_F would be, respectively,

$$R_{SM} = \frac{\partial RO_M}{\partial P_M}\frac{\bar{P}_M}{\overline{RO}_M} = 0.5880\left(\frac{4.85}{3.43}\right) = 0.831 \tag{12.41a}$$

$$R_{SF} = \frac{\partial RO_M}{\partial P_F}\frac{\bar{P}_F}{\overline{RO}_M} = 0.1238\left(\frac{4.98}{3.43}\right) = 0.180 \tag{12.41b}$$

The relative sensitivity values indicate that the rainfall during March has a greater effect on the runoff in March than on the rainfall in February. This seems very rational.

The bivariate correlations of RO_M versus P_M and P_A are 0.857 and 0.291, respectively. The relationship of Equation 12.34 does not apply because it is for the bivariate case, whereas the two regression coefficients are based on multiple regression analysis. However, the same principle applies; specifically, that correlation is a measure of importance, and in this case, the two correlation coefficients agreed with the relative sensitivities of Equations 12.40, with the rainfall in March being more important than the rainfall in February.

TABLE 12.2
Summary of Multiple Regression Analysis of French Board River Data Matrix

Variable	t	b	R_{xy}
X_1 Precipitation t	0.758	0.3699	0.638
X_2 Precipitation $t-1$	0.432	0.2108	0.292
X_3 Precipitation $t-2$	0.277	0.1348	0.105
X_4 Precipitation $t-3$	0.172	0.0838	–0.058
X_5 Runoff $t-1$	–0.174	–0.1731	0.301
X_6 Temperature t	–0.327	–0.0294	–0.202
X_7 Temperature $t-1$	–0.350	–0.0317	–0.336
X_8 Temperature $t-2$	0.002	0.0001445	–0.321
X_9 Evaporation t	0.512	0.3905	–0.123
X_{10} Evaporation $t-1$	–0.327	–0.2508	–0.303
X_{11} Evaporation $t-2$	–0.010	–0.007894	–0.345
X_{12} Wind speed t	0.021	0.008474	0.181
		Intercept = 1.827	

Example 12.5

Monthly water yield (in.) data for the French Broad River serves as the criterion variable and is regressed on 12 predictor variables, all of which are monthly values of rainfall, water yield, temperature, evaporation, or wind speed for the same month t or previous months $t-1$, $t-2$, or $t-3$. The variables are given in column 1 of Table 12.2.

The resulting regression coefficients (b), standardized partial regression coefficients (t), and bivariate predictor-criterion correlation coefficients (R) are given in Table 12.2. The correlation coefficients indicate that X_1 is the most important, while X_2, X_5, X_7, X_8, X_{10}, and X_{11} are much less important and each is of equal importance. The t values indicate that X_1 is most important with X_9 and X_2 less important, but more important than X_7, X_6, X_{10}, and X_3. Thus, the two methods provide different assessments of the relative importance of the predictor variables. If the regression model is used to represent the system, then the t values are a better measure of the importance than the R values.

12.3 TIME VARIATION OF SENSITIVITY

Many hydrologic design models compute a runoff hydrograph, which is the variation of discharge with time. Watershed and reservoir storages also vary with time. The temporal variation of discharge and storage is a key factor in hydrologic decisions. Where time is a factor in a model, the sensitivity functions can vary with time. Given that sensitivity is a measure of importance, it then follows that importance varies with time. For a hydrologic model, baseflow discharge will be sensitive to some model parameters and inputs, while peak discharges will be more sensitive to other parameters. In evaluating the rationality of a model, it is important to ensure that the temporal variations of the importance of model outputs are rational.

Example 12.4 provided constant values of both the deviation and relative sensitivities that were not time dependent. Sensitivity will actually vary depending on the values selected for the variables. This general statement implies that the sensitivity of a hydrologic model to errors in the input will vary with the state of the system (i.e., the values of the coefficients and variables). This is easily illustrated using a three-coefficient model for estimating the mean monthly temperature:

$$T = b_0 + b_1 \sin \omega M + \cos \omega M \tag{12.42}$$

in which M is time, such as the month, and ω is a scaling factor. Assume that least-squares fitting provided the following values for the fitting coefficients:

$$T = 55.97 - 10.05 \sin \omega M - 17.64 \cos \omega M \tag{12.43}$$

Taking the derivatives of the temperature with respect to the three coefficients yields

$$\frac{\partial T}{\partial b_0} = 1 \tag{12.44a}$$

$$\frac{\partial T}{\partial b_1} = \sin \omega M \tag{12.44b}$$

$$\frac{\partial T}{\partial b_2} = \cos \omega M \tag{12.44c}$$

These derivatives indicate that the sensitivity of T is not necessarily a constant. Error in b_0 is linearly related to T since a unit change in b_0 causes a unit change in T, which is evident from the sensitivity of 1 for $\partial T/\partial b_0$. However, the sensitivities of T to b_1 and b_2 are not constant and will vary with the month. When ωM equals either 0° or 180°, error in b_1 will not cause any error in T, while the effect of error in b_2 will be at its greatest. Similarly, when ωM equals either 90° or 270°, the effect of error in b_1 will be greatest while error in b_2 will have no effect on T.

This example illustrates that model sensitivity is not always constant. For the time-dependent model of this example, the sensitivity varied with time, which would be true of time-dependent rainfall-runoff models. While sensitivity analyses of hydrologic models are usually performed using the mean values of both the coefficients and the input variables, it is quite likely that the sensitivity estimates should be computed for the watershed conditions at which a design will be made. For example, if a hydrologic model is being used for design at extreme flood conditions, such as at the 100-year rainfall event, the sensitivity analysis should be performed at the conditions that would exist at that time rather than at average watershed conditions. A high soil-moisture state would be more appropriate than the average soil-moisture state. One can reasonably expect the relative sensitivities, and therefore the relative importance of the model elements, to vary with the state of the watershed. In summary, while the mathematical development of sensitivity was presented in its basic form, it is important to recognize that sensitivity analyses of hydrologic models must consider the time and spatial characteristics of watershed conditions.

12.4 SENSITIVITY IN MODEL FORMULATION

The general definition of sensitivity, Equation 12.5, indicates that the sensitivity of one factor depends, in the general case, on the magnitude of all system factors. For a dynamic system that is not in steady state, such as a watershed, the output and sensitivity functions will also be time dependent. The unit response function proposed by Nash (1957) is used here to demonstrate the importance of sensitivity in model formulation. The gamma distribution proposed by Nash (1957) has been used by others as a conceptual representation of the response of a watershed (Nash, 1959; Sarma, Delleur, and Rao, 1969):

$$h = h(n,\ K;\ t) = \frac{t^{n-1} e^{-t/K}}{K^n \Gamma(n)} \tag{12.45}$$

where n represents the number of equivalent reservoirs whose hydrologic output is the same as that of the drainage basin, K is the constant of proportionality relating outflow rate and storage, t represents the time from the impulse input, and $\Gamma(n)$ is the gamma function with argument n. Equation 12.6 can be used to derive the sensitivity functions for the parameters n and K. The sensitivity function for n is

$$S_n = \frac{\partial h}{\partial n} = \frac{t^{n-1} e^{-t/K} [\ln_e (t/K) - \Gamma'(n)/\Gamma(n)]}{K^n \Gamma(n)} \tag{12.46a}$$

where $\Gamma'(n)$ is the derivative of $\Gamma(n)$ with respect to n. The sensitivity function for K is

$$S_K = \frac{\partial h}{\partial K} = \frac{t^{n-1} e^{-t/K} (t - nK)}{K^{n+2} \Gamma(n)} \tag{12.46b}$$

It is evident from Equations 12.46a and 12.46b that both sensitivity functions depend on the values of n and K and that the sensitivity of both parameters varies with time t. In Figure 12.2, both the unit response function and the two parametric sensitivity functions are plotted as a function of time. The fact that the sensitivity function of K is larger than that of n does not imply that K is more important because these functions have different dimensions and thus cannot be compared. However, the sensitivity functions for two parameters indicate that the parameters have their greatest influence on the rising limb of the unit response function. Also, at the peak of the hydrograph, both sensitivity functions have declined to approximately one-third of their maximum sensitivity. Although such information concerning the effect

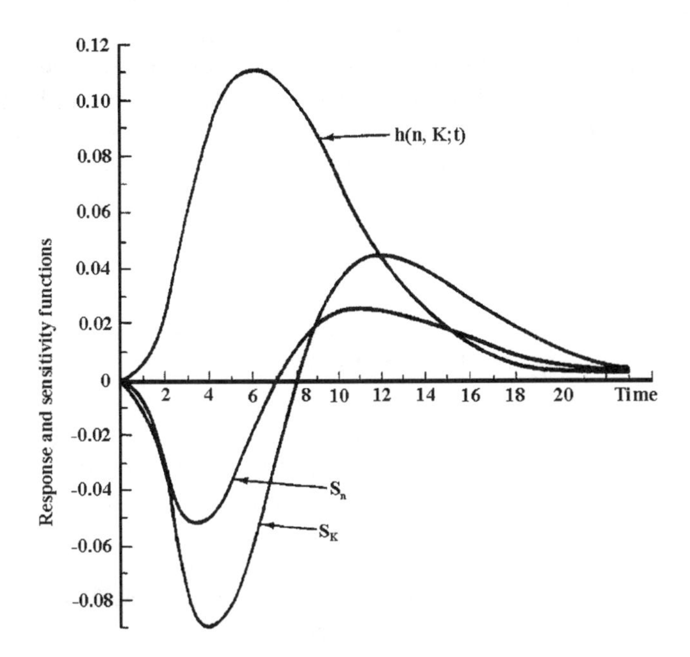

FIGURE 12.2 Nash model sensitivity: $n = 4.0$, $K = 2.0$.

of parameter variation is only qualitative, it can be of considerable value in the process of deriving relationships between the optimal parameter values and watershed or storm properties. It appears that the values of n and K are more sensitive to watershed processes that affect the rising limb of the hydrograph than to processes influential at the peak area or the recession. Thus, models that estimate values of n and K may be more effective if they use channel properties rather than soil properties.

Sensitivity functions can also suggest interrelationships between parameters. Generally speaking, it would be preferable to formulate a model with parameters that are not highly correlated. Parameters such as n and K that are highly intercorrelated may confound the optimization technique and thus prevent the identification of the true values. Failure to identify the true parameter values would adversely affect relationships derived between model parameters and physical characteristics of the watershed. The effect of parameter interaction on relationships between optimal model parameters and watershed properties can be reduced if the similarities of the parameter sensitivity functions are considered when formulating the model structure. The functions shown in Figure 12.2 show that n and K are highly correlated and better accuracy may be achieved using a model in which the parameters are less highly correlated.

Example 12.6

As a watershed urbanizes, the measured flood peaks tend to increase for a given rainfall. This creates a nonstationary flood series that needs to be adjusted before design discharges can be estimated with the measured flood series. It has been shown that urbanization has a greater effect on the more frequent events (e.g., 2-year discharge) than on the low frequency events (e.g., 100-year discharge). The effect of urbanization also depends on the amount of impervious cover. To develop a method of adjusting measured flood discharges for urbanization, it is necessary to understand the sensitivity of peak discharges to both the amount of urbanization and the magnitude of a discharge. This sensitivity led to the development of the adjustment method of Chapter 5.

Numerous reports have presented models that relate imperviousness and peak discharge. Sarma et al. (1969) provided the following equation:

$$q_p = 484.1\, A^{0.723}\, (1+I)^{1.516}\, P_E^{1.113}\, T_R^{-0.403} \tag{12.47}$$

in which A is the drainage area (mi^2), I is the fraction of imperviousness, P_E is the depth of rainfall excess (in.), T_R is the duration of rainfall excess (hr), and q_p is the peak discharge (ft^3/sec). Equation 12.47 yields a relative sensitivity of 1.5 for peak discharge as a function of imperviousness, which means that a 1% change in I yields a 1.5% change in q_p.

The 1975 version of the Soil Conservation Service (SCS) TR-55 provided a graph that related the percentage of imperviousness and a peak factor, which was a multiplication value to apply to a rural watershed discharge. The peak factor varied with the runoff curve number and varied according to values as high as 1.8. Overall, the graph suggested a relative sensitivity as high as 2.6, but generally was lower.

Rantz (1971) provided a method based on simulation results that used the percentages of both basin developed and channel sewered as predictors of the ratio of the postdevelopment to predevelopment peak discharges. The values were as large as 4 for the 2-year event to as large as 2.5 for the 100-year event. Graphs were also provided for the 5-, 10-, and 25-, and 50-year events.

Other graphs and equations that provide similar information are available in the literature. These were used to develop the adjustment graph of Figure 5.9. Since the methods did not indicate the same value of the peak index ratio for any level of imperviousness and since the methods did not always use the same predictor variables, the assessed sensitivities tended to be approximate measures of central tendencies of the different effects. The "average" sensitivities for the different levels of imperviousness and exceedance probability were plotted and then smoothed to produce Figure 5.9.

12.5 SENSITIVITY AND DATA ERROR ANALYSIS

Data used for model verification or for estimating values of model parameters invariably contain error. The magnitude and distribution of error often cannot be evaluated. For example, airspeed is often included in evaporation models as a measure of air instability. But whether airspeed measurements quantitatively measure the effect of air instability on evaporation rates is difficult to assess. However, the effects of data error from other sources can be quantitatively evaluated. For example, instrument specifications supplied by manufacturers can be used to estimate the potential error in an observed measurement due to the inaccuracy of the recording instrument. Radiometers that are expected to measure solar radiation R to within 0.5% of the true value are commercially available. For such instruments, the expected error E_R can be estimated by

$$E_R = 0.005R \tag{12.48}$$

The sensitivity of evaporation E computed with the Penman model (Penman, 1948, 1956) to change in radiation is estimated by

$$S_R = \frac{\partial E}{\partial R} = \frac{\Delta}{\Delta + \gamma} \tag{12.49}$$

where Δ is the slope of the saturation vapor pressure curve at air temperature and γ is the psychrometric constant. When the radiation is expressed in equivalent inches of evaporation, the error E_E in an estimate of daily evaporation from instrument insensitivity can be estimated by the product

$$E_E = E_R(S_R) \tag{12.50}$$

Equation 12.50 is a specific example of the deviation sensitivity of Equation 12.22. For a temperature of 70°F and incoming solar radiation of 500 ly/day the expected error in evaporation would be approximately 0.01 in., or 4% of the total evaporation. This simplified example indicates the potential value of using an analytical sensitivity function to estimate the effect of data error. For complex models, analytical solutions can not be developed. The method of parameter perturbation would require considerably more computer time to estimate the effect of data error on complex models.

Reproducibility is measure of how well different designers can estimate the value of a property at a site with the same input information and procedure. The lack of reproducibility is a source of random error. Sensitivity and error analyses are a means of examining the effect of the lack of reproducibility on a hydrologic design variable, such as peak discharge.

Example 12.7

A simple model for computing peak discharge can be used to demonstrate the use of sensitivity and error analysis. The model gives the peak discharge (q_p) as a function of the unit peak discharge (q_u) in cfs/mi² per in. of runoff, the drainage area (A) in square miles, and the depth of runoff (Q) in in.:

$$q_p = q_u A Q \tag{12.51}$$

The unit peak discharge is a function of the time of concentration (t_c) in hours; for t_c greater than 0.7 hr, the relationship is

$$q_u = 321.48\, t_c^{-0.74946} \tag{12.52}$$

The runoff depth can be estimated as

$$Q = \frac{(P - 0.2S)^2}{P + 0.8S} \tag{12.53}$$

in which P is the 24-hour rainfall depth in inches for the selected exceedance probability, and S is the maximum retention, which is given by

$$S = \frac{1000}{CN} - 10 \tag{12.54}$$

in which CN is the runoff curve number. The SCS lag method can be used to estimate the time of concentration (t_c),

$$t_c = \frac{1.67 HL^{0.8}(S+1)^{0.7}}{1900Y^{0.5}} \tag{12.55}$$

in which HL is the hydraulic length in feet and Y is the watershed slope in percent. The hydraulic length can be computed from the drainage area (A_a) in acres:

$$HL = 209A_a^{0.6} \tag{12.56}$$

Using the equations above, the following sensitivity functions can be derived for the input variables of P, A, t_c, and CN:

$$\frac{\partial q_p}{\partial P} = q_p\left(\frac{2}{P - 0.2S} - \frac{1}{P + 0.8S}\right) \tag{12.57}$$

$$\frac{\partial q_p}{\partial A_a} = \frac{q_p}{A_a}(1 + 0.48b) \tag{12.58}$$

$$\frac{\partial q_p}{\partial t_c} = \frac{bq_p}{t_c} \tag{12.59}$$

$$\frac{\partial q_p}{\partial CN} = \frac{100q_p}{CN^2}\left(\frac{-7b}{S+1} + \frac{4}{P - 0.2S} + \frac{8}{P + 0.8S}\right) \tag{12.60}$$

in which b is the exponent of Equation 12.52. The sensitivity functions of Equations 12.57 to 12.60 are a function of the peak discharge (q_p) and the input variables (A_a, CN, and P). These sensitivity functions can be used to compute relative sensitivity functions, as follows:

$$R_p = \frac{\partial q_p}{\partial P}\frac{P}{q_p} = P\left(\frac{2}{P-0.2S} - \frac{1}{P+0.8S}\right) \quad (12.61)$$

$$R_{A_a} = \frac{\partial q_p}{\partial A_a}\frac{A_a}{q_p} = 1 + 0.48b \quad (12.62)$$

$$R_{t_c} = \frac{\partial q_p}{\partial t_c}\frac{t_c}{q_p} = b \quad (12.63)$$

$$R_{CN} = \frac{\partial q_p}{\partial CN}\frac{CN}{q_p} = \frac{100}{CN}\left(\frac{-7b}{S+1} + \frac{4}{P-0.2S} + \frac{8}{P+0.8S}\right) \quad (12.64)$$

in which R_p, R_A, R_{tc}, and R_{CN} are the relative sensitivity functions for the rainfall depth, the drainage area, the time of concentration, and the curve number, respectively. These are functions only of the rainfall and curve number, with R_A and R_t being constant for a given value of b. They will vary with P and CN. The relative sensitivity functions were computed for rainfall depths of 3 to 5 in. and CN values of 65 and 80, with the results plotted in Figure 12.3. The curves suggest that the

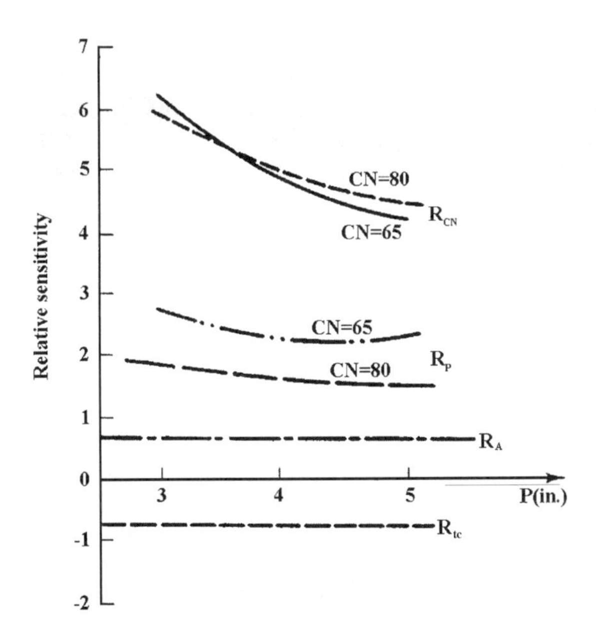

FIGURE 12.3 Relative sensitivity plots for the curve number (R_{CN}), rainfall (R_p), area (R_A), and time of concentration (R_{tc}) as a function of rainfall (P).

CN is the most important input, with the relative sensitivity not changing with a change in *CN*. Rainfall is the second most important variable, with the effect of *CN* on the relative sensitivity much greater than for the relative sensitivity of the *CN*. Drainage area and the time of concentration have similar relative sensitivities. It should be emphasized that the plots of Figure 12.3 do not indicate that error in the *CN* will have the greatest effect, only the relative error will have the greatest effect when the relative errors are equal. For unequal relative errors, the relative error in the computed discharge will vary according to the relative sensitivity functions in Figure 12.3.

The sensitivity functions of Equations 12.57 to 12.60 can be used with Equation 12.22 to perform an error analysis,

$$\Delta R = \frac{\partial q_p}{\partial X} \Delta X \tag{12.65}$$

in which X is any one of the four input variables, $\partial q_p/\partial X$ is one of the derivatives of Equations 12.57 to 12.60, and ΔX is a measure of the error. The U.S. Water Research Council (WRC) report (1981) reported coefficients of variation of 0.03, 0.02, 0.70, and 0.06 for P, A, t_c, and CN, respectively. The coefficient of variation can be converted to an average error by multiplying the coefficient of variation (C_{VX}) by the mean ($\overline{X}$) of the variable X:

$$\Delta X = C_{VX} \overline{X} \tag{12.66}$$

The effect of reproducibility errors on peak discharge can be computed with the equations above. For example, if we assume that $A_a = 500$ acres, $P = 5$ in., and $CN = 80$, we can compute $t_c = 1.498$ hr from Equation 12.55 and the derivatives of Equations 12.57 to 12.60. The values of the reproducibility error using Equation 12.65 are

$$\Delta R_P = \frac{\partial q_p}{\partial P} \Delta P = 161.9[0.03(5)] = 24 \text{ cfs} \tag{12.67}$$

$$\Delta R_A = \frac{\partial q_p}{\partial A_a} \Delta A_a = 0.687[0.02(500)] = 7 \text{ cfs} \tag{12.68}$$

$$\Delta R_{t_c} = \frac{\partial q_p}{\partial t_c} \Delta t_c = -268.6[0.7(1.498)] = -282 \text{ cfs} \tag{12.69}$$

$$\Delta R_{CN} = \frac{\partial q_p}{\partial CN} \Delta CN = 29.6[0.06(80)] = 142 \text{ cfs} \tag{12.70}$$

The values of Equations 12.67 to 12.70 indicate that the lack of reproducibility in the drainage area and rainfall cause little error in peak discharge but a very large error due to the lack of reproducibility of t_c and CN. These results suggest that drainage and stormwater management policies should be written to minimize the problem associated with errors in t_c and CN.

Example 12.8

Manning's equation is widely used for estimating the velocity (V) of flow, such as in the calculation of the time of concentration. It requires three inputs: roughness coefficient (n), hydraulic radius (R_h), and slope of the flow path (S):

$$V = \frac{1.49}{n} R_h^{2/3} S^{1/2} \tag{12.71}$$

Each of the inputs is the value of a random variable with errors that vary depending on the application. Of course, error in the inputs is transferred to the computed velocity and, subsequently, to other computed values based on the velocity.

Tables of the roughness coefficient generally give a range of values rather than a single value. The range reflects the inaccuracy of the value and varies with the land use. For example, the range for an asphalt pavement is much smaller than the range for a vegetated earthen stream. It is not unusual for the variation to be ±20% of the median value.

The slope of the flow path is computed using elevation estimates for the two ends of the flow path and the distance between the two points. If the elevations are taken from a map, it may be necessary to interpolate between topographic lines, thus introducing inaccuracy. The length of the stream must be measured from a map and error is associated with measurement. Also, the map may not reflect the full extent of the meandering, so the length measured from the map will underestimate the true length. An error of 15% due to both of these sources of uncertainty is not unreasonable.

The hydraulic radius is generally taken as the depth of flow but this is an approximation. Also, the hydraulic radius will vary over the flow length, assuming an increase in flow as the flow moves downstream. An uncertainty of ±10% seems reasonable.

The absolute sensitivity functions for the three inputs of Equation 12.71 can be computed by applying Equation 12.5:

$$\frac{\partial V}{\partial n} = \frac{-1.49\, R_h^{2/3}\, S^{1/2}}{n^2} = \frac{-V}{n} \tag{12.72a}$$

$$\frac{\partial V}{\partial S} = \frac{1}{2}\left(\frac{1.49}{n}\right) R_h^{2/3}\, S^{-1/2} = \frac{V}{2S} \tag{12.72b}$$

$$\frac{\partial V}{\partial R_h} = \frac{2}{3}\left(\frac{1.49}{n}\right) R_h^{-1/3} S^{1/2} = \frac{2V}{3R_h} \tag{12.72c}$$

The relative sensitivities equal the exponents of the variables, that is, 1, 0.5, and 0.67 for n, S, and R_h, respectively. The deviation sensitivities are useful for assessing the effects of errors in the inputs.

Equations 12.72 indicate that the absolute sensitivities and, therefore, the deviation sensitivities depend on the values of the three inputs. Thus, the sensitivities and error effects depend on the magnitudes of the three input variables. For illustrating the calculation of the range of errors in velocity that can result from errors in the inputs, consider the case where the R_h is 2.2 ft, n is 0.045, and S is 0.015 ft/ft. This gives a velocity (Equation 12.71) of 6.86 ft/sec. The expected errors in the three inputs are

$$\Delta n = 0.20(0.045) = 0.009 \tag{12.73a}$$

$$\Delta S = 0.15(0.015) = 0.00225 \tag{12.73b}$$

$$\Delta R_h = 0.10(2.2) = 0.22 \text{ ft} \tag{12.73c}$$

These expected errors result in deviation sensitivities of

$$\Delta V_n = \frac{\partial V}{\partial n} \cdot \Delta n = \frac{-6.86}{0.045}(0.009) = -1.372 \text{ ft/sec} \tag{12.74a}$$

$$\Delta V_S = \frac{\partial V}{\partial S} \cdot \Delta S = \frac{6.86(0.00225)}{2(0.015)} = 0.515 \text{ ft/sec} \tag{12.74b}$$

$$\Delta VR_h = \frac{\partial V}{\partial R_h} \cdot \Delta R_h = \frac{2(6.86)}{3(2.2)} = 0.457 \text{ ft/sec} \tag{12.74c}$$

These deviation sensitivities are somewhat different from the relative sensitivities. The relative sensitivities indicate that n is twice as important as S and one-third more important than R_h. However, because of the higher percentage of error in n, the deviation sensitivity indicates that the error in n causes error in the velocity that is more than 2.5 times greater than the error due to either the slope or the hydraulic radius. Also, even though the percentage error in S is 50% greater than the percentage error in R_h (i.e., 15% vs. 10%), the effects indicated by the deviation sensitivities show that the larger error in slope causes approximately the same error in velocity as caused by the error in R_h. This occurs because of the larger relative sensitivity of R_h.

This example illustrates that both the relative sensitivity and deviation sensitivity provide useful information, but they have different uses. Relative sensitivity is a useful measure of relative importance and where the uncertainties are unknown or equal. Deviation sensitivity analysis is useful for error analyses.

12.6 SENSITIVITY OF MODEL COEFFICIENTS

The predicted output of a model depends on the model structure, the values of the predictor (input) variables, and the coefficients. Parametric and component sensitivity analyses are useful in evaluating model structure. The relative sensitivity of inputs is useful for comparing the effects of inputs. Model coefficients are frequently derived empirically, and thus, their values are subject to error and dependent on the range of the input variables. Where coefficients are related to specific variables, sensitivity analyses of the coefficients can provide an additional assessment of the importance of model inputs.

Example 12.9

Peak discharge equations are often placed in the form of a multiple-predictor power model:

$$q_p = b_0\, x_1^{b_1}\, x_2^{b_2} \cdots x_p^{b_p} \tag{12.75}$$

The following are examples for the case where the equations require two predictor variables (Dillow, 1996):

$$\hat{q}_{p_1} = 451\, A^{0.635}\, (F+10)^{-0.266} \tag{12.76a}$$

$$\hat{q}_{p_2} = 1410\, A^{0.761}\, (F+10)^{-0.782} \tag{12.76b}$$

in which $\hat{q}_{p_1}$ and $\hat{q}_{p_2}$ are the estimated 2-year peak discharges for the Piedmont and Western Coastal Plain regions of Maryland (ft³/sec), A is the drainage area (mi²), and F is the percent forest cover.

General expressions for the sensitivity functions for the coefficients $b_i (i = 0, 1, 2 \ldots, p)$ of the multiple-predictor power model of Equation 12.75 are

$$\frac{\partial q_p}{\partial b_0} = X_1^{b_1}\, X_2^{b_2} \cdots X_p^{b_p} = \hat{q}_p / b_0 \tag{12.77a}$$

$$\frac{\partial q_p}{\partial b_i} = b_0\, X_1^{b_1} \cdots X_p^{b_p}\ \ln_e\, X_i = \hat{q}_p\ \ln_e\, X_i \qquad \text{for any } i \tag{12.77b}$$

$$\frac{\partial V}{\partial R_h} = \frac{2}{3}\left(\frac{1.}{}\right. \tag{12.77c}$$

$$\frac{\partial q_p / q_p}{\partial b_i / b_i} = \left(\frac{\partial q_p}{\partial b_i}\right)\left(\frac{b_i}{q_p}\right) = b_i\ \ln_e\, X_i \tag{12.77d}$$

$$\Delta q_p = \frac{\partial q_p}{\partial b_0}\,\Delta b_0 = \frac{q_p \Delta b_0}{b_0} \tag{12.77e}$$

$$\Delta q_p = \frac{\partial q_p}{\partial b_i}\,\Delta b_i = (q_p\ \ln_e\, X_i)\,\Delta b_i \tag{12.77f}$$

Using these general forms, the importance of the coefficients of Equations 12.75 can be assessed.

Consider, for example, the case of 50-mi² watersheds with 20% forest cover in the two regions. For the Piedmont region, the discharge is 2188 ft³/sec. For the Western Coastal Plain, the discharge is 1937 ft³/sec. For the Piedmont region, the absolute sensitivity functions are

$$\frac{\partial q_{p_1}}{\partial b_0} = \frac{2188}{451} = 4.851 \tag{12.78a}$$

$$\frac{\partial q_{p_1}}{\partial b_1} = 2188 \ \ln_e \ (50) = 8560 \tag{12.78b}$$

$$\frac{\partial q_{p_1}}{\partial b_2} = 2188 \ \ln_e \ (20+10) = 7442 \tag{12.78c}$$

The relative sensitivities are

$$\frac{\partial q_p / q_p}{\partial b_0 / b_0} = 1 \tag{12.79a}$$

$$\frac{\partial q_p / q_p}{\partial b_1 / b_1} = 0.635 \ \ln_e \ (50) = 2.48 \tag{12.79b}$$

$$\frac{\partial q_p / q_p}{\partial b_2 / b_2} = -0.266 \ \ln_e \ (30) = -0.946 \tag{12.79c}$$

The relative sensitivities indicate that the coefficient for the drainage area has a significantly greater relative effect on variation of the computed peak discharge than either the intercept coefficient or the coefficient for the forest cover term. The negative sign on the relative sensitivity of b_2 only suggests that increasing the magnitude of b_2 would decrease the peak discharge. It is the magnitude of the relative sensitivity that suggests importance. For example, two quantities, one with a relative sensitivity of +1 and the other with a value of −1, are equally important.

The values for the Piedmont region can be compared with those for the Western Coastal Plain. Table 12.3 summarizes the calculations. For the 50-mi² watershed, with a peak discharge of 1937 ft³/sec, the relative sensitivities of b_0, b_1, and b_2 are 1, 2.98, and −2.56, respectively. These indicate that the two slope coefficients are similar in importance, but have comparatively greater importance than the intercept coefficient b_0.

The sensitivities for 5-mi² watersheds with 50% forest cover are also given in Table 12.3. The results suggest that the coefficients are responsible for different effects than shown for the larger 50-mi² watershed. Specifically, in the Piedmont

TABLE 12.3
Sensitivity Analysis of Peak Discharge Equations

Region	Area (mi^2)	Forest Cover (%)	q_p	$\partial q_p/\partial b_0$	$\partial q_p/\partial b_1$	$\partial q_p/\partial b_2$	$\left(\frac{dq_p/q_p}{db_0/b_0}\right)$	$\left(\frac{dq_p/q_p}{db_1/b_1}\right)$	$\left(\frac{dq_p/q_p}{db_2/b_2}\right)$
Piedmont	50	20	2188	4.851	8560	7440	1	2.48	–0.946
	5	50	421.7	0.9350	678	1727	1	1.02	–1.09
Western Coastal Plain	50	20	1937	1.374	7578	6588	1	2.98	–2.56
	5	50	195.3	0.1385	314	800	1	1.23	–3.20

region, the three coefficients are equally important, with relative sensitivities from 1 to 1.09 (absolute values). However, for the Western Coastal Plain the coefficient for the forest cover dominates in importance.

Example 12.10

The design of irrigation canals and stream restoration projects are two examples of watershed change where streambed erosion is a significant concern. Numerous methods have been developed for estimating erosion. One class of methods assumes that significant channel degradation does not begin until the critical tractive stress (τ_c) is exceeded. Four equations that have been fit to data to compute the τ_c follow:

Schoklitsch (1914): $$\tau_c = [0.201(\gamma_s - \gamma)\lambda d^3]^{0.5} \tag{12.80}$$

Tiffany and Bentzel (1935): $$\tau_c = 0.029[(\gamma_s - \gamma)(d/M)]^{0.5} \tag{12.81}$$

Leliavsky (1955): $$\tau_c = 166d \tag{12.82}$$

Krey (1925): $$\tau_c = 0.000285(\gamma_s - \gamma)d^{1/3} \quad \text{for } 0.1 \text{ mm} < d < 3 \text{ mm} \tag{12.83a}$$

$$\tau_c = 0.076(\gamma_s - \gamma)d \quad \text{for } 6 \text{ mm} < d \tag{12.83b}$$

in which γ_s and γ are the specific weights of the soil and water, respectively; λ is a grain shape coefficient (λ = 1 for spheres, λ = 4.4 for flat grains); M is a grain distribution modulus; d is the mean grain diameter (mm); and τ_c is the critical tractive stress (kg/m^2).

Recognizing that the coefficients of these equations are not highly accurate and the difficulty of obtaining soil of a homogeneous gradation, the sensitivity of these equations to the diameter is of interest. The following are the sensitivity function $\partial\tau_c/\partial d$ of τ_c with respect to d for the four models given above:

Schoklitsch (1914): $$1.5[0.201(\gamma_s - \gamma)\lambda]^{0.5} d^{0.5} \tag{12.84}$$

Tiffany and Bentzel (1935): $$0.0145[(\gamma_s - \gamma)/M]^{0.5} d^{-0.5} \tag{12.85}$$

Leliavsky (1955): $$166 \tag{12.86}$$

Krey (1925): $$0.000095(\gamma_s - \gamma)d^{-2/3} \quad \text{for } 0.1 \text{ mm} < d < 3 \text{ mm} \tag{12.87a}$$

$$0.076(\gamma_s - \gamma) \quad \text{for } 6 \text{ mm} < d \tag{12.87b}$$

Equations 12.84 to 12.87 indicate considerable variation in the effect of grain diameter. Schoklitsch's equation indicates that the sensitivity is directly related to *d*. Tiffany's and the first part of Krey's equations indicate a negative relationship between sensitivity and *d*. Leliavsky's and the second part of Krey's equations suggest that the sensitivity is not a function of grain diameter. Which of these equations is selected for use in estimating critical shear stress may depend on the apparent rationality of the sensitivity function. If, for example, the critical shear stress should be less sensitive to *d* as *d* increases, then Equations 12.81 and 12.83 may be considered as being more rational.

Sensitivity analyses to the coefficients may also be useful since the equations are empirical and provide widely different estimates of τ_c. In addition, the relative sensitivity functions of the multivariable equations could be used to assess model rationality.

12.7 WATERSHED CHANGE

The second word of the term *watershed change* suggests a clear relationship with the term *sensitivity analysis*. The latter by definition is a measure of change. Therefore, sensitivity analysis is an especially important modeling tool when dealing with watershed change.

12.7.1 Sensitivity in Modeling Change

The four phases of modeling have been emphasized: conceptualization, formulation, calibration, and verification. When conceptualizing a model that shows watershed change, it is necessary to identify the variables that can influence the value of the criterion variable. For example, a time-of-concentration model that was to be used for evaluating change in land use would have to include roughness and length, but rainfall and slope would be less important in that watershed. However, if the effect of climate change on time of concentration was of interest, then certainly a rainfall characteristic would need to be included. In order for a model to accurately show realistic effects, it must include variables that are directly related to the change. This does not imply that variables not directly related to the change can be omitted. It is well known in modeling that when significant variables are omitted, fitting coefficients related to other predictor variables will attempt to compensate for their absence. This distorts the sensitivity of the model to the variables that are included in the model. For example, a time of concentration model may not need the rainfall variable or slope to show watershed change, but the effect of change may be inaccurate if they are omitted. Ultimately, a model should include all significant variables if effects are to be assessed with the model. Some variables can be omitted if only prediction estimates are needed.

Just as model sensitivity is important in model conceptualization, it is important in formulating models to evaluate watershed change. For example, it is generally believed that small amounts of urbanization will have a very minimal effect on peak discharge rates. Also, once a watershed is close to ultimate development, any additional development has a minor effect on discharge rates. It seems that additional development of a moderately developed watershed has the greatest effect on discharge rates. This would suggest that a logistic curve (see Figure 10.5) may be an appropriate functional form to use.

The problem of split-sample testing versus jackknife evaluation was previously discussed. Both methods have disadvantages, yet it is important to verify a model intended to show the effects of watershed change. A sensitivity analysis that involves both methods could be applied. Split-sample testing separates the data into two sets with record lengths of $n/2$. Jackknife testing also separates the data into two sets, but such that the record lengths are 1 and $n - 1$. A sensitivity analysis could be conducted that uses the same concept but separates the data into parts with record lengths of k and $n - k$, where k is varied from 1 (jackknifing) to $n/2$ (split-sampling). The sensitivity function would be the standard error or the test data versus the value of k. Such a sensitivity analysis may suggest the stability of the verification process.

12.7.2 Qualitative Sensitivity Analysis

The mathematical foundation of sensitivity suggests that sensitivity analysis is a modeling tool that requires criteria and variables that are numerically scaled on continuous or ratio scales. Actually, the concept of sensitivity analysis applies to conditions just as much as it does to variables. The concept can be applied to conditions ranked on nominal or ordinal scales. For example, the accuracy of a hydrograph model might be compared for watersheds subject to flooding from both rain- and snow-generated runoff events. Of interest might be the accuracy of peak discharge prediction for rain- and snow-generated floods. A correlation coefficient could be computed for rainfall events and another correlation coefficient for snowmelt events. If the correlations are significantly different, as suggested by a test of hypothesis on two correlation coefficients, then the conclusion is made that the model is more sensitive to one type of runoff-generating precipitation. A similar type of analysis might be made for hurricane- versus nonhurricane-generated runoff.

12.7.3 Sensitivity Analysis in Design

This chapter has concentrated on the fundamental concepts of sensitivity as a modeling tool. In addition to its important role in hydrologic modeling of watershed change, sensitivity analysis is also important in hydrologic design where watershed change is involved. Watershed change may be of either a temporal or spatial nature, but generally both types of change must be addressed when designs are made on watersheds where change has or is expected to occur. In summary, the foundations of sensitivity analysis as a modeling tool are important in both design and analysis and in addressing both spatial and temporal watershed change.

The subject of frequency analysis was introduced in Chapter 5. It assumes that the data are measured from a homogeneous system. Most watersheds, even those in rural areas, are subject to considerable spatial change over time. Afforestation can easily introduce nonhomogeneity into a sequence of annual maximum discharges. Even a decline in the hydrologic condition of forest cover can introduce nonhomogeneity into a measured flood record. Changing forest-covered land to agricultural or urban land obviously affects the applicability of frequency analysis concepts. But the threshold of watershed change that is significant is not known, which is a useful role for a sensitivity study. Also, the hydrologic effect of land use change is most likely nonlinear, which makes sensitivity analysis a useful and necessary tool for assessing the effects of urbanization on a specific watershed. A sensitivity analysis of land use changes is useful for assessing the applicable of design discharges obtained from a frequency analysis.

Unit hydrograph models are widely used in hydrologic design. Their development is generally assumed to be limited to stationary systems, not watersheds on which land cover is nonhomogeneously distributed in space and time. For such watersheds, a unit hydrograph that reflects that nonhomogeneity would be needed to reflect the actual pattern of land use. While subdividing the watershed is the usual way that nonhomogeneity is addressed, it has the inherent problem of increasing the peak discharge solely because of subdivision. The development of a unit hydrograph that reflects spatial nonhomogeneity of land cover could be the result of a sensitivity analysis and would be preferable to watershed subdivision.

Design methods are based on numerous constraining assumptions. For example, design storm methods assume that the T-year rainfall causes the T-year runoff. As another example, the kinematic wave equation for estimating travel times of sheet-flow runoff assumes that the mean depth of flow equals the flow depth at the end of the flow length. Both of these assumptions are not conceptually supportable, yet they persist. Where designs are critical, sensitivity analyses could be used to check the effect of such assumptions. If a sensitivity analysis shows that the design depends on the accuracy of the assumption, then justification for the assumption is necessary.

Hydrology makes considerable use of empirical models. Peak discharge equations, time-of-concentration models, and bed-load erosion models are a few examples. The fitted coefficients are subject to considerable variation, which depends on the quality and quantity of the data used to calibrate the model. Where the accuracy of the fitted coefficients is not known, it is wise to perform a sensitivity study of the effects of error variation in the coefficients.

12.8 PROBLEMS

12-1 Apply the definition of sensitivity to define the output of a unit hydrograph model to the rainfall input, the unit hydrograph parameters, and the loss function coefficients.

12-2 Discuss the accuracy of sensitivity estimates made with Equation 12.3 given that it is a simplification of Equation 12.2.

12-3 Discuss the effect of model nonlinearity on the accuracy of the numerical approximation of S with Equation 12.7.

12-4 Assume the transfer functions for the model of Figure 12.1 are $Y = aX$ and $Z = bY^2 + cY + d$.

Determine the following:

The component sensitivity function dZ/dY.

The model sensitivity function dZ/dX.

The parametric sensitivity function dZ/da.

12-5 Assume the transfer functions for the model of Figure 12.1 are: $Y = \lambda e^{-\lambda x}$ and

$$Z = \begin{cases} a + bY & \text{for } Y \le Y_c \\ a + bY_c + d(Y - Y_c)^f & \text{for } Y > Y_c \end{cases}$$

Determine the following:
The component sensitivity function dZ/dY.
The model sensitivity function dZ/dX.
The parametric sensitivity function $dZ/d\lambda$.

12-6 For the model of Problem 12-4, determine the relative sensitivity functions for the three sensitivities requested in Problem 12-4.

12-7 Given the following model, $y = ce^{-bt}\cos[(w^2 - b^2)^{0.5}t + \theta]$, determine the relative sensitivity functions of Y with respect to b and θ.

12-8 Given the logistic model, $Y = \frac{k}{1+e^{bx}}$, determine the relative sensitivity function of Y with respect to k, b, and X.

12-9 Given the variable-intercept power model, $Y = KX^m e^{nx}$, determine the relative sensitivity functions of Y with respect to K, m, and n.

12-10 Assume evaporation E is related to the daily air temperature (T), the average percent relative humidity (H), and the mean daily wind speed as follows:

$$E = 0.007 + 0.0012T - 0.0008H + 0.0002W$$

Compute the relative sensitivities of E with respect to T, H, and W at $T = 70°$, $H = 65\%$, and $W = 45$.

12-11 Determine the errors in E that result from errors in T, H, and W of $\Delta T = 0.5°$, $\Delta H = 3\%$, and $\Delta W = 0.5$ for the model and conditions of Problem 12-10.

12-12 The weir equation, $q = C_w Lh^{1.5}$, is used to determine the discharge over a rectangular weir of length L when the water is at the depth h above the weir, with C_w being the discharge coefficient. Compute the absolute, relative, and deviation sensitivity functions for each of the input variables: C_w, L, and h.

12-13 The orifice equation, $q = C_o A(2gh)^{0.5}$, is used to determine the discharge through an orifice of area A, when the orifice is submerged at depth h, with C_o being the orifice coefficient. Compute the absolute, relative, and deviation sensitivity function for each of the input variables C_o, A, and h.

12-14 Horton's infiltration capacity (f) equation ($f = f_c + (f_o - f_c)\, e^{-kt}$) estimates the capacity as a function of time (t), and depends on three parameters (f_o, f_c, k) of the soil type, characteristics of the overlying vegetation, and the soil moisture state of the soil. Compute the absolute, relative, and deviation sensitivity functions for each of the input variables t, k, f_c, and f_o.

12-15 The double-routing-unit hydrograph model, $q(t) = te^{-Kt}/K^2$, depends on the single fitting parameter K. Compute the sensitivity function for K and discusses the implications of temporal variation of the sensitivity to K in parameterizing the coefficient K.

12-16 Discuss the uses of the absolute, relative, and deviation sensitivity functions.

13 Frequency Analysis Under Nonstationary Land Use Conditions

Glenn E. Moglen

13.1 INTRODUCTION

Many hydrologic designs are based on estimates of flood magnitudes associated with a specified return period. Flood-frequency analyses based on data collected by stream gages can be used to determine these flood magnitudes. In the event of land use change associated with urbanization, deforestation, or changes in agricultural practices within the gaged watershed, the annual maximum time series recorded by the gage includes a trend or nonstationary component that reflects the effect of the land use change. Because urbanization typically increases the flood response of a watershed, a flood frequency analysis performed on a nonstationary time series will lead to underestimation of flood magnitudes and insufficient, underdesigned structures. Accounting for this nonstationarity is, therefore, essential for appropriate design.

13.1.1 Overview of Method

The method presented in this chapter may be used to adjust a peak-discharge time series that is nonstationary because of changing land use within the gaged watershed over the gaging period. The method has several parts. First, the method focuses on deriving a spatially sensitive time series of land use. This step requires resourcefulness and creativity on the part of the hydrologist to obtain relevant data and to organize these data into a format, most likely using making use of geographic information systems (GIS) technology, that can be readily used to generate the values necessary as input to the hydrologic model. The next step is to calibrate the hydrologic model over the gaging period being studied, while taking into account the spatially and temporally varying land use. The final step is to use the calibrated model to generate a synthetic time series of peak discharge, related to the observed time series, but adjusted to reflect a single land use condition such as the current or ultimate land use. This chapter examines the differences in derived flood-frequency behavior between the observed (nonstationary) and adjusted (stationary) peak-discharge time series.

13.1.2 Illustrative Case Study: Watts Branch

The process of accounting for nonstationarity in the flood record and ultimately performing a flood-frequency analysis based on an adjusted flood series is illustrated for a watershed in the Piedmont region of Maryland just north of Washington, D.C. This watershed, Watts Branch, has a drainage area of 3.7 square miles at the location of U.S. Geological Survey (USGS) gage station 01645200, which was active from 1958 to 1987. According to the Maryland Department of Planning's assessment as of 2000, it was composed of a mix of residential densities totaling 35% of the land area. Commercial and industrial land uses cover 23%, and other urban uses (institutional and open urban land) cover an additional 15% of the land area. A significant percentage (18%) of agricultural land remains within the watershed with the remainder (9%) made up of deciduous forest. By comparison, at the time of a 1951 aerial photograph, the rough land use distribution was 15% urban, 64% agricultural, and 21% forest. Figure 13.1 shows a comparison of the spatial distribution of these land uses in 1951 and 2000. As further evidence of the changes this watershed has undergone, see the literature focused on channel enlargement and geomorphic change (e.g., Leopold, Wolman, and Miller, 1964; Leopold, 1973; Leopold, 1994).

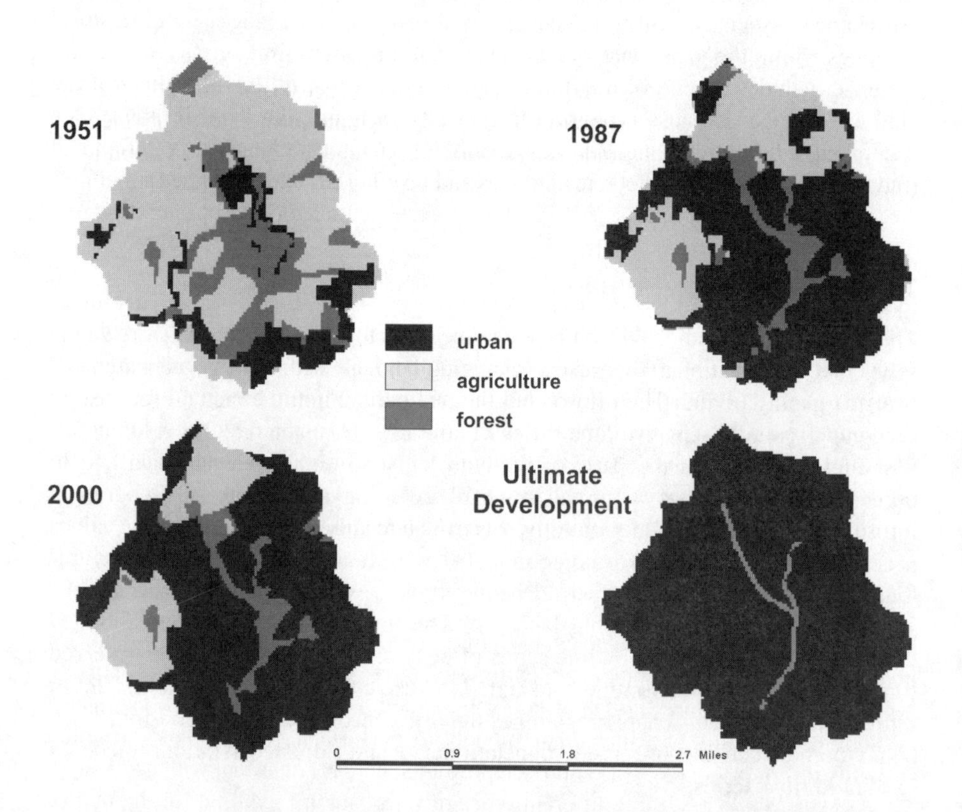

FIGURE 13.1 Spatial distribution of land use in Watts Branch watershed in 1951, 1987, 2000, and under ultimate development conditions.

The flood of record took place in 1975 with a discharge of 3400 ft^3/sec. In 1972, Hurricane Agnes, responsible for the flood of record in many neighboring watersheds, produced an annual maximum flow of 2900 ft^3/sec. The Watts Branch watershed is used throughout subsequent sections to illustrate the various phases of the modeling process.

13.2 DATA REQUIREMENTS

In a flood-frequency analysis for a stationary system, the only data required are the observed annual maximum series, $Q_{p,o}(t)$. Because of the land use change that induced nonstationarity into $Q_{p,o}(t)$, other data are required: an observed causal rainfall time series, $P(t)$, and several GIS data sets such as digital elevation models (DEMs), land use, and soils. These data sets are described in greater detail in the following sections.

13.2.1 Rainfall Data Records

Rainfall data are collected through a nationwide network of rain gages and, more recently, radar and satellite imagery. These data are archived and readily available on the Internet at a number of websites, the most accessible being the National Climatic Data Center (NCDC, 2001). This site provides free download access for point rainfall data. Data are stored in a database that is accessed through the website allowing the location and extraction of rainfall data that suits a range of selection criteria such as latitude/longitude, state/county/city name, ZIP code, or station identification number.

13.2.2 Streamflow Records

Streamflow data are collected and archived by the USGS and are similarly made available for extraction via a web-based interface that allows for a range of potential selection criteria (U.S. Geological Survey, 2001b). Data are organized and archived in two forms: daily averaged flows and the annual maximum series. In the case of the annual maximum series data, the discharge is accompanied by a field that also identifies the date of occurrence of the annual maximum. This chapter will focus on the annual maximum series and any trends that may be present in this series as a result of land use change within the gaged watershed.

13.2.3 GIS Data

Rainfall-runoff estimates of peak discharge could be developed with a range of potential models. This chapter uses Natural Resources Conservation Service (NRCS; U.S. Soil Conservation Service, 1985, 1986) methods to develop such estimates. Although the details of using other models to perform a similar analysis would certainly differ, the spirit of the approach presented here is consistent with any model.

As stated earlier, several different types of GIS data are required for the hydrologic modeling of the annual maximum discharges. First and foremost, topographic data in the form of a DEM is probably the most fundamental data type required for this analysis. The DEM serves first to determine flow paths and provide an automated delineation of the study watershed that provides an estimate of the watershed area. The DEM then allows for the estimation of slopes and times of concentration that are central to the analysis as well. These data are made available by the USGS (U.S. Geological Survey, 2001c) at several map scales. The data used in the case study presented here are derived from the 1:24,000 map scale and have a resolution of 30 meters.

NRCS methods depend heavily on the estimation of a curve number, requiring information about the area distribution of both land use and hydrologic soil type. Thus, coverages of both land use and hydrologic soil type are required. Land use coverages may be obtained from a number of sources. The USGS GIRAS (Mitchell et al., 1977) is probably the oldest, widely available data set. It tends to reflect land use of an approximately 1970s vintage. These data are now commonly distributed as part of the core data set in the BASINS model (U.S. Environmental Protection Agency, 2001a). Newer data may also exist on a more regionally varied level, such as the MRLC data set (Vogelmann et al., 1998a; Vogelmann, Sohl, and Howard, 1998b; U.S. Environmental Protection Agency, 2001b) that covers many states in the eastern United States and dates to approximately the early 1990s. Other more high-quality data sets will likely be available on an even more limited basis, perhaps varying by state, county, or municipality. Generally, the higher-quality data will reflect conditions from periods more recent than the GIRAS data mentioned earlier. Knowledge of land use from before the 1970s will likely need to be gleaned from nondigital sources such as historical aerial photography or paper maps.

Soils data are generally obtained from the NRCS. The NRCS publishes two different sets of digital soils data: STATSGO (NRCS, 2001a) and SSURGO (NRCS, 2001b). The STATSGO data are the coarser of the two and are digitized from 1:250,000 scale maps (except in Alaska where the scale is 1:1,000,000) with a minimum mapping unit of about 1544 acres. These data are available anywhere within the United States. The SSURGO data are digitized from map scales ranging from 1:12,000 to 1:63,360. SSURGO is the most detailed level of soil mapping done by the NRCS. These data are in production at this time and availability varies on a county-by-county basis. In the case study presented here, Watts Branch lies entirely within Montgomery County, Maryland, one of the counties for which SSURGO data are available.

13.3 DEVELOPING A LAND-USE TIME SERIES

The particular emphasis of this chapter is to consider the effect of changing land use on peak discharge; land use is not static, but rather continually changing in both time and space throughout the time series. A practical problem that generally arises is that the GIS data to support the modeling of peak discharge on an annual basis do not exist. In general, one has access to, at best, several different maps of land use corresponding to different "snapshots" in time. This section provides and illustrates a method to develop a land-use time series on an annual time step.

The data required to develop a land-use time series are two different maps of land use covering the extents of the watershed and spanning the same time period as the available annual maximum discharge record. Additionally, data are required that convey the history of land use development at times between the two land use snapshots. Such data are typically available in the form of tax maps. In the example provided here, the Maryland Department of Planning publishes such data (Maryland Department of Planning, 2001) that indicate tax map information at the detailed level of individual parcel locations. One of the attributes associated with these data is the date of construction of any structure on the property.

The notation $LU(x,t)$ is used here to indicate the land use across all locations in the vector x within the watershed being studied, and t is any generic time in years. The land-use time series is initialized to be the land use at time, t_1, indicated by the earlier of the two land use maps. This land use is denoted by $LU(x,t_1)$. If t_1 is earlier than approximately 1970, it is likely that the required land use data are not available digitally, but rather in the form of a paper map or aerial photograph. Such data will need to be georeferenced and then digitized into a hierarchical land-use classification scheme such as the one created by Anderson et al. (1976). Figure 13.1 shows land use over the study watershed at times $t_1 = 1951$ and $t_2 = 2000$. (For completeness, this figure also shows the watershed at an intermediate time, $t = 1987$, and at some future time corresponding to ultimate development conditions. These land use conditions are discussed later in this chapter.) Using the $LU(x,t_1)$ coverage as a starting point, the land use is then allowed to transition to $LU(x,t_2)$ in the specific year, t^*, that the tax map information indicates is the year of construction for that individual parcel. Applying this rule over all years $t_1 < t^* < t_2$ and for all parcels within the watershed allows the modeler to recreate land use change on an annual time step.

This process is illustrated in Figure 13.2, which shows a view of several rows of parcels in a subdivision over the years 1969 and 1970. The date shown within each parcel is the year in which the tax map data indicates that it was developed.

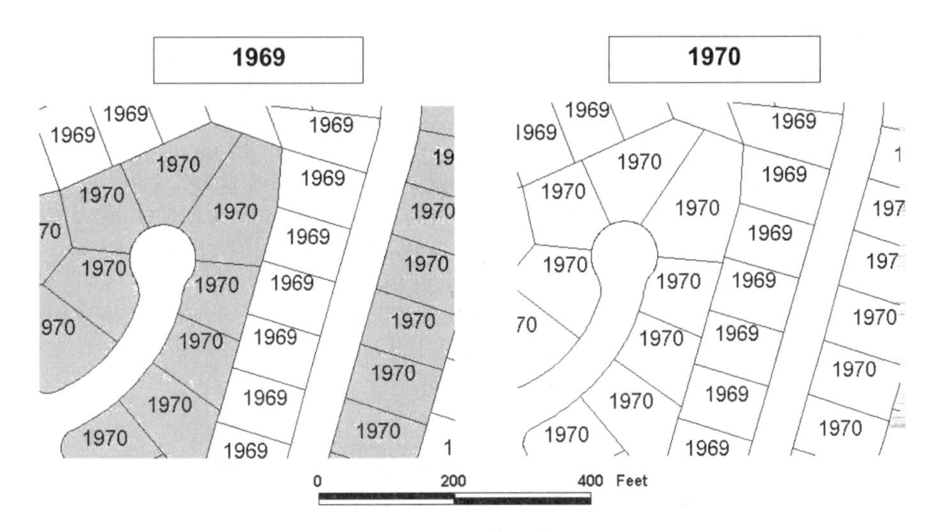

FIGURE 13.2 Parcel-level view of land-use change model. Parcels shown in white are developed and gray parcels are not developed. Note that all parcels shown become developed by 1970.

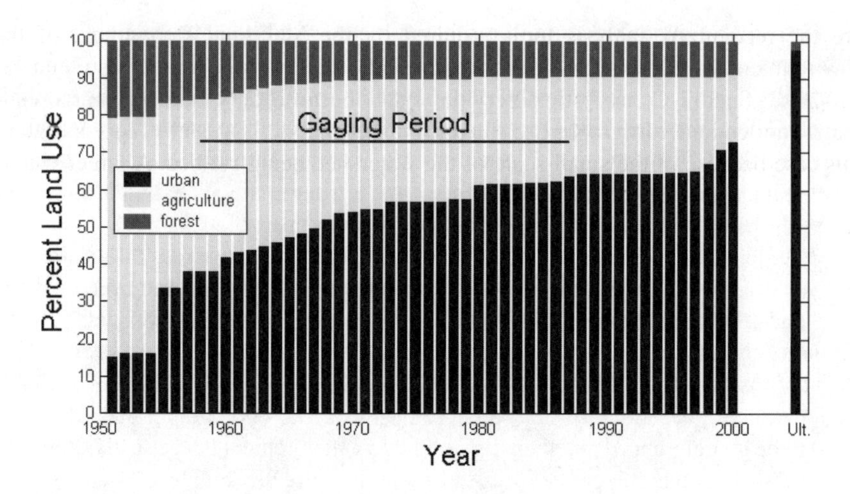

FIGURE 13.3 Aggregate land-use distribution in Watts Branch watershed over time. The bar at the far right gives the ultimate development land-use distribution.

The GIS treats all parcels shown in gray (undeveloped parcels) to remain in their original land use at time $t = t_1$, while those shown in white (developed parcels) have transitioned to their final land use at time $t = t_2$. Figure 13.3 shows the aggregate change in the land use distribution within the Watts Branch watershed between 1951 and 2000. Note that land use does not simply change linearly between these two known times, but rather it changes in an irregular fashion following the actual patterns of development as they were realized within the watershed.

13.4 MODELING ISSUES

A wide range of modeling issues confront the hydrologist performing this type of study. First and foremost is the simple selection of the model to use. Other issues include methods for calibrating the model and ultimately using the model to simulate the discharge behavior of the study watershed in a predictive sense. This section presents a discussion of these wide-ranging issues and argues for a particular series of choices throughout the modeling process, recognizing that different choices might be selected by others. The decisions presented here reflect the pragmatic needs of the engineer wishing to make use of a valuable gage record, but also recognizing the influence that urbanization has on that record.

13.4.1 SELECTING A MODEL

This study is concerned with observed and simulated peak discharges from a small watershed subject to changes in land use over time. The model that is selected must therefore predict peak discharges, be appropriate for a watershed of this size, and be sensitive to land use change in its predictions of peak discharge. The NRCS TR-55 (U.S. Soil Conservation Service, 1986) is one such model. It is selected here over the HEC-HMS model (U.S. Army Corps of Engineers, 2000) because the NRCS models are the recognized analysis tools required by the Maryland Department of the Environment for all flood discharge studies. TR-55 is chosen over the more general TR-20 (U.S. Soil Conservation Service, 1984) because only peak discharge estimates, not entire hydrographs are sought. Furthermore, this model is appropriate in this case because of the small scale of the watershed being studied. A larger watershed where reach routing is a significant part of the overall travel time, or a watershed with significant detention, would require the more sophisticated TR-20. In any case, although the details specific to the TR-55 graphical method are presented here, the general approach, which is essentially model independent, is emphasized.

The TR-55 model determines peak discharge using

$$Q_{p,s}(t) = q_u(t)AQ(t) \tag{13.1}$$

where $Q_{p,s}(t)$ is the simulated peak discharge in ft^3/sec, $q_u(t)$ is the unit peak discharge in ft^3/sec-in. of runoff, A is the area of the watershed in mi^2, and $Q(t)$ is the runoff depth in inches. The functional dependence of runoff, unit peak discharge, and simulated peak discharge on time is explicitly shown to emphasize the time-varying nature of these quantities due to changes in land use.

The runoff depth is determined from

$$Q(t) = \frac{[P(t) - I_a(t)]^2}{P(t) - I_a(t) + S(t)} = \frac{[P(t) - 0.2S(t)]^2}{P(t) + 0.8S(t)} \tag{13.2}$$

where the standard assumption is made that the initial abstraction, $I_a(t)$, is equal to 20 percent of the storage, $S(t)$. Both storage and initial abstraction are in inches units. $P(t)$ is the *causal* precipitation depth associated with the observed annual maximum discharge. This quantity is discussed in greater detail in Section 13.4.2. Storage, $S(t)$, is determined as a function of the curve number, $CN(t)$ using

$$S(t) = \frac{1000}{CN(t)} - 10 \tag{13.3}$$

The curve number is determined using a standard "look-up table" approach given the spatial overlap of land use and soils and using the NRCS-defined curve numbers (U.S. Soil Conservation Service, 1985). The time-varying nature of $CN(t)$ is due to the time-varying land use within the watershed as discussed earlier in Section 13.3.

The unit peak discharge, $q_u(t)$, is a function of two quantities, time of concentration, $t_c(t)$, and the ratio of the initial abstraction to the precipitation, $I_a(t)/P(t)$. Again, the dependence of these quantities on time is shown here explicitly. The unit peak discharge is generally determined graphically; however, this procedure is automated with the GIS using the equation

$$\log[q_u(t)] = C_0[r(t)] + C_1[r(t)] \cdot \log[t_c(t)] + C_2[r(t)] \cdot \{\log[t_c(t)]\}^2 \tag{13.4}$$

TABLE 13.1
Constants C_0, C_1, and C_2 Used in Equation 13.4 for the SCS Type II Storm Distribution

I_a/P	C_0	C_1	C_2
0.10	2.55323	–0.61512	–0.16403
0.30	2.46532	–0.62257	–0.11657
0.35	2.41896	–0.61594	–0.08820
0.40	2.36409	–0.59857	–0.05621
0.45	2.29238	–0.57005	–0.02281
0.50	2.20282	–0.51599	–0.01259

Source: U.S. Soil Conservation Service, Urban Hydrology for Small Watersheds, Technical Release 55, U.S. Department of Agriculture, Washington, DC, 1986.

where $r(t)$ is determined from

$$r(t) = \frac{I_a(t)}{P(t)} \tag{13.5}$$

and C_0, C_1, and C_2, are tabular functions of this ratio. Values of these constants for the U.S. Soil Conservation Service Type II storm distribution, which is appropriate for the study watershed are provided in Table 13.1.

For consistency, the time of concentration, $t_c(t)$, was determined in this study using the SCS lag equation (U.S. Soil Conservation Service, 1973), rather than the often-used velocity method. In this case, the dependency of the velocity method on spatially and time-varied surface roughness would be too arbitrary to characterize consistently. The lag equation's dependency on curve number, which is characterized very carefully throughout this study, was instead chosen as the basis for developing t_c estimates.

The SCS lag time, $t_l(t)$ in minutes, is determined using

$$t_l(t) = \frac{100L^{0.8}\left(\dfrac{1000}{CN(t)} - 9\right)^{0.7}}{1900Y^{0.5}} \tag{13.6}$$

where L is the longest flow path in the watershed in feet, and Y is the basin averaged slope in percentages.

The lag time is converted to a time of concentration by multiplying by a factor of 1.67 and also accounting for speedup in runoff rates due to imperviousness introduced in the urbanization process. Time of concentration is thus determined using the following equation (McCuen, 1982):

$$t_c(t) = 1.67t_l(t)\cdot\{1 - I(t)[(D_0 + D_1\cdot CN(t) + D_2\cdot CN^2(t) + D_3\cdot CN^3(t)]\} \tag{13.7}$$

TABLE 13.2
Percent Imperviousness Associated with Various Land Uses Present in Watts Branch, Maryland

Land Use	Percent Imperviousness
Low-density residential	25
Medium-density residential	30
High-density residential	65
Commercial	82
Industrial	70
Institutional	50
Open urban land	11
Cropland	0
Pasture	0
Deciduous forest	0
Mixed forest	0
Brush	0

Source: Weller, personal communication.

where $I(t)$ is the time-varying imperviousness of the watershed in percentages, and D_0, D_1, D_2, and D_3 are -6.789×10^{-3}, 3.35×10^{-4}, -4.298×10^{-7}, and -2.185×10^{-8}, respectively. Imperviousness was determined as a simple lookup function of land use based on values determined by the Maryland Department of Planning for their generalized land use data. These values are provided in Table 13.2.

13.4.2 Calibration Strategies

Rainfall data obtained from the NCDC Web site (National Climatic Data Center, 2001) mentioned in Section 13.2.1 were obtained for the Rockville, Maryland rain gage (Coop ID# 187705). Annual precipitation values were determined around a three-dimensional window centered on the date of the annual maximum flood. Because of potential time/date mismatches between the occurrence of precipitation and peak discharge, the observed precipitation associated with the peak discharge was taken to be the maximum of the sum over a 2-day window either beginning on ending on the day associated with the annual maximum discharge.

Even allowing for potential time/date mismatches, the observed precipitation is quite small on four occasions (1958, 1964, 1969, and 1981). In fact, *no* precipitation is observed to explain the annual maximum in 1969. A plausible explanation for this is the relatively small scale of the study watershed. Convectively generated rainfall is highly varied in space compared to frontal-system generated rainfall, and the rain gage, although close, is not actually within the study watershed. Because of the size of the watershed, the annual maximum discharge is likely to be associated with convective summer events rather than the frontal precipitation more common in the cooler months. The data support this hypothesis with 19 of 27 annual maxima observed in the months of June through September.

The hydrologic engineer has a number of parameters available to calibrate the above model to the observed annual maximum series. Both curve numbers and t_c values are frequent candidates for calibration in a typical analysis. Given the time series implications of this analysis, it did not make sense to adjust either of these quantities since an adjustment of a quantity might be made "up" in one year and "down" in the next. The physical basis for such an adjustment pattern is unclear. Instead, given the presumed inaccuracies in the observed precipitation record, precipitation was used as the only calibration parameter. The causal precipitation, $P(t)$, in Equation 13.2, was calibrated by setting the model outlined in Equations 13.1 through 13.7 into Equation 13.8, such that the observed and simulated annual maximum discharges were the same within a small (0.1%) tolerance.

$$\frac{|Q_{p,s}[t, P(t)] - Q_{p,o}(t)|}{Q_{p,o}(t)} < 0.001 \qquad (13.8)$$

where $Q_{p,s}[t, P(t)]$ is the simulated peak discharge in year, t, and assuming a causal precipitation depth, $P(t)$. $Q_{p,o}(t)$ is the observed peak discharge in year, t. The observed and causal precipitation were determined to have a correlation coefficient, R, of 0.67. Table 13.3 and Figure 13.4 provide a summary and comparison of the causal (simulated) and observed precipitation depths.

13.4.3 Simulating a Stationary Annual Maximum-Discharge Series

With the causal precipitation time series determined, it is a straightforward process to determine the annual maximum discharge that would have been observed had land use remained constant over the period of record of the stream gage. In fact, the only question facing the hydrologic engineer is what land use to employ in the simulation. As a side-product of the calibration process, representations of land use on an annual time step corresponding to each year from 1951 to 2000 are available. Using any *one* of these years, t^*, and the causal precipitation time series developed in the calibration step, the annual maximum series that would be observed as if the land use were fixed for that particular year, t^* can be generated. In more mathematical terms,

$$Q_{p,t^*}(t) = Q_{p,s}[t, P(t), LU(x, t^*)] \qquad (13.9)$$

For illustrative purposes, the annual maxima corresponding to the following four different land use conditions have been simulated: $Q_{p,1951}(t)$ (earliest land use), $Q_{p,1987}(t)$ (last year of the gage record), $Q_{p,2000}(t)$ ("present" conditions), and $Q_{p,ult.}(t)$ (projected ultimate condition of watershed given zoning data). These simulated peak discharges are provided in Table 13.5 and shown in Figure 13.5. (For completeness,

TABLE 13.3
Observed and Causal Precipitation Associated with Annual Maximum Discharge in Watts Branch, Maryland

Hydrologic Year	Date of Peak Discharge	P_{obs} 1 Day before Peak (in.)	P_{obs} on Day of Peak (in.)	P_{obs} 1 Day after Peak (in.)	Maximum Observed 2-Day P (in.)	Causal P (in.)
1958	07/08/1958	0.00	*0.18*[a]	*0.11*	0.29	2.41
1959	09/02/1959	0.00	*1.98*	*0.26*	2.24	1.59
1960	07/13/1960	0.00	*0.00*	*1.97*	1.97	2.58
1961	04/13/1961	*0.08*	*1.95*	0.00	2.03	1.67
1962	03/12/1962	*0.00*	*1.20*	0.00	1.20	1.47
1963	06/05/1963	—	—	—	—	2.40
1964	01/09/1964	*0.00*	*0.36*	0.00	0.36	1.62
1965	08/26/1965	0.00	*1.39*	*0.64*	2.03	2.87
1966	09/14/1966	—	—	—	—	2.57
1967	08/04/1967	0.00	*0.69*	*1.21*	1.90	2.91
1968	09/10/1968	0.00	*0.68*	*1.95*	2.63	2.54
1969	06/03/1969	0.00	0.00	0.00	0.00	2.60
1970	08/14/1970	0.00	*0.00*	*2.30*	2.930	3.84
1971	08/01/1971	0.15	*0.30*	*1.84*	2.14	1.81
1972	06/21/1972	*1.36*	*7.90*	1.36	9.26	5.34
1973	06/04/1973	0.00	*0.00*	*1.98*	1.98	3.30
1974	09/28/1974	0.00	*1.60*	*0.00*	1.60	2.05
1975	09/26/1975	*1.54*	*4.46*	0.00	6.00	5.94
1976	12/31/1975	0.19	*0.57*	*2.42*	2.99	1.73
1977	03/13/1977	0.00	*1.18*	*0.46*	1.64	1.68
1978	07/31/1978	0.00	*0.20*	*4.50*	4.70	2.82
1979	08/27/1979	0.30	*0.62*	*1.90*	2.52	4.98
1980	10/01/1979	0.08	*1.92*	*0.37*	2.29	2.49
1981	08/12/1981	*0.18*	*0.08*	0.07	0.26	2.02
1982	09/02/1982	0.00	*0.00*	*1.40*	1.40	2.33
1983	05/22/1983	—	—	—	—	2.10
1984	08/01/1984	0.00	*2.87*	*0.24*	3.11	2.52
1985	07/15/1985	*0.00*	*1.34*	0.00	1.34	2.24
1986	05/21/1986	*0.72*	*1.80*	0.02	2.52	2.67
1987	09/08/1987	*0.00*	*4.07*	0.00	4.07	2.37

[a] Numbers in italics represent a day contributing to the maximum 2-day precipitation.

the observed $[Q_o(t)]$ and simulated runoff volumes, $[Q_{1951}(t)$, $Q_{1987}(t)$, $Q_{2000}(t)$, and $Q_{ult.}(t)]$, are provided first in Table 13.4.)

With these new time series determined, it is important to clearly understand what they represent. Each of the simulated time series represents the discharges that would have been observed at the stream gage, had the land use been fixed at t^* conditions over the entirety of the gaging period. Thus, the 1951 time series, $Q_{p,1951}(t)$, represents the peak discharges observed from 1958 to 1987 adjusted to the land use

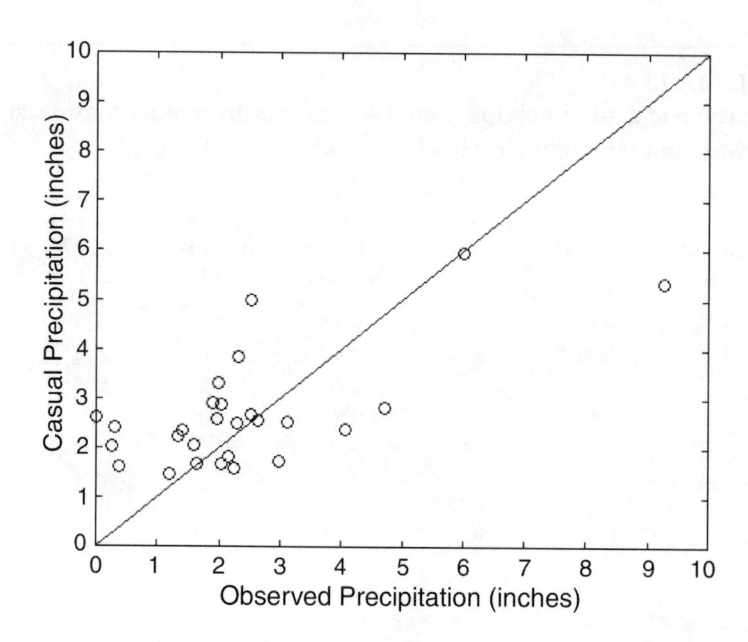

FIGURE 13.4 Comparison of observed and causal (simulated) precipitation depths associated with the annual maximum discharges in Watts Branch. The line shown corresponds to perfect agreement.

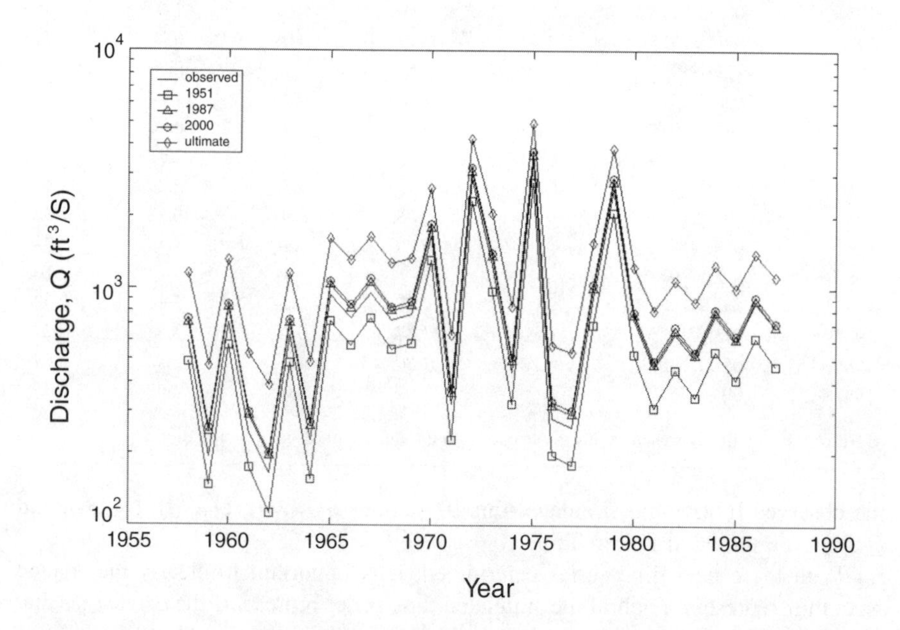

FIGURE 13.5 Time series plot showing observed and adjusted annual maximum discharges.

TABLE 13.4
Observed and Simulated Runoff Depths Associated with Annual Maximum Discharge Events in Watts Branch, Maryland

Hydrologic Year	Q_o (in.)	Q_{1951} (in.)	Q_{1987} (in.)	Q_{2000} (in.)	$Q_{ult.}$ (in.)
1958	0.76	0.70	0.81	0.82	1.02
1959	0.29	0.26	0.32	0.32	0.45
1960	0.88	0.81	0.93	0.93	1.15
1961	0.34	0.29	0.36	0.37	0.50
1962	0.25	0.21	0.27	0.27	0.38
1963	0.78	0.70	0.81	0.81	1.01
1964	0.32	0.27	0.34	0.34	0.47
1965	1.11	1.00	1.14	1.14	1.38
1966	0.90	0.80	0.93	0.93	1.14
1967	1.13	1.02	1.16	1.17	1.41
1968	0.88	0.78	0.90	0.91	1.12
1969	0.92	0.82	0.95	0.95	1.17
1970	1.85	1.72	1.89	1.90	2.19
1971	0.42	0.36	0.44	0.45	0.59
1972	3.11	2.94	3.17	3.17	3.53
1973	1.42	1.31	1.46	1.47	1.73
1974	0.56	0.49	0.58	0.58	0.76
1975	3.64	3.46	3.70	3.71	4.09
1976	0.37	0.32	0.40	0.40	0.54
1977	0.35	0.30	0.37	0.37	0.51
1978	1.07	0.97	1.10	1.11	1.34
1979	2.79	2.64	2.85	2.85	3.21
1980	0.86	0.75	0.87	0.87	1.08
1981	0.56	0.47	0.56	0.57	0.73
1982	0.75	0.65	0.76	0.76	0.96
1983	0.60	0.52	0.61	0.61	0.79
1984	0.88	0.77	0.89	0.89	1.10
1985	0.69	0.60	0.70	0.71	0.89
1986	0.98	0.86	0.99	0.99	1.22
1987	0.79	0.68	0.79	0.79	0.99

conditions present in 1951. Note that the observed discharges lie between those associated with the 1951 and 1987 land use conditions. This is to be expected since the gaging period is from 1958 to 1987. It is also of interest to note that the 1987 discharge in the $Q_{p,1987}(t)$ time series is identical to that in the observed record. In other words, under the method described here, adjustment is not made to this discharge when determining the peak discharge under 1987 land use conditions. If gaging were to have taken place in 1951 or to have continued in 2000, we would expect the discharges in both of these years under these land use conditions to also be identical to the observed record for the particular year in question. Each of these time series is valuable, because each obeys a major assumption of flood-frequency analysis that is violated by the actual, observed annual maximum series. These fixed

TABLE 13.5
Observed and Simulated Annual Maximum Discharges in Watts Branch, Maryland

Hydrologic Year	$Q_{p,o}$ (ft³/sec)	$Q_{p,1951}$ (ft³/sec)	$Q_{p,1987}$ (ft³/sec)	$Q_{p,2000}$ (ft³/sec)	$Q_{p,ult.}$ (ft³/sec)
1958	592	486	710	735	1149
1959	194	147	248	257	467
1960	710	570	822	851	1309
1961	242	175	289	299	526
1962	164	112	194	202	388
1963	630	484	706	731	1143
1964	227	156	260	270	485
1965	932	726	1031	1067	1602
1966	740	568	820	849	1306
1967	954	743	1053	1090	1633
1968	730	551	798	826	1274
1969	770	583	840	869	1334
1970	1660	1306	1791	1852	2638
1971	325	229	361	374	635
1972	2900	2334	3093	3195	4259
1973	1250	971	1355	1401	2048
1974	443	326	490	508	829
1975	3400	2776	3619	3738	4934
1976	283	197	319	330	570
1977	259	179	294	305	533
1978	916	699	995	1029	1551
1979	2600	2075	2779	2872	3863
1980	739	525	762	789	1223
1981	458	314	473	490	804
1982	641	450	660	684	1077
1983	502	346	518	536	870
1984	763	541	783	811	1254
1985	589	409	605	626	997
1986	867	615	883	914	1395
1987	686	468	686	709	1114

land-use time series are stationary; the trend of increasing discharges due to urbanization has been removed.

A few words are merited in the description of the ultimate land use condition. Zoning data were obtained from the Maryland Department of Planning spanning the area associated with the study watershed. As of 2000, the watershed is already highly urbanized; however, zoning for this watershed indicates that the remaining 18% of agricultural land and a significant proportion of the forest cover would be converted to commercial and/or residential land uses. Forest cover within 100 feet of existing streams is assumed to remain due to efforts to retain a buffer zone adjacent to streams. Land use conditions under ultimate development are reflected in the spatial distribution shown in Figure 13.1 and in the time series shown in Figure 13.3.

13.5 COMPARISON OF FLOOD-FREQUENCY ANALYSES

Hydrologic design is often based on peak discharges associated with various return periods. In stream restoration, discharges that approximate bank-full flow conditions have a return frequency of 1.5 to 2 years (Leopold, 1994). The design of bridges, culverts, detention ponds, and outlet structures may depend on frequencies ranging from 2 to 100 years depending on the structure and design specifications regulated by the state. From an engineering perspective, the consequences of the analyses presented in this chapter are on the differences in flood frequency between the various time series we have determined.

13.5.1 IMPLICATIONS FOR HYDROLOGIC DESIGN

The USGS PeakFQ program (U.S. Geological Survey, 2001a) automates the determination of the flood frequency distribution for any annual maximum time series, producing estimates consistent with Bulletin 17B (Interagency Advisory Committee on Water Data, 1982) guidelines. The four time series presented in Table 13.5 were analyzed with the results presented in Table 13.6 and Figure 13.6. The implications for hydrologic design are addressed in this section.

The main, and most obvious, consequence of performing the discharge adjustment procedure outlined here is the proliferation of peak discharge estimates as shown in Tables 13.5 and 13.6 and Figures 13.5 and 13.6. Through comparison with

TABLE 13.6
Flood Frequency Distributions for Observed and Simulated Annual Maximum Discharges in Watts Branch, Maryland

Return Period (years)	Observed Time Series (ft^3/sec)	1951 Land Use (ft^3/sec)	1987 Land Use (ft^3/sec)	2000 Land Use (ft^3/sec)	Ultimate Land Use (ft^3/sec)
2	628	461 (−166)[a]	678 (50)	702 (74)	1104 (476)
5	1215	929 (−286)	1292 (77)	1336 (121)	1952 (737)
10	1753	1373 (−380)	1864 (111)	1927 (174)	2708 (955)
25	2635	2123 (−512)	2818 (183)	2912 (277)	3929 (1294)
50	3459	2844 (−615)	3728 (269)	3852 (393)	5063 (1604)
100	4447	3727 (−720)	4836 (389)	4996 (549)	6416 (1969)
200	5625	4802 (−823)	6181 (556)	6385 (760)	8028 (2403)
500	7528	6580 (−948)	8400 (872)	8676 (1148)	10,640 (3112)

[a] Numbers in parentheses represent the difference between simulated discharges and observed discharges for the same return period.

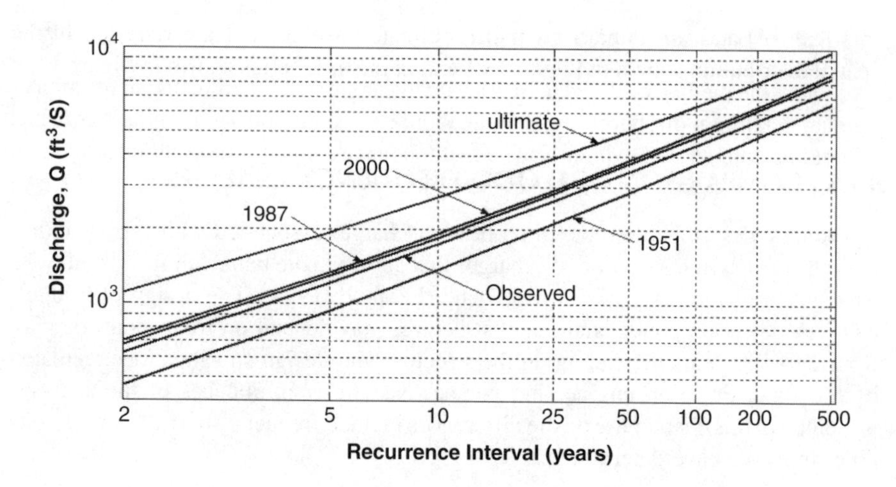

FIGURE 13.6 Observed and adjusted flood-frequency distributions.

the observed frequency distribution, the consequences of the discharge adjustment method become clear. Consider, for example, developing a hydrologic design for a bridge located at the stream gage based on a 50-year return frequency. In the absence of an adjustment method, the design discharge would be 3459 ft³/sec. Given that the gaging ended relatively recently in 1987, one might be tempted to simply use the discharges adjusted to 1987 land use conditions. In this case, the design discharge would increase by 269 ft³/sec to 3728 ft³/sec. If adjustments are made for urbanization that has taken place since the end of the gage record, the design discharge would increase to 3852 ft³/sec. In Maryland, where ultimate development determines the design discharge for new bridges, the 50-year peak discharge based on adjustment of the gage record to ultimate development land-use conditions is 5063 ft³/sec. This is 1604 ft³/sec greater than the unadjusted discharge from the gage record. The consequences on engineering design are clear: failure to adjust the observed discharge record in the face of known urbanization can lead to systematic underdesign of the structure. Depending on the structure, this underdesign might manifest itself in increased flooding, a discharge structure failing to pass the required flow and being overtopped, or in scour of a magnitude much greater than anticipated. In any case, underdesign will lead to a greater likelihood of failure of the engineered structure.

It is important to note that the adjustment procedure outlined here is of value, even if land use is relatively unchanged over the period of record of the stream gage. This is because there may still have been significant development since the gage record was collected or ultimate development conditions may represent a large degree of urbanization that was not present at the time of gaging. In both of these cases, adjusted discharges would be greater, and potentially much greater, than those associated with the observed flood-frequency distribution.

13.5.2 Assumptions and Limitations

Although the method outlined here is a valuable approach for adjusting nonstationary annual maximum discharges, it is not without its assumptions and limitations, which should be enumerated explicitly. The method, as presented, lumps all uncertainty in the modeling process into a time series of precipitation estimates that reproduce the observed peak discharge using the TR-55 method. This precipitation time series has been referred to throughout this chapter as the "causal precipitation time series." Actual differences in the hydrologic response may be due to any number of other assumptions in the TR-55 model as employed here. For instance, the curve number table that was used may not be appropriate, the antecedent conditions (assumed normal) may be dry or wet, the impervious surfaces (assumed to be directly connected to the drainage system) may instead be disconnected, subareas within the watershed may need to be considered separately, and detention storage has been ignored altogether. Design criteria instituted since the end of the gage record may mean that the ultimate land use conditions include a significant storage component; thus, peak discharges associated with ultimate development as portrayed here are much greater than would be expected in the face of stormwater management practices. If a larger watershed were considered, TR-55 would need to be replaced with the more sophisticated TR-20 or other rainfall-runoff model that both generates runoff and routes this runoff through a channel network.

Putting the hydrologic model aside for a moment, there are also assumptions and limitations associated with the land-use change model presented here. As depicted in Figures 13.1 through 13.3, the land use is only allowed to transition from one initial land use at time $t = t_1$ to another final land use at time $t = t_2$. A more robust model that would be relatively easy to incorporate would allow the landscape to transition from/to multiple land use conditions at discrete points in time. This model would then require n different snapshots in time and $n - 1$ spatially referenced data sets that convey when the conversion of each bit of landscape from condition at time $t = t_i$, to condition at time $t = t_{i+1}$, occurs. In the land use model presented earlier, there were only $n = 2$ snapshots in time and the subscript, i, took on only the value $i = 1$. The most likely limitation that the hydrologic modeler will encounter in this context is a distinct lack of land use "snapshots" at fixed points in time, and of spatially referenced data sets to convey the land use change history.

13.6 SUMMARY

This chapter has illustrated a GIS-based method to model land use change over time on an annual time step and at the parcel level of spatial detail. These land-use change data were used to drive a simple hydrologic model that is sensitive to land use change. The hydrologic model was calibrated to reproduce the observed discharges of the period of record of an actual gaged case-study watershed in Maryland. Once calibrated, the hydrologic model was then used to determine the annual maximum discharges that would have resulted in the case study watershed for fixed (rather than time-varying) land use corresponding to past, present, and future land use conditions. Using flood-frequency analysis, the fixed land-use, annual-maximum flood-frequency distributions were compared to the flood-frequency distribution for the actual observed distribution. The implications of not adjusting an annual maximum time series derived from a gaged watershed under urbanizing (nonstationary) land use conditions are that design discharges associated with any return frequency are likely to be underestimated and lead to the underdesign of a structure and a greater likelihood for that structure to ultimately fail.

13.7 PROBLEMS

13-1 You are asked to develop the hydrologic analyses necessary to build a bridge crossing at the location of the Watts Branch stream gage. The bridge piers are to be sized based on the 25-year peak flow event. Express the 1951, 1987, 2000, and ultimate development peaks for this return period as a percentage of the observed 25-year peak. Comment on the magnitudes of these percentages.

13-2 Assuming a fixed rectangular cross-sectional channel geometry that is 20 feet wide, with a slope of 0.005 ft./ft., and a Manning's roughness of 0.03,
  (a) Determine the normal depth associated with the observed, 1951, 1987, 2000, and ultimate development 100-year discharges.
  (b) If the bridge was constructed in 1987 with the deck placed to just pass the 100-year discharge at that time, by how much will it be overtopped by the 100-year event in 2000? At ultimate development conditions?

13-3 Download the PeakFQ program from the Web: *http://water.usgs.gov/software/peakfq.html* (U.S. Geological Survey, 2001a). Use the PeakFQ program and the values in Table 13.5 to verify the $Q_{p,o}$ and $Q_{p,2000}$ flood-frequency analysis values provided in Table 13.6. You will need to indicate "STA" for the skew computation option to indicate that you are using the station skew option rather than the default weighted skew option.

13-4 Let the runoff ratio time series, $R_R(t)$, be defined as:

$$R_R(t) = \frac{Q(t)}{P(t)}$$

where $Q(t)$ is the runoff depth (in inches) defined in Equation 13.2 and provided in Table 13.4 and $P(t)$ is the causal precipitation depth (in inches) also described as part of Equation 13.2 and provided in Table 13.3. Plot $R_R(t)$ for the $Q_{p,o}(t)$ and $Q_{p,2000}(t)$ time series. Perform a regression on these ratios versus time. Report the slopes that you determine from these regressions in both cases. Are these slopes consistent with what you would expect from the information contained in this chapter? Discuss.

13-5 You are asked to generalize the flood-frequency adjustment method presented in this chapter to apply to a 40-mi^2 watershed that includes several detention basins and 15 miles of channels. Describe how you would go about doing this. What additional information beyond that used in the case study presented here would be necessary?

13-6 You are given access to additional land use coverages corresponding to 1964 and 1975 conditions. Describe how the land-use change model presented in this chapter would be modified to accommodate such data. Are there any inconsistencies that might arise? Would additional data be necessary?

13-7 You are given access to an additional land use coverage corresponding to 1937 conditions. Describe how the land-use change model presented in this chapter would be modified to accommodate such data. Are there any inconsistencies that might arise? Would additional data be necessary?

13-8 This chapter makes the claim that the relatively small scale of the study watershed makes it more likely to have convective summer thunderstorms drive the annual-maximum peak flow series. Outline a small study that would use only the USGS annual-maximum flow database to determine the scale (in mi^2) at which the annual maximum shifts from summer thunderstorms to cooler-season frontal events.

Appendix A

Statistical Tables

TABLE A.1
Standard Normal Distribution

z	0.00	0.01	0.02	0.03	0.04	0.05	0.06	0.07	0.08	0.09
−3.4	0.0003	0.0003	0.0003	0.0003	0.0003	0.0003	0.0003	0.0003	0.0003	0.0002
−3.3	0.0005	0.0005	0.0005	0.0004	0.0004	0.0004	0.0004	0.0004	0.0004	0.0003
−3.2	0.0007	0.0007	0.0006	0.0006	0.0006	0.0006	0.0006	0.0005	0.0005	0.0005
−3.1	0.0010	0.0009	0.0009	0.0009	0.0008	0.0008	0.0008	0.0008	0.0007	0.0007
−3.0	0.0013	0.0013	0.0013	0.0012	0.0012	0.0011	0.0011	0.0011	0.0010	0.0010
−2.9	0.0019	0.0018	0.0018	0.0017	0.0016	0.0016	0.0015	0.0015	0.0014	0.0014
−2.8	0.0026	0.0025	0.0024	0.0023	0.0023	0.0022	0.0021	0.0021	0.0020	0.0019
−2.7	0.0035	0.0034	0.0033	0.0032	0.0031	0.0030	0.0029	0.0028	0.0027	0.0026
−2.6	0.0047	0.0045	0.0044	0.0043	0.0041	0.0040	0.0039	0.0038	0.0037	0.0036
−2.5	0.0062	0.0060	0.0059	0.0057	0.0055	0.0054	0.0052	0.0051	0.0049	0.0048
−2.4	0.0082	0.0080	0.0078	0.0075	0.0073	0.0071	0.0069	0.0068	0.0066	0.0064
−2.3	0.0107	0.0104	0.0102	0.0099	0.0096	0.0094	:0091	0.0089	0.0087	0.0084
−2.2	0.0139	0.0136	0.0132	0.0129	0.0125	0.0122	0.0119	0.0116	0.0113	0.0110
−2.1	0.0179	0.0174	0.0170	0.0166	0.0162	0.0158	0.0154	0.0150	0.0146	0.0143
−2.0	0.0228	0.0222	0.0217	0.0212	0.0207	0.0202	0.0197	0.0192	0.0188	0.0183
−1.9	0.0287	0.0281	0.0274	0.0268	0.0262	0.0256	0.0250	0.0244	0.0239	0.0233
−1.8	0.0359	0.0351	0.0344	0.0336	0.0329	0.0322	0.0314	0.0307	0.0301	0.0294
−1.7	0.0446	0.0436	0.0427	0.0418	0.0409	0.0401	0.0392	0.0384	0.0375	0.0367
−1.6	0.0548	0.0537	0.0526	0.0516	0.0505	0.0495	0.0485	0.0475	0.0465	0.0455
−1.5	0.0668	0.0655	0.0643	0.0630	0.0618	0.0606	0.0594	0.0582	0.0511	0.0559
−1.4	0.0800.8	0.0793	0.0778	0.0764	0.0749	0.0735	0.0721	0.0708	0.0694	0.0681
−1.3	0.0968	0.0951	0.0934	0.0918	0.0901	0.0885	0.0869	0.0853	0.0838	0.0823
−1.2	0.1151	0.1131	0.1112	0.1093	0.1075	0.1056	0.1038	0.1020	0.1003	0.0985
−1.1	0.1357	0.1335	0.1314	0.1292	0.1271	0.1251	0.1230	0.1210	0.1190	0.1170
−1.0	0.1587	0.1562	0.1539	0.1515	0.1492	0.1469	0.1446	0.1423	0.1401	0.1379
−0.9	0.1841	0.1814	0.1788	0.1762	0.1736	0.1711	0.1685	0.1660	0.1635	0.1611
−0.8	0.2119	0.2090	0.2061	0.2033	0.2005	0.1977	0.1949	0.1922	0.1894	0.1867
−0.7	0.2420	0.2389	0.2358	0.2327	0.2296	0.2266	0.2236	0.2206	0.2177	0.2148
−0.6	0.2743	0.2709	0.2676	0.2643	0.2611	0.2578	0.2546	0.2514	0.2483	0.2451
−0.5	0.3085	0.3050	0.3015	0.2981	0.2946	0.2912	0.2877	0.2843	0.2810	0.2776

(Continued)

TABLE A.1
Standard Normal Distribution (*Continued*)

z	0.00	0.01	0.02	0.03	0.04	0.05	0.06	0.07	0.08	0.09
–0.4	0.3446	0.3409	0.3312	0.3336	0.3300	0.3264	0.3228	0.3192	0.3156	0.3121
–0.3	0.3821	0.3783	0.3745	0.3707	0.3669	0.3632	0.3594	0.3557	0.3520	0.3483
–0.2	0.4201	0.4168	0.4129	0.4090	0.4052	0.4013	0.3974	0.3936	0.3897	0.3859
–0.1	0.4602	0.4562	0.4522	0.4483	0.4443	0.4404	0.4364	0.4325	0.4286	0.4247
0.0	0.5000	0.4960	0.4920	0.4880	0.4840	0.4801	0.4761	0.4721	0.4681	0.4641
0.0	0.5000	0.5040	0.5080	0.5120	0.5160	0.5199	0.5239	0.5279	0.5319	0.5359
0.1	0.5398	0.5438	0.5478	0.5517	0.5557	0.5596	0.5636	0.5675	0.5714	0.5753
0.2	0.5193	0.5832	0.5871	0.5910	0.5948	0.5981	0.6026	0.6064	0.6103	0.6141
0.3	0.6179	0.6217	0.6255	0.6293	0.6331	0.6368	0.6406	0.6443	0.6480	0.6517
0.4	0.6554	0.6591	0.6628	0.6664	0.6700	0.6736	0.6772	0.6808	0.6844	0.6879
0.5	0.6915	0.6950	0.6985	0.7019	0.7054	0.7088	0.7123	0.7157	0.7190	0.7224
0.6	0.7257	0.7291	0.7324	0.7357	0.1389	0.7422	0.7454	0.7486	0.7517	0.7549
0.7	0.7580	0.7611	0.7642	0.7673	0.7704	0.7734	0.7764	0.7794	0.780.23	0.7852
0.8	0.7881	0.7910	0.7939	0.7967	0.7995	0.8023	0.8051	0.8078	0.8106	0.8133
0.9	0.8159	0.8186	0.8212	0.8238	0.8264	0.8289	0.8315	0.8340	0.8365	0.8389
1.0	0.8413	0.8438	0.8461	0.8485	0.8508	0.8531	0.8554	0.8577	0.8599	0.8621
1.1	0.8643	0.8665	0.8686	0.8708	0.8729	0.8749	0.8770	0.8790	0.8810	0.8830
1.2	0.8849	0.8869	0.8888	0.8907	0.8925	0.8944	0.8962	0.8980	0.8997	0.9015
1.3	0.9032	0.9049	0.9066	0.9082	0.9099	0.9115	0.9131	0.9147	0.9162	0.9177
1.4	0.9192	0.9207	0.9222	0.9236	0.9251	0.9265	0.9279	0.9292	0.9306	0.9319
1.5	0.9332	0.9345	0.9357	0.9370	0.9382	0.9394	0.9406	0.9418	0.9429	0.9441
1.6	0.9452	0.9463	0.9474	0.9484	0.9495	0.9505	0.9515	0.9525	0.9535	0.9545
1.7	0.9554	0.9564	0.9573	0.9582	0.9591	0.9599	0.9608	0.9616	0.9625	0.9633
1.8	0.9641	0.9649	0.9656	0.9664	0.9671	0.9678	0.9686	0.9693	0.9699	0.9706
1.9	0.9713	0.9719	0.9726	0.9732	0.9738	0.9744	0.9750	0.9756	0.9761	0.9767
2.0	0.9772	0.9778	0.9783	0.9788	0.9793	0.9198	0.9803	0.9808	0.9812	0.9817
2.1	0.9821	0.9826	0.9830	0.9834	0.9838	0.9842	0.9846	0.9850	0.9854	0.9857
2.2	0.9861	0.9864	0.9868	0.9871	0.9875	0.9878	0.9881	0.9884	0.9887	0.9890
2.3	0.9893	0.9896	0.9898	0.9901	0.9904	0.9906	0.9909	0.9911	0.9913	0.9916
2.4	0.9918	0.9920	0.9922	0.9925	0.9927	0.9929	0.9931	0.9932	0.9934	0.9936
2.5	0.9938	0.9940	0.9941	0.9943	0.9945	0.9946	0.9948	0.9949	0.9951	0.9952
2.6	0.9953	0.9955	0.9956	0.9957	0.9959	0.9960	0.9961	0.9962	0.9963	0.9964
2.7	0.9965	0.9966	0.9967	0.9968	0.9969	0.9970	0.9971	0.9972	0.9973	0.9974
2.8	0.9974	0.9975	0.9976	0.9977	0.9977	0.9978	0.9979	0.9979	0.9980	0.9981
2.9	0.9981	0.9982	0.9982	0.9983	0.9984	0.9984	0.9985	0.9985	0.9986	0.9986
3.0	0.9987	0.9987	0.9987	0.9988	0.9988	0.9989	0.9989	0.9989	0.9990	0.9990
3.1	0.9990	0.9991	0.9991	0.9991	0.9992	0.9992	0.9992	0.9992	0.9993	0.9993
3.2	0.9993	0.9993	0.9994	0.9994	0.9994	0.9994	0.9994	0.9995	0.9995	0.9995
3.3	0.9995	0.9995	0.9995	0.9996	0.9996	0.9996	0.9996	0.9996	0.9996	0.9997
3.4	0.9997	0.9997	0.9997	0.9997	0.9997	0.9997	0.9997	0.9997	0.9997	0.9998

TABLE A.2
Critical Values of the t Statistic

	Level of Significance, One-Tailed								
	0.25	0.20	0.10	0.05	0.025	0.0125	0.005	0.0025	0.0005
	Level of Significance, Two-Tailed								
df	0.50	0.40	0.20	0.10	0.05	0.025	0.01	0.005	0.001
1	1.000	1.376	3.078	6.314	12.706	25.452	63.657	318.300	636.600
2	0.816	1.061	1.886	2.920	4.303	6.205	9.925	14.089	31.598
3	0.765	0.978	1.638	2.353	3.182	4.176	5.841	7.453	12.941
4	0.741	0.941	1.533	2.132	2.776	3.495	4.604	5.598	8.610
5	0.727	0.920	1.476	2.015	2.571	3.163	4.032	4.773	6.859
6	0.718	0.906	1.440	1.943	2.447	2.969	3.707	4.317	5.959
7	0.711	0.896	1.415	1.895	2.365	2.841	3.499	4.029	5.405
8	0.706	0.889	1.397	1.860	2.306	2.752	3.355	3.832	5.041
9	0.703	0.883	1.383	1.833	2.262	2.685	3.250	3.690	4.781
10	0.700	0.879	1.372	1.812	2.228	2.634	3.169	3.581	4.587
11	0.697	0.876	1.363	1.796	2.201	2.593	3.106	3.497	4.437
12	0.695	0.873	1.356	1.782	2.179	2.560	3.055	3.428	4.318
13	0.694	0.870	1.350	1.771	2.160	2.533	3.012	3.372	4.221
14	0.692	0.868	1.345	1.761	2.145	2.510	2.977	3.326	4.140
15	0.691	0.866	1.341	1.753	2.131	2.490	2.947	3.286	4.073
16	0.690	0.865	1.337	1.746	2.120	2.473	2.921	3.252	4.015
17	0.689	0.863	1.333	1.740	2.110	2.458	2.898	3.222	3.965
18	0.688	0.862	1.330	1.734	2.101	2.445	2.878	3.197	3.922
19	0.688	0.861	1.328	1.729	2.093	2.433	2.861	3.174	3.883
20	0.687	0.860	1.325	1.725	2.086	2.423	2.845	3.153	3.850
21	0.686	0.859	1.323	1.721	2.080	2.414	2.831	3.135	3.819
22	0.686	0.858	1.321	1.717	2.074	2.406	2.819	3.119	3.792
23	0.685	0.858	1.319	1.714	2.069	2.398	2.807	3.104	3.761
24	0.685	0.857	1.318	1.711	2.064	2.391	2.797	3.090	3.745
25	0.684	0.856	1.316	1.708	2.060	2.385	2.787	3.078	3.725
26	0.684	0.856	1.315	1.706	2.056	2.379	2.779	3.067	3.707
27	0.684	0.855	1.314	1.703	2.052	2.373	2.771	3.056	3.690
28	0.683	0.855	1.313	1.701	2.048	2.368	2.763	3.047	3.674
29	0.683	0.854	1.311	1.699	2.045	2.364	2.756	3.038	3.659
30	0.683	0.854	1.310	1.697	2.042	2.360	2.750	3.030	3.646
35	0.682	0.852	1.306	1.690	2.030	2.342	2.724	2.996	3.591
40	0.681	0.851	1.303	1.684	2.021	2.329	2.704	2.971	3.551
45	0.680	0.850	1.301	1.680	2.014	2.319	2.690	2.952	3.520
50	0.680	0.849	1.299	1.676	2.008	2.310	2.678	2.937	3.496

(Continued)

TABLE A.2
Critical Values of the *t* Statistic (*Continued*)

	Level of Significance, One-Tailed								
	0.25	0.20	0.10	0.05	0.025	0.0125	0.005	0.0025	0.0005
	Level of Significance, Two-Tailed								
df	0.50	0.40	0.20	0.10	0.05	0.025	0.01	0.005	0.001
55	0.679	0.849	1.297	1.673	2.004	2.304	2.669	2.925	3.476
60	0.679	0.848	1.296	1.671	2.000	2.299	2.660	2.915	3.460
70	0.678	0.847	1.294	1.667	1.994	2.290	2.648	2.899	3.435
80	0.678	0.847	1.293	1.665	1.989	2.284	2.638	2.887	3.416
90	0.678	0.846	1.291	1.662	1.986	2.279	2.631	2.878	3.402
100	0.677	0.846	1.290	1.661	1.982	2.276	2.625	2.871	3.390
120	0.671	0.845	1.289	1.659	1.981	2.273	2.621	2.865	3.381
inf	0.674	0.842	1.282	1.645	1.960	2.241	2.576	2.807	3.290

TABLE A.3
Cumulative Distribution of Chi Square

df	Probability of Greater Value												
	0.995	0.990	0.975	0.950	0.900	0.750	0.500	0.250	0.100	0.050	0.025	0.010	0.005
1	0.00	0.00	0.00	0.00	0.02	0.10	0.45	1.32	2.71	3.84	5.02	6.63	7.88
2	0.01	0.02	0.05	0.10	0.21	0.58	1.39	2.77	4.61	5.99	7.38	9.21	10.60
3	0.07	0.11	0.22	0.35	0.58	1.21	2.37	4.11	6.25	7.81	9.35	11.34	12.84
4	0.21	0.30	0.48	0.71	1.06	1.92	3.36	5.39	7.78	9.49	11.14	13.28	14.86
5	0.41	0.55	0.83	1.15	1.61	2.67	4.35	6.63	9.24	11.07	12.83	15.09	16.75
6	0.68	0.87	1.24	1.64	2.20	3.45	5.35	7.84	10.64	12.59	14.45	16.81	18.55
7	0.99	1.24	1.69	2.17	2.83	4.25	6.35	9.04	12.02	14.07	16.01	18.48	20.28
8	1.34	1.65	2.18	2.73	3.49	5.01	7.34	10.22	13.36	15.51	17.53	20.09	21.96
9	1.73	2.09	2.70	3.33	4.17	5.90	8.34	11.39	14.68	16.92	19.02	21.67	23.59
10	2.16	2.56	:3.25	3.94	4.87	6.74	9.34	12.55	15.99	18.31	20.48	23.21	25.19
11	2.60	3.05	3.82	4.57	5.58	7.58	10.34	13.70	17.28	19.68	21.92	24.72	26.76
12	3.07	3.57	4.40	5.23	6.30	8.44	11.34	14.85	18.55	21.03	23.34	26.22	28.30
13	3.57	4.11	5.01	5.89	7.04	9.30	12.34	15.98	19.81	22.36	24.74	27.69	29.82
14	4.07	4.66	5.63	6.57	7.79	10.17	13.34	17.12	21.06	23.68	26.12	29.14	31.32
15	4.60	5.25	6.27	7.26	8.55	11.04	14.34	18.25	22.31	25.00	27.49	30.58	32.80
16	5.14	5.81	6.91	7.96	9.31	11.91	15.34	19.37	23.54	26.30	28.85	32.00	34.27
17	5.70	6.41	7.56	8.67	10.09	12.79	16.34	20.49	24.7?	27.59	30.19	33.41	35.72
18	6.26	7.01	8.23	9.39	10.86	13.68	17.34	21.60	25.99	28.87	31.53	34.81	37.16
19	6.84	7.63	8.91	10.12	11.65	14.56	18.34	22.72	27.20	30.14	32.85	36.19	38.58
20	7.34	8.26	9.59	10.85	12.44	15.45	19.34	22.83	28.41	31.41	34.17	37.57	40.00

(*Continued*)

TABLE A.3
Cumulative Distribution of Chi Square

	Probability of Greater Value												
df	0.995	0.990	0.975	0.950	0.900	0.750	0.500	0.250	0.100	0.050	0.025	0.010	0.005
21	8.03	8.90	10.28	11.59	13.24	16.34	20.34	24.93	29.62	32.67	35.48	38.93	41.40
22	8.64	9.54	10.98	12.34	14.04	17.24	21.34	26.04	30.81	33.92	36.78	40.29	42.80
23	9.26	10.20	11.69	13.09	14.85	18.14	22.34	27.14	32.01	35.17	38.08	41.64	44.18
24	9.89	10.86	12.40	13.85	15.66	19.04	23.34	28.24	33.20	36.42	39.36	42.98	45.56
25	10.52	11.52	13.12	14.61	16.47	19.94	24.34	29.34	34.38	37.65	40.65	44.31	46.93
26	11.16	12.20	13.84	15.38	17.29	20.84	25.34	30.43	35.56	38.89	41.92	45.64	48.29
27	11.81	12.88	14.57	16.15	18.11	21.57	26.34	31.53	36.74	40.11	43.19	46.96	49.64
28	12.46	13.56	15.31	16.93	18.94	22.66	27.34	32.62	37.92	41.34	44.46	48.28	50.99
29	13.12	14.26	16.05	17.71	19.77	23.57	28.34	33.71	39.09	42.56	45.72	49.59	52.34
30	13.19	14.95	16.79	18.49	20.60	24.48	29.34	34.80	40.26	43.77	46.98	50.89	53.67
40	20.71	22.16	24.43	26.51	29.05	33.66	39.34	45.62	51.80	55.76	59.34	63.69	66.77
50	27.99	29.71	32.36	34.76	37.69	42.94	49.33	56.33	63.17	67.50	71.42	76.15	79.49
60	35.53	37.48	40.48	43.19	46.46	52.29	59.33	66.98	74.40	79.08	83.30	88.38	91.95
70	43.28	45.44	48.76	51.74	55.33	61.70	69.33	77.58	85.53	90.53	95.02	100.42	104.22
80	51.17	53.54	57.15	60.39	64.28	71.14	79.33	88.13	96.58	101.88	106.63	112.33	116.32
90	59.20	61.75	65.65	69.13	73.29	80.62	89.33	98.64	107.56	113.14	113.14	124.12	128.30
100	67.33	70.06	74.22	77.93	82.36	90.13	99.33	109.14	118.50	124.34	129.56	135.81	140.17

TABLE A.4
Values of F Distribution

5% in Upper Tail

	df1								
df2	**1**	**2**	**3**	**4**	**5**	**6**	**7**	**8**	**9**
1	161.40	199.50	215.70	224.60	230.20	234.00	236.80	238.90	240.50
2	18.51	19.00	19.16	19.25	19.30	19.33	19.35	19.37	19.38
3	10.13	9.55	9.28	9.12	9.01	8.94	8.89	8.85	8.81
4	7.71	6.94	6.59	6.39	6.26	6.16	6.09	6.04	6.00
5	6.61	5.79	5.41	5.19	5.05	4.95	4.88	4.82	4.77
6	5.99	5.14	4.76	4.53	4.39	4.28	4.21	4.15	4.10
7	5.59	4.74	4.35	4.12	3.97	3.87	3.79	3.73	3.68
8	5.32	4.46	4.07	3.84	3.69	3.58	3.50	3.44	3.39
9	5.12	4.26	3.86	3.63	3.48	3.37	3.29	3.23	3.18
10	4.96	4.10	3.71	3.48	3.33	3.22	3.14	3.07	3.02
11	4.84	3.98	3.59	3.36	3.20	3.09	3.01	2.95	2.90
12	4.75	3.89	3.49	3.26	3.11	3.00	2.91	2.85	2.80
13	4.67	3.81	3.41	3.18	3.03	2.92	2.83	2.77	2.71
14	4.60	3.74	3.34	3.11	2.96	2.85	2.76	2.70	2.65
15	4.54	3.68	3.29	3.06	2.90	2.79	2.71	2.64	2.59
16	4.49	3.63	3.24	3.01	2.85	2.74	2.66	2.59	2.54
17	4.45	3.59	3.20	2.96	2.81	2.70	2.61	2.55	2.49
18	4.41	3.55	3.16	2.93	2.77	2.66	2.58	2.51	2.46
19	4.38	3.52	3.13	2.90	2.74	2.63	2.54	2.48	2.42
20	4.35	3.49	3.10	2.87	2.71	2.60	2.51	2.45	2.39
21	4.32	3.47	3.07	2.84	2.68	2.57	2.49	2.42	2.37
22	4.30	3.44	3.05	2.82	2.66	2.55	2.46	2.40	2.34
23	4.28	3.42	3.03	2.80	2.64	2.53	2.44	2.37	2.32
24	4.26	3.40	3.01	2.7J3	2.62	2.51	2.42	2.36	2.30
25	4.24	3.39	2.99	2.76	2.60	2.49	2.40	2.34	2.28
26	4.23	3.37	2.98	2.74	2.59	2.47	2.39	2.32	2.27
27	4.21	3.35	2.96	2.73	2.57	2.46	2.37	2.31	2.25
28	4.20	3.34	2.95	2.71	2.56	2.45	2.36	2.29	2.24
29	4.18	3.33	2.93	2.70	2.55	2.43	2.35	2.28	2.22
30	4.17	3.32	2.92	2.69	2.53	2.42	2.33	2.27	2.21
40	4.08	3.23	2.84	2.61	2.45	2.34	2.25	2.18	2.12
60	4.00	3.15	2.76	2.53	2.37	2.25	2.17	2.10	2.04
120	3.92	3.07	2.68	2.45	2.29	2.17	2.09	2.02	1.96
inf	3.84	3.00	2.60	2.37	2.21	2.10	2.0.1	1.94	1.88

(Continued)

TABLE A.4
Values of F Distribution (*Continued*)

5% in Upper Tail

	df1									
df2	**10**	**12**	**15**	**20**	**24**	**30**	**40**	**60**	**120**	**inf**
1	241.90	243.90	245.90	248.00	249.10	250.10	251.10	252.20	253.30	254.30
2	19.40	19.41	19.43	19.44	19.45	19.46	19.47	19.48	19.49	19.50
3	8.79	8.74	8.70	8.66	8.64	8.62	8.59	8.57	8.55	8.53
4	5.96	5.91	5.86	5.80	5.77	5.15	5.72	5.69	5.66	5.63
5	4.74	4.68	4.62	4.56	4.53	4.50	4.46	4.43	4.40	4.36
6	4.06	4.00	3.94	3.87	3.84	3.81	3.77	3.74	3.70	3.67
7	3.64	3.57	3.51	3.44	3.41	3.38	3.34	3.30	3.27	3.23
8	3.35	3.28	3.22	3.15	3.12	3.08	3.04	3.01	2.97	2.93
9	3.14	3.07	3.01	2.94	2.90	2.86	2.83	2.79	2.75	2.71
10	2.98	2.91	2.85	2.77	2.74	2.70	2.66	2.62	2.58	2.54
11	2.85	2.79	2.72	2.65	2.61	2.57	2.53	2.49	2.45	2.40
12	2.75	2.69	2.62	2.54	2.51	2.47	2.43	2.38	2.34	2.30
13	2.67	2.60	2.53	2.46	2.42	2.38	2.34	2.30	2.25	2.21
14	2.60	2.53	2.46	2.39	2.35	2.31	2.27	2.22	2.18	2.13
15	2.54	2.48	2.40	2.33	2.29	2.25	2.20	2.16	2.11	2.07
16	2.49	2.42	2.35	2.28	2.24	2.19	2.15	2.11	2.06	2.01
17	2.45	2.38	2.31	2.23	2.19	2.15	2.10	2.06	2.01	1.96
18	2.41	2.34	2.27	2.19	2.15	2.11	2.06	2.02	1.97	1.92
19	2.38	2.31	2.23	2.16	2.11	2.07	2.03	1.98	1.93	1.88
20	2.35	2.28	2.20	2.12	2.08	2.04	1.99	1.95	1.90	1.84
21	2.32	2.25	2.18	2.10	2.05	2.01	1.96	1.92	1.87	1.81
22	2.30	2.23	2.15	2.07	2.03	1.98	1.94	1.89	1.84	1.78
23	2.27	2.20	2.13	2.05	2.01	1.96	1.91	1.86	1.81	1.76
24	2.25	2.18	2.11	2.03	1.98	1.94	1.89	1.84	1.79	1.73
25	2.24	2.16	2.09	2.01	1.96	1.92	1.87	1.82	1.77	1.71
26	2.22	2.15	2.07	1.99	1.95	1.90	1.85	1.80	1.75	1.69
27	2.20	2.13	2.06	1.97	1.93	1.88	1.84	1.79	1.73	1.67
28	2.19	2.12	2.04	1.96	1.91	1.87	1.82	1.77	1.71	1.65
29	2.18	2.10	2.03	1.94	1.90	1.85	1.81	1.75	1.70	1.64
30	2.16	2.09	2.01	1.93	1.89	1.84	1.79	1.74	1.68	1.62
40	2.08	2.00	1.92	1.84	1.79	1.74	i.69	1.64	1.58	1.51
60	1.99	1.92	1.84	1.75	1.70	1.65	1.59	1.53	1.47	1.39
120	1.91	1.83	1.75	1.66	1.61	1.55	1.50	1.43	1.35	1.25
inf	1.83	1.75	1.67	1.57	1.52	1.46	1.39	1.32	1.22	1.00

(Continued)

TABLE A.4
Values of F Distribution (*Continued*)

					1% in Upper Tail				
					df1				
df2	1	2	3	4	5	6	7	8	9
1	4052.00	5000.00	5403.00	5625.00	5764.00	5859.00	5928.00	5981.00	6022.00
2	98.50	99.00	99.17	99.25	99.30	99.33	99.36	99.37	99.39
3	34.12	30.82	29.46	28.71	28.24	27.91	27.67	27.49	27.35
4	21.20	18.00	16.69	15.98	15.52	15.21	14.98	14.80	14.66
5	16.26	13.27	12.06	11.39	10.97	10.67	10.46	10.29	10.16
6	13.75	10.92	9.78	9.15	8.75	8.47	8.26	8.10	7.98
7	12.25	9.55	8.45	7.85	7.46	7.19	6.99	6.84	6.72
8	11.26	8.65	7.59	7.01	6.63	6.37	6.18	6.03	5.91
9	10.56	8.02	6.99	6.42	6.06	5.80	5.61	5.47	5.35
10	10.04	7.56	6.55	5.99	5.64	5.39	5.20	5.06	4.94
11	9.65	7.21	6.22	5.67	5.32	5.07	4.89	4.74	4.63
12	9.33	6.93	5.95	5.41	5.06	4.82	4.64	4.50	4.39
13	9.07	6.70	5.74	5.21	4.86	4.62	4.44	4.30	4.19
14	8.86	6.51	5.56	5.04	4.69	4.46	4.28	4.14	4.03
15	8.68	6.36	5.42	4.89	4.56	4.32	4.14	4.00	3.89
16	8.53	6.23	5.29	4.77	4.44	4.20	4.03	3.89	3.78
17	8.40	6.11	5.18	4.67	4.34	4.10	3.93	3.79	3.68
18	8.29	6.01	5.09	4.58	4.25	4.01	3.84	3.71	3.60
19	8.18	5.93	5.01	4.50	4.11	3.94	3.77	3.63	3.52
20	8.10	5.85	4.94	4.43	4.10	3.87	3.70	3.56	3.46
21	8.02	5.78	4.87	4.37	4.04	3.81	3.64	3.51	3.40
22	7.95	5.72	4.82	4.31	3.99	3.76	3.59	3.45	3.35
23	7.88	5.66	4.76	4.26	3.94	3.71	3.54	3.41	3.30
24	7.82	5.61	4.72	4.22	3.90	3.67	3.50	3.36	3.26
25	7.77	5.57	4.68	4.18	3.85	3.63	3.46	3.32	3.22
26	7.72	5.53	4.64	4.14	3.82	3.59	3.42	3.29	3.18
27	7.68	5.49	4.60	4.11	3.78	3.56	3.39	3.26	3.15
28	7.64	5.45	4.57	4.07	3.75	3.53	3.36	3.23	3.12
29	7.60	5.42	4.54	4.04	3.73	3.50	3.33	3.20	3.09
30	7.56	5.39	4.51	4.02	3.70	3.47	3.30	3.17	3.07
40	7.31	5.18	4.31	3.83	3.51	3.29	3.12	2.99	2.89
60	7.08	4.98	4.13	3.65	3.34	3.12	2.95	2.82	2.72
120	6.85	4.79	3.95	3.48	3.17	2.96	2.79	2.66	2.56
inf	6.63	4.61	3.18	3.32	3.02	2.80	2.64	2.51	2.41

(*Continued*)

TABLE A.4
Values of F Distribution (*Continued*)

1% in Upper Tail

	df1									
df2	**10**	**12**	**15**	**20**	**24**	**30**	**40**	**60**	**120**	**inf**
1	6056.00	6106.00	6157.00	6209.00	6235.00	6261.00	6287.00	6313.00	6339.00	6366.00
2	99.40	99.42	99.43	99.45	99.46	99.47	99.47	99.48	99.49	99.50
3	27.23	27.05	26.87	26.69	26.60	26.50	26.41	26.32	26.22	26.13
4	14.55	14.37	14.20	14.02	13.93	13.84	13.75	13.65	13.56	13.46
5	10.05	9.89	9.72	9.55	9.47	9.38	9.29	9.20	9.11	9.02
6	7.87	7.72	7.56	7.40	7.31	7.23	7.14	7.06	6.97	6.88
7	6.62	6.47	6.31	6.16	6.07	5.99	5.91	5.82	5.74	5.65
8	5.81	5.67	5.52	5.36	5.28	5.20	5.12	5.03	4.95	4.86
9	5.26	5.11	4.96	4.81	4.73	4.65	4.57	4.48	4.40	4.31
10	4.85	4.71	4.56	4.41	4.33	4.25	4.17	4.08	4.00	3.91
11	4.54	4.40	4.25	4.10	4.02	3.94	3.86	3.78	3.69	3.60
12	4.30	4.16	4.01	3.86	3.78	3.70	3.62	3.54	3.45	3.36
13	4.10	3.96	3.82	3.66	3.59	3.51	3.43	3.34	3.25	3.17
14	3.94	3.80	3.66	3.51	3.43	3.35	3.27	3.18	3.09	3.00
15	3.80	3.67	3.52	3.37	3.29	3.21	3.13	3.05	2.96	2.87
16	3.69	3.55	3.41	3.26	3.18	3.10	3.02	2.93	2.84	2.75
17	3.59	3.46	3.31	3.16	3.08	3.00	2.92	2.83	2.75	2.65
18	3.51	3.37	3.23	3.08	3.00	2.92	2.84	2.75	2.66	2.57
19	3.43	3.30	3.15	3.00	2.92	2.84	2.76	2.67	2.58	2.49
20	3.37	3.23	3.09	2.94	2.86	2.78	2.69	2.61	2.52	2.42
21	3.31	3.17	3.03	2.88	2.80	2.72	2.64	2.55	2.46	2.36
22	3.26	3.12	2.98	2.83	2.75	2.67	2.58	2.50	2.40	2.31
23	3.21	3.07	2.93	2.78	2.70	2.62	2.54	2.45	2.35	2.26
24	3.17	3.03	2.89	2.74	2.66	2.58	2.49	2.40	2.31	2.21
25	3.13	2.99	2.85	2.70	2.62	2.54	2.45	2.36	2.27	2.17
26	3.09	2.96	2.81	2.66	2.58	2.50	2.42	2.33	2.23	2.13
27	3.06	2.93	2.78	2.63	2.55	2.47	2.38	2.29	2.20	2.10
28	3.03	2.90	2.75	2.60	2.52	2.44	2.35	2.26	2.17	2.06
29	3.00	2.87	2.73	2.57	2.49	2.41	2.33	2.23	2.14	2.03
30	2.98	2.84	2.70	2.55	2.47	2.39	2.30	2.21	2.11	2.01
40	2.80	2.66	2.52	2.37	2.29	2.20	2.11	2.02	1.92	1.80
60	2.63	2.50	2.35	2.20	2.12	2.03	1.94	1.84	1.73	1.60
120	2.47	2.34	2.19	2.03	1.95	1.86	1.76	1.66	1.53	1.38
inf	2.32	2.18	2.04	1.88	1.79	1.70	1.59	1.47	1.32	1.00

TABLE A.5
Critical Values of Runs Test

n2	2	3	4	5	6	7	8	9	10	11	12	13	14	15	16	17	18	19	20
n1																			
2											2	2	2	2	2	2	2	2	2
3					2	2	2	2	2	2	2	2	2	3	3	3	3	3	3
4				2	2	2	3	3	3	3	3	3	3	3	4	4	4	4	4
5			2	2	3	3	3	3	3	4	4	4	4	4	4	4	5	5	5
6		2	2	3	3	3	3	4	4	4	4	5	5	5	5	5	5	6	6
7		2	2	3	3	3	4	4	5	5	5	5	5	6	6	6	6	6	6
8		2	3	3	3	4	4	5	5	5	6	6	6	6	6	7	7	7	7
9		2	3	3	4	4	5	5	5	6	6	6	7	7	7	7	8	8	8
10		2	3	3	4	5	5	5	6	6	7	7	7	7	8	8	8	8	9
11		2	3	4	4	5	5	6	6	7	7	7	8	8	8	9	9	9	9
12	2	2	3	4	4	5	6	6	7	7	7	8	8	8	9	9	9	10	10
13	2	2	3	4	5	5	6	6	7	7	8	8	9	9	9	10	10	10	10
14	2	2	3	4	5	5	6	7	7	8	8	9	9	9	10	10	10	11	11
15	2	3	3	4	5	6	6	7	7	8	8	9	9	10	10	11	11	11	12
16	2	3	4	4	5	6	6	7	8	8	9	9	10	10	11	11	11	12	12
17	2	3	4	4	5	6	7	7	8	9	9	10	10	11	11	11	12	12	13
18	2	3	4	5	5	6	7	8	8	9	9	10	10	11	11	12	12	13	13
19	2	3	4	5	6	6	7	8	8	9	10	10	11	11	12	12	13	13	13
20	2	3	4	5	6	6	7	8	9	9	10	10	11	12	12	13	13	13	14

(*Continued*)

TABLE A.5
Critical Values of Runs Test (*Continued*)

n2	2	3	4	5	6	7	8	9	10	11	12	13	14	15	16	17	18	19	20
n1																			
2																			
3																			
4				9	9														
5			9	10	10	11	11												
6			9	10	11	12	12	13	13	13	13								
7				11	12	13	13	14	14	14	14	15	15	15					
8				11	12	13	14	14	15	15	16	16	16	16	17	17	17	17	17
9					13	14	14	15	16	16	16	17	17	18	18	18	18	18	18
10					13	14	15	16	16	17	17	18	18	18	19	19	19	20	20
11					13	14	15	16	17	17	18	19	19	19	20	20	20	21	21
12					13	14	16	16	17	18	19	19	20	20	21	21	21	22	22
13						15	16	17	18	19	19	20	20	21	21	22	22	23	23
14						15	16	17	18	19	20	20	21	22	22	23	23	23	24
15						15	16	18	18	19	20	21	22	22	23	23	24	24	25
16							17	19	19	20	21	21	22	23	23	24	25	25	25
17							17	18	19	20	21	22	23	23	24	25	25	26	26
18							17	18	19	20	21	22	23	24	25	25	26	26	27
19							17	18	20	21	22	23	23	24	25	26	26	27	27
20							17	18	20	21	22	23	24	25	25	26	27	27	28

Note: For a one-tailed test, any sample value that is equal to or smaller than the tabled value is significant at the 2.5% level. For a two-tailed test, any value that is equal to or smaller than the tabled value in the upper table or equal to or greater than the tabled value in the lower table is significant at the 5% level of significance.

TABLE A.6
Kendall's τ Statistic

n	T	P	n	T	P	n	T	P	n	T	P
3	1.000	0.167	7	1.000	0.000	9	1.000	0.000	10	1.000	0.000
	0.333	0.500		0.905	0.001		0.944	0.000		0.956	0.000
4	1.000	0.042		0.810	0.005		0.889	0.000		0.911	0.000
	0.667	0.167		0.714	0.015		0.833	0.000		0.867	0.000
	0.333	0.375		0.619	0.035		0.778	0.001		0.822	0.000
	0.000	0.625		0.524	0.068		0.722	0.003		0.778	0.000
5	1.000	0.008		0.429	0.119		0.667	0.006		0.733	0.001
	0.800	0.042		0.333	0.191		0.611	0.012		0.689	0.002
	0.600	0.117		0.238	0.281		0.556	0.022		0.644	0.005
	0.400	0.242		0.143	0.386		0.500	0.038		0.600	0.008
	0.200	0.408		0.048	0.500		0.444	0.060		0.556	0.014
	0.000	0.592	8	1.000	0.000		0.389	0.090		0.511	0.023
6	1.000	0.001		0.929	0.000		0.333	0.130		0.467	0.036
	0.867	0.008		0.857	0.001		0.278	0.179		0.422	0.054
	0.733	0.028		0.786	0.003		0.222	0.238		0.378	0.078
	0.600	0.068		0.714	0.007		0.167	0.306		0.333	0.108
	0.467	0.136		0.643	0.016		0.111	0.381		0.289	0.146
	0.333	0.235		0.571	0.031		0.056	0.460		0.244	0.190
	0.200	0.360		0.500	0.054		0.000	0.540		0.200	0.242
	0.067	0.500		0.429	0.089					0.156	0.300
				0.357	0.138					0.111	0.364
				0.286	0.199					0.067	0.431
				0.214	0.274					0.022	0.500
				0.143	0.360						
				0.071	0.452						
				0.000	0.548						

	Tail Probability on |T| for a Two-Tailed Test[a]				
	0.200	**0.100**	**0.050**	**0.020**	**0.010**
	Right-Tail Probability on T or Left-Tail Probability on $-T$ for One-Tailed Test				
n	**0.100**	**0.005**	**–0.025**	**0.010**	**0.005**
11	0.345	0.418	0.491	0.564	0.600
12	0.303	0.394	0.455	0.545	0.576
13	0.308	0.359	0.436	0.513	0.564
14	0.275	0.363	0.407	0.473	0.516
15	0.276	0.333	0.390	0.467	0.505
16	0.250	0.317	0.383	0.433	0.483
17	0.250	0.309	0.368	0.426	0.471
18	0.242	0.294	0.346	0.412	0.451
19	0.228	0.287	0.333	0.392	0.439
20	0.221	0.274	0.326	0.379	0.421
21	0.210	0.267	0.314	0.371	0.410

(Continued)

TABLE A.6
Kendall's τ Statistic (*Continued*)

	Tail Probability on |T| for a Two-Tailed Test[a]				
	0.200	**0.100**	**0.050**	**0.020**	**0.010**
	Right-Tail Probability on *T* or Left-Tail Probability on −*T* for One-Tailed Test				
n	**0.100**	**0.005**	**−0.025**	**0.010**	**0.005**
22	0.203	0.264	0.307	0.359	0.394
23	0.202	0.251	0.296	0.352	0.391
24	0.196	0.246	0.290	0.341	0.377
25	0.193	0.240	0.287	0.333	0.367
26	0.188	0.237	0.280	0.329	0.360
27	0.179	0.231	0.271	0.322	0.356
28	0.180	0.228	0.265	0.312	0.344
29	0.172	0.222	0.261	0.310	0.340
30	0.172	0.218	0.255	0.301	0.333

Note: P values are the cumulative probability for the right tail from T to its maximum value of 1. The same probability is a cumulative left-tail probability, from the minimum value minus 1 to the value of $-T$.

[a] For n greater than 10, this table segment provides the smallest value of T (largest value of $-T$) for which the right-tail (left-tail) probability for a one-sided test is less than or equal to the values shown. The same values apply to the absolute value for a two-sided test with the tail probability shown.

TABLE A.7
Critical Values for Univariate Analyses with Spearman–Conley One-Tailed Serial Correlation Test

Sample Size	Lower-Tail Level of Significance (%)					Upper-Tail Level of Significance (%)				
	10	**5**	**1**	**0.5**	**0.1**	**10**	**5**	**1**	**0.5**	**0.1**
5	−0.8	−0.8	−1.0	−1.0	−1.0	0.4	0.6	1.0	1.0	1.0
6	−0.7	−0.8	−0.9	−1.0	−1.0	0.3	0.6	0.8	1.0	1.0
7	−0.657	−0.771	−0.886	−0.943	−0.943	0.371	0.486	0.714	0.771	0.943
8	−0.607	−0.714	−0.857	−0.893	−0.929	0.357	0.464	0.679	0.714	0.857
9	−0.572	−0.667	−0.810	−0.857	−0.905	0.333	0.452	0.643	0.691	0.809
10	−0.533	−0.633	−0.783	−0.817	−0.883	0.317	0.433	0.617	0.667	0.767
11	−0.499	−0.600	−0.746	−0.794	−0.867	0.297	0.406	0.588	0.648	0.745
12	−0.473	−0.564	−0.718	−0.764	−0.836	0.291	0.400	0.573	0.627	0.727
13	−0.448	−0.538	−0.692	−0.734	−0.815	0.287	0.385	0.552	0.601	0.706
14	−0.429	−0.516	−0.665	−0.714	−0.791	0.275	0.369	0.533	0.588	0.687
15	−0.410	−0.495	−0.644	−0.692	−0.771	0.270	0.358	0.521	0.574	0.671
16	−0.393	−0.479	−0.621	−0.671	−0.754	0.261	0.350	0.507	0.557	0.656

(Continued)

TABLE A.7
Critical Values for Univariate Analyses with Spearman–Conley One-Tailed Serial Correlation Test (*Continued*)

Sample Size	Lower-Tail Level of Significance (%)					Upper-Tail Level of Significance (%)				
	10	5	1	0.5	0.1	10	5	1	0.5	0.1
17	−0.379	−0.462	−0.603	−0.650	−0.735	0.256	0.341	0.491	0.544	0.641
18	−0.365	−0.446	−0.586	−0.633	−0.718	0.250	0.333	0.480	0.530	0.627
19	−0.354	−0.433	−0.569	−0.616	−0.703	0.245	0.325	0.470	0.518	0.613
20	−0.344	−0.421	−0.554	−0.600	−0.687	0.240	0.319	0.460	0.508	0.601
21	−0.334	−0.409	−0.542	−0.586	−0.672	0.235	0.313	0.451	0.499	0.590
22	−0.325	−0.399	−0.530	−0.573	−0.657	0.230	0.307	0.442	0.489	0.580
23	−0.316	−0.389	−0.518	−0.560	−0.643	0.226	0.301	0.434	0.479	0.570
24	−0.307	−0.379	−0.506	−0.549	−0.632	0.222	0.295	0.426	0.471	0.560
25	−0.301	−0.370	−0.496	−0.537	−0.621	0.218	0.290	0.419	0.463	0.550
26	−0.294	−0.362	−0.486	−0.526	−0.610	0.214	0.285	0.412	0.455	0.540
27	−0.288	−0.354	−0.476	−0.516	−0.600	0.211	0.280	0.405	0.448	0.531
28	−0.282	−0.347	−0.467	−0.507	−0.590	0.208	0.275	0.398	0.441	0.523
29	−0.276	−0.341	−0.458	−0.499	−0.580	0.205	0.271	0.392	0.434	0.515
30	−0.270	−0.335	−0.450	−0.491	−0.570	0.202	0.267	0.386	0.428	0.507

TABLE A.8
Critical Values of Durbin–Watson Statistic for 1% and 5% Levels of Significance (α)

		Lower Critical Values					Upper Critical Values				
		Number of Predictor Variables									
α	n	1	2	3	4	5	1	2	3	4	5
1%	15	0.81	0.70	0.59	0.49	0.39	1.07	1.25	1.46	1.70	1.96
	20	0.95	0.86	0.77	0.63	0.60	1.15	1.27	1.41	1.57	1.74
	25	1.05	0.98	0.90	0.83	0.75	1.21	1.30	1.41	1.52	1.65
	30	1.13	1.07	1.01	0.94	0.88	1.26	1.34	1.42	1.51	1.61
	40	1.25	1.20	1.15	1.10	1.05	1.34	1.40	1.46	1.52	1.58
	50	1.32	1.28	1.24	1.20	1.16	1.40	1.45	1.49	1.54	1.59
	60	1.38	1.35	1.32	1.28	1.25	1.45	1.48	1.52	1.56	1.60
	80	1.47	1.44	1.42	1.39	1.36	1.52	1.54	1.57	1.60	1.62
	100	1.52	1.50	1.48	1.45	1.44	1.56	1.58	1.60	1.63	1.65
5%	15	1.08	0.95	0.82	0.69	0.56	1.36	1.54	1.75	1.97	2.21
	20	1.20	1.10	1.00	0.90	0.79	1.41	1.54	1.68	1.83	1.99
	25	1.28	1.21	1.12	1.04	0.95	1.45	1.55	1.66	1.77	1.89
	30	1.35	1.28	1.21	1.14	1.07	1.49	1.57	1.65	1.74	1.83

(*Continued*)

TABLE A.8
Critical Values of Durbin–Watson Statistic for 1% and 5% Levels of Significance (α) (*Continued*)

		Lower Critical Values					Upper Critical Values				
		Number of Predictor Variables									
α	n	1	2	3	4	5	1	2	3	4	5
	40	1.44	1.39	1.34	1.29	1.23	1.54	1.60	1.66	1.72	1.79
	50	1.50	1.46	1.42	1.38	1.34	1.59	1.63	1.67	1.72	1.77
	60	1.55	1.51	1.48	1.44	1.41	1.62	1.65	1.69	1.73	1.77
	80	1.61	1.59	1.56	1.53	1.51	1.66	1.69	1.72	1.74	1.77
	100	1.65	1.63	1.61	1.59	1.57	1.69	1.72	1.74	1.76	1.78

TABLE A.9
Critical Values of Wilcoxon Matched-Pairs Signed-Ranks (T) Test

	Two-Tailed Level of Significance		
	0.050	0.02	0.010
	One-Tailed Level of Significance		
N	0.025	0.01	0.005
6	0		
7	2	0	
8	4	2	0
9	6	3	2
10	8	5	3
11	11	7	5
12	14	10	7
13	17	13	10
14	21	16	13
15	25	20	16
16	30	24	20
17	35	28	23
18	40	33	28
19	46	38	32
20	52	43	38
21	59	49	43
22	66	56	49
23	73	62	55
24	81	69	61
25	89	77	68

TABLE A.10
Selected Values for Siegel–Tukey Test

n	*L*	*p*	*R*	*n*	*L*	*p*	*R*	*n*	*L*	*p*	*R*
	$m = 5$				$m = 6$				$m = 7$		
5	17	0.016	38	6	25	0.013	53	7	35	0.013	70
	18	0.028	37		26	0.021	52		36	0.019	69
	19	0.048	36		27	0.032	51		37	0.027	68
	20	0.075	35		28	0.047	50		38	0.036	67
	21	0.111	34		29	0.066	49		39	0.049	66
6	17	0.009	43	7	26	0.011	58	8	36	0.010	76
	18	0.015	42		27	0.017	54		37	0.014	75
	19	0.026	41		28	0.026	56		38	0.020	74
	20	0.041	40		29	0.037	55		39	0.027	73
	21	0.063	39		30	0.051	54		40	0.036	72
7	18	0.009	47	8	27	0.010	63		41	0.047	71
	19	0.015	46		28	0.015	62		42	0.060	70
	20	0.024	45		29	0.021	61	9	38	0.011	81
	21	0.037	44		30	0.030	60		39	0.016	80
	22	0.053	43		31	0.041	59		40	0.021	79
8	19	0.009	51		32	0.054	58		41	0.027	78
	20	0.015	50	9	29	0.013	67		42	0.036	77
	21	0.023	49		30	0.018	66		43	0.045	76
	22	0.033	48		31	0.025	65		44	0.057	75
	23	0.047	47		32	0.033	64	10	40	0.012	86
9	20	0.009	55		33	0.044	63		41	0.017	85
	21	0.014	54		34	0.057	62		42	0.022	84
	22	0.021	53	10	30	0.011	72		43	0.028	83
	23	0.030	52		31	0.016	71		44	0.035	82
	24	0.041	51		32	0.021	70		45	0.044	81
	25	0.056	50		33	0.028	69		46	0.054	80
10	21	0.010	59		34	0.036	68				
	22	0.014	58		35	0.047	67				
	23	0.020	57		36	0.059	66				
	24	0.028	56								
	25	0.038	55								
	26	0.050	54								
8	46	0.010	90	9	59	0.009	112	10	74	0.009	136
	47	0.014	89		60	0.012	111		75	0.012	135
	48	0.019	88		66	0.047	105		82	0.045	128
	49	0.025	87		67	0.057	104		83	0.053	127
	50	0.032	86	10	61	0.009	119				
	51	0.041	85		62	0.011	118				
	52	0.052	84		69	0.047	111				
9	48	0.010	96		70	0.056	110				
	49	0.014	95								
	50	0.018	94								

(Continued)

TABLE A.10
Selected Values for Siegel–Tukey Test (*Continued*)

n	L	p	R	n	L	p	R	n	L	p	R
		m = 5				m = 6				m = 7	
	51	0.023	93								
	52	0.030	92								
	53	0.037	91								
	54	0.046	90								
	55	0.057	89								
10	50	0.010	102								
	51	0.013	101								
	56	0.042	96								
	57	0.051	95								

Note: The table entries provide sample sizes m and n where $m \leq n$ and the extreme values in each tail (left, L; right, R) for selected cumulative probabilities (p).

TABLE A.11
Deviates (*K*) for Pearson III Distribution

	Standardized Skew (g)						
Probability	**0.0**	**0.1**	**0.2**	**0.3**	**0.4**	**0.5**	**0.6**
0.9999	−3.71902	−3.50703	−3.29921	−3.09631	−2.89907	−2.70836	−2.52507
0.9995	−3.29053	−3.12167	−2.96698	−2.80889	−2.65390	−2.50257	−2.35549
0.9990	−3.09023	−2.94834	−2.80786	−2.66915	−2.53261	−2.39867	−2.26780
0.9980	−2.87816	−2.75106	−2.63672	−2.51741	−2.39942	−2.28311	−2.16884
0.9950	−2.57583	−2.48187	−2.38795	−2.29423	−2.20092	−2.10825	−2.01644
0.9900	−2.32635	−2.25258	−2.17840	−2.10394	−2.02933	−1.95472	−1.88029
0.9800	−2.05375	−1.99913	−1.94499	−1.88959	−1.83361	−1.77716	−1.72033
0.9750	−1.95996	−1.91219	−1.86360	−1.81427	−1.76427	−1.71366	−1.66253
0.9600	−1.75069	−1.71580	−1.67999	−1.64329	−1.60574	−1.56740	−1.52830
0.9500	−1.64485	−1.61594	−1.58607	−1.55527	−1.52357	−1.49101	−1.45762
0.9000	−1.281.55	−1.27037	−1.25824	−1.24516	−1.23114	−1.21618	−1.20028
0.8000	−0.84162	−0.84611	−0.84986	−0.85285	−0.85508	−0.85653	−0.85718
0.7000	−0.52440	−0.53624	−0.54757	−0.55839	−0.56867	−0.57840	−0.58757
0.6000	−0.25335	−0.26882	−0.28403	−0.29897	−0.31362	−0.32796	−0.34198
0.5704	−0.17733	−0.19339	−0.20925	−0.22492	−0.24037	−0.25558	−0.27047
0.5000	0.00000	−0.01662	−0.03325	−0.04993	−0.06651	−0.08302	−0.09945
0.4296	0.17733	0.16111	0.14472	0.12820	0.11154	0.09478	0.07791
0.4000	0.25335	0.23763	0.22168	0.20552	0.18916	0.17261	0.15589
0.3000	0.52440	0.51207	0.49927	0.48600	0.47228	0.45812	0.44352
0.2000	0.84162	0.83639	0.83044	0.82377	0.81638	0.80829	0.79950
0.1000	1.28155	1.29178	1.30105	1.30936	1.31671	1.32309	1.32850
0.0500	1.64485	1.67279	1.69971	1.72562	1.75048	1.77428	1.79101
0.0400	1.75069	1.78462	1.81756	1.84949	1.88039	1.91022	1.93896

TABLE A.11
Deviates (*K*) for Pearson III Distribution (*Continued*)

Probability	0.0	0.1	0.2	0.3	0.4	0.5	0.6
0.0250	1.95996	2.00688	2.05290	2.09795	2.14202	2.18505	2.22702
0.0200	2.05375	2.10697	2.15935	2.21.081	2.26133	2.31084	2.35931
0.0100	2.32635	2.39961	2.47226	2.54421	2.61539	2.68572	2.75514
0.0050	2.57583	2.66965	2.76321	2.85636	2.94900	3.04102	3.13232
0.0020	2.87816	2.99978	3.12169	3.24371	3.36566	3.48737	3.60872
0.0010	3.09023	3.23322	3.37703	3.52139	3.66608	3.81090	3.95567
0.0005	3.29053	3.45513	3.62113	3.78820	3.95605	4.12443	4.29311
0.0001	3.71902	3.93453	4.15301	4.37394	4.59687	4.82141	5.04718

Probability	0.7	0.8	0.9	1.0	1.1	1.2	1.3
0.9999	−2.35015	−2.18448	−2.02891	−1.88410	−1.75053	−1.62838	−1.51752
0.9995	−2.21328	−2.01661	−1.94611	−1.82241	−1.7,0603	−1.59738	−1.49673
0.9990	−2.14053	−2.01739	−1.89894	−1.78572	−1.67825	−1.57695	−1.48216
0.9980	−2.05701	−1.94806	−1.84244	−1.74062	−1.64305	−1.55016	−1.46232
0.9950	−1..92580	−1.83660	−1.74919	−1.66390	−1.58110	−1.50114	−1.42439
0.9900	−1.80621	−1.73271	−1.66001	−1.58838	−1.51808	−1.44942	−1.38267
0.9800	−1.66325	−1.60604	−1.54886	−1.49188	−1.43529	−1.37929	−1.32412
0.9750	−1.61099	−1.55914	−1.50712	−1.45507	−1.40314	−1.35153	−1.30042
0.9600	−1.48852	−1.44813	−1.40720	−1.36584	−1.32414	−1.28225	−1.24028
0.9500	−1.42345	−1.38855	−1.35299	−1.31684	−1.28019	−1.24313	−1.20578
0.9000	−1.18347	−1.16574	−1.14712	−1.12762	−1.10726	−1.08608	−1.06413
0.8000	−0.85703	−0.85607	−0.85426	−0.85161	−0.84809	−0.84369	−0.83841
0.7000	−0.59615	−0.60412	−0.61146	−0.61815	−0.62415	−0.62944	−0.63400
0.6000	−0.35565	−0.36889	−0.38186	−0.39434	−0.40638	−0.41794	−0.42899
0.5704	−0.28516	−0.29961	−0.31368	−0.32740	−0.34075	0.35370	−0.36620
0.5000	−0.11578	−0.13199	−0.14807	−0.16397	−0.17968	−0.19517	−0.21040
0.4296	0.06097	0.04397	0.02693	0.00987	−0.00719	−0.02421	−0.04116
0.4000	0.13901	0.12199	0.10486	0.08763	0.07032	0.05297	0.03560
0.3000	0.42851	0.41309	0.39729	0.38111	0.36458	0.34772	0.33054
0.2000	0.79002	0.77986	0.76902	0.75752	0.74537	0.73257	0.71915
0.1000	1.33294	1.33640	1.33889	1.34039	1.34092	1.34047	1.33904
0.0500	1.81864	1.83916	1.85856	1.87683	1.89395	1.90992	1.92472
0.0400	1.96660	1.99311	2.01848	2.04269	2.06573	2.08758	2.10823
0.0250	2.26790	2.30764	2.34623	2.38364	2;41984	2.45482	2.48855
0.0200	2.40670	2.45298	2.49811	2.54206	2.58480	2.62631	2.66657
0.0100	2.82359	2.89101	2.95735	3.02256	3.08660	3.14944	3.21103
0.0050	3.22281	3.31243	3.40109	3.48874	3.57530	3.66073	3.74497
0.0020	3.72957	3.84981	3.96932	4.08802	4.20582	4.32263	4.43839
0.0010	4.10022	4.24439	4.38807	4.53112	4.67344	4.81492	4.95549
0.0005	4.46189	4.63057	4.79899	4.96701	5.13449	5.30130	5.46735
0.0001	5.27389	5.50124	5.72899	5.95691	6.18480	6.41249	6.63980

Probability	1.4	1.5	1.6	1.7	1.8	1.9	2.0
0.9999	−1.41753	−1.32774	−1.24728	−1.17520	−1.11054	−1.05239	−0.99990
0.9995	−1.40413	−1.31944	−1.24235	−1.17240	−1.10901	−1.05159	−0.99950

(*Continued*)

TABLE A.11
Deviates (*K*) for Pearson III Distribution (*Continued*)

Probability	1.4	1.5	1.6	1.7	1.8	1.9	2.0
0.9990	−1.39408	−1.31275	−1.23805	−1.16974	−1.10743	−1.05068	−0.99900
0.9980	−1.37981	−1.30279	−1.23132	−1.16534	−1.10465	−1.04898	−0.99800
0.9950	−1.35114	−1.28167	−1.21618	−1.15477	−1.09749	−1.04427	−0.99499
0.9900	−1.31815	−1.25611	−1.19680	−1.14042	−1.08711	−1.03695	−0.98995
0.9800	−1.26999	−1.21716	−1.16584	−1.11628	−1.06864	−1.02311	−0.97980
0.9750	−1.25004	−1.20059	−1.15229	−1.10537	−1.06001	−1.01640	−0.97468
0.9600	−1.19842	−1.15682	−1.11566	−1.07513	−1.03543	−0.99672	−0.95918
0.9500	−1.16827	−1.13075	−1.09338	−1.05631	−1.01973	−0.98381	−0.94871
0.9000	−1.04144	−1.01810	−0.99418	−0.96977	−0.94496	−0.91988	−0.89464
0.8000	−0.83223	−0.82516	0.81720	−0.80837	−0.79868	−0.78816	−0.77686
0.7000	−0.63779	−0.64080	−0.64300	−0.64436	−0.64488	−0.64453	−0.64333
0.6000	−0.43949	−0.44942	−0.45873	−0.46739	−0.47538	−0.48265	−0.48917
0.5704	−0.37824	−0.38977	−0.40075	−0.41116	−0.42095	−0.43008	−0.43854
0.5000	−0.22535	−0.23996	−0.25422	−0.26808	−0.28150	−0.29443	−0.30685
0.4296	−0.05803	−0.07476	−0.09132	−0.10769	−0.12381	−0.13964	−0.15516
0.4000	0.01824	0.00092	−0.01631	−0.03344	−0.05040	−0.06718	−0.08371
0.3000	0.31307	0.29535	0.27740	0.25925	0.24094	0.22250	0.20397
0.2000	0.70512	0.69050	0.67532	0.65959	0.64335	0.62662	0.60944
0.1000	1.33665	1.33330	1.32900	1.32376	1.31760	1.31054	1.30259
0.0500	1.93836	1.95083	1.96213	1.97227	1.9.8124	1.98906	1.99573
0.0400	2.12768	2.14591	2.16293	2.17873	2.19332	2.20670	2.21888
0.0250	2.52102	2.55222	2.58214	2.61076	2.63810	2.66413	2.68888
0.0200	2.70556	2.74325	2.77964	2.81472	2.84848	2.88091	2.91202
0.0100	3.27134	3.33035	3.38804	3.44438	3.49935	3.55295	3.60517
0.0050	3.82798	3.90973	3.99016	4.06926	4.14700	4.22336	4.29832
0.0020	4.55304	4.66651	4.77875	4.88971	4.99937	5.10768	5.21461
0.0010	5.09505	5.23353	5.37087	5.50701	5.64190	5.77549	5.90776
0.0005	5.63252	5.79673	5.95990	6.12196	6.28285	6.44251	6.60090
0.0001	6.86661	7.09277	7.31818	7.54272	7.76632	7.98888	8.21034

Probability	0.0	−0.1	−0.2	−0.3	−0.4	−0.5	−0.6
0.9999	−3.71902	−3.93453	−4.15301	−4.37394	−4.59687	−4.82141	−5.04718
0.9995	−3.29053	−3.45513	−3.62113	−3.78820	−3.95605	−4.12443	−4.29311
0.9990	−3.09023	−3.23322	−3.37703	−3.52139	−3.66608	−3.81090	−3.95567
0.9980	−2.87816	−2.99978	−3.12169	−3.24371	−3.36566	−3.48737	−3.60872
0.9950	−2.57583	−2.66965	−2.76321	−2.85636	−2.94900	−3.04102	−3.13232
0.9900	−2.32635	−2.39961	−2.47226	−2.54421	−2.61539	−2.68572	−2.75514
0.9800	−2.05375	−2.10697	−2.15935	−2.21081	−2.26133	−2.31084	−2.35931
0.9750	−1.95996	−2.00688	−2.05290	−2.09195	−2.14202	−2.18505	−2.22702
0.9600	−1.75069	−1.78462	−1.81756	−1.84949	−1.88039	−1.91022	−1.93896
0.9500	−1.64485	−1.67279	−1.69971	−1.72562	−1.75048	−1.77428	−1.79701
0.9000	−1.28155	−1.29178	−1.30105	−1.30936	−1.31611	−1.32309	−1.32850
0.8000	−0.84162	−0.83639	−0.83044	−0.82377	—0.81638	−0.80829	−0.79950
0.7000	−0.52440	−0.51207	−0.49927	−0.48600	−0.47228	−0.45812	−0.44352

TABLE A.11
Deviates (*K*) for Pearson III Distribution (*Continued*)

Probability	0.0	−0.1	−0.2	−0.3	−0.4	−0.5	−0.6
0.6000	−0.25335	−0.23763	−0.22168	−0.20552	−0.18916	−0.17261	−0.15589
0.5704	−0.17733	−0.16111	−0.14472	−0.12820	−0.11154	−0.09178	−0.07791
0.5000	0.00000	0.01662	0.03325	0.04993	0.06651	0.08302	0.09945
0.4296	0.17733	0.19339	0.20925	0.22492	0.24037	0.25558	0.27047
0.4000	0.25335	0.26882	0.28403	0.29897	0.31362	0.32796	0.34198
0.3000	0.52440	0.53624	0.54757	0.55839	0.56867	0.57840	0.58757
0.2000	0.84162	0.84611	0.84986	0.85285	0.85508	0.85653	0.85718
0.1000	1.28155	1.27037	1.25824	1.24516	1.23114	1.21618	1.20028
0.0500	1.64485	1.61594	1.58607	1.55527	1.52357	1.49101	1.45762
0.0400	1.75069	1.71580	1.67999	1.64329	1.60574	1.56740	1.52830
0.0250	1.95996	1.91219	1.86360	1.81427	1.76427	1.71366	1.66253
0.0200	2.05375	1.99973	1.94499	1.88959	1.83361	1.77716	1.72033
0.0100	2.32635	2.25258	2.17840	2.10394	2.02933	1.95472	1.88029
0.0050	2.57583	2.48187	2.38795	2.29423	2.20092	2.10825	2.01644
0.0020	2.87816	2.75706	2.63672	2.51741	2.39942	2.28311	2.16884
0.0010	3.09023	2.94834	2.80786	2.66915	2.53261	2.39867	2.26780
0.0005	3.29053	3.12167	2.96998	2.80889	2.65390	2.50257	2.35549
0.0001	3.71902	3.50703	3.29921	3.09631	2.89907	2.70836	2.52507

Probability	−0.7	−0.8	−0.9	−1.0	−1.1	−1.2	−1.3
0.9999	−5.27389	−5.50124	−5.72899	−5.95691	−6.18480	−6.41249	−6.63980
0.9995	−4.46189	−4.63057	−4.79899	−4.96.701	−5.13449	−5.30130	−5.46735
0.9990	−4.10022	−4.24439	−4.38807	−4.53112	−4.67344	−4.81492	−4.95549
0.9980	−3.72957	−3.84981	−3.96932	−4.08802	−4.20582	−4.32263	−4.43839
0.9950	−3.22281	−3.31243	−3.40109	−3.48874	−3.51530	−3.66073	−3.74491
0.9900	−2.82359	−2.89101	−2.95735	−3.02256	−3.08660	−3.14944	−3.21103
0.9800	−2.40670	−2.45298	−2.49811	−2.54206	−2.58480	−2.62631	−2.66657
0.9750	−2.26790	−2.30164	−2.34623	−2.38364	−2.41984	−2.45482	−2.48855
0.9600	−1.96660	−1.99311	−2.01848	−2.04269	−2.06573	−2.08758	−2.10823
0.9500	−1.81864	−1.83916	−1.85856	−1.87683	−1.89395	−1.90992	−1.92472
0.9000	−1.33294	−1.33640	−1.33889	−1.34039	−1.34092	−1.34047	−1.33904
0.8000	−0.79002	−0.77986	−0.76902	−0.75152	−0.74537	−0.73257	−0.71915
0.7000	−0.42851	−0.41309	−0.39729	−0.38111	−0.36458	−0.34772	−0.33054
0.6000	−0.13901	−0.12199	−0.10486	−0.08763	−0.07032	−0.05291	−0.03560
0.5704	−0.06097	−0.04397	−0.02693	−0.00987	0.00719	0.02421	0.04116
0.5000	0.11578	0.13199	0.14807	0.16397	0.17968	0.19517	0.21040
0.4296	0.28516	0.29961	0.31368	0.32740	0.34075	0.35370	0.36620
0.4000	0.35565	0.36889	0.38186	0.39434	0.40638	0.41794	0.42899
0.3000	0.59615	0.60412	0.61146	0.61815	0.62415	0.62944	0.63400
0.2000	0.85703	0.85607	0.85426	0.85161	0.84809	0.84369	0.83841
0.1000	1.18347	1.16574	1.14712	1.12762	1.10726	1.08608	1.06413
0.0500	1.42345	1.38855	1.35299	1.31684	1.28019	1.24313	1.20578
0.0400	1.48852	1.44813	1.40720	1.36584	1.32414	1.28225	1.24028
0.0250	1.61099	1.55914	1.50712	1.45507	1.40314	1.35153	1.30042

(*Continued*)

TABLE A.11
Deviates (*K*) for Pearson III Distribution (*Continued*)

Probability	–0.7	–0.8	–0.9	–1.0	–1.1	–1.2	–1.3
0.0200	1.66325	1.60604	1.54886	1.49188	1.43529	1.37929	1.32412
0.0100	1.80621	1.73271	1.66001	1.58838	1.51808	1.44942	1.38267
0.0050	1.92580	1.83660	1.74919	1.66390	1.58110	1.50114	1.42439
0.0020	2.05701	1.94806	1.84244	1.74062	1.64305	1.55016	1.46232
0.0010	2.14053	2.01739	1.89894	1.78572	1.61825	1.57695	1.48216
0.0005	2.21328	2.01661	1.94611	1.82241	1.70603	1.59738	1.49673
0.0001	2.35015	2.18448	2.02891	1.88410	1.75053	1.62838	1.51752

Probability	–1.4	–1.5	–1.6	–1.7	–1.8	–1.9	–2.0
0.9999	–6.86661	–7.09277	–7.31818	–7.54272	–7.76632	–7.98888	–8.21034
0.9995	–5.63252	–5.796.73	–5.95990	–6.12196	–6.28285	–6.44251	–6.60090
0.9990	–5.09505	–5.23353	–5.37087	–5.50701	–5.64190	–5.77549	–5.90776
0.9980	–4.55304	–4.66651	–4.77875	–4.88971	–4.99937	–5.10768	–5.21461
0.9950	–3.82798	–3.90973	–3.99016	–4.06926	–4.14700	–4.22336	–4.29832
0.9900	–3.27134	–3.33035	–3.38804	–3.44438	–3.49935	–3.55295	–3.60517
0.9800	–2.70556	–2.74325	–2.77964	–2.81472	–2.84848	–2.88091	–2.91202
0.9750	–2.52102	–2.55222	–2.58214	–2.61076	–2.63810	–2.66413	–2.68888
0.9600	–2.12768	–2.14591	–2.16293	–2.17873	–2.19332	–2.20670	–2.21888
0.9500	–1.93836	–1.95083	–1.96213	–1.97227	–1.98124	–1.98906	–1.99573
0.9000	–1.33665	–1.33330	–1.32900	–1.32376	–1.31760	–1.31054	–1.30259
0.8000	–0.70512	–0.69050	–0.67532	–0.65959	–0.64335	–0.62662	–0.60944
0.1000	–0.31307	–0.29535	–0.27740	–0.25925	–0.24094	–0.22250	–0.20397
0.6000	–0.01824	–0.00092	0.01631	0.03344	0.05040	0.06718	0.08371
0.5704	0.05803	0.01476	0.09132	0.10769	0.12381	0.13964	0.15516
0.5000	0.22535	0.23996	0.25422	0.26808	0.28150	0.29443	0.30685
0.4296	0.37824	0.38977	0.40075	0.41116	0.42095	0.43008	0.43854
0.4000	0.43949	0.44942	0.45873	0.46739	0.47538	0.48265	0.48917
0.3000	0.63779	0.64080	0.64300	0.64436	0.64488	0.64453	0.64333
0.2000	0.83223	0.82516	0.81720	0.80837	0.79868	0.78816	0.77686
0.1000	10.04144	10.01810	0.99418	0.96977	0.94496	0.91988	0.89464
0.0500	1.16827	1.13075	1.09338	1.05631	1.01973	0.98381	0.94871
0.0400	1.19842	1.15682	1.11566	1.07513	1.03543	0.99672	0.95918
0.0250	1.25004	1.20059	1.15229	1.10537	1.06001	1.01640	0.97468
0.0200	1.26999	1.21716	1.16584	1.11628	1.06864	1.02311	0.97980
0.0100	1.31815	1. 25611	1.19680	1.14042	1.08111	1.03695	0.98995
0.0050	1.35114	1.28167	1.21618	1.15477	1.09749	1.04427	0.99499
0.0020	1.37981	1.30279	1.23132	1.16534	1.10465	1.04898	0.99800
0.0010	1.39408	1.31275	1.23805	1.16974	1.10743	1.05068	0.99900
0.0005	1.40413	1.31944	1.24235	1.17240	1.10901	1.05159	0.99950
0.0001	1.41753	1.32774	1.24728	1.17520	1.11054	1.05239	0.99990

TABLE A.12
Critical Values for Kolmogorov–Smirnov One-Sample Test

	Level of Significance				
N	**0.20**	**0.15**	**0.10**	**0.05**	**0.01**
1	0.900	0.925	0.950	0.975	0.995
2	0.684	0.726	0.776	0.842	0.929
3	0.565	0.597	0.642	0.708	0.828
4	0.494	0.525	0.564	0.624	0.733
5	0.446	0.474	0.510	0.565	0.669
6	0.410	0.436	0.470	0.521	0.618
7	0.381	0.405	0.438	0.486	0.577
8	0.358	0.381	0.411	0.457	0.543
9	0.339	0.360	0.388	0.432	0.514
10	0.322	0.342	0.368	0.410	0.490
11	0.307	0.326	0.352	0.391	0.468
12	0.295	0.313	0.338	0.375	0.450
13	0.284	0.302	0.325	0.361	0.433
14	0.274	0.292	0.314	0.349	0.418
15	0.266	0.283	0.304	0.338	0.404
16	0.258	0.274	0.295	0.328	0.392
17	0.250	0.266	0.286	0.318	0.381
18	0.244	0.259	0.278	0.309	0.371
19	0.237	0.252	0.272	0.301	0.363
20	0.231	0.246	0.264	0.294	0.356
25	0.210	0.220	0.240	0.270	0.320
30	0.190	0.200	0.220	0.240	0.290
35	0.180	0.190	0.216	0.230	0.270
over 35	$\frac{1.07}{\sqrt{n}}$	$\frac{1.14}{\sqrt{n}}$	$\frac{1.22}{\sqrt{n}}$	$\frac{1.36}{\sqrt{n}}$	$\frac{1.63}{\sqrt{n}}$

TABLE A.13
Cumulative Probabilities for the Wald–Wolfowitz Runs Test

n_1	n_2	R	p	n_1	n_2	R	p	n_1	n_2	R	p	n_1	n_2	R	p
2	2	2	0.3333	2	15	2	0.0147	3	5	3	0.1429	3	13	4	0.1143
2	3	2	0.2000	2	15	3	0.1250	3	6	2	0.0238	3	14	2	0.0029
2	4	2	0.1333	2	15	4	0.3309	3	6	3	0.1071	3	14	3	0.0250
2	5	2	0.0952	2	16	2	0.0131	3	7	2	0.0167	3	14	4	0.1015
2	5	3	0.3333	2	16	3	0.1176	3	7	3	0.0833	3	15	2	0.0025
2	6	2	0.0714	2	16	4	0.3137	3	7	4	0.2833	3	15	3	0.0221
2	6	3	0.2857	2	17	2	0.0117	3	8	2	0.0121	3	15	4	0.0907
2	7	2	0.0556	2	17	3	0.1111	3	8	3	0.0667	3	15	5	0.3309
2	7	3	0.2500	2	17	4	0.2982	3	8	4	0.2364	3	16	2	0.0021
2	8	2	0.0444	2	18	2	0.0105	3	9	2	0.0091	3	16	3	0.0196
2	8	3	0.2222	2	18	3	0.1053	3	9	3	0.0545	3	16	4	0.0815
2	9	2	0.0364	2	18	4	0.2842	3	9	4	0.2000	3	16	5	0.3137
2	9	3	0.2000	2	19	2	0.0095	3	10	2	0.0070	3	17	2	0.0018
2	10	2	0.0303	2	19	3	0.1000	3	10	3	0.0455	3	17	3	0.0175
2	10	3	0.1818	2	19	4	0.2714	3	10	4	0.1713	3	17	4	0.0737
2	11	2	0.0256	2	20	2	0.0087	3	11	2	0.0055	3	17	5	0.2982
2	11	3	0.1667	2	20	3	0.0952	3	11	3	0.0385	3	18	2	0.0015
2	12	2	0.0220	2	20	4	0.2597	3	11	4	0.1484	3	18	3	0.0158
2	12	3	0.1538	3	3	2	0.1000	3	12	2	0.0044	3	18	4	0.0669
2	13	2	0.0190	3	3	3	0.3000	3	12	3	0.0330	3	18	5	0.2842
2	13	3	0.1429	3	4	2	0.0571	3	12	4	0.1297	3	19	2	0.0013
2	14	2	0.0167	3	4	3	0.2000	3	13	2	0.0036	3	19	3	0.0143
2	14	3	0.1333	3	5	2	0.0357	3	13	3	0.0286	3	19	4	0.0610

n_1	n_2	R	p	n_1	n_2	R	p	n_1	n_2	R	p	n_1	n_2	R	p
3	19	5	0.2714	4	9	4	0.0853	4	14	5	0.1206	4	19	3	0.0026
3	20	2	0.0011	4	9	5	0.2364	4	14	6	0.2735	4	19	4	0.0148
3	20	3	0.0130	4	10	2	0.0020	4	15	2	0.0005	4	19	5	0.0727
3	20	4	0.0559	4	10	3	0.0140	4	15	3	0.0049	4	19	6	0.1764
3	20	5	0.2597	4	10	4	0.0679	4	15	4	0.0266	4	20	2	0.0002
4	4	2	0.0286	4	10	5	0.2028	4	15	5	0.1078	4	20	3	0.0023
4	4	3	0.1143	4	11	2	0.0015	4	15	6	0.2487	4	20	4	0.0130
4	5	2	0.0159	4	11	3	0.0110	4	16	2	0.0004	4	20	5	0.0666
4	5	3	0.0714	4	11	4	0.0549	4	16	3	0.0041	4	20	6	0.1632
4	5	4	0.2619	4	11	5	0.1158	4	16	4	0.0227	5	5	2	0.0079
4	6	2	0.0095	4	12	2	0.0011	4	16	5	0.0970	5	5	3	0.0397
4	6	3	0.0476	4	12	3	0.0088	4	16	6	0.2270	5	5	4	0.1667
4	6	4	0.1905	4	12	4	0.0451	4	17	2	0.0003	5	6	2	0.0043
4	7	2	0.0061	4	12	5	0.1538	4	17	3	0.0035	5	6	3	0.0238
4	7	3	0.0333	4	12	6	0.3352	4	17	4	0.0195	5	6	4	0.1104
4	7	4	0.1424	4	13	2	0.0008	4	17	5	0.0877	5	6	5	0.2619
4	7	5	0.3333	4	13	3	0.0071	4	17	6	0.2080	5	7	2	0.0025
4	8	2	0.0040	4	13	4	0.0374	4	18	2	0.0003	5	7	3	0.0152
4	8	3	0.0242	4	13	5	0.1351	4	18	3	0.0030	5	7	4	0.0758
4	8	4	0.1091	4	13	6	0.3021	4	18	4	0.0170	5	7	5	0.1970
4	8	5	0.2788	4	14	2	0.0007	4	18	5	0.0797	5	8	2	0.0016
4	9	2	0.0028	4	14	3	0.0059	4	18	6	0.1913	5	8	3	0.0101
4	9	3	0.0182	4	14	4	0.0314	4	19	2	0.0002	5	8	4	0.0536

TABLE A.13
Cumulative Probabilities for the Wald–Wolfowitz Runs Test (*Continued*)

n_1	n_2	R	p	n_1	n_2	R	p	n_1	n_2	R	p	n_1	n_2	R	p
5	8	5	0.1515	5	13	4	0.0133	5	17	5	0.0276	6	7	2	0.0012
5	9	2	0.0010	5	13	5	0.0525	5	17	6	0.0823	6	7	3	0.0076
5	9	3	0.0070	5	13	6	0.1450	5	17	7	0.2281	6	7	4	0.0425
5	9	4	0.0390	5	13	7	0.3298	5	18	3	0.0007	6	7	5	0.1212
5	9	5	0.1189	5	14	2	0.0002	5	18	4	0.0047	6	7	6	0.2960
5	9	6	0.2867	5	14	3	0.0016	5	18	5	0.0239	6	8	2	0.0007
5	10	2	0.0007	5	14	4	0.0106	5	18	6	0.0724	6	8	3	0.0047
5	10	3	0.0050	5	14	5	0.0441	5	18	7	0.2098	6	8	4	0.0280
5	10	4	0.0290	5	14	6	0.1246	5	19	3	0.0006	6	8	5	0.0862
5	10	5	0.0949	5	14	7	0.2990	5	19	4	0.0040	6	8	6	0.2261
5	10	6	0.2388	5	15	2	0.0001	5	19	5	0.0209	6	9	2	0.0004
5	11	2	0.0005	5	15	3	0.0013	5	19	6	0.0641	6	9	3	0.0030
5	11	3	0.0037	5	15	4	0.0085	5	19	7	0.1937	6	9	4	0.0190
5	11	4	0.0220	5	15	5	0.0374	5	20	3	0.0005	6	9	5	0.0629
5	11	5	0.0769	5	15	6	0.1078	5	20	4	0.0033	6	9	6	0.1748
5	11	6	0.2005	5	15	7	0.2722	5	20	5	0.0184	6	10	2	0.0002
5	12	2	0.0003	5	16	3	0.0010	5	20	6	0.0570	6	10	3	0.0020
5	12	3	0.0027	5	16	4	0.0069	5	20	7	0.1793	6	10	4	0.0132
5	12	4	0.0170	5	16	5	0.0320	5	20	8	0.3252	6	10	5	0.0470
5	12	5	0.0632	5	16	6	0.0939	6	6	2	0.0022	6	10	6	0.1369
5	12	6	0.1698	5	16	7	0.2487	6	6	3	0.0130	6	10	7	0.2867
5	13	2	0.0002	5	17	3	0.0008	6	6	4	0.0671	6	11	2	0.0002
5	13	3	0.0021	5	17	4	0.0057	6	6	5	0.1753	6	11	3	0.0014

n_1	n_2	R	p	n_1	n_2	R	p	n_1	n_2	R	p	n_1	n_2	R	p
6	11	4	0.0095	6	15	4	0.0030	6	19	3	0.0001	7	9	2	0.0002
6	11	5	0.0357	6	15	5	0.0139	6	19	4	0.0012	7	9	3	0.0014
6	11	6	0.1084	6	15	6	0.0475	6	19	5	0.0065	7	9	4	0.0098
6	11	7	0.2418	6	15	7	0.1313	6	19	6	0.0238	7	9	5	0.0350
6	12	2	0.0001	6	15	8	0.2655	6	19	7	0.0785	7	9	6	0.1084
6	12	3	0.0010	6	16	3	0.0003	6	19	8	0.1706	7	9	7	0.2308
6	12	4	0.0069	6	16	4	0.0023	6	20	3	0.0001	7	10	2	0.0001
6	12	5	0.0276	6	16	5	0.0114	6	20	4	0.0009	7	10	3	0.0009
6	12	6	0.0869	6	16	6	0.0395	6	20	5	0.0055	7	10	4	0.0064
6	12	7	0.2054	6	16	7	0.1146	6	20	6	0.0203	7	10	5	0.0245
6	13	3	0.0007	6	16	8	0.2365	6	20	7	0.0698	7	10	6	0.0800
6	13	4	0.0051	6	17	3	0.0002	6	20	8	0.1540	7	10	7	0.1818
6	13	5	0.0217	6	17	4	0.0018	7	7	2	0.0006	7	11	3	0.0006
6	13	6	0.0704	6	17	5	0.0093	7	7	3	0.0041	7	11	4	0.0043
6	13	7	0.1758	6	17	6	0.0331	7	7	4	0.0251	7	11	5	0.0175
6	13	8	0.3379	6	17	7	0.1005	7	7	5	0.0775	7	11	6	0.0600
6	14	3	0.0005	6	17	8	0.2114	7	7	6	0.2086	7	11	7	0.1448
6	14	4	0.0039	6	18	3	0.0002	7	8	2	0.0003	7	11	8	0.2956
6	14	5	0.0173	6	18	4	0.0014	1	8	3	0.0023	7	12	3	0.0004
6	14	6	0.0575	6	18	5	0.0078	7	8	4	0.0154	7	12	4	0.0030
6	14	7	0.1514	6	18	6	0.0280	7	8	5	0.0513	7	12	5	0.0128
6	14	8	0.2990	6	18	7	0.0886	7	8	6	0.1492	7	12	6	0.0456
6	15	3	0.0004	6	18	8	0.1896	7	8	7	0.2960	7	12	7	0.1165

(*Continued*)

TABLE A.13
Cumulative Probabilities for the Wald–Wolfowitz Runs Test (*Continued*)

n_1	n_2	R	p	n_1	n_2	R	p	n_1	n_2	R	p	n_1	n_2	R	p
7	12	8	0.2475	7	16	6	0.0172	7	20	4	0.0003	8	10	7	0.1170
7	13	3	0.0003	7	16	7	0.0536	7	20	5	0.0018	8	10	8	0.2514
7	13	4	0.0021	7	16	8	0.1278	7	20	6	0.0075	8	11	3	0.0003
7	13	5	0.0095	7	16	9	0.2670	7	20	7	0.0278	8	11	4	0.0021
7	13	6	0.0351	7	17	4	0.0006	7	20	8	0.0714	8	11	5	0.0090
7	13	7	0.0947	7	17	5	0.0034	7	20	9	0.1751	8	11	6	0.0341
7	13	8	0.2082	7	17	6	0.0138	7	20	10	0.3060	8	11	7	0.0882
7	14	3	0.0002	7	17	7	0.0450	8	8	2	0.0002	8	11	8	0.1994
7	14	4	0.0015	7	17	8	0.1097	8	8	3	0.0012	8	12	3	0.0002
7	14	5	0.0072	7	17	9	0.2392	8	8	4	0.0089	8	12	4	0.0014
7	14	6	0.0273	7	18	4	0.0005	8	8	5	0.0317	8	12	5	0.0063
7	14	7	0.0777	7	18	5	0.0021	8	8	6	0.1002	8	12	6	0.0246
7	14	8	0.1760	7	18	6	0.0112	8	8	7	0.2145	8	12	7	0.0674
7	14	9	0.3359	7	18	7	0.0381	8	9	3	0.0007	8	12	8	0.1591
7	15	3	0.0001	7	18	8	0.0947	8	9	4	0.0053	8	12	9	0.2966
7	15	4	0.0011	7	18	9	0.2149	8	9	5	0.0203	8	13	3	0.0001
7	15	5	0.0055	7	19	4	0.0004	8	9	6	0.0687	8	13	4	0.0009
7	15	6	0.0216	7	19	5	0.0022	8	9	7	0.1573	8	13	5	0.0044
7	15	7	0.0642	7	19	6	0.0092	8	9	8	0.3186	8	13	6	0.0181
7	15	8	0.1496	7	19	7	0.0324	8	10	3	0.0004	8	13	7	0.0521
7	15	9	0.2990	7	19	8	0.0820	8	10	4	0.0033	8	13	8	0.1278
7	16	4	0.0008	7	19	9	0.1937	8	10	5	0.0134	8	13	9	0.2508
7	16	5	0.0043	7	19	10	0.3332	8	10	6	0.0479	8	14	4	0.0006

n_1	n_2	R	p	n_1	n_2	R	p	n_1	n_2	R	p	n_1	n_2	R	p
8	14	5	0.0032	8	17	8	0.0570	8	20	11	0.3322	9	12	6	0.0137
8	14	6	0.0134	8	17	9	0.1340	9	9	3	0.0004	9	12	7	0.0399
8	14	7	0.0408	8	17	10	0.2518	9	9	4	0.0030	9	12	8	0.1028
8	14	8	0.1034	8	18	4	0.0002	9	9	5	0.0122	9	12	9	0.2049
8	14	9	0.2129	8	18	5	0.000.10	9	9	6	0.0445	9	13	4	0.0004
8	15	4	0.0004	8	18	6	0.0047	9	9	7	0.1090	9	13	5	0.0022
8	15	5	0.0023	8	18	7	0.0169	9	9	8	0.2380	9	13	6	0.0096
8	15	6	0.0101	8	18	8	0.0473	9	10	3	0.0002	9	13	7	0.0294
8	15	7	0.0322	8	18	9	0.1159	9	10	4	0.0018	9	13	8	0.0789
8	15	8	0.0842	8	18	10	0.2225	9	10	5	0.0076	9	13	9	0.1656
8	15	9	0.1816	8	19	4	0.0001	9	10	6	0.0294	9	13	10	0.3050
8	15	10	0.3245	8	19	5	0.0008	9	10	7	0.0767	9	14	4	0.0003
8	16	4	0.0003	8	19	6	0.0037	9	10	8	0.1786	9	14	5	0.0015
8	16	5	0.0017	8	19	7	0.0138	9	10	9	0.3186	9	14	6	0.0068
8	16	6	0.0077	8	19	8	0.0395	9	11	3	0.0001	9	14	7	0.0220
8	16	7	0.0257	8	19	9	0.1006	9	11	4	0.0011	9	14	8	0.0612
8	16	8	0.0690	8	19	10	0.1971	9	11	5	0.0049	9	14	9	0.1347
8	16	9	0.1556	8	20	5	0.0006	9	11	6	0.0199	9	14	10	0.2572
8	16	10	0.2856	8	20	6	0.0029	9	11	7	0.0549	9	15	4	0.0002
8	17	4	0.0002	8	20	7	0.0114	9	11	8	0.1349	9	15	5	0.0010
8	17	5	0.0013	8	20	8	0.0332	9	11	9	0.2549	9	15	6	0.0049
8	17	6	0.0060	8	20	9	0.0878	9	12	4	0.0007	9	15	7	0.0166
8	17	7	0.0207	8	20	10	0.1751	9	12	5	0.0032	9	15	8	0.0418

TABLE A.13
Cumulative Probabilities for the Wald–Wolfowitz Runs Test (*Continued*)

n_1	n_2	R	p	n_1	n_2	R	p	n_1	n_2	R	p	n_1	n_2	R	p
9	15	9	0.1102	9	18	11	0.2545	10	11	4	0.0006	10	14	6	0.0036
9	15	10	0.2174	9	19	5	0.0003	10	11	5	0.0027	10	14	7	0.0122
9	16	4	0.0001	9	19	6	0.0015	10	11	6	0.0119	10	14	8	0.0367
9	16	5	0.0007	9	19	7	0.0061	10	11	7	0.0349	10	14	9	0.0857
9	16	6	0.0036	9	19	8	0.0193	10	11	8	0.0920	10	14	10	0.1775
9	16	7	0.0127	9	19	9	0.0524	10	11	9	0.1849	10	14	11	0.3062
9	16	8	0.0377	9	19	10	0.1144	10	11	10	0.3350	10	15	5	0.0005
9	16	9	0.0907	9	19	11	0.2261	10	12	4	0.0003	10	15	6	0.0025
9	16	10	0.1842	9	20	5	0.0002	10	12	5	0.0017	10	15	7	0.0088
9	16	11	0.3245	9	20	6	0.0012	40	12	6	0.0078	10	15	8	0.0275
9	17	5	0.0005	9	20	7	0.0048	10	12	7	0.0242	10	15	9	0.0673
9	17	6	0.0027	9	20	8	0.0157	10	12	8	0.0670	10	15	10	0.1445
9	17	7	0.0099	9	20	9	0.0441	10	12	9	0.1421	10	15	11	0.2602
9	17	8	0.0299	9	20	10	0.0983	10	12	10	0.2707	10	16	5	0.0003
9	17	9	0.0751	9	20	11	0.2013	10	13	4	0.0002	10	16	6	0.0018
9	17	10	0.1566	9	20	12	0.3313	10	13	5	0.0011	10	16	7	0.6065
9	17	11	0.2871	10	10	3	0.0001	10	13	6	0.0053	10	16	8	0.0209
9	18	5	0.0004	10	10	4	0.0010	10	13	7	0.0170	10	16	9	0.0533
9	18	6	0.0020	10	10	5	0.0045	10	13	8	0.0493	10	16	10	0.1180
9	18	7	0.0077	10	10	6	0.0185	10	13	9	0.1099	10	16	11	0.2216
9	18	8	0.0240	10	10	7	0.0513	10	13	10	0.2189	10	17	5	0.0002
9	18	9	0.0626	10	10	8	0.1276	10	14	4	0.0001	10	17	6	0.0013
9	18	10	0.1336	10	10	9	0.2422	10	14	5	0.0007	10	17	7	0.0048

n_1	n_2	R	p	n_1	n_2	R	p	n_1	n_2	R	p	n_1	n_2	R	p
10	17	8	0.0160	10	20	8	0.0076	11	13	7	0.0101	11	16	8	0.0118
10	17	9	0.0425	10	20	9	0.0225	11	13	8	0.0313	11	16	9	0.0317
10	17	10	0.0968	10	20	10	0.0550	11	13	9	0.0736	11	16	10	0.0757
10	17	11	0.1893	10	20	11	0.1200	11	13	10	0.1569	11	16	11	0.1504
10	17	12	0.3197	10	20	12	0.2175	11	13	11	0.2735	11	16	12	0.2665
10	18	5	0.0002	11	11	4	0.0003	11	14	5	0.0004	11	17	5	0.0001
10	18	6	0.0009	11	11	5	0.0016	11	14	6	0.0019	11	17	6	0.0006
10	18	7	0.0036	11	11	6	0.0073	11	14	7	0.0069	11	17	7	0.0025
10	18	8	0.0124	11	11	7	0.0226	11	14	8	0.0223	11	17	8	0.0087
10	18	9	0.0341	11	11	8	0.0635	11	14	9	0.0551	11	17	9	0.0244
10	18	10	0.0798	11	11	9	0.1349	11	14	10	0.1224	11	17	10	0.0600
10	18	11	0.1621	11	11	10	0.2599	11	14	11	0.2235	11	17	11	0.1240
10	18	12	0.2809	11	12	4	0.0002	11	15	5	0.0002	11	17	12	0.2265
10	19	5	0.0001	11	12	5	0.0010	11	15	6	0.0013	11	18	6	0.0004
10	19	6	0.0001	11	12	6	0.0046	11	15	7	0.0048	11	18	7	0.0018
10	19	7	0.0028	11	12	7	0.0150	11	15	8	0.0161	11	18	8	0.0065
10	19	8	0.0096	11	12	8	0.0443	11	15	9	0.0416	11	18	9	0.0189
10	19	9	0.0276	11	12	9	0.0992	11	15	10	0.0960	11	18	10	0.0478
10	19	10	0.0661	11	12	10	0.2017	11	15	11	0.1831	11	18	11	0.1027
10	19	11	0.1392	11	12	11	0.3350	11	15	12	0.3137	11	18	12	0.1928
10	19	12	0.2470	11	13	4	0.0001	11	16	5	0.0002	11	18	13	0.3205
10	20	6	0.0005	11	13	5	0.0006	11	16	6	0.0009	11	19	6	0.0003
10	20	7	0.0021	11	13	6	0.0030	11	16	7	0.0034	11	19	7	0.0013

(Continued)

TABLE A.13
Cumulative Probabilities for the Wald–Wolfowitz Runs Test (*Continued*)

n_1	n_2	R	p	n_1	n_2	R	p	n_1	n_2	R	p	n_1	n_2	R	p
11	19	8	0.0049	12	13	7	0.0061	12	16	8	0.0068	12	19	7	0.0006
11	19	9	0.0148	12	13	8	0.0201	12	16	9	0.0191	12	19	8	0.0025
11	19	10	0.0383	12	13	9	0.0498	12	16	10	0.0487	12	19	9	0.0080
11	19	11	0.0853	12	13	10	0.1126	12	16	11	0.1020	12	19	10	0.0223
11	19	12	0.1644	12	13	11	0.2068	12	16	12	0.1933	12	19	11	0.0524
11	19	13	0.2830	12	14	5	0.0002	12	16	13	0.3149	12	19	12	0.1085
11	20	6	0.0002	12	14	6	0.0011	12	17	6	0.0003	12	19	13	0.1973
11	20	7	0.0010	12	14	7	0.0040	12	17	7	0.0013	12	19	14	0.3189
11	20	8	0.0037	12	14	8	0.0138	12	11	8	0.0048	12	20	7	0.0005
11	20	9	0.0116	12	14	9	0.0358	12	17	9	0.0142	12	20	8	0.0019
11	20	10	0.0308	12	14	10	0.0847	12	17	10	0.0373	12	20	9	0.0061
11	20	11	0.0712	12	14	11	0.1628	12	17	11	0.0813	12	20	10	0.0175
11	20	12	0.1404	12	14	12	0.2860	12	17	12	0.1591	12	20	11	0.0424
11	20	13	0.2500	12	15	5	0.0001	12	17	13	0.2693	12	20	12	0.0900
12	12	5	0.0005	12	15	6	0.0007	12	18	6	0.0002	12	20	13	0.1693
12	12	6	0.0028	12	15	7	0.0027	12	18	7	0.0009	12	20	14	0.2803
12	12	7	0.0095	12	15	8	0.0096	12	18	8	0.0035	13	13	5	0.0002
12	12	8	0.0296	12	15	9	0.0260	12	18	9	0.0106	13	13	6	0.0010
12	12	9	0.0699	12	15	10	0.0640	12	18	10	0.0288	13	13	7	0.0038
12	12	10	0.1504	12	15	11	0.1286	12	18	11	0.0651	13	13	8	0.0131
12	12	11	0.2632	12	15	12	0.2351	12	18	12	0.1312	13	13	9	0.0341
12	13	5	0.0003	12	16	6	0.0005	12	18	13	0.2304	13	13	10	0.0812
12	13	6	0.0017	12	16	7	0.0018	12	19	6	0.0001	13	13	11	0.1566

n_1	n_2	R	p	n_1	n_2	R	p	n_1	n_2	R	p	n_1	n_2	R	p
13	13	12	0.2772	13	16	12	0.1396	13	19	10	0.0132	14	15	7	0.0009
13	14	5	0.0001	13	16	13	0.2389	13	19	11	0.0324	14	15	8	0.0036
13	14	6	0.0006	13	17	6	0.0002	13	19	12	0.0714	14	15	9	0.0107
13	14	7	0.0024	13	17	7	0.0007	13	19	13	0.1365	14	15	10	0.0291
13	14	8	0.0087	13	17	8	0.0027	13	19	14	0.2353	14	15	11	0.0642
13	14	9	0.0236	13	17	9	0.0084	13	20	7	0.0002	14	15	12	0.1306
13	14	10	0.0589	13	17	10	0.0234	13	20	8	0.0010	14	15	13	0.2247
13	14	11	0.1189	13	17	11	0.0535	13	20	9	0.0033	14	16	6	0.0001
13	14	12	0.2205	13	17	12	0.1113	13	20	10	0.0100	14	16	7	0.0006
13	15	6	0.0004	13	17	13	0.1980	13	20	11	0.0254	14	16	8	0.0024
13	15	7	0.0016	13	17	14	0.3215	13	20	12	0.0575	14	16	9	0.0073
13	15	8	0.0058	13	18	6	0.0001	13	20	13	0.1138	14	16	10	0.0207
13	15	9	0.0165	13	18	7	0.0005	13	20	14	0.2012	14	16	11	0.0470
13	15	10	0.0430	13	18	8	0.0019	13	20	15	0.3200	14	16	12	0.1007
13	15	11	0.0906	13	18	9	0.0061	14	14	6	0.0004	14	16	13	0.1804
13	15	12	0.1753	13	18	10	0.0175	14	14	7	0.0015	14	16	14	0.2986
13	15	13	0.2883	13	18	11	0.0415	14	14	8	0.0056	14	17	7	0.0004
13	16	6	0.0002	13	18	12	0.0890	14	14	9	0.0157	14	17	8	0.0016
13	16	7	0.0010	13	18	13	0.1643	14	14	10	0.0412	14	17	9	0.0051
13	16	8	0.0040	13	18	14	0.2752	14	14	11	0.0871	14	17	10	0.0149
13	16	9	0.0117	13	19	7	0.0003	14	14	12	0.1697	14	17	11	0.0355
13	16	10	0.0316	13	19	8	0.0014	14	14	13	0.2198	14	17	12	0.0779
13	16	11	0.0695	13	19	9	0.0045	14	15	6	0.0002	14	17	13	0.1450

TABLE A.13
Cumulative Probabilities for the Wald–Wolfowitz Runs Test (*Continued*)

n_1	n_2	R	p	n_1	n_2	R	p	n_1	n_2	R	p	n_1	n_2	R	p
14	17	14	0.2487	14	20	11	0.0153	15	17	8	0.0009	15	19	14	0.1251
14	18	7	0.0002	14	20	12	0.0368	15	17	9	0.0031	15	19	15	0.2109
14	18	8	0.0011	14	20	13	0.0763	15	17	10	0.0095	15	19	16	0.3286
14	18	9	0.0035	14	20	14	0.1432	15	17	11	0.0237	15	20	8	0.0003
14	18	10	0.0108	14	20	15	0.2387	15	17	12	0.0546	15	20	9	0.0010
14	18	11	0.0266	15	15	6	0.0001	15	17	13	0.1061	15	20	10	0.0034
14	18	12	0.0604	15	15	7	0.0006	15	17	14	0.1912	15	20	11	0.0094
14	18	13	0.1167	15	15	8	0.0023	15	17	15	0.3005	15	20	12	0.0237
14	18	14	0.2068	15	15	9	0.0070	15	18	7	0.0001	15	20	13	0.0512
14	18	15	0.3221	15	15	10	0.0199	15	18	8	0.0006	15	20	14	0.1014
14	19	7	0.0002	15	15	11	0.0457	15	18	9	0.0021	15	20	15	0.1766
14	19	8	0.0007	15	15	12	0.0974	15	18	10	0.0067	15	20	16	0.2831
14	19	9	0.0025	15	15	13	0.1749	15	18	11	0.0173	16	16	7	0.0002
14	19	10	0.0079	15	15	14	0.2912	15	18	12	0.0412	16	16	8	0.0009
14	19	11	0.0202	15	16	7	0.0003	15	19	13	0.0830	16	16	9	0.0030
14	19	12	0.0471	15	16	8	0.0014	15	18	14	0.1546	16	16	10	0.0092
14	19	13	0.0942	15	16	9	0.0046	15	18	15	0.2519	16	16	11	0.0228
14	19	14	0.1720	15	16	10	0.0137	15	19	8	0.0004	16	16	12	0.0528
14	19	15	0.2776	15	16	11	0.0328	15	19	9	0.0014	16	16	13	0.1028
14	20	7	0.0001	15	16	12	0.0728	15	19	10	0.0047	16	16	14	0.1862
14	20	8	0.0005	15	16	13	0.1362	15	19	11	0.0127	16	16	15	0.2933
14	20	9	0.0018	15	16	14	0.2362	15	19	12	0.0312	16	17	7	0.0001
14	20	10	0.0058	15	17	7	0.0002	15	19	13	0.0650	16	17	8	0.0006

n_1	n_2	R	p	n_1	n_2	R	p	n_1	n_2	R	p	n_1	n_2	R	p
16	17	9	0.0019	16	19	15	0.1594	17	18	10	0.0027	17	20	15	0.0955
16	17	10	0.0062	16	19	16	0.2603	17	18	11	0.0075	17	20	16	0.1680
16	11	11	0.0160	16	20	8	0.0002	17	18	12	0.0194	17	20	17	0.2631
16	17	12	0.0385	16	20	9	0.0006	17	18	13	0.0422	18	18	8	0.0001
16	17	13	0.0718	16	20	10	0.0020	17	18	14	0.0859	18	18	9	0.0005
16	17	14	0.1465	16	20	11	0.0058	17	18	15	0.1514	18	18	10	0.0017
16	11	15	0.2397	16	20	12	0.0153	17	18	16	0.2495	18	18	11	0.0050
16	18	8	0.0004	16	20	13	0.0345	17	19	8	0.0001	18	18	12	0.0134
16	18	9	0.0013	16	20	14	0.0716	17	19	9	0.0005	18	18	13	0.0303
16	18	10	0.0042	16	20	15	0.1300	17	19	10	0.0018	18	18	14	0.0640
16	18	11	0.0113	16	20	16	0.2188	17	19	11	0.0052	18	18	15	0.1171
16	18	12	0.0282	16	20	17	0.3297	17	19	12	0.0139	18	18	16	0.2004
16	18	13	0.0591	17	17	8	0.0003	17	19	13	0.0313	18	18	17	0.3046
16	18	14	0.1153	17	17	9	0.0012	17	19	14	0.0659	18	19	9	0.0003
16	18	15	0.1956	17	17	10	0.0041	17	19	15	0.1202	18	19	10	0.0011
16	18	16	0.3091	17	17	11	0.0109	11	19	16	0.2049	18	19	11	0.0034
16	19	8	0.0002	17	17	12	0.0272	17	19	17	0.3108	18	19	12	0.0094
16	19	9	0.0008	17	17	13	0.0572	17	20	9	0.0003	18	19	13	0.0219
16	19	10	0.0029	17	17	14	0.1122	17	20	10	0.0012	18	19	14	0.0479
16	19	11	0.0080	17	17	15	0.1907	17	20	11	0.0036	18	19	15	0.0906
16	19	12	0.0207	17	17	16	0.3028	17	20	12	0.0100	18	19	16	0.1606
16	19	13	0.0450	17	18	8	0.0002	17	20	13	0.0233	18	19	11	0.2525
16	19	14	0.0908	17	18	9	0.0008	17	20	14	0.0506	18	20	9	0.0002

(Continued)

TABLE A.13
Cumulative Probabilities for the Wald–Wolfowitz Runs Test (*Continued*)

n_1	n_2	R	p	n_1	n_2	R	p	n_1	n_2	R	p	n_1	n_2	R	p
18	20	10	0.0007	19	20	13	0.0109								
18	20	11	0.0023	19	20	14	0.0255								
18	20	12	0.0066	19	20	15	0.0516								
18	20	13	0.0159	19	20	16	0.0981								
18	20	14	0.0359	19	20	17	0.1650								
18	20	15	0.0701	19	20	18	0.2610								
18	20	16	0.1285	20	20	10	0.0003								
18	20	11	0.2088	20	20	11	0.0009								
18	20	18	0.3182	20	20	12	0.0029								
19	19	9	0.0002	20	20	13	0.0075								
19	19	10	0.0007	20	20	14	0.0182								
19	19	11	0.0022	20	20	15	0.0380								
19	19	12	0.0064	20	20	16	0.0748								
19	19	13	0.0154	20	20	17	0.1301								
19	19	14	0.0349	20	20	18	0.2130								
19	19	15	0.0683	20	20	19	0.3143								
19	19	16	0.1256												
19	19	17	0.2044												
19	19	18	0.3127												
19	20	9	0.0001												
19	20	10	0.0005												
19	20	11	0.0015												
19	20	12	0.0044												

TABLE A.14
Critical Values for the Kolmogorov–Smirnov Two-Sample Test

	One-Tailed		Two-Tailed	
n	5%	1%	5%	1%
3	3			
4	4		4	
5	4	5	5	5
6	5	6	5	6
7	5	6	6	6
8	5	6	6	7
9	6	7	6	7
10	6	7	7	8
11	6	8	7	8
12	6	8	7	8
13	7	8	7	9
14	7	8	8	9
15	7	9	8	9
16	7	9	8	10
17	8	9	8	10
18	8	10	9	10
19	8	10	9	10
20	8	10	9	11
21	8	10	9	11
22	9	11	9	11
23	9	11	10	11
24	9	11	10	12
25	9	11	10	12
26	9	11	10	12
27	9	12	10	12
28	10	12	11	13
29	10	12	11	13
30	10	12	11	13
35	11	13	12	
40	11	14	13	

TABLE A.15
Critical Values for the Pearson Correlation Coefficient for the Null Hypothesis H_o: $p = 0$ and the One-Tailed Alternative H_a: $p > 0$ and the Two-Tailed Alternative H_a: $p \neq 0$

	Level of Significance for a One-Tailed Test					
	0.1000	0.0500	0.0250	0.0100	0.0050	0.0005
	Level of Significance for a Two-Tailed Test					
Degrees of Freedom	0.2000	0.1000	0.0500	0.0200	0.0100	0.0010
1	0.9511	0.9877	0.9969	0.9995	0.9999	1.0000
2	0.8001	0.9000	0.9500	0.9800	0.9900	0.9990
3	0.6871	0.8053	0.8783	0.9343	0.9587	0.9912
4	0.6083	0.7293	0.8114	0.8822	0.9172	0.9741
5	0.5509	0.6694	0.7545	0.8329	0.8745	0.9508
6	0.5068	0.6215	0.7067	0.7888	0.8343	0.9249
7	0.4716	0.5823	0.6664	0.7498	0.7976	0.8982
8	0.4428	0.5495	0.6319	0.7154	0.7646	0.8721
9	0.4187	0.5214	0.6020	0.6850	0.7348	0.8471
10	0.3980	0.4972	0.5760	0.6581	0.7079	0.8233
11	0.3801	0.4762	0.5529	0.6339	0.6836	0.8010
12	0.3645	0.4574	0.5324	0.6120	0.6614	0.7800
13	0.3506	0.4409	0.5139	0.5922	0.6411	0.7604
14	0.3383	0.4258	0.4973	0.5742	0.6226	0.7419
15	0.3272	0.4124	0.4821	0.5577	0.6055	0.7247
16	0.3170	0.4000	0.4683	0.5425	0.5897	0.7084
17	0.3076	0.3888	0.4556	0.5285	0.5750	0.6932
18	0.2991	0.3783	0.4438	0.5154	0.5614	0.6853
19	0.2914	0.3687	0.4329	0.5033	0.5487	0.6652
20	0.2841	0.3599	0.4227	0.4921	0.5368	0.6524
21	0.2774	0.3516	0.4133	0.4816	0.5256	0.6402
22	0.2711	0.3438	0.4044	0.4715	0.5151	0.6287
23	0.2652	0.3365	0.3961	0.4623	0.5051	0.6177
24	0.2598	0.3297	0.3883	0.4534	0.4958	0.6073
25	0.2545	0.3233	0.3809	0.4451	0.4869	0.5974
26	0.2497	0.3173	0.3740	0.4372	0.4785	0.5880
27	0.2452	0.3114	0.3673	0.4297	0.4706	0.5790
28	0.2408	0.3060	0.3609	0.4226	0.4629	0.5703
29	0.2365	0.3009	0.3550	0.4158	0.4556	0.5620
30	0.2326	0.2959	0.3493	0.4093	0.4487	0.5541
40	0.2018	0.2573	0.3044	0.3578	0.3931	0.4896
60	0.1650	0.2109	0.2500	0.2948	0.3248	0.4078
120	0.1169	0.1496	0.1779	0.2104	0.2324	0.2943
250	0.0808	0.1035	0.1230	0.1453	0.1508	0.2038

Appendix B

Data Matrices

TABLE B.1
Pond Creek Annual Maximum Flood Series (1945–1968)

1	2000	2	1740	3	1460	4	2060	5	1530
6	1590	7	1690	8	1420	9	1330	10	607
11	1380	12	1660	13	2290	14	2590	15	3260
16	2490	17	3080	18	2520	19	3360	20	8020
21	4310	22	4380	23	6220	24	4320		

TABLE B.2
Nolin River Annual Maximum Discharge Flood Series (1945–1968)

4390	3550	2470	6560	5170	4720	2720
5290	6580	548	6840	3810	6510	8300
7310	1640	4970	2220	2100	8860	2300
4280	7900	5500				

TABLE B.3
Sediment Yield Data

X_1	X_2	X_3	X_4	Y
0.135	4.000	40.000	0.000	0.140
0.135	1.600	40.000	0.000	0.120
0.101	1.900	22.000	0.000	0.210
0.353	17.600	2.000	−17.000	0.150
0.353	20.600	1.000	−18.000	0.076
0.466	70.000	23.000	0.000	2.670
0.466	32.200	57.000	0.000	0.610
0.833	41.400	4.000	−12.000	0.690
0.085	14.900	22.000	19.000	2.310
0.085	17.200	2.000	13.000	2.650
0.085	30.500	15.000	12.000	2.370
0.193	6.300	1.000	0.000	0.510
0.167	14.000	3.000	16.000	1.420
0.235	2.400	1.000	44.000	1.650
0.448	19.700	3.000	−16.000	0.140
0.329	5.600	27.000	0.000	0.070
0.428	4.600	12.000	0.000	0.220
0.133	9.100	58.000	0.000	0.200
0.149	1.600	28.000	0.000	0.180
0.266	1.200	15.000	−7.000	0.020
0.324	4.000	48.000	0.000	0.040
0.133	19.100	44.000	0.000	0.990
0.133	9.800	32.000	0.000	0.250
0.133	18.300	19.000	8.000	2.200
0.356	13.500	58.000	0.000	0.020
0.536	24.700	62.000	0.000	0.030
0.155	2.200	27.000	0.000	0.036
0.168	3.500	24.000	0.000	0.160
0.673	27.900	22.000	0.000	0.370
0.725	31.200	10.000	−5.000	0.210
0.275	21.000	64.000	0.000	0.350
0.150	14.700	37.000	0.000	0.640
0.150	24.400	29.000	0.000	1.550
1.140	15.200	11.000	−7.000	0.090
1.428	14.300	1.000	−20.000	0.170
1.126	24.700	44.000	0.000	0.170
0.492	31.000	0.000	−14.000	0.660

Note: X_1 = climate variable (average annual precipitation (in.) divided by average annual temperature (°F); X_2 = watershed slope (%); X_3 = coarse soil particle index (% of soil particles coarser than 1 mm in surface 5.1 cm of soil); X_4 = soil aggregation index (indicator of aggregation of clay-sized particles in surface 5.1 cm of soil); Y = sediment yield (acre-ft per square mile per year)

TABLE B.4
Back Creek Annual Flood Series (1929–1931, 1939–1973)

8750	4060	6300	3130	4160	6700	3880	8050	4020
1600	4460	4230	3010	9150	5100	9820	6200	10700
3880	3420	3240	6800	3740	4700	4380	5190	3960
5600	4670	7080	4640	6680	8360	5210	18700	15500
22400	536							

TABLE B.5
Annual Snowmelt Runoff Data, Sevier River at Hatch

X_1	X_2	X_3	X_4	X_5	X_6	Y
2.8	23.1	32.6	21.5	46.4	11.9	88.0
5.0	8.9	6.9	2.3	15.3	2.8	16.7
3.1	14.1	16.4	11.8	28.4	4.8	35.4
3.4	11.5	11.6	8.8	19.9	3.7	17.5
2.6	9.4	9.6	0.0	19.4	0.0	21.0
2.3	11.6	16.4	7.3	21.9	0.3	40.2
3.6	18.3	17.9	13.8	30.3	8.4	62.8
5.3	7.4	7.4	5.1	13.5	0.0	13.8
2.4	10.7	8.6	4.1	17.0	2.2	17.4
2.5	11.8	9.0	4.3	18.1	3.8	18.2
3.3	19.7	24.7	14.4	33.0	5.3	53.8
3.6	7.0	1.7	0.0	8.8	0.0	18.9
2.5	9.8	8.2	5.3	13.1	2.1	29.1
2.3	12.0	13.4	8.9	19.9	2.3	51.9
4.5	9.9	12.9	3.4	23.2	1.0	39.4
4.2	9.0	9.7	2.1	19.7	0.0	55.7
5.4	13.8	18.3	10.4	25.1	4.8	56.4
4.8	24.2	34.4	24.6	42.0	10.6	107.1
6.5	9.0	4.3	1.1	14.8	0.5	22.9
3.2	10.7	8.6	5.9	17.3	2.8	24.3
3.6	6.3	2.0	0.0	16.3	0.0	20.9
4.2	17.6	23.8	17.1	35.2	9.8	91.3
5.7	6.6	6.2	0.0	15.0	0.7	16.6
2.6	14.3	14.1	8.2	23.2	3.5	37.4
2.9	14.2	14.4	9.9	21.1	5.2	25.2

Note: X_1 = monthly baseflow (×1000 acre-feet), October; X_2 = snow water equivalent (in.), April 1, Castle Valley; X_3 = snow water equivalent (in.), April 1, Duck Creek; X_4 = snow water equivalent (in.), April 1, Harris Flat; X_5 = snow water equivalent (in.), April 1, Midway Valley, X_6 = snow water equivalent (in.), April 1, Panguitch Lake, X_7 = total streamflow (×1000 acre-feet) at Hatch, April 1 to July 31.

TABLE B.6
Project Cost Data (70 Projects)

X_1	X_2	X_3	X_4	X_5	X_6	X_7	X_8	X_9	X_{10}	Y
10	0.01	0.50	20	36	39	70	2963	52	52	103996
10	0.10	0.70	32	36	51	136	7100	164	164	191311
10	2.00	0.50	29	27	48	153	3400	120	89	151499
10	0.30	0.50	18	36	45	85	2200	52	35	90341
10	0.30	0.50	19	24	45	57	2600	51	26	96291
10	0.90	0.40	17	15	60	290	1900	133	133	84576
5	1.10	0.30	41	15	60	329	6040	296	296	156546
5	1.40	0.30	23	15	60	278	3410	250	250	82947
3	0.80	0.55	55	15	36	502	6631	174	122	125063
3	0.10	0.50	100	24	96	560	9999	503	503	841601
3	0.10	0.50	14	24	66	109	2450	90	90	141600
5	2.10	0.50	24	18	66	370	4092	210	105	131931
5	0.50	0.50	19	18	84	290	4415	245	245	236006
5	1.20	0.50	12	24	108	1270	4440	847	254	271064
5	3.40	0.50	39	21	60	267	4152	147	147	82055
5	1.00	0.50	35	24	84	520	3738	307	61	126308
15	0.30	0.51	32	12	48	169	3552	116	116	166532
5	2.60	0.40	125	12	66	258	7931	217	217	349543
10	2.20	0.50	160	12	27	90	7360	30	24	179454
15	1.10	0.50	25	12	108	1940	2786	1485	891	341758
5	1.50	0.45	16	15	36	70	2640	48	48	81140
15	3.20	0.70	40	15	30	81	5180	20	20	89013
5	1.60	0.60	35	15	36	64	4020	39	39	82347
5	1.10	0.45	44	12	54	145	7255	87	87	98684
5	0.60	0.45	36	15	54	105	5388	80	56	124142
10	1.60	0.60	30	15	96	410	2225	181	181	63980
1	0.25	0.45	31	15	30	42	3382	70	70	26585
1	0.30	0.45	31	15	30	12	1704	20	20	11759
1	0.25	0.46	29	15	36	18	3294	30	30	29918
10	1.90	0.38	18	15	48	148	1653	79	28	34462
10	0.90	0.40	18	15	54	171	1965	92	50	47029
10	1.60	0.38	15	15	24	77	1285	38	38	56644
10	1.90	0.35	15	15	48	136	2607	76	76	39049
5	4.80	0.60	30	15	36	50	1160	44	44	53811
5	2.50	0.60	15	15	18	23	1300	18	18	25636
10	1.50	0.40	24	48	66	334	2029	144	144	55540
15	4.50	0.55	33	15	24	60	2465	11	11	58885
15	5.00	0.55	21	15	24	188	2830	14	14	61884
5	2.30	0.45	17	15	36	85	2050	21	21	51807
2	2.50	0.50	12	15	27	53	1371	40	36	24250
2	1.80	0.50	33	15	36	77	4441	96	96	51116
10	2.00	0.20	23	12	27	48	6400	93	56	34067
10	3.60	0.50	8	12	27	55	640	19	19	13348
5	1.40	0.35	19	18	54	214	2900	52	52	79339

TABLE B.6
Project Cost Data (70 Projects)(*Continued*)

X_1	X_2	X_3	X_4	X_5	X_6	X_7	X_8	X_9	X_{10}	Y
5	0.90	0.50	20	15	54	142	2781	48	48	59610
5	0.80	0.50	17	15	48	100	3169	59	59	39330
5	1.70	0.50	8	18	24	53	1641	31	31	18307
5	1.10	0.50	23	18	54	163	4778	111	111	70742
5	2.10	0.50	13	21	24	45	1779	40	40	28092
5	2.20	0.50	23	21	54	186	2600	101	61	47961
3	0.40	0.51	27	15	42	53	2720	70	33	58991
3	1.50	0.50	19	15	30	46	1570	90	90	30901
5	2.10	0.50	25	18	27	56	2823	85	85	46448
3	1.50	0.65	27	24	30	33	1760	34	34	43109
5	0.30	0.50	30	12	36	28	2755	33	33	34433
5	1.50	0.50	31	12	18	15	2392	30	30	14844
5	0.50	0.50	17	12	27	15	1608	19	19	8169
5	0.40	0.50	16	12	21	7	1698	12	12	11788
5	1.00	0.40	36	15	54	172	2730	69	69	43130
10	2.00	0.50	9	30	45	122	1295	56	18	47740
50	0.70	0.50	10	15	24	21	900	23	13	2541
10	0.01	0.50	16	15	45	38	2300	57	57	63937
10	0.01	0.50	14	15	39	87	3940	37	37	69180
10	0.15	0.50	12	12	42	31	2115	111	44	35822
10	0.20	0.50	12	18	42	67	1992	109	31	49492
10	1.00	0.50	4	18	33	60	1054	38	4	14653
10	0.01	0.45	9	21	36	34	1330	19	19	37378
10	2.40	0.70	10	36	48	200	882	96	73	23778
5	1.00	0.45	9	30	48	47	1190	47	19	12997
5	0.40	0.45	7	18	30	23	1600	16	16	29336

Note: X_1 = return period (years); X_2 = slope (%); X_3 = runoff coefficient; X_4 = number of inlets; X_5 = smallest diameter (in.); X_6 = largest diameter (in.); X_7 = outlet capacity (cfs); X_8 = length of sewers (ft); X_9 = total area (acres); X_{10} = developed area (acres); Y = project cost ($).

TABLE B.7
Monthly Rainfall (*P*) and Runoff (*RO*) Depths (in.) for Chattooga River

P	*RO*	*P*	*RO*	*P*	*RO*	*P*	*RO*	*P*	*RO*	*P*	*RO*
Year 1		**Year 4**		**Year 7**		**Year 10**		**Year 13**		**Year 16**	
3.51	1.00	3.04	1.15	13.93	5.83	4.42	2.06	4.52	4.13	4.38	1.71
4.72	1.26	2.55	1.27	2.91	3.45	4.81	1.78	3.10	1.79	6.09	2.09
10.33	4.00	7.22	2.78	5.58	3.83	4.13	1.84	0.94	1.55	6.65	2.98
11.92	6.36	10.59	3.10	7.75	4.50	5.12	2.65	6.83	2.54	0.578	3.47
4.83	3.63	7.55	5.40	8.15	6.59	3.10	2.44	10.61	7.03	5.77	4.90
7.40	4.28	3.47	4.40	8.42	5.14	11.00	7.07	4.28	6.14	4.75	4.02
4.57	4.34	8.34	6.37	3.27	6.05	7.02	3.56	9.33	4.25	6.97	5.24
2.62	3.10	5.69	3.07	2.88	3.46	4.39	3.48	8.42	5.49	5.72	4.43
4.24	2.30	7.32	3.65	4.37	2.67	10.05	2.97	3.14	3.28	6.68	5.24
3.68	1.40	1.92	2.38	4.98	2.39	1.13	3.92	3.32	2.34	4.83	2.52
4.23	1.13	2.17	1.43	8.40	3.09	2.73	2.39	7.84	2.55	8.44	4.16
0.44	0.65	8.31	1.76	8.87	2.42	3.13	1.75	4.67	1.86	8.86	3.91
Year 2		**Year 5**		**Year 8**		**Year 11**		**Year 14**		**Year 17**	
0.31	0.55	6.58	2.97	5.92	3.37	0.33	1.36	6.14	2.61	3.12	3.07
3.54	0.84	11.82	5.01	1.40	2.15	6.61	1.96	5.61	4.71	5.99	3.63
8.41	2.05	6.79	5.19	3.21	1.97	4.42	2.50	4.35	3.12	5.24	3.41
2.68	1.70	4.92	4.51	3.54	2.31	10.44	5.45	4.20	4.11	3.93	3.38
8.97	3.54	6.33	4.21	10.97	4.91	5.98	4.10	4.86	3.43	3.07	3.05
4.63	3.08	6.19	4.56	6.42	5.42	12.46	7.38	3.98	3.96	5.14	3.44
6.70	4.48	9.10	5.87	5.60	5.17	11.65	8.80	3.74	2.64	4.97	3.53
8.52	5.07	5.04	6.08	3.23	3.38	3.33	5.55	6.77	2.93	3.84	2.43
5.26	2.88	2.40	2.87	9.45	3.53	4.54	2.84	7.60	6.41	7.95	3.00
9.53	2.96	12.76	4.65	6.32	2.94	6.83	2.42	9.19	4.87	7.72	2.01
3.39	2.38	5.69	2.61	8.98	3.40	8.19	2.57	16.01	5.25	5.83	2.15
2.90	1.26	1.23	1.52	2.65	2.57	6.16	3.14	3.52	4.29	2.88	1.32
Year 3		**Year 6**		**Year 9**		**Year 12**		**Year 15**		**Year 18**	
1.44	1.19	0.41	1.36	1.59	1.71	10.69	8.49	5.80	3.30	9.85	2.43
2.80	1.04	3.36	1.14	5.81	2.25	4.25	3.25	6.0.1	3.59	4.79	2.89
1.95	1.02	3.08	1.36	14.85	7.57	7.00	5.09	11.01	6.66	3.51	2.03
2.23	0.86	5.96	2.64	7.88	5.81	4.00	4.62	5.34	5.89	7.93	3.38
12.77	4.20	5.92	2.51	6.90	4.86	7.16	5.57	1.40	3.24	8.15	4.87
5.30	3.45	7.29	3.07	7.89	5.90	8.23	6.27	7.82	4.73	7.23	5.33
9.69	4.73	7.27	4.28	6.07	7.16	4.91	5.36	4.59	4.24	3.27	3.68
2.90	2.80	11.88	5.25	2.01	3.52	4.66	4.41	4.81	3.22	4.09	3.24
2.70	1.75	1.19	5.10	6.79	3.50	6.52	4.37	3.90	3.44	6.84	2.21
8.11	1.97	9.50	3.24	4.88	1.89	3.48	3.20	4.71	2.05	8.66	2.72
2.72	1.07	3.23	1.99	4.54	1.51	3.41	2.61	1.42	1.54	3.91	4.36
6.00	0.94	8.20	2.99	5.55	1.29	6.16	2.13	5.73	1.53	6.84	2.20

References

Anderson, D.G., Effects of Urban Development on Floods in Northern VA, Water Supply Paper 2001-C, U.S. Geological Survey, Reston, VA, 1970.

Anderson, J.R., Hardy, E.E., Roach, J.T., and Witmer, R.E., A Land Use and Land Cover Classification System for Use with Remote Sensor Data, Professional Paper 964, U.S. Geological Survey, Reston, VA, 1976.

Carter, R.W., Magnitude and Frequency of Floods in Suburban Areas, Professional Paper 424-B, B9-B11, U.S. Geological Survey, Reston, VA, 1961.

Conley, L.C. and McCuen, R.H., Modified critical values for Spearman's test of serial correlation, *J. Hydrologic Eng.*, 2(3), 133–135, 1997.

Crawford, N.H. and Linsley, R.K., Digital Simulation in Hydrology: Stanford Watershed Model IV, Technical Report No. 39, Stanford University, Palo Alto, CA, 1964.

Davis, J.C., *Statistics and Data Analysis in Geology*, John Wiley & Sons, New York, 1973.

Dawdy, D.R. and O'Donnell, T., Mathematical models of catchment behavior, *ASCE J. Hydrologic Div.*, 91(HY4), 123–137, 1965.

Dillow, J.J.A., Technique for Estimating Magnitude and Frequency of Peak Flows in Maryland. USGS-WRI-95-4154, U.S. Geological Survey, Towson, MD, 1996.

Draper, N.R. and Smith, H., *Applied Regression Analysis*, John Wiley & Sons, New York, 1966.

Dunne, T. and L.B. Leopold, *Water in Environmental Planning*, W.H. Freeman, San Francisco, 1978.

Gibbons, J.D., *Nonparametric Methods for Quantitative Analysis*, Holt, Rinehart & Winston, New York, 1976.

Gilbert, R.O., *Statistical Methods for Environmental Pollution Monitoring*, Van Nostrand Reinhold, New York, 1987.

Haan, C.T., *Statistical Methods in Hydrology*, Iowa State University Press, Ames, 1977.

Helsel, D.R. and Hirsch, R.M., *Statistical Methods in Water Resources*, Elsevier, Amsterdam, 1992.

Hirsch, R.M., Alexander, R.B., and Smith, R.A., Selection of methods for the detection and estimation of trends in water quality, *Water Resour. Res.*, 27, 803–813, 1991.

Hirsch, R.M., Slack, J.R., and Smith, R.A., Techniques of trend analysis for monthly water quality data, *Water Resour. Res.*, 18, 107–121, 1982.

Hufschmidt, M. M. and Fiering, M. B., *Simulation Techniques for Design of Water-Resource Systems*, Harvard University Press, Cambridge, MA, 1966.

Inman R.L. and Conover W.J., *A Modern Approach to Statistics*, John Wiley & Sons, New York, 1983.

Interagency Advisory Committee on Water Data, Guidelines for Determining Flood Flow Frequency: Bulletin 17B, U.S. Geological Survey, Office of Water Data Coordination, Reston, VA, 1982.

James, L.D., Using a digital computer to estimate the effects of urban development on flood peaks, *Water Resour. Res.*, 1, 223–234, 1965.

James, L.D., Watershed Modeling: An Art or a Science? Paper presented at Winter Meeting, American Society of Agricultural Engineers, Chicago, 1970.

Karlinger, M.R., and J.A. Skrivan, Kriging Analysis of Mean Annual Precipitation, Powder River Basin, Montana and Wyoming, USGS/WRD/WRI/81-050, U.S. Geological Survey, Tacoma, WA, May 1981.

Kendall, M.G., *Rank Correlation Methods*, Charles Griffin, London, 1975.

Krey, H., Grenzen der Öbertragbarkeit der Versuchsergebnisse, *Z. Angew., Math. Mech.*, 5, 6, 1925.

Leliavsky, S., *Introduction to Fluvial Hydraulics*, Constable, London, 1955.

Leopold, L.B., *A View of the River*, Harvard University Press, Cambridge, MA, 1994.

Leopold, L.B, M.G. Wolman, and J.P. Miller, *Fluvial Processes in Geomorphology*, W.H. Freeman, San Francisco, 1964.

Leopold, L.B., River channel change with time — an example, *Bull. Geol. Soc., Amer.* 84, 1845–1860, 1973.

Maass, A. et al., *Design of Water-Resource Systems*, Harvard University Press, Cambridge, MA, 1962.

Mann, H.B., Nonparametric test against trend, *Econometrica*, 13, 245–259, 1945.

Mann, H.B. and Whitney, D.R., On a test of whether one or two random variables is stochastically larger than the other, *Ann. Math. Stat.*, 18, 50–60, 1947.

Maryland Department of Planning, MdProperty View, Available at *http://www.op.state.md.us/data/mdview.htm*, 2001.

McCuen, R.H., *A Guide to Hydrologic Analysis Using SCS Methods*, Prentice-Hall, Englewood Cliffs, NJ, 1982.

McCuen, R.H., *Microcomputer Applications in Statistical Hydrology*, Prentice-Hall, Englewood Cliffs, NJ, 1993.

McCuen, R.H., Effect of historic information on the accuracy of flood estimates, *J. Floodplain Manage.*, 1, 15–34, 2000.

McNemar, Q., *Psychological Statistics*, 4th ed., John Wiley & Sons, New York, 1969.

Mendenhall, W. and Sincich, T., *Statistics for Engineering and the Sciences*, 3rd ed., Dellen Publishing, San Francisco, 1992.

Miller, I. and Freund, J.E., *Probability and Statistics for Engineers*, Prentice-Hall, Englewood Cliffs, NJ, 1965.

Mitchell, W.B., Guptill, S.C., Anderson, K.E., Fegeas, R.S., and Hallam, C.A., GIRAS—A Geographic Information Retrieval and Analysis System for Handling Land Use and Land Cover Data, Professional Paper 1059, U.S. Geological Survey, Reston, VA, 1997.

Mosteller, F. and Tukey, J.W., *Data Analysis and Regression*, Addison-Wesley, Reading, MA, 1977.

Nash, J.E., The form of instantaneous unit hydrographs, *J. Int. Assoc. Sci. Hydrol.*, 45, 114–121, 1957.

Nash, J.E., Systematic determination of unit hydrograph parameters, *J. Geophys. Res.*, 64 111–115, 1959.

National Climatic Data Center, NCDC: Locate Weather Observation Station Record, available at *http://lwf.ncdc.noaa.gov/oa/climate/stationlocator.html*, 2001.

Natural Environmental Research Council Report, Washington, D.C., 1975.

Natural Resources Conservation Service, NRCS State Soil Geographic (STATSGO) Data Base, available at *http://www.ftw.nrcs.usda.gov/statsgo.html*, 2001a.

Natural Resources Conservation Service, NRCS Soil Survey Geographic (SSURGO) Data Base, available at *http://www.ftw.nrcs.usda.gov/ssurgo.html*, 2001b.

Penman, H.L., Natural evaporation from open water, bare soil, and grass, *Proc. Royal Soc. (London)*, A, 193, 120–145, 1948.

Penman, H.L., Estimating evaporation, *Trans. Amer. Geophysic. Union*, 37(1), 43–46, 1956.

Pilon, P.J. and Harvey, K.D., Consolidated Frequency Analysis (CFA), Version 3.1 Reference Manual, Environment Canada, Ottawa, 1992.

Rantz, S.E., Suggested Criteria for Hydrologic Design of Storm-Drainage Facilities in the San Francisco Bay Region, California, Open File Report, U.S. Geological Survey, Menlo Park, CA, 1971.

Rosner, B., On the detection of many outliers, *Technometrics*, 17, 221–227, 1975.

Rosner, B., Percentage points for a generalized ESD many-outlier procedure, *Technometrics*, 25, 165–172, 1983.

Sarma, P.G.S., J.W. Delleur, and A.R. Rao, A Program in Urban Hydrology Part II, Tech. Report No. 9, Water Research Center, Purdue University, Lafayette, IN, October 1969.

Sauer, V.B., Thomas, W.O. Jr., Stricker, V.A., and Wilson, K.V., Flood Characteristics of Urban Watersheds in the United States, Water-Supply Paper 2207, U.S. Geological Survey, Reston, VA, 1983.

Schoklitsch, A., *Über Schleppkraft und Geschiebebewegung*, Engelmann, Leipzig, 1914.

Siegel, S., *Nonparametric Statistics for the Behavioral Sciences*, McGraw-Hill, New York, 1956.

Spencer, C.S. and McCuen, R.H., Detection of outliers in Pearson type III data, *J. Hydrologic Eng.*, 1, 2–10, 1996.

Taylor, C.H. and Loftis, J.C., Testing for trend in lake and ground water quality time series, *Water Resour. Bull.*, 25, 715–726, 1989.

Tiffany, J.B. and Bentzel, C.B., Sand mixtures and sand movement in fluvial models, *Trans. ASCE*, 100, 1935.

U.S. Army Corps of Engineers, HEC-HMS Technical Reference Manual, Hydrologic Engineering Center, U.S. Army Corps of Engineers, Davis, CA, 2000.

U.S. Environmental Protection Agency, BASINS data by hydrologic unit code: BASINS data download, available at *http://www.epa.gov/OST/BASINS/gisdata.html*, 2001a.

U.S. Environmental Protection Agency, Multi-resolution land characteristics—MRLC homepage, available at *http://www.epa.gov/mrlc*, 2001b.

U.S. Geological Survey, PeakFQ, available at *http://water.usgs.gov/software/peakfq.html*, 2001a.

U.S. Geological Survey, Surface-water data for the nation, available at *http://water.usgs.gov/nwis/sw*, 2001b.

U.S. Geological Survey, U.S. GeoData, available at *http://edcwww.cr.usgs.gov/doc/edchome/ndcdb/ndcdb.html*, 2001c.

U.S. Soil Conservation Service, A Method for Estimating Volume and Rate of Runoff in Small Watersheds, TP-149, U.S. Department of Agriculture, Washington, DC, 1973.

U.S. Soil Conservation Service, *Computer Program for Project Formulation*, Technical. Release 20, U.S. Department of Agriculture, Washington, DC, 1984.

U.S. Soil Conservation Service, National Engineering Handbook, Supplement A, Section 4, Hydrology, Chapter 10, U.S. Department of Agriculture, Washington, DC, 1985.

U.S. Soil Conservation Service, *Urban Hydrology for Small Watersheds*, Technical Release 55, U.S. Department of Agriculture, Washington, DC, 1986.

U.S. Water Resources Council (WRC), Estimating Peak Flow Frequencies for Natural Ungaged Watersheds, Hydrology Comm., Washington, D.C., 1981.

Vogel, R.M., The probability plot correlation coefficient test for normal, lognormal, and Gumbel distributional hypotheses, *Water Resour. Res.*, 23(10), 2013, 1986.

Vogelmann, J.E., Sohl, T., Howard, S.M., and Shaw, D.M., Regional land cover characterization using Landsat Thematic Mapper data and ancillary data sources, *Environ. Monitoring Assessment*, 51, 415–428, 1998a.

Vogelmann, J.E., Sohl, T., and Howard, S.M., Regional characterization of land cover using multiple sources of data, *Photogrammetric Eng. Remote Sensing*, 64, 45–57, 1998b.

Wallis, J.R., Matalas, N.C., and Slack, J.R., Just a moment, *Water Resour. Res.*, 10, 211–219, 1974.

Zoch, R.T., On the relation between rainfall and stream flow, *Mon. Weather Rev.*, 62, 315–322, 1934.

Index

A

B

C

D

E

N

O

P

R

S

T

U

V

W